Springer Water

The book series Springer Water comprises a broad portfolio of multi- and inter-disciplinary scientific books, aiming at researchers, students, and everyone interested in water-related science. The series includes peer-reviewed monographs, edited volumes, textbooks, and conference proceedings. Its volumes combine all kinds of water-related research areas, such as: the movement, distribution and quality of freshwater; water resources; the quality and pollution of water and its influence on health; the water industry including drinking water, wastewater, and desalination services and technologies; water history; as well as water management and the governmental, political, developmental, and ethical aspects of water.

More information about this series at https://link.springer.com/bookseries/13419

Vladimir Pešić · Djuradj Milošević · Marko Miliša
Editors

Small Water Bodies of the Western Balkans

Springer

Editors
Vladimir Pešić
Department of Biology, Faculty of Sciences
University of Montenegro
Podgorica, Montenegro

Djuradj Milošević
Department of Biology and Ecology
University of Niš
Niš, Serbia

Marko Miliša
Department of Biology, Faculty of Science
University of Zagreb
Zagreb, Croatia

ISSN 2364-6934 ISSN 2364-8198 (electronic)
Springer Water
ISBN 978-3-030-86477-4 ISBN 978-3-030-86478-1 (eBook)
https://doi.org/10.1007/978-3-030-86478-1

This Springer imprint is published by the registered company Springer Nature Switzerland AG
The registered company address is: Gewerbestrasse 11, 6330 Cham, Switzerland

Contents

Chapter 1
Springs as Essential Water Sources for Dependent Ecosystems in Karst

Zoran Stevanović, Želimir Pekaš, Aleksandra Maran Stevanović, Romeo Eftimi, and Milan Radulović

Abstract The karst covers some 30% of the territory of former Yugoslavia and similarly Western Balkans countries and represent fully contrasted terrain. On one side there are dry hilly mountainous terrains with deep groundwater table and terrestrial ecosystems adapted to these circumstances while on another, in the foothills and along erosional bases are often located strong karst springs which originate many perennials and sinking rivers or recharge the lakes. These rich karst aquifers are the main source for water supply of local inhabitants and aquatic ecosystems. In this "classical karst" region where the Dinaric Mountains arch represents dominant geological structure there are numerous landscapes, features and biotopes under different kinds of protection, from those included in UNESCO's World Heritage list to natural monuments and areas protected by national legislatives. They are inhabited by specific biocenoses including underground endemic species such as famous olm *Proteus anguinus*. Although rich in surface and groundwater resources in comparison to rest of Europe water demands are permanently increasing as result of tourism expansion, intensive agriculture and urbanization. This is why during recession and drought periods lack of sufficient water may cause restrictions in potable water supply along with water shortage for dependent ecosystems. Establish a proper balance between abstraction for domestic water supply and water provision to dependent ecosystems is one of the main challenges and requests which had been included in

Z. Stevanović (✉)
Faculty of Mining & Geology, Department of Hydrogeology, Centre for Karst Hydrogeology, University of Belgrade, Djušina 7, 11000 Belgrade, Serbia

Ž. Pekaš
Hrvatske Vode, Zagreb, Croatia
e-mail: zelimir.pekas@voda.rs

A. M. Stevanović
Natural History Museum, Njegoševa 51, 11000 Belgrade, Serbia
e-mail: amaran@nhmbeo.rs

R. Eftimi
Independent Researcher, Tirana, Albania

M. Radulović
Faculty of Civil Engineering, University of Montenegro, Podgorica, Montenegro

V. Pešić et al. (eds.), *Small Water Bodies of the Western Balkans*, Springer Water,
https://doi.org/10.1007/978-3-030-86478-1_1

several recently prepared River Basin Management Plans of concerned countries. In certain areas engineering solutions and interventions may improve water situation and ensure more water available to the consumers. Improper solid waste management and lack of wastewater treatment plants represent other main threats to the quality of spring water.

Keywords Springs · Karst · Water supply · Dependent ecosystem · Ecological flow · Aquifer regulation

1.1 Introduction

There are certain disputes about the boundaries of the Western Balkans but in line with the official definition of the European Union (EU) [1], there are five countries, namely: Albania, Bosnia and Herzegovina, North Macedonia, Montenegro and Serbia, plus the territory of Kosovo.[1] However, due to geographical and geological similarities and especially presence of karst, this region should be evaluated by including also two other ex-Yugoslavian countries—Croatia and Slovenia.

Entire region for long geological time was under Tethys Ocean and then after retreat of the sea water it had been exposed to uplifting and intensive Alpine orogenesis. The Alpine orogenic belt consists of several branches: the largest is the Dinarides, extending from Carso (the area between Italy and Slovenia) over Croatia and Bosnia and Herzegovina to the western part of North Macedonia and the Vardar Valley. The orientation of the system is NW–SE, parallel to the Adriatic Sea. Other branches of Alpine orogenic belt are the Carpathians, extending over eastern Serbia and the Karnian Alps extending over northern Slovenia (Fig. 1.1).

The karst is a specific environment, resulted from dissolution of carbonate rocks caused by the two factors: erosional—mechanical water activity and corrosional—chemical water activity [2–4].

The Dinarides is also region where a scientific discipline karstology was born at the end of the nineteenth century as a result of the studies of Jovan Cvijić [5]. The Dinaric karst geology, hydrology and geomorphology were also studied by Herak [6], Mijatović [7], Bonacci [8], Milanović [9], Bonacci et al. [10], Stevanović et al. [11] and many other authors who helped to better understand this "classical" highly developed karst, one of the world's largest [12].

Intensive Alpine tectonic activities have resulted in the creation of a complex system of faults and fractures that now act as preferential groundwater flow paths. Moreover, climatic conditions, particularly the successions of wet and warm periods, have significantly contributed to karstification. And finally, a very specific relief had been formed, both on the surface and underground. The stream network, in most of carbonate rock complex, was often completely degraded, consists of temporary or

[1] This designation is without prejudice to positions on status, and in line with UNSCR 1244 and the ICJ Opinion on the Kosovo Declaration of Independence.

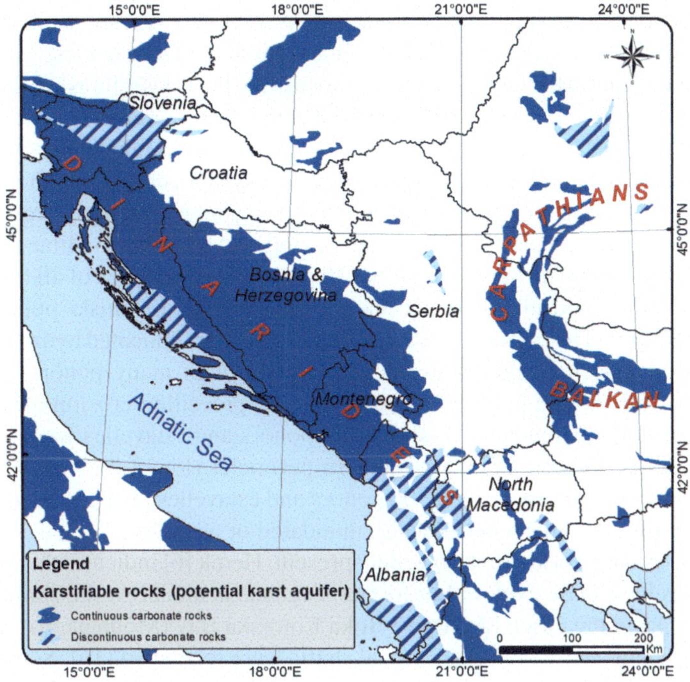

Fig. 1.1 Karst distribution along the Western Balkans' countries (extract from the WOKAM map [14], inputs and karst boundaries provided by Stevanović et al. [11])

sinking streams while only major streams such as Cetina, Zrmanja, Krka, Neretva, Zeta, Vjosa have perennial character.

The large karstic springs are extended all across the region and ensure potable water supply from time immemorial. During Roman time many cities were established just in their vicinity [13]. The results of recently completed project the World Karst Aquifer Maps, WOKAM [14] confirm that region has the densest concentration of large springs, which minimal discharge exceeding 2 m^3/s. They are of crucial importance not only for ensuring water supply of local population and dependent ecosystems, but also for industrial and touristic development. However, a large variation of springs discharge throughout the year and minimal values during periods of peak demands cause water shortage in certain areas and restrictions in water supply imposed by local waterworks.

1.2 Karst and Karst Aquifers Distribution in the Dinarides

The Dinaric karst is a mountainous region with a prevalence of highly karstified rocks and large karstic poljes and valleys formed in tectonic depressions. The karst is almost entirely carbonate (limestones and dolomites), its thickness is often over 1000 m and it is mostly of Mesozoic age [11]. The development of the Dinaric karst was gradual. Herak [6] stated that at the end of the Triassic or during the Lower Jurassic, the Triassic carbonate rocks were first exposed to the impact of water circulation processes. At the end of the Upper Cretaceous and during the Paleocene (Laramian tectonic phase), intensive uplifting and folding took place in Alpine system, during which most of the carbonate rocks were tectonised. Along with uplifting intensive faulting resulted with creation of large tectonic depressions—karstic poljes. According to Milanović [15] in the Dinaric karst there are approximately 130 poljes. The total area of all these poljes is about 1,350 km^2, and among them is the world's largest—Livanjsko polje covering an area of 380 km^2. The poljes are characterized by very complicated hydrogeological relations: drainage of surface water is achieved through many ponors. These are frequently located in the polje areas nearest to the prevailing erosion base [11]. In the Nikšićko polje (Montenegro), about 880 ponors and estavelles were identified, 851 of which are located along its southern perimeter (Fig. 1.2). In Popovo Polje (Herzegovina) there are more than 500 ponors and estavelles [15]. Some of them are lakes or swamps, the others periodically inundated or even dry.

The underground forms are also widely present. Herak [6] indicates that more than 12,000 caves have been explored in former Yugoslavia alone. At the Kameno more ("Stone sea") and the Orjen Mt. above Boka Kotorska Bay (Montenegro), within an area of only 8 km^2 more than 300 vertical shafts were registered [16]. Some of them were speleologically surveyed to depths of 200–350 m.

The karst aquifer recharges from precipitation and waters percolating from numerous sinking rivers. Depending on locality, morphology and karstification properties the average infiltration rate can be assessed to vary from between 50 and 80% of the precipitation [11, 17]. In the Dinaric region of ex-Yugoslavia there are 230 springs with a minimal discharge of over 100 l/s, while about 100 springs have a minimal discharge of over 500 l/s [18]. In the Albanian karst there are roughly 110 springs with an average discharge exceeding 100 l/s [17]. Milanović [19] noted that "only through three huge springs along the Neretva Valley and Adriatic coast (Buna

Fig. 1.2 Specific cylindric structures for preventing water sinking (left) and cylindric dam located in southern perimeter of Vrmac part of Nikšićko polje (Montenegro)

Fig. 1.3, Bunica and Ombla) and a few spring zones in the Kotor Bay, more than 150 m^3/s is discharged annually into the Adriatic Sea directly or indirectly through the Neretva River".

Four capital cities of the Western Balkans receive drinking water from the karstic aquifers, Sarajevo, Podgorica, Skopje and Tirana, respectively.

In the countries Montenegro, Bosnia and Herzegovina, Albania, each citizen has more than 5,000 m^3 of water available annually. This amount is ten times higher than the level which in certain UN documents is indicated as limit for "water stressed countries". In an average hydrological year, each inhabitant of Montenegro has 21,395 m^3 of water available, but utilises just 1.18% of this volume. In Bosnia and Herzegovina the utilisation rate is even lower, below 1%. Citizens of Croatia and Albania also use less than 5% of water that is available to them per capita [20]. The above figures are closely linked to the distribution of karst terrains, high permeability of karst aquifer systems, and highly effective infiltration of rainfall into the ground.

Although the Western Balkans and Dinaric region in particular is one of the most water-rich regions in the world, due to its specific water regime and the behaviour of the karst, local population and ecosystems often suffers from water shortages. In

Fig. 1.3 The spring Buna, one the world's largest concerning maximal discharge of 380 m^3/s (minimal is 2.95 m^3/s)

Fig. 1.4 The water cistern and dry plateau in high mountain of Durmitor (Montenegro)

the mountains high above the sea level and the erosional base, groundwater table is often very deep, with no available surface waters. The only way to provide water to the local villagers and their livestock is to build cisterns and specific intakes for collecting rainwater (Fig. 1.4).

Several areas and sites of the Dinaric karst are recognized and protected by international and national legislative acts. The most important are the four UNESCO World Heritage sites: the Plitvice Lakes in Croatia, the Škocjan Caves in Slovenia, the Durmitor National Park in Montenegro, and Ohrid Lake in North Macedonia which is shared with Albania. The areas that are on the UNESCO tentative list are: the Velebit Mountain (Croatia), Classical Karst (Slovenia) and the Tara National Park (Serbia). Several karstic poljes including Skadar Lake as the largest in the Balkans, are protected under the Ramsar List [21]. However, the region as a whole has never been placed under protection. This deficiency has been recognized in the report of Paul Williams [22], who recommended to the International Union of Conservation of Nature (IUCN) to initiate the process of recognition of the Dinaric karst as a World Heritage area.

1.3 Vulnerability of Karst Aquifers and Variability of Springs Discharge

Of all the aquifer systems, karst aquifers are characterised by the most dynamic regime: the water table fluctuates, and the springs' discharge or water chemistry can vary from one day or hour to the next [3, 4, 23]. It is caused by high porosity and well-developed channels in karstified rocks. During the dry season and low aquifer water table, water circulation in the karst system is characterized by a slow movement of aquifer waters. The water waves labeled with dye take two- to five-fold less time to travel the same distance during a season of high hydrologic activity [15]. According to Milanović [15] based on numerous tracing experiments conducted in Eastern Herzegovina, the average flow velocity varies in a wide range of 0.002–55.2 cm/s.

Based on 380 conducted experiments Komatina [18] concluded that the frequency of groundwater velocities in Dinaric karst is as follows: in 70% of cases from 0 to 5 cm/s; in 20% of cases 5 to 10 cm/s; and in 10% of cases more than 10 cm/s. A large fluctuation in the water table is common in the region [19, 24]. For instance, the water level can change by 312 m during a period of 183 days as an example in the observation borehole Z-3 in the Nevesinjsko Polje (personal communication of Ž. Zubac). In the Cetina River basin the maximum recorded water table increase was 3.17 m/h.

Large variation of discharge is typical characteristic of many springs. For these that are of gravity type this is almost a rule. Such examples are several springs located along the Boka Kotorska Bay in Montenegro [16] where the ratio $Q_{max} : Q_{min}$ can be over 1 : 100.000. They discharge a few hundreds m^3/s in maximum, but in low water seasons are completely dry or functioning as submarine springs (vruljas, Fig. 1.5).

Ascending springs with extended siphons and with deeper water storage characterised by more stable discharge. A special kind of sources are "eyes" (locally "oka") sublacustrine springs along the Skadar Lake shoreline [25]. One of the deepest eye is the Raduš spring, which discharges 66 m below sea level [26].

Typical example of seasonal and daily variation can be seen at hydrograph of Rijeka Crnojevića spring in the Skadar basin of Montenegro (Fig. 1.6). For selected period of hydrologic year 1987–88 ratio $Q_{max} : Q_{min}$ is 1 : 87 (0.9–78.4 m^3/s), while very quick response on fallen rains is also evident.

The presented hydrogeological settings are the reason why karst aquifers are characterized by a low attenuation capacity and are highly vulnerable to pollution. Many

Fig. 1.5 Sopot spring discharge from cave orifice above sea level. Yield is larger than 100 m^3/s (left). During low water season only submarine discharge exists, location of flux is indicated by arrow (right)

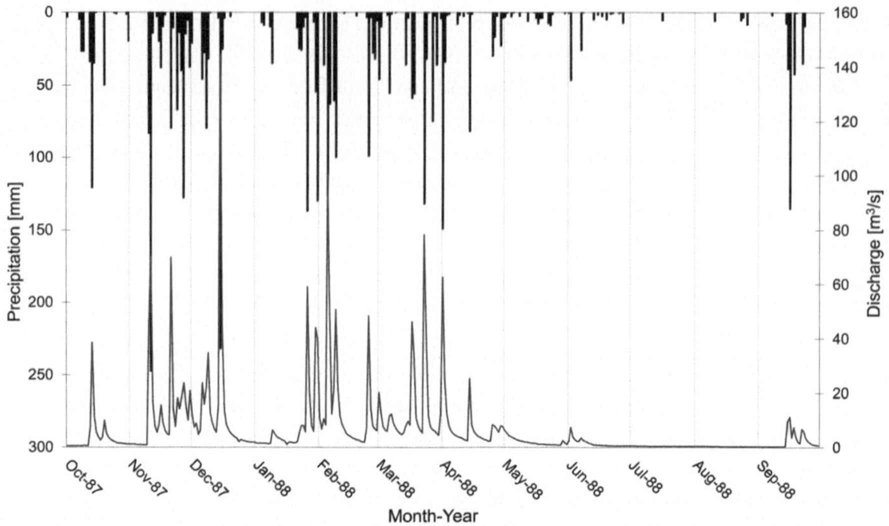

Fig. 1.6 Hydrograph of Rijeka Crnojevića spring discharge (hydrometric station "Brodska njiva") with precipitation bar graph (climatological station Cetinje) (after Živaljević [27]; reprinted from Radulović [26], Springer copyrights)

methods have been developed in water practice to assess their intrinsic vulnerability. Methods such as DRASTIC [28] or EPIK [29] are just some of those that are applied the most in water management practice. Stevanović & Marinović [30] have developed a new method SODA for regional assessment of aquifer vulnerability, which has been applied for assessment of groundwater component in the prepared River Basin Management Plans in Bosnia & Herzegovina (Sava River) and Montenegro (Adriatic and Danube river basins).

1.4 Ecosystems, Water Demands and Ecological Flows

Karst aquifers are the essential source of water supply for dependent ecosystems, both terrestrial and aquatic, as well as, surficial and underground (hypogean). In the aquatic, where the survival of organisms directly depends on water, any drought has a significant negative impact, while the drying out of a surface water body directly leads to the extinction of the existing plants and animals.

Zalewski et al. [31] promoted the term "ecohydrology" by explaining that hydrological variability significantly influences biotic structure and activity. Bonacci et al. [10] stated that subterranean karst ecosystems are extremely sensitive to environmental changes, namely more than most others. It is simply the result of variable karst hydrology and groundwater regime. Karst ecohydrology should therefore help in understanding and explaining the sensitivity and stresses of ecosystems to hydrological extremes such as droughts and floods.

An underground ecosystem in karst except bat colonies and wild animals as casual visitors, including endemic species especially adapted to the cave environment (troglobites). In the latter group are various insects and aquatic animals such as the famous olm *Proteus anguinus*, which inhabits exclusively the Dinaric karst. Precious springwater of karst aquifers [32] is rich in micro and macro organisms—diatoms, bryophytes, vascular plants, mollusks, ostracods, stoneflies, etc. [33–36]. The rich biodiversity of the entire Western Balkans has been studied for many years and by many researchers, but with particular intensity from of the mid-nineteenth century [37]. Regional studies of rarely distributed and endangered plants and animals can be found among others in the works of Karaman [38, 39], Sket [40], Schneider-Jacoby et al. [41], Albrecht and Wilke [42], Petrović et al. [43], Talevski et al. [44], Pešić et al. [45, 46] Delić et al. [47].

These rich literature sources suggest that some of the species are completely adapted to the difficult circumstances of surviving in karst. As such, the stygophilic freshwater small fish Gaovica (*Delminichthys ghetaldii*) which inhabit Popovo Polje in Eastern Herzegovina spent most of its life in the underground cavities, linked to ponors, estavelles and springs. In the past, before regulation of the Trebišnjica the largest sinking river of Europe, Gaovica lives in temporary lake of Popovo polje during periods of floods, while goes underground only after draining of polje via numerous ponors in periods of drought. Nowadays, its natural life cycle has been completely disrupted by the lack of appearance of an annual surface lake in which to breed and exchange genes.

The Eastern Herzegovina region is also well-known by systems of cascade poljes and regional and local underground flows oriented towards the Adriatic Sea as regional erosional base (Fig. 1.7). In these closed depressions the surface flows can be temporary or very short, extended from the spring located on one margin of

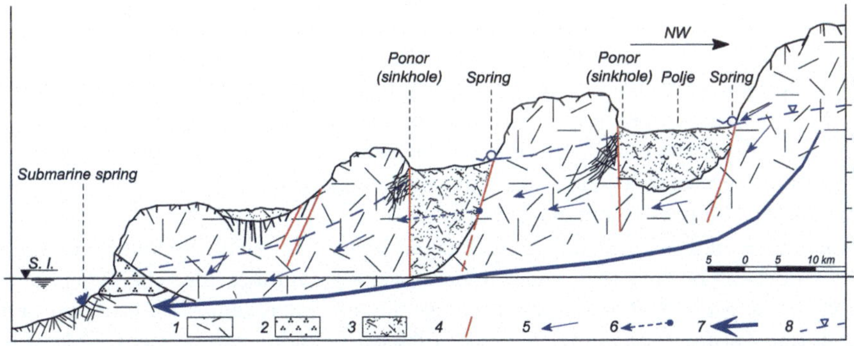

Fig. 1.7 Schematic typical cross-section through the karstic poljes of the Dinarides (after Mijatović [7], modified by Stevanović. Springer copyrights). Legend: 1. karstified Mesozoic limestone; 2. flysch barrier; 3. porous aquifer of polje; 4. fault; 5. groundwater flow; 6. direction of flow around the barrier; 7. regional flow; 8. groundwater table

Fig. 1.8 Ključka Rijeka in Cerničko polje from the spring (Vilina pećina) to ponor (photo courtesy of Petar Milanović)

polje to ponor on another, whereas disappeared waters reappears on the springs of lower positioned polje. As such, Milanović [19] states that Zalomka River in Gatačko polje is flowing 213 days in average hydrological year, but there is not any flow for the rest of year. Milanović [19] also describes short rivers of Vrijeka in Dabarsko polje (0.043—25 m^3/s) and Ključka rijeka in Cerničko polje which flowing only on distance between spring and ponor of 300 m (Fig. 1.8).

Estavelles have specific hydraulic mechanism, functioning as springs or ponors depending on pressure within aquifer system. One of the typical and largest in the region is Gornjepoljski vir in Nikšićko polje, Montenegro [25]. The local life habitat just follows fluctuations of water table moving up- or downstream, but is important persistent presence of water. But, adaptation to mechanism of some other karst water features such as intermittent (ebb and flow) springs is much more difficult. Such kind of spring named Mukavica (not very distant from Gornjepoljski vir), during drought periods is rapidly flowing or stopping to flow in frequency of about an hour [25]. Some other springs have no regular ebb and flow frequency, but incidentally may stop to discharging for certain period of time. Such unstable regime and yield interruptions characterises Šavnička glava the source of short Šavnička River at the Durmitor Mt. foothills (Montenegro) [48].

Adaptation of flora and fauna to changed hydrogeological, hydraulic, climate environment can be very long, measurable in geological timescale. The Quaternary period was especially dynamic concerning climatic interchanges—glaciation and interglacial warming stadiums. In terms of karstification, groundwater table was deepening, many previous surface rivers disappeared and become dry and blind valleys, and many new springs activated as drainage points of aquifer systems. Milanović [19] explains such paleo river Vala, which extends between Zavala in Trebišnjica basin and Slano at Adriatic Coast, and represented the main outflow of

Popovo polje in the geological past. An assessment of paleohydrographic network of Boka Kotorska Bay in post-glacial period, which now is completely missing, is presented by Stevanović [49].

Except natural factors, which influencing discharge of springs (floods, droughts, climate variations), the major threats for ecosystems are artificial—building the dam and reservoirs and non-controlling water extraction.

In contrast to water shortage, the high discharge of the springs and rivers during winter or springtime causes continual seasonal flooding of cultivated land in depressions and karst poljes, and this was the main reason why large projects to regulate river flows were initiated in all the countries in the region after WW II and many of these were implemented during the 1960–1970s. Today many streams are dammed, and their waters are utilized by hydropower plants. The major dams and reservoirs have been built on the Cetina, Neretva, Trebišnjica, Zeta, and Drini rivers [11]. Dinaric karst becomes even a referent area for the successful completion of dams in karst, very problematic media from the point of view of water losses [23]. About 2/3 of total existing hydro power facilities of the concerned countries are located in the Dinaric karst area, which plays significant role in these countries' economies. However, change of type of aquatorium and microclimate, conversion of fast flowing rivers into stagnant reservoirs, most commonly without constructed fish passages, has serious implications on local aquatic, but also terrestrial ecosystems preventing migratory paths of the aquatic fauna.

When it comes to water extraction, requirement for sustainable water resources management comprises needs to define water demands of dependent eco-systems. A large part of the watercourses in the karst areas start their flow at karst springs but abstracting most of these waters for public water supply or irrigation may cause the lack of water downstream, particularly during the dry summer periods when the demand for water is the highest. In EU, the term "ecological flow" and conception of its definition is prescribed in Guidance Document No. 31 "Ecological flows in the implementation of the Water Framework Directive" [50]. The experiences of Croatia and Bosnia & Herzegovina in this domain can be considered as relevant for other countries, as well.

The ecological flow (EF) in terms of biological minimum in Croatia is defined as "the flow which ensures the survival and development of biocenoses in a river as a biotope" [51]. According to the Croatian Water Law [52], the ecological flow is the quality, quantity and timing of surface water and groundwater required to sustain the functions and processes of freshwater and estuarine ecosystems and the human livelihoods and wellbeing that depend on these ecosystems. However, the obligation to implement and the control of implementation of the EFs is implemented mostly for the projects related to the use of water for energy needs.

The following approaches are most frequently used for the assessment of EFs:

I. EF as the minimum average monthly discharge with a 95% probability of occurrence,

II. EF as an indicator that depends on the characteristics of the watercourse and its banks.

(a) For the watercourses with extremely steep banks—application of Matthey's formulas [53]:

- For the watercourses or parts of watercourses with an average discharge up to 5 m^3/s with the application of Matthey's three formulas for EF expressed as Q_E:

$$Q_E = \frac{15 \cdot Q_{300}}{(\ln Q_{300})^2} \geq 50\ (1/s),$$

$$Q_E = \frac{15 \cdot Q_{347}}{(\ln Q_{347})^2}\ (1/s),$$

$$Q_E = 0.35 \cdot Q_{347}\ (1/s),$$

of which the one with the strictest criterion is applied.

- For larger watercourses (watercourses with an average discharge of more than 5 m^3/s):

$$Q_E = 0.25 \cdot Q_{347} + 75\ (1/s)$$

In that process, both the inputs about the characteristic discharges (obtained based on the analysis of duration curves) and the results are presented in the «l/s» dimension.

It needs to be noted that some hydrologists [54] for the same formula instead of the characteristic discharge of the 300th day from the average duration curve (Q_{374}) take ($Q_{80\%}$), with that parameter implying the discharge of the 80% duration at the average duration curve, i.e. the 292nd day discharge. On the 347th day the appropriate value expressed in percentage terms is the 95% duration which is in much more widespread use in expert hydrological terminology and is also closer to the standard methods of determining the characteristic values of the average duration and frequency curves. For that reason, the above-mentioned expressions for Matthey's formulas are used in such a way that the characteristic discharges from the duration curve are defined as the above-mentioned % duration.

(b) For the watercourses with moderately steep banks and watercourses with indented banks, among the hydrological methods it is recommended to use [53]:

- The method defining the biological minimum (EF) as the average minimum annual discharge defined as an arithmetic mean of the annual absolute minimum discharges recorded in the analysed period

$$Q_E = Q_{MIN(SR)}\ (1/s)$$

III. EF as an alarming limit value

$$Q_E = 0.2 Q_{80\%}$$

According to Bonacci [53], an alarming limit value represents the lowest current discharge value below which the value shall not fall:

$Q_{80\%}$ is the discharge of 80% duration on the average duration curve.

Since the EF definition methodology has not been established in Croatia yet for sources at which water is abstracted for water supply, the above alarming EF limit value is most often used as the baseline. It as a rule gives the lowest values of the discharge that needs to be ensured in the watercourse downstream of the water supply abstraction point. Since the current use of water for potable water supply needs is mostly significantly higher, ensuring the alarming value as a rule represents a step forward compared to the state so far.

In addition to the hydrological methods of the EF definition, the use of biological (habitat modelling) and holistic methods is also on the rise. The current trends support a holistic approach which includes interdisciplinary groups of experts aiming at sustainable solutions.

The Ordinance on the definition of ecological flow in Bosnia & Herzegovina (B&H) introduces the two levels of assessment [55]:

(1) I Level—General assessment of EF for all water bodies by use of hydrology methods, and
(2) II Level—Special assessment of EF for water bodies in designated protected areas.

The hydrology method considers the three possible components for calculating EF: average minimal flow, average flow and decade averaging flow.

The study on special assessment must includes analysis of ecosystem (processes, habitats, species), valuing criteria for assessing success, and targets to be achieved. Determined EF in protected area must ensure normal functioning of ecosystem as a whole and endangered species in particular, as well as maintenance of required water quality. Monitoring stations should be located upstream and next to the intake, downstream and next to the intake, upstream at the section of maximal water tail of the reservoir, and downstream next to the dam site (Fig. 1.9).

1.5 How to Regulate Karst Aquifer Drainage?

In complex and sensitive karst aquifers the main challenge for many waterworks is to ensure water supply and to avoid restrictions or total interruptions in water provision during summer–autumn critical periods. The minimal discharges of the springs and the accordingly reduced minimal river flows during recession periods result in water deficit not only for water consumers but also for dependent eco systems. If this situation can be improved? The general answer is yes, but not everywhere, and only under restrictive circumstances.

If aquifer is well karstified and has adequate storage in its deeper parts it is often possible, just as it is in the case of open water reservoirs, to regulate and manage ecological flow downstream by various engineering interventions [56]. "Engineering

Fig. 1.9 Mratinje dam site on the Piva River (Montenegro) during construction in 1980, when derivation tunnel evacuated all river water from riverbed downstream to enable construction works

regulation of karst aquifers implies water intake or other structures that aim to create physical conditions for controlling and exploiting the water reserves in a controlled manner by changing the natural regime. This is, therefore, an artificial intervention and application which the standard tapping structures do not permit" [56]. The aim of engineering regulation of an aquifer is to enable tapping of the necessary water volume during certain periods (usually periods of increased water demands) by being able to count on sufficient water replenishment (during the following wet seasons). It is true, however, that this aim is not achievable everywhere, and these two possible scenarios exist [57]:

1. Replenishment potential is sufficient to cover a "loan" over a short period of time (within the same or the next hydrologic cycle)
2. The deficit cannot be compensated, and a water table decline is envisaged.

The first scenario considers temporary over-pumping via drilled wells (Fig. 1.10), water pumping from deep siphons, or constructed galleries. If pumped additional amount of water is equally or proportionally shared between pipeline and riverbed, the temporary reduction of natural springflow is not a problem, but there is benefit of both, humans and ecosystem. But, if ecological flow downstream is not ensured or pumped amount cannot be compensated during next floods, this situation leads to over-extraction and deterioration of environment.

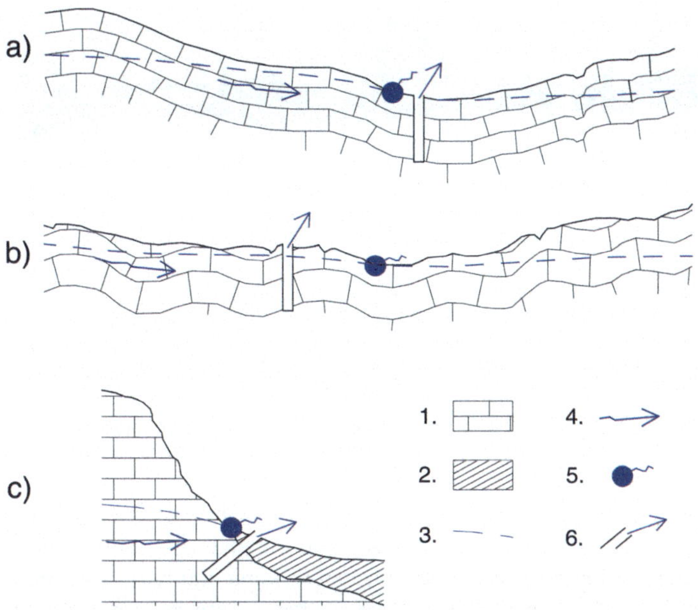

Fig. 1.10 Controlling the groundwater flow in drainage area. **a** vertical well downstream (beneath) the spring; **b** vertical well upstream (above) the spring; **c** inclined well drilled through impervious cover. Legend: 1. karst aquifer; 2. impervious rocks; 3. groundwater table; 4. groundwater flow direction; 5. spring; 6. discharge from well (after Stevanović [56], Springer copyrights)

Although natural karst water quality is generally very good, there is high risks of pollution due to high vulnerability of aquifers and fast migration of pollutants. Improper solid waste management and lack of wastewater treatment plants are the main threats to the quality of spring water. Pumping water from deeper aquifer parts may have positive effects on water quality. Stratification of water chemistry within aquifer system is possible even in dynamic system such karst is. Therefore, lower positioned and stagnant water layers characterised by longer interaction rocks-water and better self-purification capacity [58].

The key factor to ensure sustainable development of aquifer system in dynamic karst aquifers is monitoring [59]. The Fig. 1.11 shows the transboundary (Albania-Montenegro) river of Cijevna at its mouth to the Morača River. At this important river which part is under protection in status of natural monument there is not any active hydrological station [60]. This river is regularly dying out at lower sections and proper monitoring should be a key element for understanding level of human impact (pumping for irrigation from the river and from connected aquifer) on its hydrology regime.

With exception of Slovenia and Croatia in the Dinaric region monitoring of groundwater is far from satisfactory. In Slovenia in the year 2017, a total of 27 observation points were located at springs, or at streams in head watershed, in the proximity of springs. In karstic terrains of Croatia some 48 observation points (dominantly

Fig. 1.11 Cijevna River at its mouth to Morača River (Montenegro) in July 2018

springs including those that are tapped, or streams in the vicinity of the spring sites) are included in the quantitative monitoring and the majority of observation stations are automated. In Serbia just five are located in karst aquifers. A similar situation, with a very limited number of water sites in karst aquifers monitored by national institutions despite their wide extension, can be found in Bosnia & Herzegovina, Montenegro and Albania [61]. However, recently prepared River Basin Management Plans in these countries envisage the expansion of the groundwater monitoring network [62].

1.6 Conclusions

The "classical karst" of Dinarides as part of the Western Balkans is characterised by abundant water reserves, but variable in space and time. The emblem of the Dinarides are large karst springs with their most dense concentration at global level. However, low discharge of springs and rivers during recession summer–autumn periods is one of the main threats to local ecosystems, especially those that are aquatic.

As rich in groundwater reserves the Dinaric karst is also rich in biodiversity, which includes endemic species especially adapted to the cave or spring environment.

The question of how much spring water can be utilised for public water supply and how much water should be left in riverbeds as ecological flow is a question that is commonly discussed among water engineers and ecologists. The best option would be to regulate the regime of aquifer system and enable over-abstraction for temporary periods to provide more water to both, humans and biodiversity.

The following mitigation measures are advisable:

1. Improving the minimal flow by means of artificial regulation of aquifers' regimes,

2. Introducing systematic monitoring of water quantity and quality, as well as the status of biocenoses,
3. Reducing water pollution through extensive construction of sanitary landfills and facilities, as well as wastewater treatment plants,
4. Raising the awareness of the local population regarding the importance of water resources and protection of valuable ecosystems.

References

1. EC EU web site. https://ec.europa.eu/trade/policy/countries-and-regions/regions/western-balkans/. Accessed 20 May 2021
2. Bakalowicz M (2005) Karst groundwater: a challenge for new resources. Hydrogeol J 13:148–160. https://doi.org/10.1007/s10040-004-0402-9
3. Kresic N (2013) Water in Karst. Management, vulnerability and restoration. McGraw Hill, New York
4. Stevanović Z (2015) Characterization of karst aquifer. In: Stevanović Z (ed) Karst aquifers—characterization and engineering. Springer Intern. Publ., Switzerland, pp 47–126. https://doi.org/10.1007/978-3-319-12850-4
5. Cvijić J (1893) Das Karstphänomen. Versuch einer morphologischen Monographie. Geographischen Abhandlung, Wien V(3):218–329
6. Herak M (1972) Karst of Yugoslavia. In: Herak M, Stringfield VT (eds) Karst: important Karst regions of the Northern Hemisphere. Elsevier Publishing Company, Amsterdam, pp 25–83
7. Mijatović B (1984) Hydrogeology of the Dinaric Karst. Intern. Contrib. to Hydrogeology, IAH, vol 4. Heise, Hannover, p 254
8. Bonacci O (1987) Karst hydrology; with special reference to the Dinaric Karst. Springer-Verlag, Berlin, p 184
9. Milanović P (2005) Water potential in south-eastern Dinarides. In: Stevanović Z, Milanović P (eds) Water resources and environmental problems in Karst CVIJIĆ 2005, Spec. ed. FMG. Belgrade, pp 249–257
10. Bonacci O, Pipan T, Culver D (2009) A framework for karst ecohydrology. Environ Geol 56(5):891–900. https://doi.org/10.1007/s00254-008-1189-0
11. Stevanović Z, Kukurić N, Pekaš Ž, Jolović B, Pambuku A, Radojević D (2016) Dinaric Karst aquifer—one of the world's largest transboundary systems and an ideal location for applying innovative and integrated water management, In: Stevanović Z, Krešić N, Kukurić N (eds) Karst without boundaries. CRC Press/Balkema, EH Leiden, Taylor & Francis Group, London, pp 3–25
12. Ford D, Williams P (2007) Karst hydrogeology and geomorphology. Wiley, Chichester
13. Stevanović Z (2010) Major springs of southeastern Europe and their utilization. In: Kresic N, Stevanović Z (eds) Groundwater hydrology of springs: engineering, theory, management, and sustainability. Elsevier Inc., Amsterdam, pp 391–412
14. Goldscheider N, Zhao CH, Auler A, Bakalowicz M, Broda S, Drew D, Hartmann J, Jiang G, Moosdorf N, Stevanović Z, Veni G (2020) Global distribution of carbonate rocks and karst water resources. Hydrogeol J 28(5):1661–1677. https://doi.org/10.1007/s10040-020-02139-5
15. Milanović P (1981) Karst hydrogeology. Water Resources Publications, Littleton, CO, p 434
16. Milanović S (2007) Hydrogeological characteristics of some deep siphonal springs in Serbia and Montenegro karst. Environ Geol 51(5):755–759
17. Eftimi R (2010) Hydrogeological characteristics of Albania, AQUA-mundi, AM01012, vol 1, pp 79–92

18. Komatina M (1984) Hydrogeologic features of the Dinaric Karst. In: Mijatović B (ed) Hydrogeology of the Dinaric Karst. Intern. Contrib. to Hydrogeology, IAH, vol 4. Heise, Hannover, pp 55–73
19. Milanović P (2006) Karst of eastern Herzegovina and Dubrovnik littoral. ASOS, Belgrade, p 362
20. Stevanović Z (2021) Karst aquifers of Southeast Europe—essential and rich resource of potable waters, Recueil des Rapports du Comité pour le karst et spéléologie, Acad. Serbe des Sci. et des Arts, vol 11, Belgrade, pp 53–68
21. Gunn J (2021) Karst groundwater in UNESCO protected areas: a global overview. Hydrogeol J 297–314
22. Williams P (2008) World heritage caves and karst. IUCN, Gland, p 57
23. Milanović P (2000) Geological engineering in karst. Zebra Publishing Ltd., Belgrade, p 347
24. Pekas Z, Jolovic B, Radojevic D, Pambuku A, Stevanović Z, Kukurić N, Zubac Z (2012) Unstable regime of Dinaric karst aquifers as a major concern for their sustainable utilization. In: Proceedings of 39 IAH congress, CD Publications, Niagara Falls
25. Radulović M (2000) Karst hydrogeology of Montenegro. Spec. issue of geological Bulletin, vol XVIII. Spec. ed. Geol. Survey of Montenegro, Podgorica, p 271
26. Radulović MM, Radulović M, Stevanović Z, Sekulić G, Radulović V, Burić M, Novaković D, Vako E, Blagojević M, Dević N, Radojević D (2015) Hydrogeology of the Skadar Lake basin (Southeast Dinarides) with an assessment of considerable subterranean inflow. Environ Earth Sci 74(1):71–80. https://doi.org/10.1007/s12665-015-4090-7
27. Živaljević R (1991) Hydrological analysis of karst water flow—Crnojevića River case study (in Serbian). Doctoral Dissertation, University of Montenegro, Podgorica
28. Aller L, Bennet T, Lehr J, Petty R, Hackett G (1987) DRASTIC: a standardized system for evaluating ground water pollution potential using hydrogeologic settings. EPA, Chicago, Illinois, U.S
29. Dörfliger N, Zwahlen F (1997) EPIK: a new method for outlining of protection areas in karstic environment. In: Günay G, Johnson I (eds) Karst water and environmental impacts. Balkema, Rotterdam, pp 117–123
30. Stevanović Z, Marinović V (2020) A methodology for assessing the pressures on transboundary groundwater quantity and quality—experiences from the Dinaric karst. Geol Croat 73(2):107–118. https://doi.org/10.4154/gc.2020.08
31. Zalewski M, Janauer GS, Jolankai G (eds) (1997) Ecohydrology—a new paradigm for the sustainable use of aquatic resources. IHP UNESCO technical documents in hydrology No. 7, Paris, p 58
32. White W (2010) Springwater geochemistry. In: Krešić N, Stevanović Z (eds) Groundwater hydrology of springs: engineering, theory, management and sustainability. Elsevier, Burlington, pp 231–268
33. Cantonati M, Gerecke R, Bertuzzi E (2006) Springs of the Alps, sensitive ecosystems to environmental change: from biodiversity assessments to long-term studies. In: Lami A, Boggero A (eds) Ecology of high altitude aquatic systems in the Alps. Developments of hydrobiology. Hydrobiologia 59–96
34. Cantonati M, Bertuzzi E, Spitale D (2007) The spring habitat: biota and sampling methods. Monografie del Museo Tridentino Scienze Naturali 4. Museo Tridentino di Scienze Naturali, Trento, Italy
35. Springer EA, Stevens LE, Anderson D, Parnell RA, Kreamer D, Flora S (2008) A comprehensive springs classification system: integrating geomorphic, hydrogeochemical, and ecological criteria. In: Stevens LE, Meretsky VJ (eds) Arid land springs in North America: ecology and conservation. University of Arizona Press, Tucson, pp 49–75
36. Stevens LE, Springer EA, Ledbetter DJ (2011) Inventory and monitoring protocols for springs ecosystems. Springs Stewardship Institute, Museum of North Arizona, Flagstaff. http://docs.springstewardship.org/PDF/Springs_Inventory_Protocols_110602.pdf. Accessed 16 March 2021
37. Lučić I (2009) History of studies on Dinaric Karst, using the example of Popovo Polje. Doctoral Dissertation, University of Nova Gorica, Nova Gorica

38. Karaman G (1974) Catalogus Faunae Jugoslaviae, Crustacea Amphipoda (Contribution to the Knowledge of the Amphipoda 60)—Consil. Academ. Scient. SFRY Academia Scient. et Art. Slovenica, Ljubljana 3(3):1–44
39. Karaman G (2012) Further investigations of the subterranean genus Niphargus Schiödte, 1849 (fam. Niphargidae) in Serbia. (Contribution to the Knowledge of the Amphipoda 264). Agric For (Podgorica) 58(2):45–64
40. Sket B (1994) Distribution patterns of some subterranean Crustacea in the territory of the former Yugoslavia. Hydrobiologia 287(1):65–75
41. Schneider-Jacoby M, Schwarz UP, Sackl P, Dhora D, Saveljić D, Štumberger B (2006) Rapid assessment of the ecological value of the Bojana-Buna delta (Albania/Montenegro). Euronatur, Radolfzell, p 102
42. Albrecht C, Wilke T (2008) Ancient Lake Ohrid: biodiversity and evolution; Patterns and Processes of Speciation in Ancient Lakes, Developments in Hydrobiology, Springer, Netherlands. Hydrobiologia 615:103–140
43. Petrović D, Stešević D, Vuksanović S (2008) Material for the Red Book of the flora of Montenegro. In: Proceedings of III Intern. Symp. of ecologists of the Republic of Montenegro. Natura Montenegrina 7:605–631
44. Talevski T, Milošević D, Marić D, Caković D, Talevska M, Talevska A (2009) Biodiversity of Ichthyofauna from Lake Prespa, Lake Ohrid and Lake Skadar. Biotechnol Biotechnol Equip 23(2):400–404
45. Pešić V, Karaman GS, Kostianoy AG (eds) The Skadar/Shkodra lake environment. Springer, Berlin, Heidelberg
46. Pešić V, Paunović M, Kostianoy A (eds) The rivers of Montenegro. In: The handbook of environmental chemistry, vol 93. Springer, Cham
47. Delić T, Kapla A, Colla A (2020) Orogeny, sympatry and emergence of a new genus of Alpine subterranean Trechini (Coleoptera: Carabidae). Zool J Linn Soc 189(4):1217–1231
48. Stevanović Z, Blagojević M (eds) (2021) Hydrogeology and climate changes impact on aquifer systems of Drina River basin. Ministry of Agriculture, Forestry and Water Management of Montenegro, Podgorica, p 315
49. Stevanović Z (2021) The impact of climate changes on groundwater. In: Proceedings of the Intern. Conf. "100 years of Milanković's theory on climate changes", 18–19 Nov. 2020. Soc. "Milutin Milanković", Belgrade, pp 157–176
50. EU Commission (2015) Ecological flows in the implementation of the Water Framework Directive, Guidance document No 31, Brussels
51. Hrvatske vode (2009) Strategy for water management (in Croatian), Zagreb
52. Official Gazette of the Republic of Croatia, 66/2019: Water Law (Zakon o vodama), Zagreb
53. Mišetić S, Pavlin Ž (2004) Approach to define ecological flow in the Republic of Croatia. In: Proceedings of the seminar "High and Low Waters" (in Croatian), Soc. of Civil Eng. Zagreb and Croat. Hydrol Soc., Zagreb, pp 205–221
54. Bonacci O (2003) Ecohydrology of water resources and open streams (in Croatian), Građevinsko-arhitektonski fakultet Sveučilišta u Splitu i IGH, p 487
55. Official Gazette (SN) FBiH 4/2013 (2013) Ordinance on the definition of ecological flow (Pravilnik o načinu određivanja ekološki prihvatljivog protoka)
56. Stevanović Z (2015) Engineering regulation of karstic springflow to improve water sources in critical dry periods. In: Stevanović Z (ed) Karst aquifers—characterization and engineering. Springer Intern. Publ., Switzerland, pp 490–530. https://doi.org/10.1007/978-3-319-12850-4
57. Stevanović Z (2009) Utilization and regulation of springs, In: Kresic N, Stevanović Z (eds) Groundwater hydrology of springs: engineering, theory, management and sustainability. Elsevier Inc., BH, Burlington-Oxford, pp 339–388
58. Vasić Lj, Stevanović Z, Milanović S, Petrović B (2014) Attenuation of bacteriological contaminants in karstic siphons and relative barrier purifiers—case examples from Carpathian karst in Serbia. In: Andreo B, Carrasco F, Duran JJ, Jimenez P, LaMoreaux JW (eds) Hydrogeological and environmental investigations in karst systems, environmental earth sciences, vol 1. Springer, Cham, pp 449–456

59. Milanović S, Vasić Lj (2015) Monitoring of karst groundwater. In: Stevanović Z (ed) Karst aquifers—characterization and engineering. Springer Intern. Publ., Switzerland, pp 47–126
60. Blagojević M, Stevanović Z, Radulović MM, Marinović V, Petrović B (2020) Transboundary groundwater resource management: needs for monitoring the Cijevna River Basin (Montenegro–Albania). Environ Earth Sci Springer 79:74. https://doi.org/10.1007/s12665-020-8809-8
61. Eftimi R (2020) Karst and karst water resources of Albania and their management. Carbonate Evaporites 35:1–14. https://doi.org/10.1007/s13146-020-00599-0
62. Stevanović Z, Maran-Stevanović A (2021) Monitoring as the key factor for sustainable use and protection of groundwater in karst environments. Sustainability 13:5468. https://doi.org/10.3390/su13105468

Chapter 2
Small Standing-Water Ecosystems in the Transitional Temperate Climate of the Western Balkans

Maria Špoljar, Spase Shumka, Orhideja Tasevska, Tea Tomljanović, Aleksandar Ostojić, Anita Galir Balkić, Jasna Lajtner, Bledar Pepa, Tvrtko Dražina, and Ivančica Ternjej

Abstract Small standing-water ecosystems (SWE, i.e. ponds, lakes, reservoirs), natural or anthropogenic origin, dominate in the global landscape, contributing to the high diversity of habitats and species as well as environmental heterogeneity. Water chemistry, morphometry, climate and the level of human activities are extremely

M. Špoljar (✉) · J. Lajtner · T. Dražina · I. Ternjej
Faculty of Science, Department of Biology, University of Zagreb, Rooseveltov trg 6, 10000 Zagreb, Croatia
e-mail: maria.spoljar@biol.pmf.hr

J. Lajtner
e-mail: jasna.lajtner@biol.pmf.hr

T. Dražina
e-mail: tvrtko.drazina@biol.pmf

I. Ternjej
e-mail: ivancica.ternjej@biol.pmf.hr

S. Shumka
Faculty of Biotechnology and Food, Agricultural University of Tirana, Kodër Kamëz, SH1, 1000 Tirana, Albania

O. Tasevska
Department of Zooplankton, Hydrobiological Institute in Ohrid, Naum Ohridski 50, 6000 Ohrid, North Macedonia
e-mail: orhidejat@hio.edu.mk

T. Tomljanović
Faculty of Agriculture, Department of Fisheries, Apiculture, Wildlife Management and Special Zoology, University of Zagreb, Svetošimunska 25, 10000 Zagreb, Croatia
e-mail: ttomljanovic@agr.hr

A. Ostojić
Faculty of Science, University of Kragujevac, Radoja Domanovića 12, 34000 Kragujevac, Serbia
e-mail: aleksandar.ostojic@pmf.kg.ac.rs

A. G. Balkić
Department of Biology, Josip Juraj Strossmayer University of Osijek, Ul. Cara Hadriana 8/A, 31000 Osijek, Croatia
e-mail: agalir@biologija.unios.hr

V. Pešić et al. (eds.), *Small Water Bodies of the Western Balkans*, Springer Water,
https://doi.org/10.1007/978-3-030-86478-1_2

important in the shaping of the functioning of SWE. The transitional climate in the Western Balkans region represents a collision of different impacts and ecotone from colder- to warmer-temperate and Mediterranean climates. Peculiarities of the SWE ecosystem functioning highlight ecology of understudied aquatic biocoenoses, especially zooplankton component, and its comprehensive abiotic–biotic interplays. Results revealed high invertebrates diversity, functional traits of biocoenotic components, and their trophic network. SWE are recognized as resilient ecosystems, capable to recover, however within limited conditions. The present survey points out recognition of ecological and social importance of SWE, and contribute to the conservation and sustainability of these important and vulnerable ecosystems.

Keywords Biodiversity · Zooplankton: Rotifera · Cladocera · Copepoda · Macroinvertebrates · Fish · Alternative states transparent versus turbid · Ecosystem functioning

2.1 Introduction

Small standing water ecosystems (SWE) mainly related to shallow water bodies (SWB), of either natural or artificial origin, perennial or temporary, dominate the global landscape, including pools, ponds, small lakes, and wetlands. Their share is estimated at 30% of the standing water surface of the Earth [1, 2]. SWE are defined variously regarding an area and depth, e.g.: (1) as shallow (relative depth < 3 m) lentic water bodies, with a surface range of 10^3–10^6 m^2 [3]; (2) as shallow (<20 m deep) and small lentic water bodies ranging in surface between 1 m^2 and several hectares (1 ha = 10,000 m^2), ≈10 ha [4, 5]. In this survey, ecosystem functioning will be presented in 28 SWB and take into consideration exclusively permanent standing water bodies in the extended frame related to area and depth, thus, ranged <1 ha up to 35 ha and in-depth <1 m up to 16 m, respectively. The majority of the studied SWB fits with morphometry into the core interpretation by Scheffer [3] related to the area (400–10,000 m^2) and depth (relative depth <3 m), 17 and 22 SWB, respectively (Fig. 2.1 and Table 2.1; [6–10]). According to scarce data, the share of standing water on the Western Balkan in the national territory is <3% in Albania, and ≈0.45% in Croatia, where freshwater SWE significantly contribute.

A broad spectrum of origin could be recognized among SWB: natural depression, flow-through lake/riverine lake, glacial, flood-plane, but lately many artificial SWE arise, e.g.: gravel pits, fish ponds, reservoirs, urban ponds and lakes, mainly with strong anthropogenic impact and heavily modified (Table 2.1). Nine of studied SWE are situated within a natural protected area, and a majority of them are without any protection and legislation, and this ratio unfortunately, reflect the worldwide interplay between protected/controlled and threatened SWE [6–10]. From the second half of twentieth century to-date awareness of the importance and conservation of SWB

B. Pepa
Faculty of Technical Sciences, Vlora University, 9400 Vlora, Albania

Fig. 2.1 Map of studied SWE across Western Balkans (up) and heterogeneity of different small waterbodies: (the second row, left to right) glacial Valamara Lake (A8), transparent Jankovac Reservoir with submerged macrophyte *Hippuris vulgaris* (Cro3), turbid Krapina Oxbow without macrophytes (Cro5). *United Nations Security Council Resolution 1244. Photo by S. Shumka and M. Špoljar

increased as reflected in the establishment of: (i) Ramsar convention related to the protection of wetland habitats (1971); (ii) Natura 2000 network in the European Union; and (iii) European Pond Conservation Network (EPCN, established in 2004).

SWE are of great importance providing multiple ecosystem services, e.g., important in irrigation and other agricultural activities, aquaculture, primarily as fish ponds and are destinations for numerous sports and recreational activities, habitat and landscape heterogeneity and species diversity, particularly invertebrates, and crucial habitat for amphibians. Although SWE are worldwide abundant, and their ecological and civilization importance is confirmed, generally are neglected and overlooked by social and scientific communities, i.e., scientific monitoring, policy and consequently in conservation. Also, they are exposed to various anthropogenic impacts manifested in eutrophication as anthropogenically enhanced nutrient enrichment, an inflow of toxic effluents, backfilling with various solid waste and soil, hydro-morphological degradation and allochthonous/invasive species introduction [11]. More than 50% of SWE in this survey is of artificial origin (i.e., reservoirs), and some of them are under heavy anthropogenic stressors (municipal effluents, agricultural leaching, addition of fertilizers in fish ponds). Thus, SWE joined to the global problem of habitats loss resulting in loss of biodiversity, as consequence of deterioration in the ecosystem functioning. All these ecosystems are distinguished by significant diversity of sensitive habitats and species, which so far are understudied and remain extremely vulnerable due to significant anthropogenic impact [12–14].

Table 2.1 Morphometry and biotic features related to the studied SWE across Western Balkans

	SWE hydronim	Abbreviation	Origin/type of SWE	Protected area (PA)	Area (m^2)	Depth (Z_{max}, m)	Macrophyte coverage (%)	Macrophyte type	Fish presence
Small and shallow waterbodies	Lake of Rajca	A11	Artificial/reservoir	National Park Shebenik-Jabllanica	1800	9	5	S	0
	Lake of Dragan	A4	Natural/glacial	National Park Shebenik-Jabllanica	600	3	90	S	0
	Lake i Zi (Black)	A5	Natural/glacial	Nature Park Korab Korritnik	4800	11	35	S	1
	Lake of Valamara	A8	Natural/glacial		800	14	0	0	0
	Sutla1	Cro1	Natural backwater		1984	3	40	S	1
	Sutla2	Cro2	Natural backwater		2632	4	0	0	1
	PapukR1	Cro3	Artificial/reservoir	Nature Park Papuk	8736	1.9	95	S	1
	PapukR2	Cro4	Artificial/reservoir	Nature Park Papuk	6630	1.95	98	S	1
	Škrčev kut	Cro5	Artificial/oxbow		3600	1.6	0	F	1
	Krapina oxbow1	Cro6	Artificial/oxbow		5550	4	5.5	E	1
	Krapina oxbow2	Cro7	Artificial/oxbow		3888	3	7.8	E	1
	Pond Mali Sakadaš	Cro8	Natural/floodplain lake	Nature Park Kopački rit	4112	0.726		E	1
	Pond I	Sr1	Artificial/gravel pit		894	0.88	30	E	1
	Pond II	Sr2	Artificial/gravel pit		1883	1.5	50	S	1
	Pond III	Sr3	Artificial/gravel pit		1621	0.783	20	S	1

(continued)

Table 2.1 (continued)

	SWE hydronim	Abbreviation	Origin/type of SWE	Protected area (PA)	Area (m^2)	Depth (Z_{max}, m)	Macrophyte coverage (%)	Macrophyte type	Fish presence
	Pond IV	Sr4	Artificial/oxbow		429	1.16	5	E	1
	Pond V	Sr5	Artificial/oxbow		6341	2.13	50	E	1
Large waterbodies	Lake of Pernaska	A1	Natural lake		324,700	6	10	S	1
	Reservoir of Poloska	A10	Artificial/reservoir		72,000	9	30	E	1
	Lake of Dega	A2	Natural lake	Nature monument	308,600	4.5	85	S	1
	Lake of Dorbi	A3	Natural lake		105,400	5	20	E	1
	Reservoir of Fushe Studa	A5	Artificial/reservoir	National Park Shebenik-Jabllanica	245,000	14	5	S	1
	Lake of Dushku	A7	Natural/glacial		224,000	16	45	E	1
	Reservoir of Pretusha	A9	Artificial/reservoir		351,000	12	10	S	1
	Lake Sakadaš	Cro9	Natural/floodplain Lake	Nature Park Kopački rit	127,143	4.87	3	S	1
	Konče 1	NM1	Artificial/reservoir		20,000	6	20	S	1
	Konče 3	NM2	Artificial/reservoir		19,000	6	20	S	1
	Pond VI	Sr6	Artificial/oxbow		14,206	1.83	5	S	1

2.2 Environmenal Features of Small Waterbodies in the Temperate Climate

Small water bodies compared to larger lentic freshwater ecosystems and deep lakes are characterized by a lower area/perimeter ratio that highlights considerable impact of ecotone zone to their metabolism and functioning [15]. SWE are considered polymictic waterbodies, with the horizontal distribution of abiotic and biotic factors [3, 16] Alternative Stable States Theory, ASS [17], includes alteration of SWE between transparent and turbid state of the water. A fundamental role in the transition between those states is phosphorous concentration, where exceeding the threshold over 100 μg P L^{-1} can cause alteration from favourable transparent into the undesirable turbid state. The ratio between the two types of primary producers, macrophytes *vs.* phytoplankton is known to be related to transparency *vs.* turbidity, i.e., macrophytes maintain a transparent state, opposite to the phytoplankton domination in turbid lakes [3, 18, 19]. Submerged macrophytes, with a complex habitus architecture particularly act as ecosystem engineers by modifying abiotic conditions (increase dissolved oxygen concentrations and light availability, and reduce nutrient concentrations and sediment resuspension) and biotic interactions [19–23]. Both, biological and water quality elements should be in equilibrium to maintain sustainable SWE ecosystem functioning (Fig. 2.2).

The transparent state is generally featured by environmental conditions and biotic interplays: lower total phosphorous concentration, submerged macrophytes coverage, abundant large-sized algivore zooplankton (cladoceran *Daphnia*), lower abundance of planktivore fish and top-down control in the ecosystem over predatory

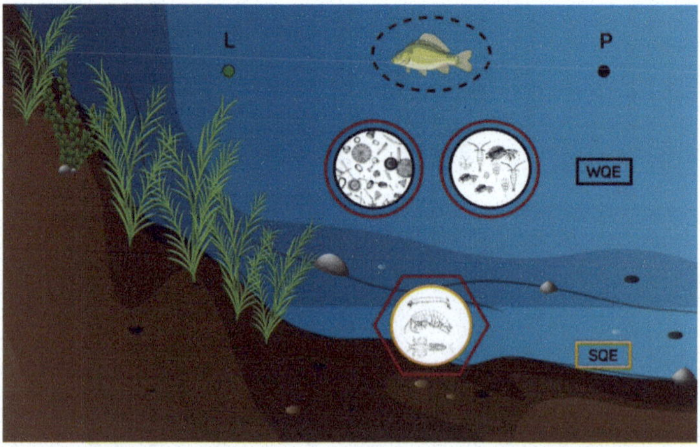

Fig. 2.2 Key aquatic chemical and biological elements—in the assessment of the functioning of SWE requested are analyses of: water quality elements (WQE), sediment quality elements (SQE), primary producers (phytoplankton, submerged macrophytes), primary consumers (zooplankton) and predators, including both visual (fish) and tactile (macroinvertebrates) in pelagial (P) and littoral (L)

fish. Submerged macrophytes in the littoral zone enable the shelter to the large-sized zooplankters against fish during the day. For many zooplankters littoral zone with macrophytes is not an adequate surviving solution, because there encounter macroinvertebrate, such as tactile predators [24, 25]. Also, many studies suggested the existence of different alternative states at the same trophic conditions, where the absence of macrophyte coverage and presence of cyprinids could cause a shift into a turbid state [22, 25].

In the turbid state different scenario prevailed: bottom-up control governed ecosystem functioning, affected by higher concentrations of nutrients (phosphorous, nitrogen) and reflect in increased phytoplankton biomass (indicated often through chlorophyll *a* concentration), reduced or absent macrophytes, or eventually the presence of floating or emerged macrophytes coverage, abundant small-sized detritivore zooplankton (i.e., cladoceran *Bosmina*, rotifer *Keratella, Brachionus*), increased abundance of planktivore fish and reduced predatory fish. In such environmental condition does not exist strong control of phytoplankton, thus productivity causes eutrophication and deteriorated ecosystem functioning. ASS obtains principles of SWB functioning in the colder temperate climate in NW Europe, while on the lower geographic latitudes, with the higher impact of warmer Mediterranean climate SWE deviated from that pattern [26, 27]. That fact is very important to notice in the global climate changes with a pronounced effect of warming, and thus predict and prevent such scenario in temperate climate. Even within temperate climate across latitudinal gradient differences in environmental conditions could be recently observed [28], as higher chlorophyll *a* and phosphorus values in northern waterbodies (e.g., Poland) opposite to higher oxygen concentration and transparency in southern waterbodies (e.g., Croatia).

Results conducted in presented SWE suggest nutrient concentrations as main drivers in the ecosystem functioning reflected in increased turbidity, the concentration of dissolved and particulate organic matter, as well as chlorophyll *a* [6, 8, 10, 25, 29]. Share of the macrophyte coverage in the SWE surface, the complexity of macrophyte architecture and fish size structure particularly impact zooplankton component [6, 7, 25, 30]. Most of the studied SWE are confirmed as eu- to hypertrophic, in different ASS, except transparent Cro3 and Cro4 which are in slightly mesotrophic condition. Regardless of the presented, the switch between transparent *vs.* turbid state could be triggered by various impacts, and accordingly control of entire water quality elements could enhance survival of diverse SWE and their biodiversity.

2.3 Zooplankton Assemblage in Temperate Shallow Ecosystems

Zooplankton organisms, with main groups: Rotifera, Cladocera and Copepoda (Fig. 2.3), contribute significantly to the aquatic ecosystem diversity, play a pivotal role between primary producers and higher trophic levels, and are also known as very

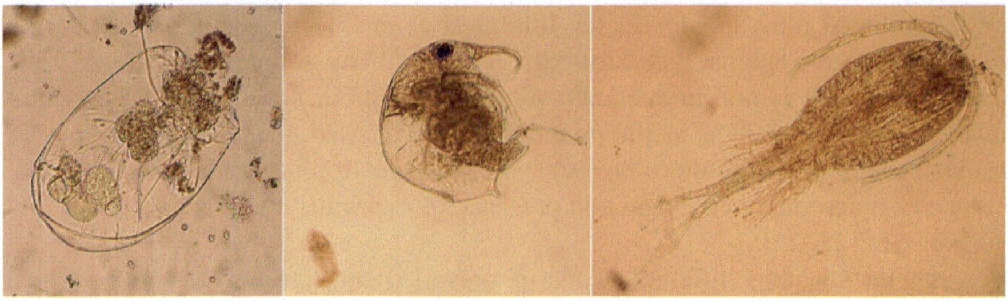

Fig. 2.3 Main representatives of the zooplankton *Asplanchna priodonta* (eurytopic predatory rotifer), *Bosmina longirostris* (eurytopic detritivore cladoceran) and *Thermocyclops oithnoides* (omnivorous copepod) Photo by A. Galir Balkić

sensitive indicators of the environmental alterations [31–37]. Zooplankton indicator value arises from rapid growth, numerous populations and quick response to deteriorated environmental conditions, and application of zooplankton traits as a tool in the ecological assessment should be developed for small-shallow ecosystems [6, 38].

Many indices based on zooplankton traits, e.g. diversity, abundance, biomass, body size, functional, phylogenetic, ecological, and feeding guilds are commonly used in ecological status assessment [28, 30, 39, 40]. Zooplankton, as primary consumers in lentic ecosystems, mediates in food network between the top-down and bottom-up factors, and is less studied than phytoplankton, macroinvertebrates and fish, particularly due to it is omitted among the biological quality elements from the Water Framework Directive [41] in the European Union. Hence, more efforts should be input to elucidate zooplankton trophic importance in the ecosystem balance and water quality assessment.

Overview of zooplankton assemblage across 28 SWE of Western Balkan (A4–A11 SWE in Albania were only qualitatively checked) was selected during the summer period as optimal for macrophyte and zooplankton growth [42]. Overall 155 zooplankton species were identified, 112 in Croatia, 84 in Serbia, 36 in Albania, and 23 in North Macedonia (Table 2.2). In zooplankton diversity rotifer share 80% (120 species), while distinctly lower contributed Cladocera and Copepoda, 16% and 6%, 25 and 16 species, respectively (Table 2.2). Many species recorded in this survey as limited to specific region, are previously recorded across the studied region, but probably in different seasons and/or different localities [7, 35, 43–48].

Zooplankton assemblage among SWE differed moderately but significant (Analysis of similarities, ANOSIM $r = 0.33$, $p < 0.005$), and indicate notable differences between heavily anthropogenic stressed SWE (Sr 1–6; ANOSIM $r = 0.30–0.97$, $p = < 0.005$) in comparison to others, less exposed to human activities or in the protected area (Fig. 2.4). Dense macrophyte coverage remarkably support higher zooplankton diversity (ANOSIM $r = 0.9$, $p < 0.001$), and it is with compliance with the zooplankton study in ponds in Croatia [7, 25] and in the central-western region of Poland [11]. In that assessment, in 278 ponds in NW Poland 355 zooplankton species

Table 2.2 Zooplankton diversity in 28 SWE across Western Balkans with species functional traits related to habitat and feeding guilds (FFG). Abbreviations: Croatia (Cro), Albania (A), North Macedonia (NM), Serbia (Sr)

	Habitat	FFG	Cro	A	NM	Sr		Habitat	FFG	Cro	A	NM	Sr
Rotifera							**Rotifera**						
Anuraeopsis fissa (Gosse 1851)	P	D	*	*		*	*Kellicottia longispina* (Kellicott 1879)	P	D		*	*	
Ascomorpha ecaudis (Perty 1850)	P	A	*			*	*Keratella cohlearis* (Gosse 1851)	P	D	*	*	*	*
Ascomorpha ovalis (Bergendahl 1982)	P	A	*			*	*Keratella quadrata* (Müller 1786)	P	D	*	*		*
Ascomorpha saltans (Bartsch 1870)	P	A	*				*Keratella tecta* (Gosse 1851)	P	D	*			*
Asplanchna priodonta (Gosse 1850)	P	P	*	*	*	*	*Keratella testudo* (Ehrenberg 1832)	P	D				*
Asplanchnopus multiceps (Schrank 1793)	P	P	*				*Keratella valga* (Ehrenberg 1834)	P	D				*
Bdelloidea	L/B	D	*	*	*	*	*Lecane acus* (Harring 1913)	L	D	*			
Brachionus angularis (Gosse 1851)	P	D	*	*	*		*Lecane bulla* (Gosse 1851)	L	D	*	*		*
Brachionus angularis bidens (Plate 1886)	P	D				*	*Lecane closterocerca* (Schmarda 1859)	L	D	*	*	*	*
Brachionus budapestanensis (Daday 1885)	P	D				*	*Lecane cornuta* (Muller 1786)	L	D	*			*
Brachionus calyciflorus (Pallas 1766)	P	D				*	*Lecane flexilis* (Gosse 1886)	L	D	*			*
Brachionus caudatus (Barrois and Daday 1894)	P	D	*				*Lecane furcata* (Murray 1913)	L	D				*
Brachionus diversicornis (Daday 1883)	P	D	*			*	*Lecane hamata* (Stokes 1896)	L	D				*
Brachionus falcatus (Zacharias 1898)	P	D	*		*	*	*Lecane inermis* (Bryce 1892)	L	D	*			*

(continued)

Table 2.2 (continued)

	Habitat	FFG	Cro	A	NM	Sr		Habitat	FFG	Cro	A	NM	Sr
Brachionus forficula (Wierzejski 1891)	P	D				*	*Lecane ludwigi* (Eckstein 1883)	L	D				*
Brachionus patulus (Müller 1786)	P	D				*	*Lecane luna* (Müller 1776)	L	D	*	*	*	*
Brachionus quadridentatus (Hermann 1783)	L	D	*			*	*Lecane lunaris* (Ehrenberg 1832)	L	D	*	*	*	*
Brachionus quadridentatus melheni (Barrois and Daday 1894)	P	D				*	*Lecane nana* (Murray 1913)	L	D				*
Brachionus urceolaris (Müller 1773)	P	D				*	*Lecane pyriformis* (Daday 1905)	L	D				*
Cephalodella auriculata (Muller 1773)	L	O	*				*Lecane quadridentata* (Ehrenberg 1832)	L	D	*			*
Cephalodella catelina (Müller 1786)	L	P				*	*Lepadella patella* (Müller 1786)	L	D	*			*
Cephalodella elegans (Myers 1924)	L	P	*				*Lepadella rhomboides* (Gosse 1886)	L	D				*
Cephalodella gibba (Ehrenberg 1832)	L	O	*	*		*	*Lophocaris oxysternon*	L	D	*			
Cephalodella tenuiseta (Burn 1890)	L	P				*	*Lophocaris salpina* (Ehrenberg 1834)	L	D	*			
Cephalodella ventripes (Dixon-Nuttall 1901)	L	P				*	*Macrochaetus collinsi* (Gosse 1867)	L	D	*			
Collotheca campanulata (Dobie 1849)	P	D	*				*Monommata caudata* (Myers 1930)	L	D				*
Coulrella adriatica (Ehrenberg 1831)	L	D	*			*	*Monommata dentata* (Wulfert 1940)	L	D				*

(continued)

Table 2.2 (continued)

	Habitat	FFG	Cro	A	NM	Sr		Habitat	FFG	Cro	A	NM	Sr
Colurella obtusa (Glascott 1893)	L	D	*			*	*Monommata grandis* (Tessin 1890)	L	D	*			
Colurella uncinata (Müller 1773)	L	D	*			*	*Monommata longiseta* (Müller 1786)	L	D	*			
Cupelopagis vorax (Leidy 1857)	P	P	*				*Mytilina mucronata* (Müller 1773)	L	D	*			*
Dicranophorus caudatus	L	P	*				*Mytilina ventralis brevispina* (Ehrenberg 1830)	L	D	*			*
Dicranophorus forcipatus (Müller 1786)	L	P	*				*Mytilina videns* (Levander 1894)	L	D	*			
Dicranophorus grandis (Ehrenberg 1832)	L	P	*				*Notholca acuminata* (Ehrenberg 1832)	P	D	*			*
Elosa worrallii (Lord 1891)	L	D	*				*Notholca squamula* (Muller 1786)	P	D	*			
Encentrum aquila (Gosse 1887)	L	D	*				*Notommata* spp.	L	O	*			
Encentrum felis (Muller 1773)	L	D	*				*Platyas quadricornis* (Ehrenberg 1832)	P	D	*			*
Encentrum mustella (Milne 1885)	L	D	*				*Plationus patulus* (Muller 1786)	L	D	*			*
Epiphanes senta (Muller 1773)	P	D	*				*Pleurotrocha petromyzon* (Ehrenberg 1830)	L	O	*			
Euchlanis calpidia (Myers 1930)	L	D	*				*Polyarthra dolichoptera* (Idelson 1925)	P	D				*
Euchlanis deflexa (Gosse 1851)	L	D	*				*Polyarthra remata* (Skorikov 1896)	P	D				*
Euchlanis dilatata (Ehrenberg 1832)	L	D	*	*		*	*Polyarthra dolichoptera + vulgaris*	P	A	*			
Euchlanis meneta (Myers 1930)	L	D	*				*Polyarthra vulgaris* (Carlin 1943)	P	A		*	*	
Euchlanis oropha (Gosse 1887)	L	O	*				*Pompholyx complanata* (Gosse 1851)	P	D	*		*	*
Euchlanis phryne (Myers 1930)	L	D	*				*Pompholyx sulcata* (Hudson 1885)	P	D	*	*		

(continued)

Table 2.2 (continued)

	Habitat	FFG	Cro	A	NM	Sr		Habitat	FFG	Cro	A	NM	Sr
Euchlanis triquetra (Ehrenberg 1838)	L	D	*				*Ptygura* spp.	P	D	*			
Filinia longiseta (Ehrenberg 1834)	P	D	*	*			*Scaridium longicauda* (Müller 1786)	L	D	*			*
Filinia opoliensis (Zacharias 1898)	P	D			*		*Squatinella mutica tridentata* (Ehrenberg 1832)	L	D	*			*
Filinia passa (Müller 1786)	P	D				*	*Squatinella rostrum* (Schmarda 1846)	L	D				*
Gastropus stylifer (Imhof 1891)	P	A	*	*	*	*	*Synchaeta pectinata* (Ehrenberg 1832)	P	A	*	*		*
Harringia eupoda (Gosse 1887)	P	P	*				*Synchaeta* sp.	P	A				*
Hexarthra fennica (Levander 1892)	P	D	*				*Testudinella mucronata* (Gosse 1886)	L	D	*			
Hexarthra mira (Hudson 1871)	P	D	*	*	*	*	*Testudinella parva* (Ternetz 1892)	L	D				*
Itura aurita (Ehrenberg 1830)	L	D	*				*Testudinella patina* (Hermann 1783)	L	D				*
Rotifera							**Cladocera**						
Testudinella reflexa (Gosse 1887)	L	D	*				*Diaphanosoma brachyurum* (Liévin 1848)	P	A		*		
Trichocerca capucina (Wierzejski and Zacharias 1893)	P	O	*	*			*Graptoleberis testudinaria* (Fischer 1851)	L	D				*
Trichocerca cylindrica (Imhof 1891)	P	O	*				*Moina brachiata* (Jurine 1820)	P	D	*			
Trichocerca dixon-nuttalli (Jennings 1903)	P	A				*	*Picripleuroxus laevis* (Sars 1862)	L	D				*
Trichocerca gracilis (Tessin 1890)	P	O				*	*Pleuroxus striatus* (Schödler 1862)	L	D	*			
Trichocerca iernis (Gosse 1887)	P	A				*	*Pleuroxus truncatus* (Muller 1785)	L	D	*			

(continued)

Table 2.2 (continued)

	Habitat	FFG	Cro	A	NM	Sr		Habitat	FFG	Cro	A	NM	Sr
Trichocerca porcellus (Gosse 1851)	L	O				*	*Pleuroxus uncinatus* (Baird 1850)	L	D	*			
Trichocerca pussila (Jennings 1903)	P	A				*	*Scapholeberis mucronata* (Muller 1776)	L	A	*			*
Trichocerca rattus minor (Müller 1776)	P	A				*	*Sida crystallina* (Müller 1776)	L	A				*
Trichocerca similis (Wierzejski 1893)	P	A	*		*	*	*Simocephalus expinosus* (De Geer 1778)	L/P	A	*			
Trichocerca tigris (Muller 1786)	P	A	*				*Simocephalus serrulatus* (Koch 1841)	L/P	A	*			
Trichocerca sp. (Lamarck 1801)	P	A			*		*Simocephalus vetulus* (Muller 1776)	L/P	A	*	*		*
Trichotria pocillum (Müller 1776)	L	O	*										
Trichotria tetractis (Ehrenberg 1830)	L	O				*							
Cladocera							**Copepoda**						
Alona affinis (Leydig 1860)	L	D	*				*Acanthocyslops vernalis* (Fischer 1853)	L/P	A/P	*			*
Alona costata (Sars 1862)	L	D	*			*	*Arctodiaptomus salinus* (Daday 1885)				*		
Alona guttata (Sars 1862)	L	D	*				*Cyclops sp.*	P	A/P	*	*	*	
Alona quadrangularis (Müller 1776)	L/P	D				*	*Cyclops abysorum* (Sars 1863)	P	A/P		*		
Alonella excisa (Fischer 1854)	L	D	*				*Eucyclops serrulatus* (Fischer 1851)	P	A/P		*		
Alonella nana (Baird 1843)	L	D	*				*Macrocyclops albidus* (Jurine 1820)	L	A/P	*			
Bosmina longirostris (Müller 1785)	P	D	*	*	*	*	*Macrocyclops fuscus* (Jurine 1820)	L	A/P	*			
Ceriodaphnia quadrangula (Muller 1785)	L	A	*	*	*		*Mesocyclops leuckarti* (Claus 1857)	P	A/P	*	*		

(continued)

Table 2.2 (continued)

	Habitat	FFG	Cro	A	NM	Sr		Habitat	FFG	Cro	A	NM	Sr
Ceriodaphnia reticulata (Jurine 1820)	L	A		*		*	*Thermocyclops crassus* (Fischer 1853)	L	A/P				*
Chydorus sphaericus (Muller 1776)	L/P	D	*	*	*	*	*Thermocyclops oithonoides* (Sars 1863)	P	A/P	*			
Daphnia cuculata (Sars 1862)	P	A		*	*		*nauplii*		A	*	*	*	*
Daphnia longispina (Müller 1875)	P	A	*	*			*copepodites*		A/P	*	*	*	*
Daphnia pulex (Leydig 1860)	P	A	*										

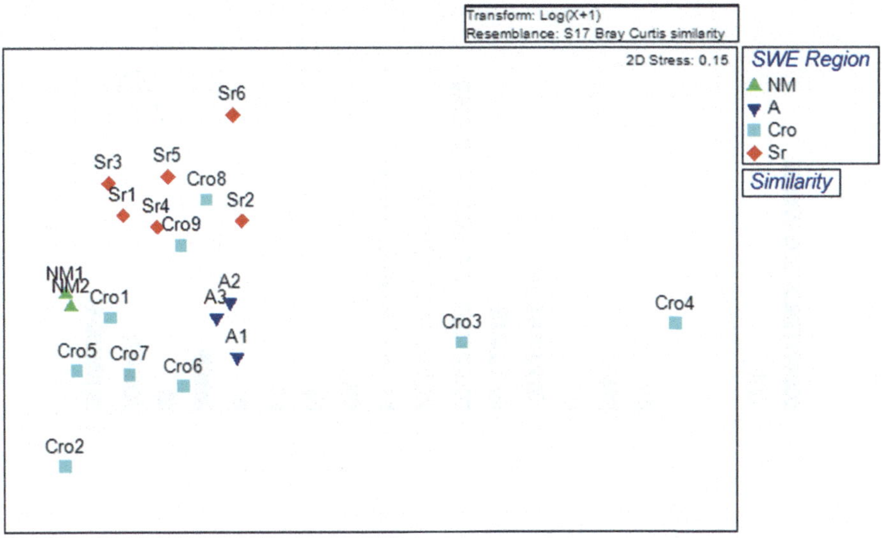

Fig. 2.4 Zooplankton assemblage (di)similarity plot indicating environmental similarity based on regional as well as ecological specificity, and separating Cro3 and Cro4 with low anthropogenic impact against other SWE (Non-metric Multidimensional Scaling, NMDS)

were identified (258 rotifer, 63 cladoceran, 34 copepod species), while within another group of 27 ponds reciprocally less diversity of 130 zooplankton species (93 rotifer, 24 cladoceran, 13 copepod species), was noted. Hereby, besides abiotic and biotic parameters, diversity depends on many factors included in the sampling strategy, e.g. spatial and temporal occasions, sampler/net, number and heterogeneity of SWE.

According to results, widespread species in SWE of Western Balkans are: among rotifers—*Asplanchna priodonta* (Gosse 1850), *Gastropus stylifer* (Imhof 1891), *Hexarthra mira* (Hudson 1871), *Keratella cohlearis* (Gosse 1851) in pelagal, and Bdelloidea, *Cephalodella* spp., *Lecane closterocerca* (Schmarda 1859), *Lecane luna* (Müller 1776) and *Lecane lunaris* (Ehrenberg 1832) in littoral; among cladocerans—*Bosmina longirostris* (O. F. Müller 1785) in pelagial and *Chydorus sphaericus* (O. F. Muller 1776) in littoral; among copepods—larval stages, nauplii and copepodites, *Acanthocyslops vernalis* (Fischer 1853) and *Mesocyclops leuckarti* (Claus 1857) in pelagial (Table 2.2).

Zooplankton abundance ranged from 1000 up to 12,000 Ind. L^{-1}, with rotifers share from 40 up to 95%. Extremely low abundance ($\approx$20 Ind. L^{-1}) was recorded in slightly mesotrophic SWE, Cro3 and Cro4, in the Nature Park Papuk, Croatia (Fig. 2.5). Algivores, large-sized cladocerans were present in low abundance and share, except in Cro3 and Cro4, 72 and 79%, respectively. Namely, large-sized cladocerans are important as efficient grazers over phytoplankton, and suggesting a balanced and transparent ecosystem. Small-sized cladocerans are mainly detritivores

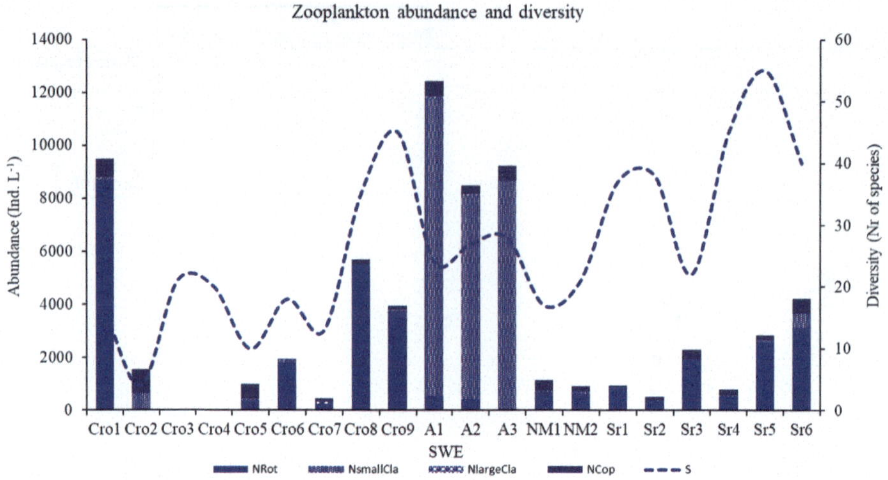

Fig. 2.5 Total zooplankton diversity (number of species) and abundance (Ind. L^{-1}) of Rotifera, small- and large-sized Cladocera and Copepoda. Abbreviation: S—number of species

and indicate increased organic matter and its decay, reflected in increased concentration of suspended, particulate organic matter (Fig. 2.5, SWE A1 to A3). Very rare knowledge related to glacial SWE of Balkan Peninsula indicates that in these lakes, Valamara and i Zi dominated Cyclopoida, *Cyclops abysorum* (Sars 1863), *Eucyclops serrulatus* (Fischer 1851) and *A. vernalis*, and Calanoida, *Arctodiaptomus salinus* (Daday 1885).

Within SWE of Western Balkan's zooplankton assemblage showed dissimilarity from 77 to 81%. Peculiarities for SWE in Albania reveal the prevalence of small-sized cladocerans, e.g. *Alona rectangula* (Sars 1861), *Chydorus sphaericus* (O. F. Muller 1776) and copepod, *M. leuckarti* (Table 2.3). In SWE in Croatia dominated rotifers, but generally in lower abundance than in other SWE, except *Polyarthra* spp. (*P. dolichoptera* + *vulgaris*) and *Trichocerca similis* (Wierzejski 1893) occurred in significantly higher abundance. In SWE in North Macedonia distinct assemblage was caused by a higher abundance of detritivore rotifers *H. mira*, *Brachionus angularis* (Gosse 1851) and *K. cochlearis.* Domination of algivore rotifer *Polyarthra remata* (Skorikov 1896) featured SWE in Serbia. In the survey of Cladocera diversity, its gradient declined from the northern (in Croatia 18 species) to southern (Albania 8, North Macedonia 4 species) SWE. This finding follows results of inter climate [49], and intraclimate [28] studies, confirming less tolerance of cladocerans to a higher temperature, even in a slight gradient along the temperate climate of Western Balkan.

Among cladocerans and copepods almost 50% of species were algivores, while among rotifers only 10% species. These species are planktonic, inhabit pelagial. Among rotifers, many species from genera are detritivores and abundant in pelagial (*Keratella, Hexarthra, Filinia*). Otherwise, the majority of detritivore species inhabit

Table 2.3 (Di)similarity of zooplankton assemblage among SWE (similarity percentages, SIMPER analysis) regarding geographical distance

	Average dissimilarity = 81%			Average dissimilarity = 76%	
Species	A versus Cro	Contrib%	Species	NM versus A	Contrib%
Alona rectangula (Sars 1861)	>	5.54	*Alona rectangula* (Sars 1861)	<	5.45
Chydorus sphaericus (Muller 1776)	>	4.62	*Hexarthra mira* (Hudson 1871)	>	4.44
Mesocyclops leuckarti (Claus 1857)	>	4.24	*Brachionus angularis* (Gosse 1851)	>	3.98
Diaphanosoma brachyurum (Liévin 1848)	>	3.90	*Mesocyclops leuckarti* (Claus 1857)	<	3.94
Kellicottia longispina (Kellicott 1879)	>	3.49	*Chydorus sphaericus* (Muller 1776)	<	3.77
copepodites	>	3.26			
Polyarthra spp.	<	2.10			
	Average dissimilarity = 78%			Average dissimilarity = 77%	
Species	A versus Sr	Contrib%	Species	NM versus Cro	Contrib%
Alona rectangula (Sars 1861)	>	4.48	*Hexarthra mira* (Hudson 1871)	>	6.64
Polyarthra dolichoptera (Idelson 1925)	<	3.36	*Polyarthra vulgaris* (Carlin 1943)	>	4.72
Mesocyclops leuckarti (Claus 1857)	>	3.23	*Brachionus angularis* (Gosse 1851)	>	4.33
Diaphanosoma brachyurum (Liévin 1848)	>	2.94	*Ceriodaphnia quadrangula* (Müller 1785)	>	4.16
Polyarthra remata (Skorikov 1896)	<	2.81	*Gastropus stylifer* (Imhof 1891)	>	3.90
			Keratella cochlearis (Gosse 1851)	>	3.88

(continued)

Table 2.3 (continued)

	Average dissimilarity = 80%			Average dissimilarity = 80%	
Species	Cro versus Sr	Contrib%	Species	NM versus Sr	Contrib%
Polyarthra dolichoptera (Idelson 1925)	<	4.23	*Hexarthra mira* (Hudson 1871)	>	4.19
Anuraeopsis fissa (Gosse 1851)	<	3.73	*Polyarthra dolichoptera* (Idelson 1925)	<	4.06
Polyarthra remata (Skorikov 1896)	<	3.72	*Polyarthra vulgaris* (Carlin 1943)	>	3.43
Trichocerca pusilla (Jennings 1903)	<	1.91	*Keratella cochlearis* (Gosse 1851)	>	3.36
Trichocerca similis (Wierzejski 1893)	>	1.62	*Polyarthra remata* (Skorikov 1896)	<	3.11

the littoral. A huge share of littoral rotifer species in the zooplankton assemblage reflects the heterogeneity of habitat, caused by complex macrophyte architecture, as well as the significant impact of benthic-water coupling particularly prominent in SWE. The advantage of applying of functional traits approach in ecological assessments (e.g. feeding type, habitat selection) is in comprehensive insight into the food web and ecosystem functioning [6, 50], whilst in certain conditions taxon-based approach reveals more distinctions affected by environmental conditions [28].

2.4 Macroinvertebrates Interplays in the Littoral Zone of Shallow Water Ecosystems

Macroinvertebrates, mainly comprised of insect larvae, oligochaetes, gastropods and bivalves, are feeding on periphyton, algae and detritus, plant tissues or are predators that pray over other invertebrates. Complexity and heterogeneity of sediment, food subsidies, as well as macrophyte coverage affect macroinvertebrate assemblage in the littoral zone of SWE [7, 24, 25, 51, 52]. Sediment in the deeper zone is more uniform, as well as macroinvertebrate assemblage mainly consists of oligochaetes. In this survey, results from a few SWE (Cro 1 to Cro 5) are available related to macroinvertebrates, which are mainly considered as tactile predators over zooplankton. In slightly mesotrophic, and macrophyte covered SWE (Cro 3, 4) were recorded approximately 30 species, comprising mainly Diptera, Ephemeroptera, Odonata larvae, Coleoptera and Heteroptera adults, Amphipoda, and Gastropoda [53, 54].

In the eutrophic, complex submerged macrophyte covered SWE (Cro 1) approximately 50 macroinvertebrate taxa were determined, where gastropods, water mites and insect larvae were the most prominent consumers of epiphyton maintain light conditions and macrophyte stands. In hypertrophic floating-leaved SWE twice less ($\approx$20) taxa were identified. Namely, Cro 5 was floating-leaved covered in 2012, while after intense precipitation in 2013 water level increased, and macrophyte coverage was reduced (Table 2.1). Hence, water mites (35%) prevailed among the hornwort (*Ceratophyllum*) coverage in Cro1, adult hemipterans (23%) overcome in Cro2, and in Cro5 dipteran larvae constituted a considerable share (51%) in the littoral ecotone zone, with very sparsely developed emergent macrophytes or their absence. Moreover, very important is possibility of refuge against predators in the sediment, due to results indicated an increased presence of buried planktonic crustaceans in the conditions of the increased abundance of macroinvertebrate predators, mainly dipteran and dragonfly larvae in the littoral zone [24, 25]. Food web modelling indicated zooplankton and epiphyton as crucial food subsidies for invertebrate predators among complex submerged macrophyte than among the floating-leaved macrophyte stands (Fig. 2.6). Namely, this is in concordance with results that macrophyte stands are not always in the function of zooplankton shelter. Moreover, among macrophyte stems zooplankters encounter their macroinvertebrate predators.

2.5 The Fish Component in Small Waterbodies

The ecological importance of fish in the ecosystem is multiple, as they connect trophic webs, and provide carbon, phosphorous and nitrogen circulation from water to the atmosphere [55]. Fish, as visual predators, strongly affect zooplankton assemblage, and are a key factor in the food network and ecosystem balance, as well as in the alteration of transparent to turbid state [25, 49]. The specific biogeography setting of Western Balkan Countries, from one side makes quite complicated the current spatial analyses of its fish assemblage, while from the other side it is a significant European hotspot with a high fish diversity. Following Abell et al. [56] and the biogeographic units, within this area include: Dneister-Lower Danube, Dalmatia, Southeast Adriatic Drainage, Ionian Drainage and Vardar. Similarly, in other water bodies in the Balkans, the cyprinids occur in the freshwater habitat of SWE as dominant group, >70% in assemblage, including endemic species. Thus, they present 59% of the total species in the Balkan freshwater fish fauna [57]. This is due to the high presence of non-native fish species and allochthonous introduction for aquaculture and sport fishing purposes.

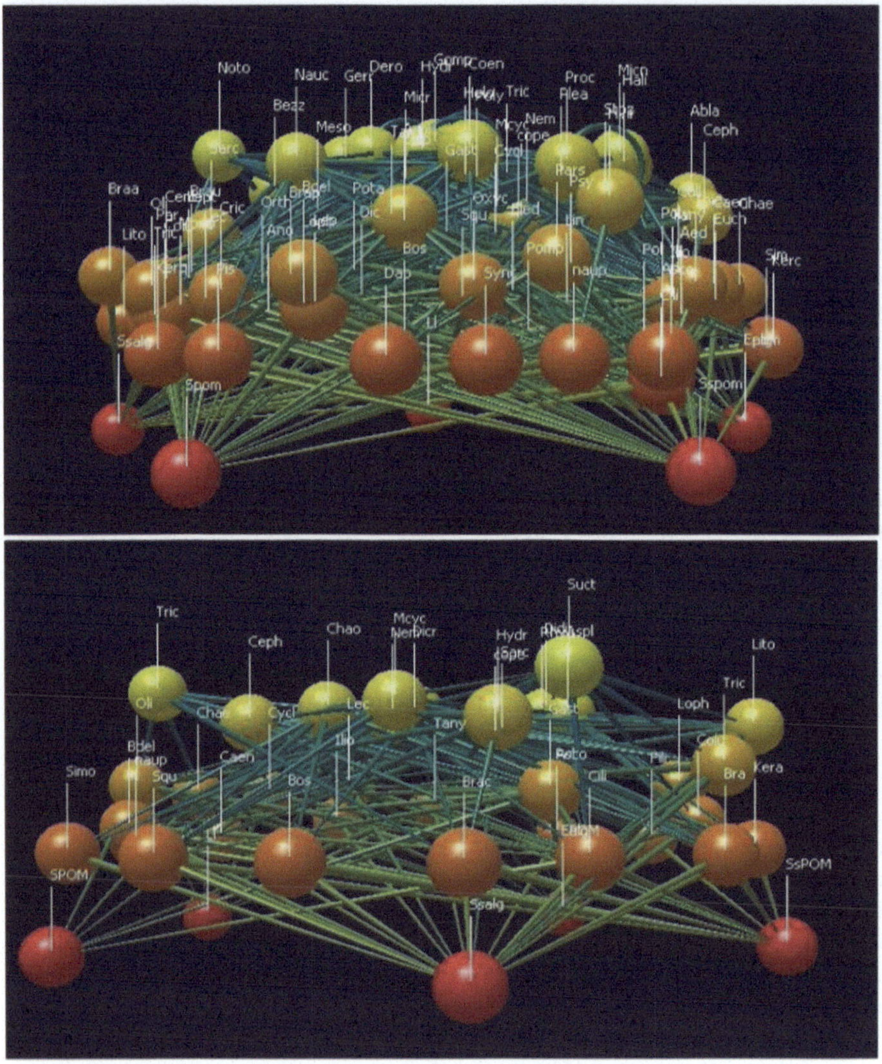

Fig. 2.6 Relaxed niche model of the food webs based on the original species in the littoral zone of submerged macrophyte covered Sutla backwater, Cro 1 (top), and floating-leaved covered Skrcev kut oxbow, Cro5 (below). Spheres represent trophic species and elongated cones represent feeding links. The position of the taxa vertically corresponds to their trophic level (TL) with basal taxa (primary producers and detritus) shown at the bottom of the network in red, and uppermost taxa at the top in yellow. Taken from [24]

Different elements need to be taken into consideration for the rich fish biodiversity in the Balkans: (a) the state is considered to have numerous relicts of the Plio-Pleistocene Peonic Lake [58]; (b) following the hypotheses of Economidis and Bănărescu [59] the Euro-Siberian taxa of East-Asian distributed southward to Europe and West Asia through temporary river connections in Oligocene and Miocene epochs that [60] bringing in the Balkans old Siberian lineages and a second wave accrued at Pliocene–Pleistocene boundary through northern dispersal that involved different river systems.

Species richness: The spatial heterogeneity of countries with temperate water bodies has a direct influence on current richness of their species. The fish species diversity of the temperate SWE has been heavily affected by human intervention and they depend on the river basin. After the 1970s, there was an intensive intervention through aquaculture development and promoting mostly Chinese stocks that were used even in natural lakes (including Lake Dushku, a glacial one part of our study). Various river systems of Albania are reported to harbour more than 80 species of fish, with numerous non-native ones [61]. The native species in SWE include bleak *Alburnus scoranza* (Heckel and Kner 1858), schneider *Alburnoides bipunctatus* (Bloch 1782), Prespa barbell *Barbus prespensis* (Karaman 1924), Ohrid spined loach *Cobitis ohridana* (Karaman 1928), gudgeon *Gobio* sp., Albanian roach *Pachychilon pictum* (Heckel and Kner 1858), chub *Squalius platyceps* (Zupančič, Marić, Naseka and Bogutskaya 2010), etc.

Data related to fish diversity in SWB were gathered for several decades. Fishes are taxonomically discerned in 11 families, where most species (25) are from Cyprinidae family. From family Cobitidae and Percidae 3 species, and other families one or two species are determined. There are 12 allochthonus fish species of which 3 are invasive. All allohtonus fishes are eurytopic by their ecological requirements with diverse feeding preferences (Fig. 2.7 and Table 2.4).

Non-native fish species: Allochthonous fish species can change the structure of communities and habitats, as well as the food web and nutrient circulation, which can fully ruin the functioning of the ecosystem. Whereas Poljakov et al. [62] provided information on the presence of introduced *Gambusia holbrooki* (Girard 1859), almost 40 years later Rakaj [61] reported the presence of altogether a total of 20 non-native species for Albania. The ten-fold increase in the abundance of exotic species was due to the importing of species for aquaculture, especially East Asian cyprinids and several salmonids [12].

The mostly wide spread aliens include *Pseudorasbora parva* (Temminck and Schlegel 1842), *G. holbrooki*, *Carassius gibelio* (Bloch 1782), *Aristichthys nobilis* (Richardson 1845), *Hypophthalmichtys molitrix* (Valenciennes 1844), *Ctenopharyngodon idella* (Valenciennes 1844), *Parabramis pekinensis* (Basilewsky 1855), *Lepomis gibbosus* (Linnaeus 1758), *Oncorhynchus mykiss* (Walbaum 1792), *Tinca tinca* (Linnaeus 1758), etc. The presence of non-native species as *C. gibelio, Esox lucius* (Linnaeus 1758), *P. parva, Gimnocephalus cernua* (Linnaeus 1758), *Ameiurus*

Fig. 2.7 Dominant fish species in SWE of Western Balkans (from the top): *Alburnoides bipunctatus, Barbus prespensis*, *Cobitis ohridana,* and *Squalius platyceps*. Photo by S. Shumka

nebulosus (Lesueur 1819), *Lepomis gibbosus* (Linnaeus 1758), etc. are reported by Talevski et al. [63] for the North Macedonian man-made water bodies.

Introduced allochthonous fish species can harm the biodiversity of an area's ecosystems. The most important is the negative impact of allochthonous freshwater fish on native species through competition for food and habitat, the transmission of diseases and parasites and hybridization [64]. Thus, the pumpkinseed (*L. gibbosus*) by its aggressive behaviour has squeezed many native species or reducing its abundance. This interlay was studied in eutrophic Bara Lake (Eastern Central Croatia) in 2015. A total of nine species is recorded in that lake, among them 55% are

Table 2.4 Fish species present in SWE and functional traits according to distribution (Cro—Croatian, NM—North Macedonia, A—Albania), fish status (A—allochthonus, I—invasive), ecological requirements (L—limnophilic, R—rheophilic, E—eurytopic) and feeding type (inse—insects, inve—invertebrates, omni—omnivore, pisc—piscivore, herb—herbivore, zoopl—zooplankton)

Family	Species	Distribution	Fish status	Ecological requirements	Feeding type
Acheilognathidae	European bitterling *Rhodeus amarus* (Bloch 1782)	WB		E	herb/inve
	Bitterling *Rhodeus sericeus* (Pallas 1776)	Srb		E	omni
Centrarchidae	Pumpkinseed *Lepomis gibbosus* (Linnaeus 1758)	WB	A, I	E	inse/inve
	Largemouth Bass *Micropterus salmoids* (Lacépède 1802)	Cro	A	E	pisc/inve
Cobitidae	Danubian Loach *Cobitis elongatoides* (Bacescu and Maier 1969)	Cro		E	inse/inve
	Ohrid spined loach *Cobitis ohridana* (Karaman 1928)	Alb		R	inse/inve
	Spined loach *Cobitis taenia* (Linnaeus 1758)	Cro/Srb		E	inv
	Weatherfish *Misgurnus fossilis* (Linnaeus 1758)	Cro/Alb		E	inve/inse
Cottidae	Bullhead *Cottus gobio* (Linnaeus 1758)	Cro		R	inve/inse
Cyprinidae	Common bream *Abramis brama* (Linnaeus 1758)	Cro/Alb		E	inve/inse
	Bleak *Alburnus alburnus* (Linnaeus 1758)	Cro/Srb		E	inve/inse
	Schneider *Alburnoides bipunctatus* (Bloch 1782)	Cro		E	inve/inse
	Bleak *Alburnus scoranza* (Heckel and Kner 1858)	Alb		E	inve/inse
	Asp *Aspius aspius* (Linnaeus 1758)	Cro		E	inve/inse/pisc
	Zope *Ballerus ballerus* (Linnaeus 1758)	Cro		E	inve/inse
	Southern Barbel *Barbus balcanicus kotli* (Tsigenopoulos, Rab and Berrebi 2002)	Cro		E	omni
	Prespa barbell *Barbus prespensis* (Karaman 1924)	Alb		E	omni

(continued)

Table 2.4 (continued)

Family	Species	Distribution	Fish status	Ecological requirements	Feeding type
	Krupatica *Blicca bjoerkna* (Linnaeus 1758)	Cro		E	inve/herb
	Crucian carp *Carassius carassius* (Linnaeus 1758)	Cro/Alb		L	omni
	Prussian carp *Carassius gibelio* (Bloch 1782)	WB	A	E	omni
	Grass Carp *Ctenopharingodon idella* (Valenciennes 1844)	Cro/Alb	A	E	herb/inve
	Common Carp *Cyprinus carpio* (Linnaeus 1758)	WB		E	omni
	Gudgeon *Gobio obtusirostris* (Valenciennes 1842)	Cro		E	inve/inse
	Silver carp *Hypophthalmichthys molitrix* (Valenciennes 1844)	Cro/Alb	A	E	zoopl
	Bighead Carp *Hypophthalmichthys nobilis* (Richardson 1845)	Cro/Alb	A	E	zoopl/herb
	Sunbleak *Leucaspius delineatus* (Heckel 1843)	Cro		E	zoopl/inse
	Dace *Leuciscus leuciscus* (Linnaeus 1758)	Cro		R	inse/pisc/herb
	Ide *Leuciscus idus* (Linnaeus 1758)	Cro		E	omni
	Albanian roach *Pachychilon pictum* (Heckel and Kner 1858)	Alb		E	omni
	Stone moroko *Pseudorasbora parva* (Temminck and Schlegel 1842)	WB	A, I	E	zoopl/inve
	Common roach *Rutilus rutilus* (Linnaeus 1758)	Cro/Srb		E	zoopl/herb/inve
	Rudd *Scardinius erythrophthalmus* (Linnaeus 1758)	WB		L	zoopl/herb
	Chub *Squalius cephalus* (Linnaeus 1758)	Cro/Srb		R	omni

(continued)

Table 2.4 (continued)

Family	Species	Distribution	Fish status	Ecological requirements	Feeding type
	Chub *Squalius platyceps* (Zupančič, Marić, Naseka and Bogutskaya 2010)	WB		E	omni
	Tench *Tinca tinca* (Linnaeus 1758)	Cro		L	zoopl/inve
Xenocyprididae	White amur bream *Parabramis pekinensis* (Basilewsky 1855)	A	A	E	inse/inve
Esocidae	Pike *Esox luciuss* (Linnaeus 1758)	WB		E	inve/pisc
Ictaluridae	Black Bullhead *Ameiurus melas* (Rafinesque 1820)	Cro	A	L	omni
	Brown bullhead *Ameiurus nebulosus* (Lesueur 1819)	WB	A	E	omni
Leuciscidae	Common nase *Chondrostoma nasus* (Linnaeus 1758)	Cro/Srb		E	omni
Percidae	Ruffe *Gymnocephalus cernua* (Linnaeus 1758)	Cro/NM		E	inve/inse
	European perch *Perca fluviatilis* (Linnaeus 1758)	WB		E	inse/inve/pisc
	Pikeperch *Sander lucioperca* (Linnaeus 1758)	Cro/Alb		E	zoopl/pisc
Salmonidae	Rainbow Trout *Oncorhynchus mykiss* (Walbaum 1792)	Cro	A	E	inse/inve/pisc
	Black Sea trout *Salmo trutta* (Linnaeus 1758)	Cro		E	inse/pisc
Siluridae	Wels catfish *Silurus glanis* (Linnaeus 1758)	WB		E	pisc
Poeciliidae	Eastern mosquitofish *Gambusia holbrooki* (Girard 1859)	Cro/Alb	A, I	E	inse

invading species and 45% are native species, where pumpkinseed dominated and shared 51% of the total fish abundance. This fish species, (*L. gibbosus*) coexisted with five autochthonous species: *Perca fluvialitis* (Linnaeus 1758), *Abramis brama* (Linnaeus 1758), *Rutilus rutilus* (Linnaeus 1758), *Leuciscus idus* and *E. lucius* and two allochthonous species: *Sander lucioperca* (Linnaeus 1758) and *P. parva*. Relatively higher growth coefficient and positive allometric growth of pumpkinseed in the Bara Lake compared to other fish species, indicates outstanding abilities of adaptation, resistance and reproduction in the new environment [65]. Also, in some waters *Ameiurus melas* (Rafinesque 1820) has severely squeezed other fish species, so such waters have become uninteresting to anglers.

Conservation challenges: The SWB as a part of freshwater habitats are facing serious threats, pressure and activities mostly accelerated due to intensive human intervention and land-use changes. With this regard, the habitat loss is a dominating feature. Further on, a significant threat is caused due to interventions in the hydraulic regime of the lake including temperate and glacial SWE, particularly 1990s. Along with habitat disturbance, the biological invasion is one of the causes of biodiversity decline. There is a consensus that climate change amplify existing threats causing significant alterations in the hydrological, thermal and biotic elements of freshwater ecosystems [66]. Also, aquaculture can be a threat to the natural environment and its biodiversity. Aquaculture projects, that are not done with the appropriate sensitivity to the environment can cause direct water pollution, destruction of wetlands and losses of flood areas. Therefore, a balanced relationship between these seemingly opposing categories is a prerequisite for successful management [67].

Fishes are considered to be the most endangered group of vertebrates today because of their connection with the aquatic ecosystem in which they live and due to the great human pressure on their habitats. The greatest impact on fish communities has human interventions that lead to the degradation of their habitat, regulation and pollution of watercourses, reclamation and transferred and foreign species [68]. Climate change will certainly cause an increase in water temperature, a decrease in the growth of individuals and more frequent floods and droughts. Therefore, there will be a change in the assemblage of fish in the freshwater ecosystems that will adapt to these changes [69].

2.6 Conclusions

Beside that the SWE are very abundant in the Western Balkans they remain the least studied component of the water ecosystems and at the current policy are almost excluded from water management planning and practices. In this review we identify the research priorities to undertake adequate protection of SWE and recommend relevant actions necessary for the more intense integration of these ecosystems into the management of catchments and landscapes. Based on our survey apparently at the present situation the impact of human activities is greater on SWE than on larger ones. Further, due to hydrological networks the ecosystem services and good quality of large freshwater bodies are directly connected with the status of SWE including tributaries, as well as entire system needs in the integrated approach. The biotic components as zooplankton organisms, macroinvertebrates and fish community contribute significantly to the aquatic ecosystem diversity, playing a role in their ecosystem functioning and directly reflecting the environmental alterations caused due to impact of human activities. This review revealed that SWE as a part of freshwater habitats is facing serious threats, pressure and activities mostly accelerated due to intensive human intervention and land-use changes. To protect small water bodies, there is now an urgent requisite to raise awareness about their ongoing alterations and their multiple beneficial functions to society. Further on there is a strong need for management and policy recognition of their importance for preserving a healthy and diverse aquatic environment. Finally, findings from this survey contribute to the new insight for improving current knowledge on SWE functioning in the Western Balkans through highlighting and syntheses of current data on their crucial role in supporting biodiversity and ecosystem services. Undoubtedly, there is a requisite to stimulate studies on a regional and wider perspective establishing and sharing the information through filling gaps through interdisciplinary approaches.

References

1. Céréghino R, Biggs J, Oertli B, Declerck S (2008) The ecology of European ponds: defining the characteristics of a neglected freshwater habitat. Hydrobiologia 597:1–6
2. Mullins ML, Doyle RD (2019) Big things come in small packages: why limnologists should care about small ponds. Acta Limnol Bras 31:e105. https://doi.org/10.1590/S2179-975X4119
3. Scheffer M (1998) Ecology of Shallow Lakes. Chapman & Hall, London, p 357
4. Rosset V, Angelibert S, Arthaud F, Bornette G, Robin J, Wezel A, Vallod D, Oertli B (2014) Is eutrophication really a major impairment for small waterbody biodiversity? J Appl Ecol 51:415–425
5. Bolpagni R, Poikane S, Laini A, Bagella S, Bartoli M, Cantonati M (2019) Ecological and conservation value of small standing-water ecosystems: a systematic review of current knowledge and future challenges. Water 11:402. https://doi.org/10.3390/w11030402
6. Špoljar M, Dražina T, Habdija I, Meseljević M, Grčić Z (2011) Contrasting zooplankton assemblages in two oxbow lakes with low transparencies and narrow emergent macrophyte belts (Krapina River, Croatia). Int Rev Hydrobiol 96:175–190
7. Špoljar M, Dražina T, Šargač J, Borojević K, Žutinić P (2012) Submerged macrophytes as a habitat for zooplankton development in two reservoirs of a flow-through system (Papuk Nature Park, Croatia). Ann Limnol Int J Lim EDP Sci 48:161–175
8. Tasevska O, Špoljar M, Gušeska D, Kostoski G (2017) Summer Rotifer assemblages in three reservoirs in the Republic of Macedonia. Acta Zool Bulg suppl 8:77–84
9. Mancinelli G, Mali S, Belmonte G (2019) species richness and taxonomic distinctness of zooplankton in ponds and small lakes from Albania and North Macedonia: the role of bioclimatic factors. Water 11:2384. https://doi.org/10.3390/w11112384
10. Stamenković O, Simić V, Stojković Piperac M, Dj M, Simić S, Ostojić A, Đorđević N, Čerba D, Petrović A, Jenačković Gocić D, Đurđević A, Koh M, Buzhdygan OY (2021) Direct, water-chemistry mediated, and cascading effects of human-impact intensification on multitrophic biodiversity in ponds. Aquat Ecol 55:187–214. https://doi.org/10.1007/s10452-020-09822-5
11. Kuczyńska-Kippen N (2020) Biodiversity of zooplankton in polish small water bodies. In: Korzeniewska E, Harnisz M (eds) Polish river basins and lakes part II—the handbook of environmental chemistry 87. Springer, Berlin, pp 56–76
12. Shumka S, Paparisto A, Grazhdani S (2008) Identification of non-native freshwater fishes in Albania and assessment of their potential threats to the national biological freshwater diversity. In: BALWOIS conference, Ohrid, Republic of Macedonia, 21–31 May 2008

13. Shuka L, Çullaj A, Shumka S, Miho A, Duka S, Bachofen R (2009) Response of drinking-water reservoir ecosystems to anthropogenic impacts in Albania: trends of interrelationship. J Int Environ Appl Sci 4:478–486
14. Skarbøvik E, Shumka S, Mukaetov D, Udaya Sekhar U (2010) Harmonised monitoring of Lake Macro Prespa as a basis for integrated water resources management. Irrig Drain Syst 24:223–238. https://doi.org/10.1007/s10795-010-9099-1
15. Wetzel RG (2001) Limnology lake and reservoir ecosystems. Academic Press, San Diego, p 1006
16. Kalff J (2002) Limnology. In: Inland water ecosystems. Prentice Hall, Upper Saddle River, New Jersey, p 608
17. Scheffer M, Hosper SH, Meijer M-L, Moss B, Jeppesen E (1993) Alternative equilibria in shallow lakes. Trends Ecol Evol 8:275–279
18. Zhang C, Liu H, Gao X, Zhang H (2016) Modeling nutrients, oxygen and critical phosphorus loading in a shallow reservoir in China with a coupled water quality—macrophytes model. Ecol Indic 66:212–219
19. Špoljar M, Zhan C, Dražina T, Zhao G, Lajtner J, Radonić G (2017) Development of submerged macrophyte and epiphyton in a flow-through system: assessment and modelling predictions in interconnected reservoirs. Ecol Indic 75:145–154
20. Bukvic-Ternjej I, Kerovec M, Mihaljević Z, Tavčar V, Mrakovčić M, Mustafić P (2001) Copepod communities in karstic mediterranean lakes along the easteren Adriatic coast. Hydrobiologia 453(454):325–333
21. Galir Balkić A, Ternjej I, Bogut I (2018) Impact of habitat heterogeneity on zooplankton assembly in a temperate river-floodplain system. Environ Monit Assess 190:143–143
22. Zhang C, Huang Y, Špoljar M, Zhang W, Kuczyńska-Kippen N (2018) Epiphyton dependency of macrophyte biomass in shallow reservoirs and implications for water transparency. Aquat Bot 150:46–52
23. Kozak A, Celewicz-Gołdyn S, Kuczyńska-Kippen N (2019) Cyanobacteria in small water bodies: the effect of habitat and catchment area conditions. Sci Total Environ 646:1578–1587
24. Špoljar M, Lajtner J, Dražina T, Malekzadeh-Viayeh R, Radanović I, Zrinščak I, Fressl J, Matijašec D (2017) Disentangling food webs interactions in the littoral of temperate shallow lakes. Nat Croat 26:145–166
25. Špoljar M, Dražina T, Lajtner J, Duić Sertić M, Radanović I, Wallace R, Matulić D, Tomljanović T (2018) Zooplankton assemblage in four temperate shallow waterbodies in association with habitat heterogeneity and alternative states. Limnologica 71:51–61
26. Scheffer M, van Nes EH (2007) Shallow lakes theory revisited: various alternative regimes driven by climate, nutrients, depth and lake size. In: Gulati GD, Lammens E, De Pauw N, Van Donk E (eds) Shallow lakes in a changing world. Springer, Dordrecht, pp 455–466
27. Moustaka-Gouni M, Michaloudi E, Sommer U (2014) Modifying the PEG model for Mediterranean lakes—no biological winter and strong fish predation. Freshw Biol 59:1136–1144
28. Kuczyńska-Kippen N, Špoljar M, Zhang C, Pronin M (2020) Zooplankton functional traits as a tool to assess latitudinal variation in the northern-southem temperate European regions during spring and autumn seasons. Ecol Indic 117:106629. https://doi.org/10.1016/j.ecolind.2020.106629
29. Galir Balkić A, Ternjej I, Katanić N (2019) Alteration in microcrustacean secondary production and herbivory effects between the River Danube and its floodplain lake. Hydrobiologia 836:185–196
30. Špoljar M, Tomljanović T, Dražina T, Lajtner J, Štulec H, Matulić D, Fressl J (2016) Zooplankton structure in two interconnected ponds: similarities and differences. Croat J Fish 74:6–13

31. Ejsmont-Karabin J (2012) The usefulness of zooplankton as lake ecosystem indicators: Rotifer trophic state index. Pol J Ecol 60:339–350
32. Špoljar M (2013) Microaquatic communities as indicators of environmental changes in lake ecosystems. J Eng Res 1:29–42
33. Kuczyńska-Kippen N (2005) On body size and habitat selection in rotifers in a macrophye-dominated lake Budzyńskie, Poland. Aquat Ecol 39:447–454
34. Kuczyńska-Kippen N (2018) The use of bdelloids in reference to rotifer biocoenotic indices as an indicator of the ecological state of small field water bodies: the effect of macrophytes, shading and trophic state of water. Ecol Indic 89:576–583
35. Tasevska O, Špoljar M, Guševska D, Kostoski G, Patcheva S, Veljanoska Sarafiloska E (2017) Zooplankton in ancient and oligotrophic Lake Ohrid (Europe) in association with environmental variables. Croat J Fish 75:95–103
36. Yu R, Yang R, Zhang C, Špoljar M, Kuczyńska-Kippen N, Sang G (2020) Vine copula-based modeling for identification of multivariate water pollution risk in an interconnected river system network. Water 12:2741. https://doi.org/10.3390/w12102741
37. Stamou G, Katsiapi M, Moustaka-Gouni M, Michaloudi E (2021) The neglected zooplankton communities as indicators of ecological water quality of Mediterranean lakes. Limnetica 40:359–373. https://doi.org/10.23818/limn.40.24
38. Jeppesen E, Nõges P, Davidson TA, Haberman J, Nõges T, Blank K, Lauridsen TL, Sřndergaard M, Sayer C, Laugaste R, Johansson LS, Bjerring R, Amsinck SL (2011) Zooplankton as indicators in lakes: a scientific-based plea for including zooplankton in the ecological quality assessment of lakes according to the European Water Framework Directive (WFD). Hydrobiologia 676:279–297
39. Ostojić AM, Radojević ID, Galir Balkić A (2015) Zooplankton community along the Sava River. In: Milačič R, Scancar J, Paunović M (eds) The Sava River. The handbook of environmental chemistry, vol 31. Springer-Verlag GmbH, Berlin, pp 317–334
40. Stamou G, Katsiapi M, Moustaka-Gouni M, Michaloudi E (2019) Trophic state assessment based on zooplankton communities in Mediterranean lakes. Hydrobiologia 844:83–103
41. Water Framework Directive (2000) European Commission, 2000. Directive 2000/60/EC of the European Parliament and of the Council of 23 October 2000 establishing a framework for community action in the field of water policy. OJEC L37:1–72
42. Beklioglu M, Meerfhoff M, Sřndergaard M, Jeppesen E (2011) Eutrophication and restoration of shallow lakes from a old temperate to a warm Mediterranean and a (sub)tropical climate. In: Ansari AA, Singh GS, Lanza GR, Rast W (eds) Eutrophication: causes, consequences and control, 1st edn. Springer, Dordrecht, pp 91–129
43. Shumka S, Miho A (2006) Data on plankton community zooplankton and phytoplankton in Drini course. In: BALWOIS conference II, Ohrid, Republic of Macedonia, 23–25 May 2006
44. Ostojić A (2010) Contribution to the knowledge on the zooplankton investigation of Balkan Peninsula—first 100 years of zooplankton research on the territory of Serbia. Kragujevac J Sci 32:117–140
45. Ostojić AM, Obradović MP, Radoičić DJ (2012) Faunistic composition of zooplankton in the Šumarice lake (Kragujevac, Serbia). Kragujevac J Sci 34:87–96
46. Ostojić AM, Radojević ID, Obradović MP, Radoičić DJ (2012) A checklist of freshwater zooplankton on the territory of Kragujevac City (Serbia) and it's surrounding areas. Water Resour Manag 2:35–42
47. Špoljar M, Habdija I, Primc-Habdija B, Sipos L (2005) Impact of environmental variables and food availability on Rotifer assemblage in the karstic Barrage Lake Visovac (Krka River, Croatia). Int Rev Hydrobiol 90:555–579

48. Shumka S, Špoljar M, Tasevska O (2018) The Zooplankton of Lake Skadar/Shkodra: species diversity and abundance. In: Pešić V, Karaman G, Kostianoy Arey G (eds) The Skadar/Shkodra Lake environment. Springer Nature Switzerland AG, pp 239–254
49. Meerhoff M, Iglesias C, De Mello FT, Clemente JM, Jensen E, Lauridsen TL, Jeppesen E (2007) Effects of habitat complexity on community structure and predator avoidance behaviour of littoral zooplankton in temperate versus subtropical shallow lakes. Freshw Biol 52:1009–1021
50. Litchman E, Ohman MD, Kiørboe T (2013) Trait-based approaches to zooplankton communities. J Plankton Res 35:473–484. https://doi.org/10.1093/plankt/fbt019
51. Miliša M, Habdija I, Primc-Habdija B, Radanović I, Matoničkin Kepčija R (2006) The role of flow velocity in the vertical distribution of particulate organic matter on moss-covered travertine barriers of the Plitvice Lakes (Croatia). Hydrobiologia 553:231–243
52. Sertić Perić M, Matoničkin Kepčija R, Radanović I, Primc B, Habdija I (2020) Freshwater reefs as mesohabitats for the assessment of diel invertebrate drift patterns. Nat Croat 29:185–203
53. Špoljar M, Šneller D, Miliša M, Lajtner J, Sertić Perić M, Radanović I (2012) Entomofauna of submerged macrophyte stands in reservoirs (Papuk Nature Park). Entomol Croat 16:7–20
54. Špoljar M, Dražina T, Lajtner J, Radanović I, Ternjej I, Mitić B, Hruševar D (2014) Influence of environmental factors on lake biocenoses and travertine barriers in Papuk Nature Park. Faculty of Science, University of Zagreb (in Croatian), Zagreb
55. Ćaleta M, Buj I, Mrakovčić M, Mustafić P, Zanella D, Marčić Z, Duplić A, Mihinjač T, Katavić I (2015) Endemic fishes of Croatia. Croatian Environment Agency, p 115
56. Abell R, Thieme M, Revenga C (2008) freshwater ecoregions of the world: a new map of biogeographic units for freshwater biodiversity conservation. Bioscience 58:403–414
57. Oikonomou A, Leprieur F, Leonardos ID (2014) Biogeography of freshwater fishes of the Balkan Peninsula. Hydrobiologia 738:205–220
58. Francke A, Wagner B, Leng MJ, Rethemeyer J (2013) A Late Glacial to Holocene record of environmental change from Lake Dojran (Macedonia, Greece). Clim Past 9:481–498
59. Economidis PS, Bănărescu PM (1991) The distributions and origins of freshwater fishes in the Balkan Peninsula, especially in Greece. Int Rev Gesamten Hydrobiol 76:257–284
60. Barbieri R, Zogaris S, Kalogianni E, Stoumboudi MT, Chatzinikolaou Y, Giakoumi S, Kapakos Y, Kommatas D, Koutsikos N, Tachos V, Vardaka L, Economou AN (2015) Freshwater fishes and lampreys of Greece: an annotated checklist; monographs on marine sciences No. 8. Hellenic Centre for Marine Research, Athens, Greece, p 133
61. Rakaj N (1995) Iktiofauna e Shqiperise (Ichthyofauna of Albania). Shtëpia Botuese “Libri Universitar”, Tirana, p 410
62. Poljakov GD, Filipi N, Basho K, Hysenaj A (1958) Pesquit e Shqiperise (Fishes of Albania). Mihal Duri, Tirana, p 230
63. Talevski T, Milosevic D, Trajchevski B, Taleska M, Lesoska L, Pejoski E (2019) The impact of alien fish species on native fish population from Lake Dojran and Lake Tikvesh (Aegean catchment area). In: Third international conference on zoology, zoonoses and epidemiology, Hissar, Bulgaria, 21–23 October 2019
64. Piria M, Simonović P, Kalogianni E, Vardakas L, Koutsikos N, Zanella D, Ristovska M, Apostolou A, Adrović A, Mrdak D, Tarkan S, Milošević D, Zanella L, Joy MK (2018) Alien freshwater fish species in the Balkans—vectors and pathways of introduction. Fish Fish (Oxf) 19:138–169
65. Khanom M, Abdur RM, Ahsan N, Siddiqui N, Tomljanović T (2020) Morphological characteristics, growth and age structure of allochthonous fish pumpkinseed, Lepomis gibbosus in Bara Lake, Croatia. Bulg J Agric Sci 26:213–222
66. Markovic D, Carrizo S, Karcher O, Walz A, David J (2017) Vulnerability of European freshwater catchments to climate change. Glob Chang Biol 23:3567–3580. https://doi.org/10.1111/gcb.13657

67. Beardmore JA, Mair GC, Lewis RI (1997) Biodiversity in aquatic systems in relation to aquaculture. Aquac Res 28:829–839
68. Kirchhofer A, Müller D (eds) (2012) Conservation of endangered freshwater fish in Europe. Birkhäuser, Basel, p 356
69. Rijnsdorp AD, Peck MA, Engelhard GH, Möllmann C, Pinnegar JK (2009) Resolving the effect of climate change on fish populations. ICES J Mar Sci 66:1570–1583

Chapter 3
Conservation Value and Habitat Diversity of Fluvial Lakes and Gravel Pits in River-Floodplain Systems

Dušanka Cvijanović

Abstract This study aimed to review habitat diversity and conservation value of small fluvial and gravel pit lakes along river-floodplain systems in the Western Balkans. Aquatic habitat types were presented according to the EU Habitats Directive and the Bern Convention, while the physical habitat properties were evaluated following the standard approach for hydromorphological assessment of lakes in Europe. Habitats considered as conservation priorities are present in both types of small water bodies, covering a wide range of water trophic gradient and hydromorphological conditions. This phenomenon, coupled with macrophytes' physical structural heterogeneity, creates extremely diverse habitats for other aquatic communities. Hence, fluvial and gravel pit lakes along the river-floodplains can be considered as hot spots of aquatic habitat diversity in the Western Balkans. However, these ecosystems are subjected to various human pressures such as alteration of hydrological regime, artificial land use of the riparian zone, sediment erosion, shoreline modification, and eutrophication in the entire region.

Keywords Lakes · Ponds · Wetlands · Floodplains · Habitats · Macrophytes · Hydromorphology

3.1 Introduction

Floodplains are transitions between land and freshwater ecosystems, with hydrology, soil, and biotic conditions strongly influenced by stream water and flooding [1]. They provide a range of important ecosystem services such as nutrient and water retention, flood control, carbon sequestration, a diversity of aquatic and wetland habitats. Fluvial processes are considered the third most important category of lake-forming mechanisms. Floodplain lakes are formed by a combination of groundwater infiltration, flood-season inundation by river exceeding bank top conditions [2]. Pulsing lateral connectivity of freshwater habitats in the floodplains depends on the river

D. Cvijanović (✉)
Faculty of Sciences, University of Novi Sad, Trg Dositeja Obradovića 3, 21000 Novi Sad, Serbia
e-mail: dusanka.cvijanovic@dbe.uns.ac.rs

V. Pešić et al. (eds.), *Small Water Bodies of the Western Balkans*, Springer Water,
https://doi.org/10.1007/978-3-030-86478-1_3

hydrological regime. These specific conditions create a unique mosaic of aquatic habitats of different successional stages [3].

Fluvial lakes and ponds, as well as gravel pit lakes along river floodplains (Figs. 3.1, 3.2 and 3.3), were recognized in previous studies as sites of high conservation interest in the Western Balkans [4, 5]. The majority of these lentic ecosystems could be classified as small water bodies (SWB), having a depth of less than 20 m and an area up to 10 ha [4–6]. In such lotic systems, hydromorphological attributes and aquatic vegetation traits play an important role in structuring other aquatic communities [6, 7]. Microhabitat heterogeneity and physical diversity generated by macrophytes provide shelter and nesting places for aquatic animals. Macrophytes are important component of food webs and nutrient cycles [7–9]. Also, macrophytes interact with the physical environment by changing and establishing hydromorphological conditions as pioneer species at the early beginning of habitat formation [10].

SWB in floodplains are crucial ecosystems in maintaining macrophyte diversity and production in riverine ecosystems [11]. The frequency with which these SWB are connected to the main river channel during flood flows may influences their macrophyte richness. The highest species richness in backwaters is typically found at a low connectivity rate. On the other hand, the highest plant biomass is found at low and medium connectivity with the main channel. Therefore, the naturalness of the shore zone and hydrological regime of small water bodies in floodplains should be considered as a priority in the Western Balkans.

Even artificial floodplain SWB may contribute to aquatic biodiversity [6, 12]. Pander et al. [12] showed that fluvial habitats in nature-like fish passes can provide important juvenile fish habitats, which are scarce in the main river channel. Gravel pit lakes were also recognized as a potentially valuable refuge for wildlife in the Western Balkans and other parts of Europe [6, 13–15].

Fig. 3.1 Fluvial pond Agla (Koviljsko-Petrovaradinski Rit wetland, Serbia). Photo by N. Lukičić

Fig. 3.2 Gravel pit lake along the Nera River Floodplain. Photo by D. Cvijanović

Regardless of their mode of formation, SWB in floodplains are susceptible to habitat alterations, land use change, and eutrophication worldwide, due to their disproportionate perimeter/area ratio [16].

Floodplains and flood-prone areas are relevant for several European policies: the Water Framework Directive [17], the Habitats Directive [18] and Birds Directives [19], and the Floods Directive [20]. While the Habitats Directive is the primary legal instrument driving species and habitat protection in the EU, the Bern Convention [21] is considered an important frame for the protection of biodiversity in Central and Eastern Europe [22]. Apart from Croatia, other countries in the Western Balkans are not part of the EU and therefore are still in one of the preparation phases of implementation of the Habitats Directive.

This study aimed to evaluate habitat diversity, and conservation value of small fluvial and gravel pit lakes along the river floodplain systems in the Western Balkans. The chapter includes a review of previously published vegetation data [4–6, 23–44]. Aquatic habitat types were presented according to the Habitats Directive and the Bern Convention. Physical habitat properties were evaluated following the standard approach for hydromorphological assessment of lakes in Europe, using the LHMS (Lake Habitat Modification) and the LHQA (Lake Habitat Quality) scores [45, 46] Conservation value was calculated based on macrophyte vegetation trait [4, 47–49].

Fig. 3.3 Kazuski Dunavac (Gornje Podunavlje wetland, Serbia), July 2010, stands of *Trapa natans* L. Photo by D. Cvijanović

3.2 River-Floodplains in the Western Balkans

Serbia and Croatia are among the countries with the highest percentage of area covered by floodplains in Europe, after Hungary, Netherlands, and Liechtenstein [1]. In total, floodplains make about 12% (26,892 km^2) of the Western Balkans area.

The main river systems with significant floodplain area in the Western Balkans are the Danube, the Tisza, the Sava, the Drina, the Drava, the Southern Morava, the Vjosë/Aoös, the Vardar, the Lim, and the Zeta, as well as tributaries of the Neretva and the Trebišnjica rivers.

One of the most important floodplain areas in the region, concerning its conservation value is the UNESCO Cross-border Biosphere Reserve 'Mura–Drava–Danube', also known as Europe's Amazon. It includes wetlands Gornje Podunavlje (Serbia), Kopački Rit (Croatia), and Gemenc (Hungary) [50]. The Gornje Podunavlje floodplain area, with a status of the special nature reserve in Serbia, is a marsh complex along 36 km of the Danube River, with a unique mosaic of aquatic, wetland, and terrestrial ecosystems. Many rare and threatened plant species and their communities are supported, as well as vulnerable habitats [51]. The Kopački Rit wetland area is located in north-eastern Croatia, in the floodplain formed by the confluence of the Danube and the Drava River [25]. The wetland includes small fluvial lakes, marshes, wet grasslands, reed beds, riverine forests, numerous channels, oxbow lakes, and fishponds. Important flooded areas along the Sava river are Lonjsko Polje and Obedska Bara. The Lonjsko Polje Nature Park follows a floodplain along the Sava River in Croatia, which runs south-eastwards towards the border with Bosnia and Herzegovina. It is mainly composed of alluvial forests, but also wet meadows and pastures. This combination of habitats and their regular flooding and high groundwater levels provide the conditions for rich biodiversity [52].

Obedska Bara is a seasonally inundated area of the Sava River floodplain in Serbia, with marshes, ponds, and seasonally flooded meadows surrounded by oak, willow, and poplar forests [43, 52]. This is the largest wetland in Serbia with near-natural water level balance and variability.

Moreover, the Vjosë/Aoös river system in Albania is characterized by unique, widely undisturbed, and maintained fluvial dynamics throughout its course, from the headwaters in Greece to the delta in southern Albania [29, 53]. Along the entire watercourse, relatively deep stagnant pools with fine sediment deposition and residence time suitable for macrophyte development are dominated by *Chara* spp., and *Potamogeton nodosus*.

According to European Aggregates Association (UEPG) [54], the estimated production of gravel and sand in the Western Balkans in 2018 was 36 tons, which was excavated from 726 sites. All excavation activities in the EU follow the provisions outlined in Article 6 of the Habitats Directive to ensure that these activities do not adversely affect the integrity of Natura 2000 sites [55]. Other EU environmental Directives are also directly relevant to the NEEI industry (Non-energy mineral extraction industry). The SEA (Strategic Environmental Assessment) Directive [56] have a purpose to ensure that the environmental consequences of certain plans and programmes are identified, assessed, and taken into account during their preparation and before their adoption. While the SEA process operates at the level of public plans and programs, Environmental Impact Assessments (EIA) [57] operate at the level of individual public and private projects.

Although the environmental legislation strictly requires re-cultivation of extracted fields in the area studied, there are no legislation guidelines for pit lake planning or designs to enhance passive remediation of water quality or habitat re-naturalization [6]. The gravel extraction in the Western Balkans is usually performed illegally, or with poorly defined rules [6].

Damnjanović et al. [6] recognized hydromorphological predictors, important for different macrophyte assemblages in gravel pit lakes along the lower course of the river Drina. They stressed that an optimal and careful selection of gravel excavation procedures and landscaping techniques following biodiversity requirements may help convert gravel excavation fields to an ecologically sustainable state. These lake attributes and their selected values may be used for early-design phases of future gravel extraction in the floodplain areas. According to Damnjanović et al. [6], two gravel pit types should be excavated within the single extraction area to support pioneering charophyte vegetation and vegetation of typical eutrophic lowland floodplain lakes as well. The first hydromorphological gravel pit type, suitable for stonewort species, should be positioned up to 100 m from the river main channel, having a surface area up to 1000 m^2 and a relative depth ratio >5%. The second gravel pit type should be located about 300 m from the river main channel, with preferable maximal depth in the range 3–4 m (at least 2 m depth), and a lake surface area between 10,000 and 20,000 m^2 (at least 4000 m^2). The relative depth ratio may vary but should be less than 5%. Generally, all sites should be designed with minimal impact to the riparian and shore zones. The actions proposed by Damnjanović et al. [6] addressed the goals of the Priority Area 6 of the European Union Strategy for the Danube Region [58], which suggests the work on establishing green infrastructure and the process of restoration of at least 15% of degraded ecosystems in the region.

In contrast to the Western Balkans, many sand and gravel extraction areas in the rest of Europe were subsequently designated as Natura 2000 sites and are contributing to the conservation of habitat types and species of EU interest [59]. For example, the creation of wetland habitats at one of the largest sands and gravel extraction sites in the UK (Needingworth Quarry) is making a substantial contribution to achieving the national targets for reedbeds and the bittern (*Botaurus stellaris*), a Birds Directive Annex I species that had undergone major declines in the UK. Moreover, all waters containing submerse stonewort communities in Germany, including secondary waters such as flooded gravel pits, are being protected following the EU Habitats Directive [60, 61].

3.3 Aquatic Habitats in Fluvial and Gravel Pit Lakes in the Western Balkans—Conservation Significance

According to the literature review, two habitat types listed in the Habitats Directive were recorded in SWB along the river floodplains in the Western Balkans: (i) hard oligo-mesotrophic waters with benthic vegetation of *Chara* spp.; and (ii) habitats of natural eutrophic lakes with *Magnopotamion* or *Hydrocharition*-type vegetation. Habitats considered as conservation priorities by the Bern Convention are present in both types of small water bodies (fluvial lakes and gravel pits) (Table 3.1). In total, 32 aquatic plant communities are present in the area studied.

Hard oligo-mesotrophic waters with benthic vegetation of *Chara* spp. (Habitats Directive code 3140) are more confined to gravel pit lakes than to fluvial ones, but probably due to lack of data. For instance, many other stonewort species were previously detected in ponds along the Danube floodplain—Special Nature Reserve Gornje Podunavlje such as *Chara tenuispina* A. Braun, *Chara vulgaris* L., *Chara baueri* A. Braun, *Nitella confervacea* (Brébisson) A. Braun ex Leonhard, *Nitella gracilis* (J. E. Smith) C. Agardh, *Nitella syncarpa* (Thuillier) Chevallier, *Nitella capillaris* (Krock.) J. Groves & Bull.-Webst., *Nitella mucronate* (A. Braun) Miq. and *Tolypella prolifera* (Ziz ex A. Braun) Leonhardi [62, 63], but no vegetation data are available.

According to Damnjanović et al. [6], the presence of stonewort species in gravel pit lakes along the Drina river floodplain, especially *Chara globularis*,

Table 3.1 Aquatic habitats in fluvial and gravel pit lakes in the Western Balkans

The Bern convention priority habitat	Association name [5]	Species name	Fluvial lakes	Gravel pits
The Habitats Directive code 3140-Hard oligo-mesotrophic waters with benthic vegetation of *Chara* spp.				
C1.25 Charophyte submerged carpets in mesotrophic waterbodies	*Charetum contrariae* Corillion 1957	*Chara contraria* A. Braun ex Kutz.	–	+
	Charetum globularis Corillion 1949 nom. mut. Prop.	*Chara globularis* Thuill.	–	+
	Nitellopsietum obtusae Dambska 1961	*Nitellopsis obtusa* (Desv. in Loisel.) J. Groves	–	+
The Habitats Directive code 3150-Natural eutrophic lakes with *Magnopotamion*or *Hydrocharition*-type vegetation				
C1.223 Floating Stratiotes aloides rafts	*Stratiotetum aloidis* Miljan 1933	*Stratiotes aloides* L.	+	–
C1.33 Rooted submerged vegetation of eutrophic waterbodies	*Potametum pectinati* Carstensen ex Hilbig 1971	*Potamogeton pectinatus* L.	+	+
	Potametum perfoliate Miljan 1933	*Potamogeton perfoliatus* L.	–	+
	Potametum trichoidis Tüxen 1974	*Potamogeton trichoides* Cham. & Schltdl.	+	–
	Najadetum marinae Fukarek 1961	*Najas marina* L.	+	+
	Najadetum minoris Ubrizsy 1961	*Najas minor* All	+	+
	Potametum crispi von Soó 1927	*Potamogeton crispus* L.	+	–
	Potametum lucentis Hueck 1931	*Potamogeton lucens* L.	+	–
	Myriophylletum verticillati Gaudet ex Šumberová in Chytrý 2011	*Myriophyllum verticillatum* L.	+	+
	Potamo pectinati-Myriophylletum spicat Rivas Goday 1964	*Myriophyllum spicatum* L.	+	+

(continued)

Table 3.1 (continued)

The Bern convention priority habitat	Association name [5]	Species name	Fluvial lakes	Gravel pits
C1.32 Free-floating vegetation of eutrophic waterbodies	*Salvinio natantis-Spirodeletum polyrhizae* Slavnić 1956	*Spirodela polyrhiza* (L.) Schleiden, *Salvinia natans* (L.) All	+	–
	Lemno-Spirodeletum polyrhizae Koch 1954	Spirodela polyrhiza (L.) Schleiden, Lemna minor L.	+	–
	Lemnetum trisulcae den Hartog 1963	*Lemna trisulca* L.	+	
	Lemno minoris-Riccietum fluitantis Šumberová et Chytrý in Chytrý 201	*Riccia fluitans* L.	+	–
	Lemno-Utricularietum Soó 1947	*Lemna minor* L., *Utricularia vulgaris* L.	+	–
	Ricciocarpetum natantis Tüxen 1974	*Ricciocarpus natans* (L.) Corda	+	–
	Lemnetum gibbae Miyawaki et J. Tüxen 1960	*Lemna gibba* L.	+	–
	Lemno gibbae-Wolffietum arrhizae Slavnić 1956	*Wolffia arrhiza* (L.) Horkel ex Wimm.	+	–
	Hydrocharitetum morsus-ranae van Langendonck 1935	*Hydrocharis morsus-ranae* L.	+	–
	Ceratophylletum demersi Corillion 1957	*Ceratophyllum demersum* L. subsp *demersum*	+	+
C1.3413 *Hottonia palustris* beds in shallow water	*Hottonietum palustris* Sauer 1947	*Hottonia palustris* L.	+*	–
Not Applicable	*Nymphaeetum albae* Vollmar 1947	Nymphaea alba L., Nuphar lutea (L.) Sm	+	–
	Nymphoidetum peltatae Bellot 1951	*Nymphoides peltata* (S G Gmelin) O Kuntze	+	–
	Polygonetum natantis Soó 1927	*Polygonum amphibium* L.	+	+
	Trapetum natantis Kárpáti 1963	*Trapa natans* L.	+	–

(continued)

Table 3.1 (continued)

The Bern convention priority habitat	Association name [5]	Species name	Fluvial lakes	Gravel pits
	Potametum graminei Lang 1967	*Potamogeton gramineus* L.	+	–
	Nymphaeo albae-Nupharetum luteae Nowiński 1927	*Nuphar lutea* (L.) Sm	+	+
	Potametum denso-nodosi de Bolós 1957	*Potamogeton nodosus* Poiret	+	+
	Potametum natantis Hild 1959	*Potamogeton nodosus* Poiret	–	+

Chara vulgaris and *Nitella gracilis*, was related to their ability to colonize newly created habitat patches and act as pioneer species [64]. These newly created habitats (aged 4–14 years), without domination of other submerged macrophytes, having a mesotrophic character and high-water transparency provided favourable conditions for charophyte submerged carpets. In Bosnia and Herzegovina, hard oligo-mesotrophic waters with benthic vegetation of *Chara vulgaris* and *Chara globularis* were recorded in pit lakes around Tuzla [65]. These artificial water bodies have similar physico-chemical and hydromorphological conditions as those described in this study. However, they are not part of any river floodplain and so were not included in the analysis.

Surface and subsurface vegetation of free-floating hydrocharids, duckweeds, ferns, liverworts, and bladderworts habitats were exclusively associated with fluvial lakes in the area studied. In general, these vegetation communities are confined to shallow littoral zones of eutrophic standing waters and consist of floating mats of various mixtures of duckweeds *Salvinia natans*, *Spirodela polyrhiza*, *Lemna minor*, and some emergent associates [6].

Stands with *Nuphar lutea,* which were rather monodominant in the area studied were recorded in mesotrophic gravel pits, as well as in eutrophic fluvial lakes [6]. Rare stands with both water lily species were recorded only in natural floodplain lentic water bodies. Other rooted species with floating leaves, which are eutrophication tolerant [53] are confined to fluvial lakes.

Both types of floodplain SWB host rooted submerged vegetation of eutrophic water bodies (Bern Convention code C1.33), dominated by pondweeds of genus Potamogeton. Biomass produced by this vegetation type is smaller than that produced by the water lily vegetation type, although quite often the dominant species occupy the entire water column [66]. This vegetation type is recognized as a conservation priority by the Bern Convention.

Potentially, other submerged habitats can be found in the area studied. In eutrophic mine-pit lakes around Zenica (Bosnia and Herzegovina), a submerged community dominated by *Potamogeton filiformis* Pers was recorded [67], belonging to

3150 Natural eutrophic lakes with *Magnopotamion* or *Hydrocharition*-vegetation type (C1.33 Rooted submerged vegetation of eutrophic waterbodies) [67]. Also, in the digital database of aquatic vegetation in Serbia [31], there are stored data for the community dominated by *Potamogeton acutifolius* Link along the river wetlands (Obedska bara). However, these vegetation stands were not recognized in the revised classification of aquatic vegetation in Serbia [5]. Furthermore, aquatic communities characterized with *Potamogeton pusillus* L., and *Zannichellia palustris* L. were recorded in Skadar/Shkodra lake, suggesting that these habitats might be found in floodplains as well [27]. According to the checklist of vascular flora in Kopački Rit national park (Croatia) [25], species *Littorella uniflora* (L.) Asch. and *Limosella aquatica* L. find their suitable conditions there. Therefore, the vegetation of amphibious plants in shallow, oligotrophic to mesotrophic water bodies with *Littorella uniflora* (C3.4 Species-poor beds of low-growing water-fringing or amphibious vegetation or C1.1 Permanent oligotrophic lakes, ponds, and pools, but no data is available), can be potentially found there, together with community dominated by *Limosella aquatica* (C3.51 Euro-Siberian dwarf annual amphibious swards, 3130). A community dominated by *Hottonia palustris* was recorded by Panjković [68] in Gornje Podunavlje wetland area, but this vegetation unit was not recognized by the classification analysis according to Cvijanović et al. [5] in the area studied. This habitat type is of high conservation interest according to Bern Convention (C1.3413 *Hottonia palustris* beds in shallow water, [68]). Habitat type characterized with *Hottonia palustris* was recorded in Kopački Rit [23] as well, but recent vegetation data are not available [25]. Moreover, Šegota et al. [28] reviewed the status of the rare aquatic plant *Groenlandia densa* (L.) Fourr. in the Western Balkans. This species was recorded in wetland ecosystems along the Sava river, around Skadar/Shkodra lake in Albania [26], and in artificial water supply reservoirs, but there is a lack of vegetation data including it.

Conservation indices that took into account rarity and designation status of macrophyte species were significantly higher for gravel pit lakes compared to fluvial ones, suggesting that gravel pit lakes in the Western Balkans can be considered as hotspots of rare aquatic flora and habitats (Fig. 3.4). On the other hand, fluvial lakes along the river floodplains support a high number of species and macrophyte assemblages (Figs. 3.4, 3.5 and 3.6, Table 3.1). Together, these two lentic types create an important refuge and source of aquatic plants' reproductive materials, which may positively influence the biodiversity of the main river channel and downstream wetland areas [69].

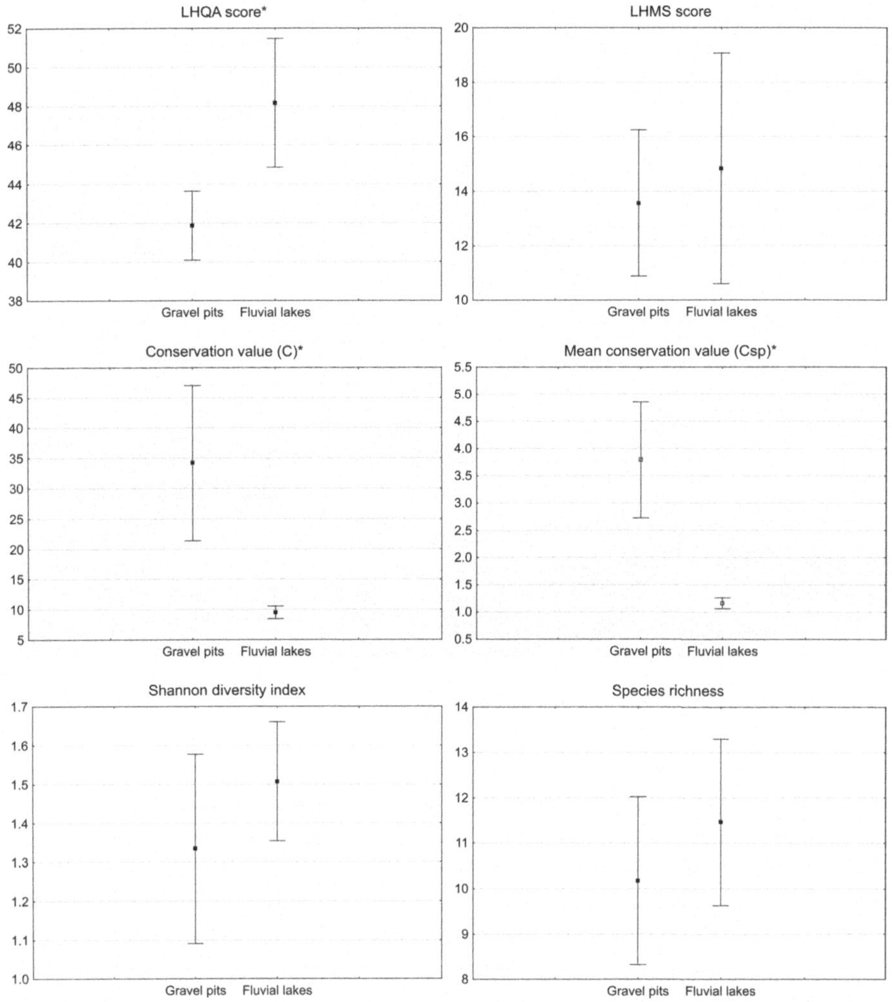

Fig. 3.4 Range and mean values of hydromorphological variables and diversity and conservation indices in floodplain fluvial lakes and gravel pits in the Western Balkans. Asterisk symbol '*' indicate statistically significant difference between gravel pits and fluvial lakes (T-test, $p < 0.05$). LHQA—Lake Habitat Quality Assessment score; LHMS—Lake Habitat Modification score

3.4 Comparison of Physico-Chemical and Hydromorphological Properties of Fluvial and Gravel Pit Lakes

The distribution of aquatic habitats between floodplain fluvial and gravel pit lakes in the Western Balkans is in line with their hydromorphological and physico-chemical conditions (Figs. 3.4 and 3.7). The mean values for all analyzed water quality vari-

Fig. 3.5 Arkanj pond (Koviljsko-Petrovaradinski Rit wetland, Serbia), July 2010, stands of water lily *Nymphaea alba* L. Photo by D. Cvijanović

ables differed significantly between lake types, but for the total organic carbon, dissolved oxygen, biological oxygen demand, and total suspended solids there was no overlap of range values. While the gravel pits may be classified as oligo-mesotrophic lakes, all fluvial lakes have eutrophic conditions [70]. Significantly higher content of the total suspended solids and electrical conductivity in fluvial lakes imply that erosional processes play an important role in water quality conditions in these ecosystems.

Hydromorphological score, which reflects physical habitat diversity (LHQA) were in the same range for both types of floodplain lentic systems as lakes throughout Europe [71], but was significantly higher for fluvial lakes than in gravel pits (Fig. 3.7). The reason for this was higher riparian vegetation complexity and longevity, the higher naturalness and diversity of riparian land cover types, the higher value of total macrophyte cover, and diversity of macrophyte functional groups in the fluvial lakes compared to the gravel pits [6, 35]. Generally, the LHMS values covered a range of moderately altered sites [45, 72–74, 75]. Although there is no significant difference between the two groups of lakes against the LHMS score, still, the in-lake use (human pressures) and the shore-line erosion were slightly higher in fluvial lakes in comparison to the gravel pits [6, 35].

Fig. 3.6 Kazuski Dunavac (Gornje Podunavlje wetland area, Serbia), October 2010, dense mats of invasive species *Azolla filiculoides* Lam. Photo by D. Cvijanović

According to previous studies [6, 35, 75], the most significant water quality variables for macrophyte assemblages in both types of floodplain SWB are oxygen saturation, total organic carbon, electroconductivity, total alkalinity. On the other hand, the importance of naturalness of riparian land-cover types, the extent and diversity of littoral habitat features (tree roots, woody debris, and overhanging vegetation), the sediment and hydrological regimes were found to be important common hydromorphological predictors for macrophytes in both types of SWB [6, 35, 75]. In general, river hydrological alterations are the most frequent and significant pressures affecting and threatening the river floodplain ecosystem in the Western Balkans [6, 29, 33, 44]. Cvijanović et al. [76] found that frequency of flooding periods during vegetation season may affect macrophyte vegetation structure probably by direct biomass removal in gravel pit lakes along the Drina river floodplain. In the same study, the frequency of the spring flooding events influenced macrophyte diversity and richness by changing trophic conditions.

Additionally, hydrological regime and water physical properties such as temperature may be affected by climate changes [77, 78], which further may accelerate the spreading of invasive species [79]. Natural eutrophic lakes of the Danube floodplain have a strong seasonal character in terms of species composition and dominance [35]. According to Laketić [35], macrophyte vegetation at these sites was characterised by stands of submerged pondweed species after the spring–summer flood

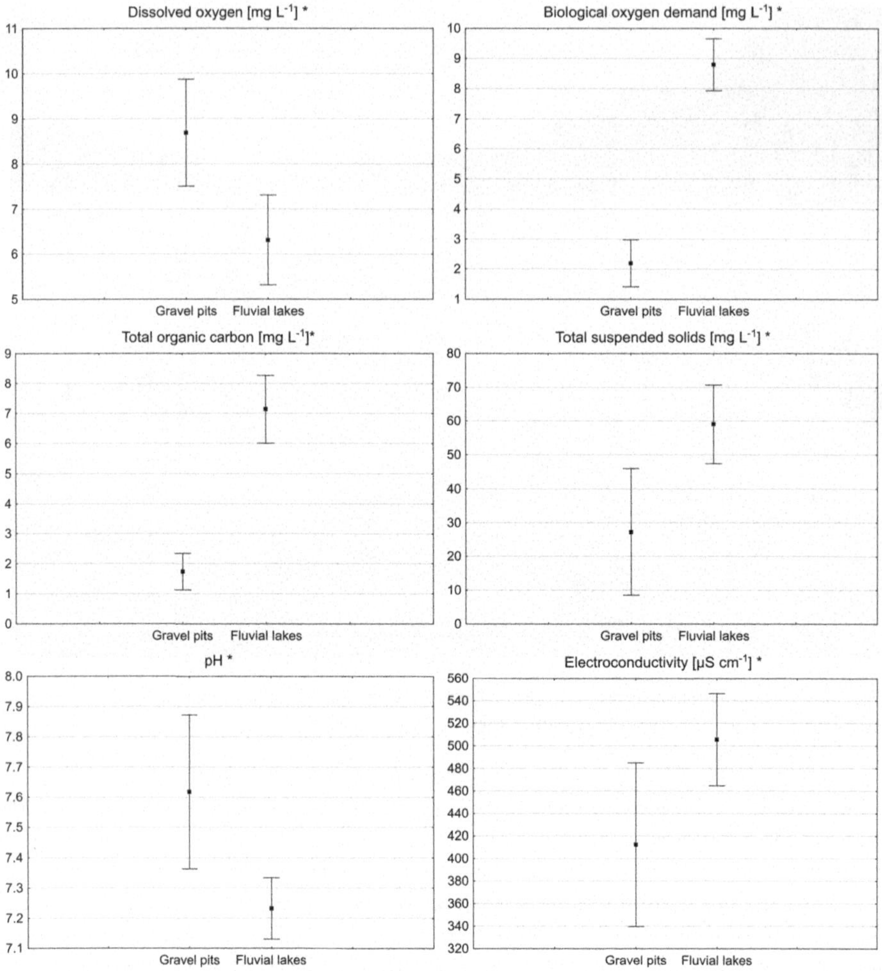

Fig. 3.7 Range and mean values of physico-chemical variables in floodplain fluvial lakes and gravel pits in the Western Balkans. Asterisk symbol '*' indicate statistically significant difference between gravel pits and fluvial lakes (T-test, $p < 0.05$)

in 2010, which usually occurs in the period from March to July. During the end of July and August 2010, macrophyte cover was dominated by duckweed species, water chestnut, and water-lilies (Fig. 3.5). After the fall mixing of water of these relatively shallow systems, which usually occurs during September, native species had almost vanished and dense stands of invasive species *Azolla filiculoides* Lam. had completely occupied these lakes (Fig. 3.6) [35].

In general, both lake types of SWB in floodplain systems of the Western Balkans are subjected to the significant spreading of aquatic invasive species: *Azolla filiculoides*, *Vallisneria spiralis* L., *Elodea canadensis* Michx., *Elodea nuttallii* (Planch.)

H. St. John, *Paspalum paspaloides* [36–38, 80–83]. According to Andjelkovic et al. [82], the Sava and Danube rivers together with the irrigation canal network of the Danube–Tisa–Danube Hydrosystem are important invasion corridors in the region. Although some other aquatic invasive species were recorded in the area studied, such as *Pistia stratiotes* L. (Begej river, Serbia [84]) and *Egeria densa* Planch. and *Myriophyllum heterophyllum* Michx (Small water bodies in the Neretva river Delta system, Croatia [85, 86]), their distributions are still localized.

3.5 Conclusions

Fluvial and gravel pit lakes along the river floodplains can be considered as hot spots of aquatic habitat diversity in the Western Balkans, following the EU Habitats Directive and the Bern Convention. Although these two lake systems may support the same habitat types, such as submerged rooted and free-floating vegetation, as well as the rooted stands with floating leaves, still, some assemblages are confined to one of these systems. Charophyte submerged carpets in mesotrophic water bodies were found to be associated with gravel pit lakes, while the free-floating duckweed communities were exclusively found in eutrophic fluvial lakes. Therefore, these two kinds of floodplain SWB cover a wide trophic gradient. This phenomenon, coupled with macrophytes' physical structural heterogeneity, creates extremely diverse habitats for other aquatic communities. However, these small water bodies are subjected to various human pressures such as alteration of hydrological regime, artificial land use of the riparian zone, sediment erosion, shoreline modification, and eutrophication in the entire region.

Acknowledgements This study was partly supported by the Rufford grant No. 28388 (Toward Cost-Effective UAV-Assisted Multimetric System for Detection of Freshwater Patches of High Conservation Value within the Danube Floodplain in Serbia) and by the Project of Ministry of Education, Science and Technological Development of the Republic of Serbia (grant number 43002 (451-03-68/2020-14/200125, 451-03-68/2020-14/200124).

References

1. European Environmental Agency (2020) Floodplain statistics viewer. https://www.eea.europa.eu/data-and-maps/data/data-viewers/floodplain-areas. Accessed 01 May 2021
2. Cohen AS (2003) Paleolimnology: the history evolution of lake systems. Oxford University Press, New York
3. Amoros C, Bornette G (2002) Connectivity biocomplexity in waterbodies of riverine floodplains. Freshw Biol 47(4):761–776
4. Radulović S, Laketić D, Teodorović I (2011) A botanical classification of standing waters in Serbia its application to conservation. Aquat Conserv Mar Freshw Ecosyst 21(6):510–527

5. Cvijanović DL, Lakušić DV, Živković MM, Novković MZ, Anđelković AA, Pavlović DM, Vukov DM, Radulović SB (2018) An overview of aquatic vegetation in Serbia. Tuexenia 38:269–286
6. Damnjanović B, Novković M, Vesić A, Živković M, Radulović S, Vukov D, Anđelković A, Cvijanović D (2019) Biodiversity-friendly designs for gravel pit lakes along the Drina River floodplain (the Middle Danube Basin, Serbia). Wetl Ecol Manag 27(1):1–22
7. Thomaz SM, Cunha ERD (2010) The role of macrophytes in habitat structuring in aquatic ecosystems: methods of measurement, causes consequences on animal assemblages' composition biodiversity. Acta Limnol Bras 22(2):218–236
8. Mendonça R, Kosten S, Lacerot G, Mazzeo N, Roland F, Ometto JP, Paz EA, Bove CP, Bueno NC, Gomes JHC, Scheffer M (2013) Bimodality in stable isotope composition facilitates the tracing of carbon transfer from macrophytes to higher trophic levels. Hydrobiologia 710(1):205–218
9. Meerhoff M, Mazzeo N, Moss B, Rodríguez-Gallego L (2003) The structuring role of free-floating versus submerged plants in a subtropical shallow lake. Aquat Ecol 37(4):377–391
10. O'Briain R, Shephard S, Coghlan B (2017) Pioneer macrophyte species engineer fine-scale physical heterogeneity in a shallow lowland river. Ecol Eng 102:451–458
11. Keruzoré AA, Willby NJ, Gilvear DJ (2013) The role of lateral connectivity in the maintenance of macrophyte diversity production in large rivers. Aquat Conserv Mar Freshw Ecosyst 23(2):301–315
12. Pander J, Nagel C, Geist J (2021) Integration of constructed floodplain ponds into nature-like fish passes supports fish diversity in a heavily modified water body. Water 13(8):1018
13. Santoul F, Figuerola J, Green AJ (2004) Importance of gravel pits for the conservation of waterbirds in the Garonne river floodplain (southwest France). Biodivers Conserv 13(6):1231–1243
14. Blaen PJ, MacDonald MA, Bradbury RB (2016) Ecosystem services provided by a former gravel extraction site in the UK under two contrasting restoration states. Conserv Soc 14(1):48–56
15. Søndergaard M, Lauridsen TL, Johansson LS, Jeppesen E (2018) Gravel pit lakes in Denmark: chemical biological state. Sci Total Environ 612:9–17
16. Bolpagni R, Poikane S, Laini A, Bagella S, Bartoli M, Cantonati M (2019) Ecological conservation value of small standing-water ecosystems: a systematic review of current knowledge future challenges. Water 11(3):402
17. Council of the European Communities (2000) Directive 2000/60/EC of the European Parliament the Council of 23rd October 2000 establishing a framework for community action in the field of water policy. Off J Eur Communities L327:1–73
18. European Commission (1992) Council directive 92/43/EEC of 21 May 1992 on the conservation of natural habitats and of wild fauna and flora. Off J Eur Union L206:7–50
19. European Commission (2009) Council directive 2009/147/EC on the conservation of wild birds (codified version). Off J Eur Communities L20:7–25
20. European Commission (2007) Directive 2007/60/EC of the European Parliament of the council of 23 October 2007 on the assessment the management of flood risks. Off J L288:28
21. Council of the European Communities (1979) Bern convention—convention on the conservation of European Wildlife Natural Habitats. Bern, 19.IX.1979
22. Epstein Y (2013) The habitats directive Bern convention: synergy dysfunction in public international EU law. Geo Int'l Envtl L Rev 26:139
23. Topić J (1989) Vegetation of the special zoological reserve of Kopački rit. Hydrobiologia 182(2):149–160
24. Ozimec S, Prlić D, Rožac V (2018) Aquatic wetland vegetation of the Kopački Rit Nature Park, Croatia. In: The 26th international scientific symposium deltas and wetlands (Book of Abstracts), vol 5. Tulcea, p 16
25. Rožac V, Prlić D, Ozimec S (2018) The vascular flora of Kopački rit Nature Park (Croatia). Acta Biol Slov 61(2):47–70

26. Kashta L (2007) Aquatic macrophytes in the Lake Shkodra-River Buna wetlands complex. In: Pinna M, Uricchio VF, Aresta M, Bsset A (eds) Rivers and citizens. Cross-border experiences in environmental protection sustainable development. Lecce, pp 82–93
27. Hadžiablahović S (2018) The diversity of the flora vegetation of Lake Skadar/Shkodra. In: Pešić V, Karaman G, Kostianoy A (eds) The Skadar/Shkodra Lake Environment. The Handbook of Environmental Chemistry, vol 80. Springer, Cham, pp 203–238
28. Šegota V, Zlatković B, Vukov D, Alegro A, Koletić N, Vuković N, Rimac A (2019) Status assessment of the rare aquatic plant *Groenlandia densa* (L.) Fourr. (Potamogetonaceae) in the Western Balkans. Bot Lett 166(2):125–133
29. Schiemer F, Beqiraj S, Drescher A, Graf W, Egger G, Essl F, Frank T, Hauer C, Hohensinner S, Miho A, Meulenbroek P (2020) The Vjosa River corridor: a model of natural hydromorphodynamics a hotspot of highly threatened ecosystems of European significance. Landsc Ecol 35(4):953–968
30. Lasić A, Jasprica N (2017) Plant communities of the two oligotrophic karstic Trebižat Lištica rivers testing the assessment of ecological status based on macrophytes. In: Gračan R, Matoničkin Kepčija R, Miliša M, Ostojić A (eds) Book of abstracts of the second symposium on freshwater biology. Croatian Association of Freshwater Ecologists, Zagreb, p 47
31. Cvijanović DL, Živković MM, Novković MZ, Anđelković AA, Radulović SB, Lakušić DV (2016) The digital database of aquatic semiaquatic vegetation in Serbia. Zbornik Matice srpske za prirodne nauke 131:145–155
32. Kraradžić B, Bulić Z, Jarić S, Mitrović M, Pavlović P (2020) Vegetation in ravine habitats of Montenegro. In: Pešić V, Paunović M, Kostianoy AG (eds) The rivers of Montenegro (The handbook of environmental chemistry). Springer, Cham, pp 201–229
33. Talevska M, Ohrid RM (2010) Influence of human activities on submerged vascular macrophytes alongside Cm Drim River. In Proceedings of the 4th Balwois conference on water, climate environment
34. Radulović S, Laketić D, Vukov D (2010) A riverside tale: assessment of altered habitat effects on macrophyte assemblage on the river Tamiš. Serbia Arch Biol Sci 62(4):1163–1174
35. Laketić DL (2013) Fitocenološka klasifikacija vegetacije jezerskog tipa u Srbiji. Dissertation, University of Belgrade
36. Alegro A, Šegota V, Rimac A, Hršak V, Vuković N, Koletić N, Stanković I (2016) Macrophyte vegetation of rivers in Croatia—diversity ecological indication potential. In: Rešetnik I, Ljubešić Z. Book of abstracts of the 5th Croatian botanical symposium with international participation. Hrvatsko botaničko društvo, Zagreb, p 6
37. Glasnović P, Novak Š, Behrič S, Fujs N (2015) Towards a checklist of the vascular flora of the Neretva River Delta (Croatia). Nat Croat 24(2):163–190
38. Rimac A, Alegro A, Šegota V, Koletić N, Stanković I, Bogdanović S, Vuković N (2021) Distribution habitat characteristics of *Vallisneria spiralis* L. in Croatia. Hacquetia 20(1):7–18
39. Shuka L, Malltezi J, Mersinllari M, Vardhami I (2008) May. Dynamics of vegetation cover of Prespa Lakes its watershed (Albanian side). In: Proceedings of the third international scientific conference on Balkan Water Observation Information System (BALWOIS), pp 27–31
40. Mullaj A, Vangjeli J, Peçi D, Imeri A (2007) General considerations on the flora vegetation of Albania's rivers. In: Pinna M, Uricchio VF, Aresta M, Bsset A (eds) Rivers and citizens. Cross-border experiences in environmental protection sustainable development. Lecce, pp 16–25
41. Šumberová K (2011) Stratiotetum aloidis Miljan 1933. In: Chytrý M (ed) Vegetace České republiky. 3. Vodní a mokřadní vegetace [Vegetation of the Czech Republic 3. Aquatic wetland vegetation]. Academia, Praha, pp 89–93
42. Alegro A, Bogdanović S, Brana S, Jasprica N, Katalinić A, Kovačić S, Nikolić T, Milović M, Pandža M, Posavec-Vukelić V, Randić M (2010) Botanički važna područja Hrvatske, Školska knjiga d.d. Zagreb
43. Igić R, Polić D, Borišev M, Janauer GA (2004) Aquatic macrophytes of Krstonošića okno (Obedska bara swamp). In: Teodorović I, Radulovic S, Bloesh J (eds) Limnological reports 35. Proceedings of the 35th conference. Novi Sad, pp 463–468

44. Vukov D, Igić R, Rućando M, Radulović S (2012) Diversity of vascular hydrophytes in the Zasavica river (Serbia): changes after thirteen years. Arch Biol Sci 64(4):1607–1617
45. Rowan JS, Carwardine J, Duck RW, Bragg OM, Black AR, Cutler MEJ, Soutar I, Boon PJ (2006) Development of a technique for lake habitat survey (LHS) with applications for the European Union Water Framework Directive. Aquat Conserv Mar Freshw Ecosyst 16(6):637–657
46. Boon P, Argillier C, Boggero A, Ciampittiello M, England J, Peterlin M, Radulović S, Rowan J, Soszka H, Urbanič G (2019) Developing a standard approach for assessing the hydromorphology of lakes in Europe. Aquat Conserv Mar Freshw Ecosyst 29(4):655–669
47. Oertli B, Joye DA, Castella E, Juge R, Cambin D, Lachavanne JB (2002) Does size matter? The relationship between pond area biodiversity. Biol Conserv 104(1):59–70
48. Linton S, Goulder R (2000) Botanical conservation value related to origin management of ponds. Aquat Conserv Mar Freshw Ecosyst 10(2):77–91
49. Rosset V, Simaika JP, Arthaud F, Bornette G, Vallod D, Samways MJ, Oertli B (2013) Comparative assessment of scoring methods to evaluate the conservation value of pond small lake biodiversity. Aquat Conserv Mar Freshw Ecosyst 23(1):23–36
50. Ramsar Sites (2021) Wetlands of international importance—Serbia. https://www.ramsar.org/wetland/serbia. Accessed 01 May 2021
51. Stojanović V, Đorđević J, Lazić L, Stamenković I, Dragićević V (2014) The principles of sustainable development of tourism in the special nature reserve Gornje Podunavlje and their impact on the local communities. Acta Geogr Slov 54(2):391–400
52. Petrović MD, Pavić D, Marković SB, Mészáros M, Jovičić A (2016) Comparison estimation of the values in wetland areas: a study of Ramsar sites Obedska bara (Serbia) Lonjsko polje (Croatia). Carpathian J Earth Environ 11(2):367–380
53. Schiemer F, Drescher A, Hauer C, Schwarz U (2018) The Vjosa River corridor: a riverine ecosystem of European significance. Acta ZooBot Austria 155:1–40
54. UEPG (2021) European Aggregates Association. Estimates of aggregates production—2018 data. https://uepg.eu/pages/figures. Accessed 28 April 2021
55. Commission E (2011) Guidance on undertaking new non-energy extractive activities in accordance with Natura 2000 requirements. Publications Office of the European Union, Luxembourg
56. European Commission (2001) Directive 2001/42/EC on the evaluation of the effects of certain plans programmes on the environment. Off J Eur Communities L197:30–37
57. Council of the European Communities (1985) Directive 85/337/EEC on the assessment of the effects of certain public private projects on the environment, as amended in 1997 (97/11/EC), 2003 (2003/35/EC) 2009 (2009/31/EC). Off J Eur Communities L175:40–48
58. European Commission (2010) COM (2010) 715/4, Communication from the commission to the European parliament, the Council, the European economic social committee and the Committee of the regions. European Union Strategy for Danube Region, Brussels
59. Olmeda C, Barov B (2019) Non-energy mineral extraction in relation to Natura 2000. Case studies. Project report Contract N 07.0202/2017/766291/SER/D.3 for the European Commission. https://ec.europa.eu/environment/nature/natura2000/management/docs/NEEI%20case%20studies%20-%20Final%20booklet.pdf. Accessed 01 May 2021
60. Federal Agency for Nature Conservation (2021) Hard oligo-mesotrophic waters with benthic vegetation of Chara spp. https://www.bfn.de/en/lrt/natura-2000-code-3140.html. Accessed 01 May 2021
61. Seelen LM, Teurlincx S, Bruinsma J, Huijsmans TM, van Donk E, Lürling M, de Senerpont Domis LN (2021) The value of novel ecosystems: disclosing the ecological quality of quarry lakes. Sci Total Environ 769:144294. https://doi.org/10.1016/j.scitotenv.2020.144294
62. Trbojević IS, Predojević DD, Šinžar-Sekulić JB, Nikolić NV, Jovanović IM, Subakov-Simić GV (2019) Charophytes of Gornje Podunavlje ponds: revitalization process aspect. Zbornik Matice srpske za prirodne nauke 136:123–131
63. Trbojević I, Milovanović V, Subakov Simić G (2020) The discovery of the rare Chara baueri (Charales, Charophyceae) in Serbia. Plants 9(11):1606

64. Vesić A, Blaženčić J, Šinžar-Sekulić J (2016) Ecological preferences of charophytes in Serbia in relation to habitat type other aquatic macrophytes. Plant Biosyst 150(3):490–500. https://doi.org/10.1080/11263504.2014.976293
65. Kamberović J, Barudanović S (2012) Algae macrophytes of mine pit lakes in the wider area of Tuzla, Bosnia Herzegovina. Nat Croat 21(1):101–118
66. Šumberová K (2011) Vegetace vodních rostlin zakořeněných ve dně (Potametea). Vegetation of aquatic plants rooted in the bottom. In: Chytrý M (ed) Vegetace České republiky. 3. Vodní a mokřadní vegetace [Vegetation of the Czech Republic 3. Aquatic wetland vegetation]. Academia, Praha, pp 101–247
67. Barudanovicl S, Masicl E (2013) State of the mine-pit lakes in the wider area of Zenica-Doboj Region. In: Proceedings of the 24th international scientific-expert-conference of agriculture food industry. University of Sarajevo, Faculty of Agriculture Food Sciences, Sarajevo, 25–28 September 2013, pp 550–555
68. Panjković B (2005) Akvatična i semiakvatična vegetacija Apatinskog i Monoštorskog rita. Dissertation, University of Novi Sad
69. Lorenz AW, Feld CK (2013) Upstream river morphology riparian land use overrule local restoration effects on ecological status assessment. Hydrobiologia 704(1):489–501
70. Leuschner C, Ellenberg H (2017) Ecology of central European non-forest vegetation: coastal to Alpine, natural to man-made habitats. In: Leuschner C, Ellenberg H (eds) Vegetation ecology of central Europe, vol 2. Springer, Cham
71. McGoff E, Aroviita J, Pilotto F, Miler O, Solimini AG, Porst G, Jurca T, Donohue L, Sandin L (2013) Assessing the relationship between the Lake Habitat Survey littoral macroinvertebrate communities in European lakes. Ecol Indic 25:205–214
72. Jurca T, Donohue L, Laketic D, Radulovic S, Irvine K (2012) Importance of the shoreline diversity features for littoral macroinvertebrate assemblages. Fundam Appl Limnol 180(2):175
73. McGoff E, Irvine K (2009) A test of the association between Lake Habitat Quality Assessment macroinvertebrate community structure. Aquat Conserv Mar Freshw Ecosyst 19(5):520–533
74. Damnjanović B (2019) Ekologija i konzervaciona vrednost vodene vegetacije šljunkara u plavnom području reke Drine. Dissertation, University of Novi Sad
75. Damnjanović B, Novković M, Živković M, Anđelković A, Pavlović D, Radulović S, Cvijanović D (2016) The interdependency of hydromorphology, water quality macrophyte vegetation in establishing conservation targets for eutrophic temperate lakes (the Middle Danube Basin, Serbia). In: Book of abstracts the International bioscience conference 6th international PSU—UNS bioscience conference—IBSC, 19–21 September 2016, Novi Sad, p 69
76. Cvijanović D, Damnjanović B, Novković M, Vesić A, Živković M, Anđelković A, Vukov D, Radulović S (2018) The role of hydrological regime in structuring macrophyte assemblages in gravel pit lakes along the Drina River floodplain (Serbia). In: Botanica Serbica, 42 (supplement 1) 7th Balkan botanical congress, 10–14 September 2018, Novi Sad, p 82
77. Basarin B, Lukić T, Pavić D, Wilby RL (2016) Trends multi-annual variability of water temperatures in the river Danube. Serbia Hydrol Process 30(18):3315–3329
78. Basarin B, Lukić T, Matzarakis A (2016) Quantification assessment of heat cold waves in Novi Sad, Northern Serbia. Int J Biometeorol 60(1):139–150
79. Kopeć D, Ratajczyk N, Wolańska-Kamińska A, Walisch M, Kruk A (2014) Floodplain forest vegetation response to hydroengineering climatic pressure—a five decade comparative analysis in the Bzura River valley (Central Poland). For Ecol Manag 314:120–130
80. Barina Z, Somogyi G, Pifkó D, Rakaj M (2018) Checklist of vascular plants of Albania. Phytotaxa 378(1):1–339
81. Cvijanović D, Damnjanović B, Novković M, Vesić A, Živković M, Anđelković A, Radulović S (2018) The role of invasive hydrophytes in structuring macrophyte assemblages in the newly formed gravel pit lakes. In: Book of abstracts joint ESENIAS DIAS scientific conference 8th ESENIAS Workshop. Management sharing of IAS data to support knowledge based decision making at regional level, 26–28 September 2018, Bucharest, p 48
82. Kočić A, Horvatić J, Jelaska SD (2014) Distribution morphological variations of invasive macrophytes *Elodea nuttallii* (Planch.) H. St. John *Elodea canadensis* Michx in Croatia. Acta Bot Croat 73(2):437–446

83. Anđelković A, Zivković MM, Cvijanović DL, Novković M, Marisavljević D, Pavlović D, Radulović S (2016) The contemporary records of aquatic plants invasion through the Danubian floodplain corridor in Serbia. Aquat Invasions 11(4):381–395
84. Živković MM, Anđelković AA, Cvijanović DL, Novković MZ, Vukov DM, Šipoš ŠŠ, Ilić MM, Pankov NP, Miljanović BM, Marisavljević DP, Pavlović, DM (2019) The beginnings of *Pistia stratiotes* L. invasion in the lower Danube delta: the first record for the province of Vojvodina (Serbia). BioInvasions Rec 8(2): 218–229
85. Jasprica N, Lasić A, Hafner D, Bratoš Cetinić A (2017) *Myriophyllum heterophyllum* Michx. (Haloragaceae) u Hrvatskoj. Nat Croat 26(1):99–103
86. Rimac A, Stanković I, Alegro A, Gottstein S, Koletić N, Vuković N (2018) The Brazilian elodea (Egeria densa Planch.) invasion reaches Southeast Europe. BioInvasions Rec 7(4):381–389

Chapter 4
Fountains—Overlooked Small Water Bodies in the Urban Areas

Dubravka Čerba and Ladislav Hamerlík

Abstract Urban and suburban areas represent a specific environment supporting the development of particular terrestrial and aquatic communities. Excluding streams and rivers, most water bodies in the urban areas are man-made, created for various purposes (e.g. decoration, recreation, gravel excavation, irrigation). Rather extreme ecosystems are the city fountains that have been usually overlooked in limnological studies. Despite the extreme environmental variables, especially water pH, temperature, turbulence, disinfection and cleaning activities, which make them inhospitable for many organisms, fountains do sustain a development of specific invertebrate communities, particularly insects, including a dipteran family Chironomidae. Studies dedicated to this unique man-made ecosystem are scarce. Those carried out indicate that fountains can harbour both very diverse chironomid communities (e.g., in Europe), but they can also be extremely species poor (e.g., in South America). The typical fountain community consists of common taxa with very wide ecological requirements together with rare species living in specific habitats, such as water distribution systems, hygropetric- and terrestrial habitats. The many new records found in fountains of countries with well-known chironomid faunas (e.g., Denmark and Czech Republic) further emphasize the importance and peculiarity of fountains. City fountains, as extreme habitat as they are, still represent very important shallow water bodies acting as a source of biodiversity in the urban areas and deserve more attention. In this chapter we (1) review the literature dealing with fountain biota and (2) bring new data from this specific habitat in the Balkans (Osijek and Varaždin, Croatia).

D. Čerba (✉)
Department of Biology, Josip Juraj Strossmayer University of Osijek, Osijek, Croatia
e-mail: dcerba@biologija.unios.hr

L. Hamerlík
Faculty of Natural Sciences, Department of Biology and Ecology, Matej Bel University, Banská Bystrica, Slovakia

Institute of Zoology, Slovak Academy of Sciences, Bratislava, Slovakia

L. Hamerlík
e-mail: ladislav.hamerlik@umb.sk

V. Pešić et al. (eds.), *Small Water Bodies of the Western Balkans*, Springer Water,
https://doi.org/10.1007/978-3-030-86478-1_4

Keywords Urban aquatic habitat · Osijek · Croatia · Chironomidae · Diptera

4.1 Introduction

> For fountains, they are a Great Beauty and Refreshment, but Pools mar all, and make the Garden unwholesome, and full of Flies and Frogs.
>
> —Sir Francis Bacon, Of Gardens, 1625.

People love water fountains. During an unbearably hot summer day in a city centre, it is hard to overlook groups of people, both tourists and locals, attracted by public fountains enjoying their refreshing action. These architectonic elements bring spirit and diversity to the urban environment. They are valuable for their history, practical use, splendour, symbolism and artistic effects along with the relaxation they provide [1, 2]. City fountains also have local climatic and health effects. The water vapor sprayed into the atmosphere reduces the ambient air temperature and contributes to its purification by removing the dirt and dust from the air [3]. At the same time, as a downside, fountains and ornamental pools are important elements in the epidemiological chain and can be sources of free-living amoebae and other microorganisms including respiratory bacterial pathogens and thus may pose a public health risk [3–5]. The common use of water fountains by humans and animals (especially birds) may contribute to the transmission of pathogenic microorganisms and also to spreading of epidemiologically important insects, such as mosquitoes [6].

From the ecological perspective, fountains represent a specific environment that can play an important role in the maintenance of terrestrial and aquatic biodiversity in the urban space. Research dedicated to this unique ecosystem is scarce and indicate that fountains can harbour both very diverse insect communities [7–9, 11], but they can also be extremely species poor [10]. The typical fountain community consists of a mixture of species that usually do not co-occur in natural water bodies, i.e. common taxa with very wide ecological requirements together with rare species living in specific habitats, such as water distribution systems, hygropetric- and terrestrial habitats. The many new faunistic records [11–13] further emphasize the importance and peculiarity of fountains as habitats.

The structure of city fountains, especially classical forms, includes one or more shallow pools of water with various ways of refill (Figs. 4.1, 4.2). The small surface area and water depth correspond to that of shallow water bodies, and their substrate structure resembles the stony littoral of lakes and ponds.

In this chapter we aim at (1) reviewing the knowledge dealing with environmental settings and biota of fountains and (2) bringing new data from this specific habitat in the Balkans (Osijek and Varaždin, Croatia).

Fig. 4.1 Examples of the variability of European water fountains. A baroque fountain in Olomouc, Czech Republic (**a**) and modern fountains of different sizes and types from Copenhagen, Denmark (**b**–**d**). Photo: L. Hamerlík

4.2 Brief History of Limnological Research of Fountains

City fountains have started to attract the attention of limnologists as study objects not long ago and thus our knowledge on their role as habitats, in ecological sense, in the urban space is rather incomplete. Some studies were marginally dealing with fountains but, to our knowledge, the first study focusing on the limnology of city fountains was that of Rieradevall and Cambra [7] who studied 10 fountains in Barcelona (along with artificial city ponds). This research was rather complex, dealing with a wide array of environmental variables and biota, including phyto- and zooplankton, and phyto- and zoobenthos (for more information see Table. 4.1), and found mainly species typical for lentic habitats of small volume, with a cosmopolitan distribution and without special ecological requirements.

The following fountain studies were focusing on Chironomidae (Diptera) communities in different European [8, 9] and South American cities [10]. These studies found a very unique composition of chironomid communities in fountains consisting of common species with wide geographical distribution, tap-water pests, hygropetric and semi-terrestrial species. The results indicated that distance to colonization source, fountain proximity and regional factors shape the fountain community. Oboňa et al. [11] carried out a comprehensive limnological survey of fountains in a Slovakian

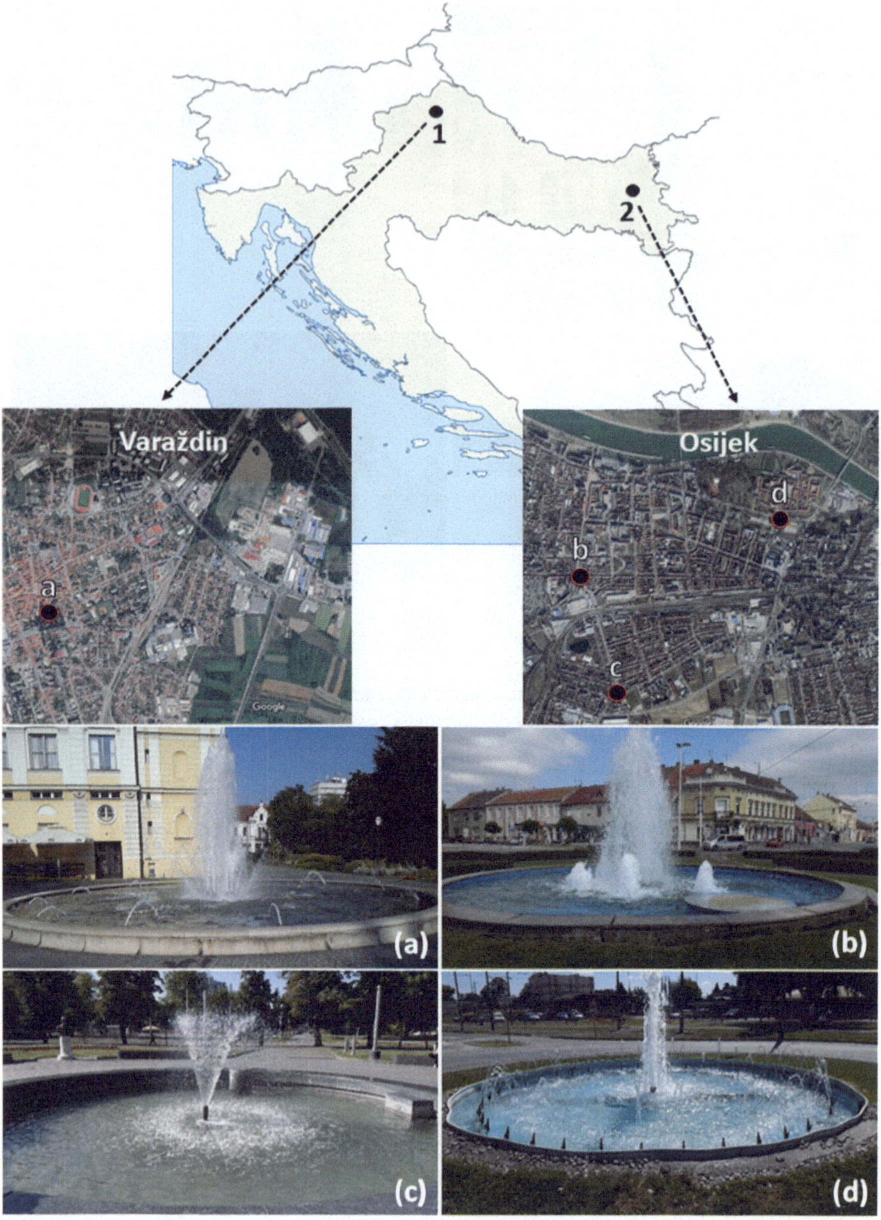

Fig. 4.2 View of the studied fountains in Croatia. Varaždin: fountain in front of the Croatian National Theatre building, HNK (**a**). Osijek: Spar (**b**), Rondel (**c**) and Mačkamama (**d**). Photo: M. Kresonja (**a**) and D. Čerba (**b**–**d**)

Table 4.1 Overview of the known limnological research of city fountains. Research is ordered according to the year of publication. Entirely faunistic research and first record papers are not included

Country	City	No. of fountains	Taxonomic groups	No. of taxa recorded	Author
Spain	Barcelona	10	Cyanobacteria, Chlorophyta, Bacillariophyceae, Ciliata, Turbellaria, Nematoda, Gastropoda, Oligochaeta, Ostracoda, Insecta (Diptera, Ephemeroptera, Heteroptera, Odonata, Trichoptera)	229 algae 91 zoobenthos (of which 30 Chironomidae)	Rieradevall and Cambra [7]
Czech Republic	Olomouc	6	Diptera (Chironomidae)	40	Hamerlík and Brodersen [8]
Denmark	Copenhagen	6	Diptera (Chironomidae)	19	Hamerlík and Brodersen [8]
Colombia	Bogotá	6	Diptera (Chironomidae)	7	Hamerík et al. [10]
Ecuador	Quito	10	Diptera (Chironomidae)	7	Hamerík et al. [10]
Sweden	Lund	4	Diptera (Chironomidae)	16	Bukvová and Hamerlík [9]
Slovakia	Prešov	6	Rotifera, Nematoda, Cladocera, Copepoda, Ostracoda, Insecta (Coleoptera, Diptera, Ephemeroptera, Heteroptera)	66 in total, Non-insects 38 (Rotifera 23), Insecta 28	Oboňa et al. [11]
Croatia	Osijek	3	Diptera (Chironomidae)	29	Kresonja [14]
	Varaždin	1	Diptera (Chironomidae)	11	Kresonja [14]

city, including zooplankton and zoobenthos and found significant seasonal changes in the composition of the fountain fauna. Fountain research has also resulted in many new faunistic records even for countries with well-studied chironomid fauna, such as Denmark [12], Czech Republic [13] or Slovakia [11]. Relatively comprehensive limnological research of the Balkan fountains has started only recently in Croatia [14] and it is still in its infancy. For brief summary of fountain research see Table 4.1.

4.3 Environmental Features of the Fountains

Water fountains are very simple ecosystems regarding their habitat heterogeneity. The simplest ones, without a reservoir, only consist of hygropetric surfaces (Fig. 4.1c) occasionally with small patches of filamentous algae [7]. Those more complex ones with a reservoir and splashing water (Fig. 4.1a, d) usually also resemble a uniform bedrock, without macrophytes and different size fractions of the bottom substrate. In some cases, patches of periphyton and algae may cover the fountain reservoir's inner walls and accumulations of organic matter may occur here and there [8–10].

Size-wise, fountains can be compared to small ponds, not exceeding hundreds of square meters but are usually much smaller, up to several tens of square meters. Obviously, they are also shallow, normally not deeper than one meter [11], resembling stony littoral of lakes and ponds.

Despite their extreme physical simplicity and similarity, even from the small amount of data on environmental variables, it is obvious that fountains epitomize a variable environment both within and among sites. Water temperatures of the few European fountains studied varied between 10.5 and 28 °C (18.3 °C on average) [7, 8, 11, 14], while temperature in the same fountain could differ by 11.5 °C during the growing season [8]. Conductivity was in general relatively high, ~ 1700 $\mu S\ cm^{-1}$ on average and varied between 556 and 2920 $\mu S\ cm^{-1}$. Fountains in southern Europe tend to have higher conductivity than in the central Europe [7, 11]. Water pH was circumneutral, ranging from 6.5 to 9 (7.4 on average) [7, 11, 14]. There is very little reliable information on the oxygen content and that shows great variability among sites. Water in the surveyed fountains was usually oversaturated all around the season [11, 14] but in Barcelona some of them had oxygen content as low as 1.6 $mg\ L^{-1}$ [7].

4.4 Biota of the Fountains

Even though fountains are understudied habitats, it is obvious that they harbour complex and diverse aquatic communities comparable with shallow water bodies, i.e. ponds and shallow lakes. The very few comprehensive studies focusing on the whole community and not only on a particular taxonomic group have recorded many higher taxa, such as Cyanobacteria, Chlorophyta, Bacillariophyceae, Ciliata, Turbellaria, Rotifera, Nematoda, Gastropoda, Oligochaeta, Ostracoda, Cladocera, Copepoda and Insecta (Coleoptera, Diptera, Ephemeroptera, Heteroptera, Odonata, Trichoptera) [7, 11].

Fountains without reservoir contain species poor benthic fauna with species of short life cycle and semi-terrestrial adaptations. The characteristic taxa of such fountains are ciliates (*Euplotes*), Nematoda and Diptera, especially Ceratopogonidae, Chironomidae (genus *Limnophyes*) and Psychodidae (particularly on filamentous algae) [7]. Fountains with water reservoirs harbour more diversified communities, most likely due to richer habitat heterogeneity. In addition to the above taxa inhabiting the hygropetric zone, detritovors dominated, such as *Chironomus*, *Cricotopus* gr. *sylvestris* (Chironomidae), Aelosomatidae, Naididae, Tubificidae (Oligochaeta), Ostracoda, Nematoda, Hydrobiidae and *Lymnaea* (Gastropoda). Predators were rare, represented by Turbellaria [7]. As much as 229 algal taxa were identified in the fountains of Barcelona with diatoms and green algae as the best represented groups. The phytobenthos was more diverse than the phytoplankton [7]. To summarize, the presence of a reservoir with an adequate amount of water, analogous to a shallow water body, represents certain hydrological stability that is necessary for the development of a complex invertebrate community, i.e. sufficient water for the oviposition and completing the metamorphosis cycle of insects, periphytic community development, primary production, detritus creation and deposition, etc.

All fountain studies highlighted the dominance of the Diptera species in the community, especially those of the Chironomidae family. The reason for that may be the biology of this insect group. Aquatic insects of temporary habitats cope with loss of water from their habitat by means of physiological tolerance, migration and life history modification [15]. It seems, however, that the dominant insects living in seasonal water bodies persist because of repeated colonization from nearby permanent habitats, rather than by physiological adaptations [16]. For the most part, these species exhibit high power of dispersal, rapid growth, short life-span, small size, opportunistic feeding and poor competitive capabilities [17], i.e. features that are typical for Chironomidae.

Dispersal plays a key role in maintaining populations in fragmented habitats; it can counteract local extinction and accelerate colonization of new habitats [18, 19]. In general, chironomids are weak fliers commonly being caught up to hundreds of metres from the source [20, 21], but can easily be transported long distances by air currents [22, 28]. Some species found commonly in the fountains, e.g. those of the *Chironomus* genus, are considered to have strong dispersal ability [20]. On the other hand, even though larvae of *Cricotopus* (also common in all studied fountains) have frequently been recorded in new available water bodies, there is no indication that adults fly more than other genera of the subfamily [28] and thus to have superior dispersal capacity. Consequently, the study of dispersal ability of particular species is of primary importance in understanding the mechanism of colonization of fountains and other shallow urban water bodies.

Up to 80 chironomid taxa, mainly species, were found in 37 fountains of six European countries (Table 4.2). Moreover, as much as 23 chironomid species were

Table 4.2 Chironomidae recorded in 37 fountains of seven European cities

	N Europe		C Europe		S Europe		
	Denmark	Sweden	Czech Rep.	Slovakia	Spain	Croatia	Croatia
	Copenhagen	Lund	Olomouc	Prešov	Barcelona	Osijek	Varaždin
Tanypodinae							
Ablabesmyia monilis (Linnaeus, 1758)	–	+	+	–	–	–	–
Macropelopia nebulosa (Meigen, 1804)	+	+	+	–	–	–	–
Procladius (*P.*) *choreus* (Meigen, 1804)	+	+	+	–	–	–	+
Procladius (*P.*) *flavifrons* Edwards, 1929	+	–	+	–	–	–	–
Procladius sp. A	–	–	+	–	–	–	–
Tanypus sp.	–	–	–	–	+	–	–
Thienemannimyia pseudocarnea Murray, 1976	–	–	+	–	–	–	–
Zavrelimyia barbatipes (Kieffer, 1911)	–	–	–	+	–	–	–
Zavrelimyia melanura (Meigen 1804)	–	–	+	–	–	–	–
Zavrelimyia nubila (Meigen, 1830)	–	+	+	+	–	–	–
Orthocladiinae							
Acricotopus lucens (Zetterstedt, 1850)	+	–	+	+	–	+	–
Bryophaenocladius subvernalis (Edwards, 1929)	+	–	+	–	–	–	–
Bryophaenocladius sp.	–	–	–	–	–	+	+
Corynoneura gr. *scutellata*	+	+	+	–	–	–	–
Corynoneura spp.	–	–	–	–	–	+	–
Cricotopus (*C.*) *bicinctus* (Meigen, 1818)	–	–	+	–	–	+	+
Cricotopus (*I.*) *ornatus* (Meigen, 1818)	+	–	+	–	–	+	–

(continued)

Table 4.2 (continued)

	N Europe		C Europe		S Europe		
	Denmark	Sweden	Czech Rep.	Slovakia	Spain	Croatia	Croatia
	Copenhagen	Lund	Olomouc	Prešov	Barcelona	Osijek	Varaždin
Cricotopus (*I.*)*reversus* Hirvenoja, 1973 / *intersectus* (Staeger, 1839)	–	+	+	–	–	+	–
Cricotopus (*I.*) *sylvestris* (Fabricius, 1794) / *trifasciatus* (Meigen, 1810)	+	+	+	+	+	+	+
Cricotopus (*C.*)*triannulatus* (Macquart, 1826) / *curtus* Hirvenoja, 1973	–	–	–	–	–	+	–
Cricotopus (*C.*)*vierriensis* Goetghebuer, 1935	–	–	–	–	–	+	+
Cricotopus spp.	–	–	+	–	+	–	+
Eukiefferiella claripennis (Lundbeck, 1898)	–	+	+	–	–	–	–
Limnophyes sp.	–	–	–	–	+	–	–
Metriocnemus eurynotus (Holmgren, 1883)	+	–	–	–	–	–	–
Metriocnemus gr. *hygropetricus* *	–	–	–	–	–	–	–
Nanocladius (*N.*)*dichromus* (Kieffer, 1906)	–	–	–	–	–	+	–
Nanocladius (*N.*)*rectinervis* (Kieffer, 1911)	–	–	+	–	–	–	–
Orthocladius (*Eud.*) *fuscimanus* (Kieffer, 1908)	+	+	+	+	–	–	+
Orthocladius (*Euo.*)*thienemanni* Kieffer, 1906	–	–	+	–	–	–	–
Orthocladius (*O.*)*glabripennis* (Goetghebuer,1921)	–	–	–	–	–	+	+
Orthocladius (*O.*) cf. *oblidens* (Walker, 1856)	–	–	+	–	–	–	–
Orthocladius (*O.*)*rhyacobius* Kieffer 1911	–	–	+	–	–	–	–
Orthocladius (*O.*) *rubicundus* (Meigen, 1818)	–	–	+	–	–	–	+

(continued)

Table 4.2 (continued)

	N Europe		C Europe		S Europe		
	Denmark	Sweden	Czech Rep.	Slovakia	Spain	Croatia	Croatia
	Copenhagen	Lund	Olomouc	Prešov	Barcelona	Osijek	Varaždin
Orthocladius spp.	–	–	–	–	–	+	–
Paralimnophyes/Limnophyes sp.	+	–	+	–	–	–	–
Psectrocladius (*P.*)*barbimanus* (Edwards, 1929)	+	–	–	–	–	–	–
Psectrocladius (*P.*)*brehmi* Kieffer, 1923	–	–	–	–	–	+	–
Psectrocladius (*P.*)*limbatellus* (Holmgren, 1869)	+	+	+	+	–	–	–
Psectrocladius (*P.*) cf.*sordidellus* (Zetterstedt, 1838) / *ventricosus* Kieffer, 1925	+	–	–	–	–	–	–
Pseudosmittia sp.	+	–	–	–	–	–	–
Pseudosmittia Pe2	–	–	–	–	–	+	–
Rheocricotopus (*P.*)*chalybeatus* (Edwards, 1929)	–	–	–	–	–	+	–
Synorthocladius semivirens (Kieffer, 1909)	–	–	+	–	–	–	–
Thienemanniella cf. *vittata* (Edwards, 1924)	–	–	+	–	–	–	–
Thienemanniella sp.	–	–	–	+	–	–	–
Orthocladiinae unidentified 1 *	–	–	–	–	–	–	–
Orthocladiinae unidentified 2 *	–	–	–	–	–	–	–
Chironominae							
Chironomus spp.	+	+	+	+	+	+	+
Cladopelma cf.*virescens* (Meigen, 1818)	–	–	–	–	–	+	–

(continued)

Table 4.2 (continued)

	N Europe		C Europe		S Europe		
	Denmark	Sweden	Czech Rep.	Slovakia	Spain	Croatia	Croatia
	Copenhagen	Lund	Olomouc	Prešov	Barcelona	Osijek	Varaždin
Cladopelma sp.	–	–	–	–	+	–	–
Cryptochironomus obreptans (Walker, 1856)	–	–	+	–	–	–	–
Dicrotendipes nervosus (Staeger, 1839)	–	–	+	–	–	+	–
Dicrotendipes notatus (Meigen 1818)	–	–	+	–	–	–	–
Dicrotendipes sp. *	–	–	–	–	–	–	–
Endochironomus tendens (Fabricius, 1775)	–	–	–	+	–	–	–
Glyptotendipes (*G.*) *barbipes* (Meigen, 1804)	–	–	+	–	–	–	–
Glyptotendipes (*G.*) *pallens* (Meigen, 1804)	–	–	–	+	–	–	–
Goeldichironomus cf. *holoprasinus* (Goeldi, 1905)*	–	–	–	–	–	–	–
Kiefferulus tendipediformis (Goetghebuer, 1921)	–	–	+	+	–	+	–
Microchironomus sp.	–	–	–	–	+	–	–
Microtendipes rydalensis (Edwards, 1929)	–	–	+	–	–	–	–
Parachironomus gr. *arcuatus*	–	+	+	–	–	+	–
Parachironomus frequens (Johannsen, 1905)	–	–	–	–	–	+	–
Parachironomus cf.*supparilis* (Edwards, 1931) *	–	–	–	–	–	–	–
Parachironomus sp.	–	–	–	–	+	–	–
Polypedilum cultellatum (Goetghebuer, 1931)	–	–	+	–	–	+	–

(continued)

Table 4.2 (continued)

	N Europe		C Europe		S Europe		
	Denmark	Sweden	Czech Rep.	Slovakia	Spain	Croatia	Croatia
	Copenhagen	Lund	Olomouc	Prešov	Barcelona	Osijek	Varaždin
Polypedilum sp.	–	–	+	–	–	+	–
Cladotanytarsus lepidocalcar Krueger, 1938	–	–	–	–	–	+	–
Micropsectra atrofasciata agg.	–	+	–	–	–	–	–
Micropsectra cf.*lindrothi* Goetghebuer, 1931	+	+	+	–	–	–	–
Micropsectra notescens (Walker, 1856)	–	–	–	+	–	–	–
Paratanytarsus bituberculatus (Edwards, 1929)	–	–	+	–	–	–	–
Paratanytarsus grimmii (Schneider 1885)	+	–	+	–	–	+	–
Paratanytarsus inopertus (Walker, 1856)	–	–	–	–	–	+	–
Paratanytarus laccophilus (Edwards, 1929)	–	+	–	–	–	–	–
Rheotanytarsus rhenanus Klink, 1983	–	–	–	–	–	+	–
Tanytarsus mendax Kieffer, 1925	+	–	+	–	–	+	–
Tanytarsus usmaensis Pagast, 1931/*nigricollis* Goetghebuer, 1939	–	+	–	–	–	–	–
Tanytarsus sp.	–	–	–	+	+	+	+
Total number of taxa	19	16	40	13	30#	29	11

Taxa marked by asterisk (*) occurred only in the South American fountains
#The complete list of taxa recorded in the Spanish fountains was not available, thus our list only consists of taxa mentioned in the text of the publication

collected in a single fountain, which is fascinating, given the temporary and uniform character of fountains.

The trophic structure of fountain invertebrates seems to be rather similar. Collectors slightly dominated over scrapers and predators; filter-feeders, if present, represented a marginal part of the assemblages [8, 10]. The dominance of collectors demonstrates the overall flexibility of chironomids in feeding mode and indicates that even small patches and a thin layer of soft sediment in the reservoir may be sufficient for them to colonize such uniform habitats.

Regarding the preference to waterflow, a wide range of species from limnobionts to rheobionts were recorded in fountains, but the difference in colonization sources (i.e. running versus standing water) was clearly reflected in the assemblage structure of particular fountains. While in the fountains with a significant running water source rheobionts represented a significant part of the assemblage, in fountains in a city with scattered lake/pond sources they were not recorded at all. Water in fountains is usually well mixed and rich in oxygen and offers acceptable conditions for the coexistence of both rheo- and limnophilic species.

Even though there were considerable differences in water temperature (both ranges and means) among fountains, the community composition did not reflect it. Temperature optima for 26 taxa recorded in fountains (see Fig. 4.5 in Hamerlík and Brodersen [8]) showed that cold as well as warm water species may coexist in them (Fig. 4.5). The estimated optima ranged from 9 to 23 °C, indicating that temperature is not an important factor determining the species distributions in fountains.

4.5 Fountains of the Balkans

We continued the research of this fascinating urban habitat and ecological niche in the Balkans; for starter in the Pannonian Ecoregion, with the prospects of widening our research across the peninsula, especially in the vicinity of the "ecologically interesting" water bodies.

The main subject of our research was a Dipteran family—Chironomidae, for the reasons mentioned above. Chironomid exuviae were sampled by applying a standard method of skimming the accumulated matter in the surface layer of water with a sampling net, mash size of 300 μm [23]. Exuviae in the fountains were sampled in a circular way across the whole surface of the fountain reservoir.

During a seven-month research period, three fountains were sampled one to three times a month in different areas of Osijek, a city in east Croatia, whilst in Varaždin, in the north-western part of Croatia, one fountain was sampled in April and June 2016. For location and view of fountains see Fig. 4.2. Both cities have temperate oceanic climate with warm summers, also known as „Cfb " in the Köppen climate classification [24], and lie on the right bank of a big lowland river (Drava), which together with artificial lakes and ponds in the cities' area, represent a colonization source of Chironomidae for the fountains. Fountains can be considered as temporary shallow water bodies and as such represent perfect microhabitats for resting and egg deposit sites for insects in urban areas. Since there are several man-made water bodies of various sizes in Osijek and Varaždin, the fountains are important "stops" and recolonization sources playing an important role in insect diversity in the cities. However, there is more to the fountains than simple urban pools or sinks. The water current coming from nozzles and sprinklers necessitate additional classification since those disturbances simulate riffle or current of streams, depending on the shape and size of the fountain. These conditions contribute to the successful colonization by rheophils and rheobionts from the lotic systems as well, which is in this case the lower reach of the Drava river.

Out of dozen fountains scattered across Osijek, the following three were most appropriate for biodiversity research: Spar, Rondel and Mačkamama (Fig. 4.2b–d). All other fountains had either drainage built in such way that nothing remained to be sampled or were built as a part of a pedestrian area, or were just not operating in the year of the research. In the city of Varaždin only one fountain was suitable for sampling (HNK, Fig. 4.2a). Rondel, Spar and HNK are made of various types of rocks, such as marble, basalt, sandstone or concrete, whilst Mačkamama is made of metal.

All fountains are filled with water at the beginning of the spring (March/April) and are functioning till the end of October, what notably influences colonization success and community development. Maintenance, cleaning or disinfection represent further challenges for the biota trying to inhabit these man-made shallow water bodies.

The four studied fountains harboured great diversity for an artificial habitat located in the middle of urban area. We identified 33 Chironomidae taxa, members of three subfamilies: Tanypodinae, Orthocladiinae and Chironominae (tribes Chironomini and Tanytarsini) (Fig. 4.3). Diversity of chironomids recorded in the fountains is rather similar, and varies from 11 taxa (HNK in Varaždin) to 18 taxa (Mačkamama) (Fig. 4.3). Most likely, this is a result of similar hydromorphological conditions

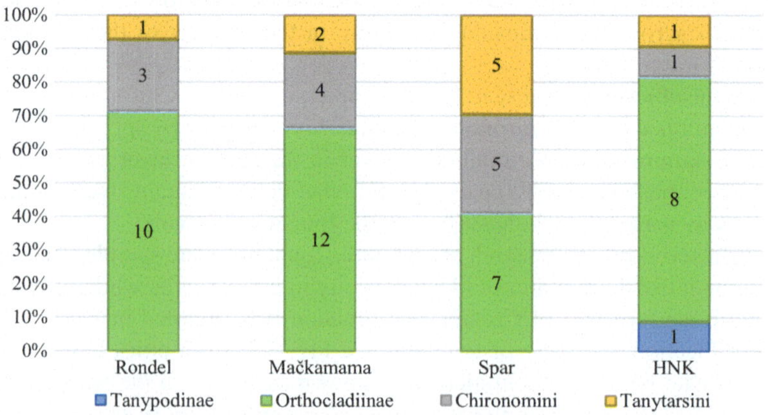

Fig. 4.3 Percentage rate of Chironomidae diversity (no. of taxa indicated by numbers in the chart) within subfamily/tribe in the fountains studied in Osijek (Rondel, Mačkamama, Spar) and Varaždin (HNK)

(area, depth, shape of profile, water regime) of the fountain reservoirs, which simulate similar living conditions to that of ponds, but with somewhat extreme abiotic parameters and strong central nozzles, what on the other hand contributes to aeration of the pool of water. The dominance of Orthocladiinae (Fig. 4.3.) was previously recorded during the research of other European fountains [8], in contrary to South American research findings with records of equal ratio of Orthocladiinae and Chironominae [10]. If we consider fountains as small temporary water bodies, the dominance of Orthocladiinae was in some way expected [25], since their smaller body sizes, good colonization traits, shorter life-cycles and fast reproduction rates, can make them more adapted to this kind of habitat [8, 26, 27]. Furthermore, these fountains are located in the vicinity of diverse natural or man-made aquatic habitats and other urban green spaces what is deeming to be necessary for repeated colonization process.

Most frequently sampled were cosmopolitan, widely distributed species with good colonizing potential, such as *Cricotopus sylvestris* and *C. intersectus* [28], accompanied with a representative of the *Chironomus* genus, a very adaptable and resilient taxon (Fig. 4.4). In slow sand filter beds, also specific temporary aquatic habitats, on different occasions researchers found *Cricotopus sylvestris* as dominating species. Jointly, in different seasons and sites, most abundant were *Procladius sagittalis* and *Tanytarsus mendax* [26, 29]. With the exception of *Tanytarsus* sp., representatives of the Tanytarsini tribe were found only in Osijek, and most of them were recorded in the Spar fountain, including *Tanytarsus* cf. *mendax*. The chironomid community composition of Croatian fountains was in accordance with the ones found in previously studied fountains [8, 9, 11, 13], represented by the aforementioned fountain chironomids along with *Orthocladius fuscimanus* [30, 31] and *Paratanytarsus grimmii*. The latter is an interesting "urban" chironomid found in water supply systems and fountains, displaying parthenogenetic reproduction abilities as adaptation to these kind

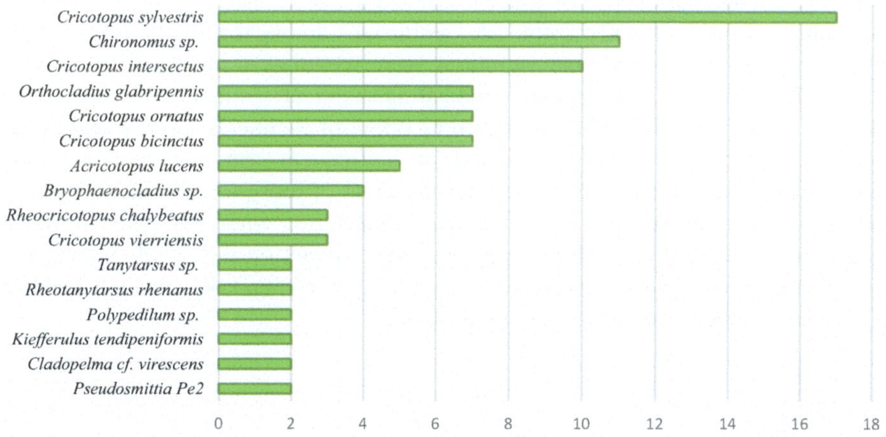

Fig. 4.4 Chironomid taxa most frequently recorded during the research of the city fountains in Osijek and Varaždin (minimum two sampling occasions)

of environments [8, 28]. *Orthocladius fuscimanus* is a hygropetric species utilizing a thin water layer residual on the walls and in cracks of fountains for egg disposal and larval development [9, 30, 31].

Notable findings also include semi-terrestrial chironomids *Bryophaenocladius* cf. *virgo* and *Pseudosmittia* Pe2, which we found in late October, at the end of the season. *Pseudosmittia* Pe2 was found in the Szentendrei-Duna in Hungary [32] and most likely found its way from the Danube to Drava River and Osijek. Previously it was thought to be restricted to Great Britain and the Iberian Peninsula. Contrary to semi-terrestrial or madicolous species, rheophilic species are not often found in water bodies with conditions typical for fountains. According to Moller Pillot [33], *Rheocricotopus chalybeatus* hasn't yet been recorded in temporary ponds, however larvae prefer stones as microhabitat what might prove to be suitable substrate if the fountain has some kind of continuous water current and is close enough to the river. Or maybe it just got stranded on the way to the suitable ovipositioning place. Exuviae of *Procladius choreus* were found only in the Varaždin fountain. Oftentimes *Procladius* inhabits temporary and man-made water bodies, including fountains [8, 25, 34].

Chironomidae are a very resilient and adaptable insect group, however, even in those genera which don't have suitable physiological tolerances to survive desiccation or chemical pollution, they compensate with higher colonization rates from surrounding permanent water bodies [15]. Even though we do not have substantial number of fountains to draw indisputable conclusion, it is indicative that we recorded a typical rheobiont *R. chalybeatus* only in the fountain closest to the river Drava. For insects inhabiting this type of habitats, r-strategy is often a solution to this problem [17].

When fountains are not, or only scarcely cleaned (e.g. once a month [11]), periphytic communities develop, primary production is high and higher amounts of organic matter can be deposited. Such fountains have more similarities with natural shallow aquatic systems and harbour complex invertebrate communities [7, 11].

Environmental parameters of water oscillated in accordance with the season and atmospheric conditions, and did not vary significantly among the fountains (Fig. 4.5). The most conspicuous seasonal variations were recorded in electric conductivity (ranging from 713 and 2450 $\mu S\ cm^{-1}$) indicating chemical treatment of the fountains. Also, on several occasions there was a strong odour of chemicals and sometimes we could see that the fountains were cleaned. Those frequent treatments with algicides and larvicides in order to prevent the development of mosquito larvae, seemingly had no effect to the success of chironomid colonization when compared to other aquatic insects, but it could have had a negative impact on the development of a more complex and diverse invertebrate community. We found only a small number of exuviae of other dipterans and not constantly.

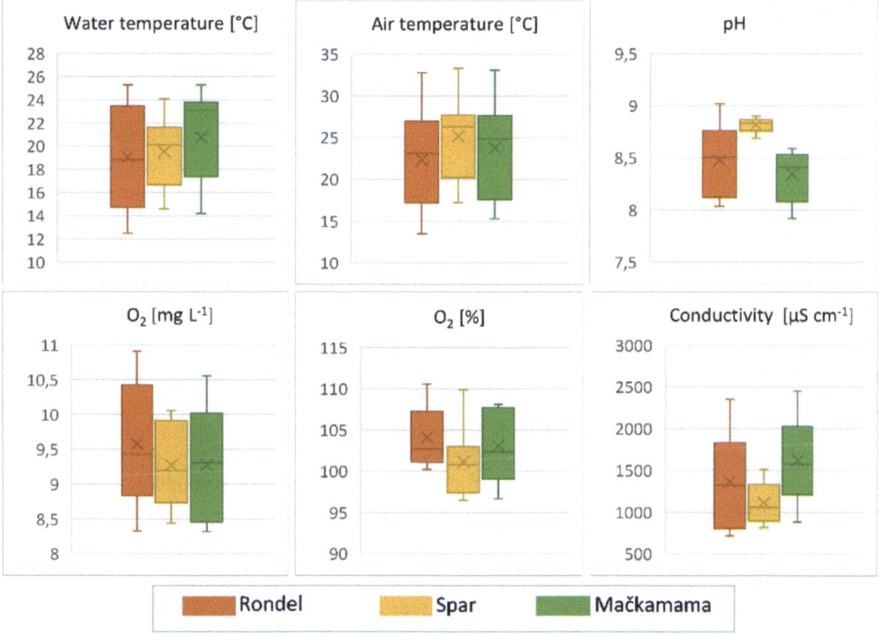

Fig. 4.5 Basic parameters of the fountain water, measured during sampling in Osijek. Box and whisker plots are showing the min–max values (whiskers), first and third quartile (box), and median and mean (line and x-sign) of the data set

4.6 Conclusion

Fountains are unique aquatic habitats, shallow water bodies that represent islands of biodiversity in urban areas. Even though cities can have other man-made or natural waters, fountains harbour specific communities and should not be overlooked in ecological studies. Furthermore, those very special features that separate them from other urban habitats enable us to further study the "true nature" of chironomid (and other invertebrate) species, their ecology and adaptability. Close and continuous research can give us new insights on how certain parameters (e.g. temperature vs. oxygen) influence the species' life cycle, behaviour or feeding habits, as we have already noticed that certain "textbook" rules do not always apply. The gained knowledge can additionally contribute to greater understanding of the observations from natural habitats, but also to monitor how global changes influence both natural and urban aquatic ecosystems.

Acknowledgements We would like to thank Matija Kresonja, Viktorija Ergović and Matej Šag, and all other colleagues who helped in the realization of our fountain research. This work was supported by the „Institutional financing" of the Department of Biology, Josip Juraj Strossmayer University of Osijek and by the Slovak Research and Development Agency under contract no. APVV-16-0236.

References

1. Hynynen A, Juuti P, Katko T (2012) Water fountains in the worldscape. International Water History Association and KehräMedia Inc
2. Juuti PS, Antoniou GP, Dragoni W, El-Gohary F, De Feo G, Katko TS, Rajala RP, Zheng XY, Drusiani R, Angelakis AN (2015) Short global history of fountains. Water 7:2314–2348
3. Burkowska-But A, Walczak M (2013) Microbiological contamination of water in fountains located in the city of Torun, Poland. Ann Agric Environ Med 20:645–648
4. Erci̇ E, Abay E, Akarsu GA, Karahan ZC (2013) Investigation of free-living Amoebae and respiratory bacterial pathogens in water samples taken from recreational fountains and ornamental pools in Ankara, Turkey. HealthMED 7:1158–1167
5. Nascimento M, Rodrigues JC, Reis L, Nogueira I, Carvalho PA, Brandão J, Duarte A, Jordao L (2016) Pathogens in ornamental waters: a pilot study. Int J Environ Res Public Health 13:216
6. Schaffner F, Kaufmann C, Hegglin D, Mathis A (2009) The invasive mosquito *Aedes japonicus* in Central Europe. Med Vet Entomol 23:448–451
7. Rieradevall M, Cambra J (1994) Urban freshwater ecosystems in Barcelona. Int Verlgung Für Theor Und Angew Limnol: VerhEn 25:1369–1372
8. Hamerlík L, Brodersen KP (2010) Non-biting midges (Diptera: Chironomidae) from fountains of two European cities: micro-scale island biogeography. Aquat Insects 32:67–79
9. Bukvová D, Hamerlík L (2015) Non-biting midges (Diptera, Chironomidae) in the fountains of Lund, SW Sweden. Entomol Tidskr 136:87–92
10. Hamerlík L, Jacobsen D, Brodersen KP (2011) Low species richness of non-biting midges (Diptera: Chironomidae) in Neotropical artificial urban water bodies. Urban Ecosyst 14:457–468
11. Oboňa J, Demková L, Smolák R, Dominiak P, Ščerbáková S (2017) Invertebrates in overlooked aquatic ecosystem in the middle of the town. Period Biol 119:47–54
12. Hamerlík L, Brodersen KP, Biba S (2010) First records of the non-biting midges *Orthocladius* (*Eudactylocladius*) *fuscimanus* (Kieffer) and *Paratanytarsus grimmii* (Schneider) (Diptera: Chironomidae) for Denmark, with notes on their ecology and distribution in artificial habitats. Stud Dipterol 17(1/2)
13. Hamerlík L (2007) Chironomidae (Diptera) from fountains new for Czech Republic. Lauterbornia 61:137–140
14. Kresonja M (2018) Urbana vodena staništa, zanemareni izvori bioraznolikosti–usporedba zajednica trzalaca (Chironomidae, Diptera) u fontanama, bajerima i rijeci. MSc Thesis, Josip Juraj Strossmayer University of Osijek. Department of Biology
15. Williams DD (1996) Environmental constraints in temporary fresh waters and their consequences for the insect fauna. J N Amn Benthol Soc 15:634–650
16. Batzer DP, Resh VH (1992) Macroinvertebrates of a California seasonal wetland and responses to experimental habitat manipulation. Wetlands 12:1–7
17. McLachlan A (1993) Can two species of midge coexist in a single puddle of rain-water? Hydrobiologia 259:1–8
18. Monaghan MT, Spaak P, Robinson CT, Ward JV (2002) Population genetic structure of 3 alpine stream insects: influences of gene flow, demographics, and habitat fragmentation. J N Am Benthol Soc 21:114–131
19. Lake PS, Bond N, Reich P (2007) Linking ecological theory with stream restoration. Freshw Biol 52:597–615
20. Delettre YR, Morvan N (2000) Dispersal of adult aquatic Chironomidae (Diptera) in agricultural landscapes. Freshw Biol 44:399–411
21. Bitušík P, Svitok M, Novikmec M, Trnková K, Hamerlík L (2017) Biological recovery of acidified alpine lakes may be delayed by the dispersal limitation of aquatic insect adults. Hydrobiologia 790:287–298
22. Heino J (2013) Does dispersal ability affect the relative importance of environmental control and spatial structuring of littoral macroinvertebrate communities? Oecologia 171:971–980

23. Ferrington LC, Blackwood MA, Wright CA, Crisp NH, Kavabaugh JL, Schmidt FJ (1991) A protocol for using surface-floating pupal exuviae of chironomidae for rapid bio assessment of changing water quality. Sediment and stream water quality in a changing environment: trends and explanation. vol 203, pp 181–190
24. Filipčić A (1998) Klimatska regionalizacija Hrvatske po W. Koeppenu za standardno razdoblje 1961.-1990. u odnosu na razdoblje 1931.-1960. Acta Geographica Croatica 33:7–15
25. Bazzanti M, Seminara M, Tamorri C (1989) A note on Chironomids (Diptera) of temporary pools in the National Park of Circeo, Central Italy. Hydrobiol Bull 23:189–193
26. Wotton RS, Armitage PD, Aston K, Blackburn JH, Hamburger M, Woodward CA (1992) Colonization and emergence of midges (Chironomidae: Diptera) in slow sand filter beds. Neth J Aquat Ecol 26:331–339
27. Čerba D, Mihaljević Z, Vidaković J (2010) Colonisation of temporary macrophyte substratum by midges (Chironomidae: Diptera). Ann Limnol Int J Limnol 46:181–190
28. Armitage PD, Cranston PS, Pinder LCV (1995) The Chironomidae: biology and ecology of nonbiting midges. Chapman and Hall, London, p 572
29. Hirabayashi K, Matsuzawa M, Yamamoto M, Nakamoto N (2004) Chironomid fauna (Diptera, Chironomidae) in a filtration plant in Japan. J Am Mosq Control Assoc 20:74–82
30. Cranston PS (1984) The taxonomy and ecology of *Orthocladius* (*Eudactylocladius*) *fuscimanus* (Kieffer), a hygropetric chironomid (Diptera). J Nat Hist 18:873–895
31. Vaillant F (1956) Récherches sur la faune madicole (hygropetrique s. l.) de France, de Corse et d'Afrique de Nord. Mémoires du Muséum national d'Histoire naturelle 11:1–258
32. Móra A, Farkas A (2012) The Chironomidae (Diptera) fauna of the Szentendrei-Duna, Hungary. Acta Biol Debrecina Suppl Oecol Hung 28:129–140
33. Moller Pillot HKM (2013) Chironomidae Larvae, Vol. 3. Orthocladiinae: biology and ecology of the aquatic Orthocladiinae, vol 3. KNNV Publishing, Zeist, p 312
34. Abellan P, Sanchéz-Fernández D, Millán A, Botella F, Sánchez-Zapata JA, Giménez A (2006) Irrigation pools as macroinvertebrate habitat in a semi-arid agricultural landscape (SE Spain). J Arid Environ 67:255–269
35. McLachlan A (1993) Can two species of midge coexist in a single puddle of rain-water? Hydrobiologia 259:1–8

Chapter 5
Temporary Ponds in Mediterranean Islands: Oases of Biodiversity

Tvrtko Dražina, Maria Špoljar, and Marko Miliša

Abstract The development of tourism in the Mediterranean and the abandonment of traditional extensive agriculture lead to the succession of several different habitat types. One of the most endangered habitats in Mediterranean region are temporary ponds. Despite small size, these ephemeral waterbodies are recognized as reservoirs of biodiversity. Shallow ponds are often only freshwater habitats on islands. In this chapter we will mainly focus on biotic interactions among macrophytes, zooplankton, macrozoobenthos and fish. With this approach we will try to give guidelines for conservation and restoration in order to prevent succession and devastation of ponds.

Keywords Pond ecology · Macrophytes · Zooplankton · Macrozoobenthos · Fish · Conservation

5.1 Introduction

Ponds can be defined as small (1 m^2 to max 5 ha), natural or man-made shallow waterbodies (normally fresh water, but occasionally brackish), which holds water for at least three months of the year or more [1]. This is just one of the possible definitions found in limnological scientific literature, but the story about these ephemeral habitats is much more complex. De Meerr et al. [2] for example, distinguished ponds and pools: the difference is that ponds permanently hold water whereas pools dry out, either regularly, yearly or every few years. Sahuquillo and Miracle [3] divided ponds into five categories, according to the hydroperiod: (I) temporary short pools (with hydroperiod < 3 months), (II) temporary intermediate pools (with hydroperiod from

T. Dražina (✉) · M. Špoljar · M. Miliša
Faculty of Science, Division of Biology, University of Zagreb, Rooseveltov trg 6, 10 000 Zagreb, Croatia
e-mail: tvrtko.drazina@biol.pmf.hr

M. Špoljar
e-mail: maria.spoljar@biol.pmf.hr

M. Miliša
e-mail: marko.milisa@biol.pmf.hr

V. Pešić et al. (eds.), *Small Water Bodies of the Western Balkans*, Springer Water,
https://doi.org/10.1007/978-3-030-86478-1_5

3 to 6 months), (III) temporary long pools (hydroperiod with more than 6 months, but with annual summer drying), (IV) semi-permanent ponds (dry every few years) and (V) permanent ponds, fed by springs. Although there is no satisfactory formal definition of ponds, these habitats are numerous in all terrestrial environments, with occurrence from polar areas to tropical rainforests. Despite their small surface area, ponds are extremely rich in biodiversity terms, especially within a region or at the landscape scale [1]. Their high contribution to ß diversity (i.e. regional diversity) comes from the facts that ponds often strongly differ in species composition amongst each other, and more important, ponds harbour species that are specific to these unique ephemeral habitats [4].

In Mediterranean region of Europe, ponds are the most common inland freshwater lentic habitats. The development of tourism in this area and the abandonment of traditional extensive agriculture lead to the succession of several different habitat types, including ponds. Thus, they are one of the priority habitat types for conservation, according to the EU Habitats Directive [5].

Mediterranean area of Croatia is part of Dinaric karst, known as world hotspot for surface and subterranean freshwater biodiversity [6, 7]. Surface waters in karstic areas are rather scarce, especially on islands. Adriatic islands are very poor regarding inland waters, and only a few of them have lakes (islands of Cres, Krk and Pag), but most of them have ponds. Lakes on Krk and Cres island were investigated in details, from hydro-morphology to biology [8–15]. Apart from this data from permanent surface island waterbodies, there is almost no data about temporary ponds. Most of investigation was conducted by enthusiasts and the NGO sector, and the data are scattered in technical reports, occasional publications or in abstracts of scientific conferences [16–22]. Nevertheless, all these scattered data indicate diverse and specific fauna and flora of island ponds. For example, Temunović and Šerić Jelaska [22] recorded in the ponds of the island of Lastovo species *Cybister tripunctatus africanus* Laporte, 1835 (Coleoptera, Dytiscidae), for the first time in Croatian fauna. Some detailed information of Odonata faunistic composition are known for several Adriatic islands [16, 21, 23, 24]. Kučinić et al. [25] published some information concerning caddisfly fauna from four Adriatic islands (Cres, Krk, Pag and Hvar).

In this chapter, we provide some new information about this type of habitat based on recently collected data. We will mainly focus on zooplankton and their biotic interactions in order to propose some specific conservation measures.

5.2 Invertebrates and Their Adaptations to Temporary Ponds

Due to ephemeral state, ponds are habitats with harsh abiotic conditions. They are often exposed to high variations of physico-chemical factors, and to high variations of hydroperiod, ranging from waterbodies lasting less than 3 months to stable, permanent ponds [3]. Only some groups of aquatic organisms can survive in this environment with such extremes. Zooplankton is such assemblage, as they are permanent aquatic residents, found in all types of habitats. Freshwater zooplankton (Rotifera, Cladocera and Copepoda) have a wide range of adaptive strategies, and they are important as early colonizers of temporary water bodies. Also, they display diverse reproductive strategies, in order to overcome harsh condition and drought. Rotifers of class Monogononta reproduce by cyclical parthenogenesis (heterogony): asexual reproduction is predominant, but sexual reproduction occurs episodically and products are resting eggs, dormant stage with thick shells [26]. Another rotifer class, Bdelloidea, reproduce entirely by asexual parthenogenesis, and anhydrobiosis is widespread strategy in order to survive unfavorable conditions [26]. Cladocerans, similarly to monogononts, usually show cyclical parthenogenesis and produce resting eggs enclosed in an ephippium, which allow them survival of dry period [27]. Second group of freshwater plankton crustaceans, copepods, display sexual reproduction. Copepod dormancy occurs in various ontogenetic stages, as resting eggs, juvenile and adult encystment, or arrested development of free-swimming nonencysted copepodids or adults [27]. Group of permanent aquatic residents that occur in only in temporary ponds are large branchiopods from orders Anostraca and Notostraca. They survive the dry periods as resting eggs that are resistant to drought and can remain viable for many years.

Also, ponds are often important habitats for different insects' orders (Ephemeroptera, Odonata, Hemiptera, Coleoptera and Diptera). They are temporary residents of these habitats, and they have different adaptation, in order to survive potential dry phase: from species capable of some dispersal that oviposit on water in spring and then aestivate or overwinter in the dry basin in various stages of the life, to species that have well-developed powers of dispersal that leave the disappearing pool and pass the dry phases in permanent waters. Some of these animals subsequently return to oviposit in the temporary ponds, when favorable hydrological conditions return.

5.3 Biotic and Abiotic Interactions in Mediterranean Ponds: A Case Study from Adriatic Islands (Croatia)

5.3.1 Study Site and Sampling Protocol

Macrophytes, zooplankton, macrozoobenthos and fish were sampled from 20 small ponds situated on three Adriatic islands (Croatia, Fig. 5.1): Rab (RAB1–RAB2), Dugi otok (DO1–DO10) and Korčula (KOR1–KOR8). The samples were obtained in spring period, from May to early June 2017. The investigated pond were extremely small waterbodies with the average diameter 12 ± 4 m, and average depth of 1.2 ± 0.5 m. The coordinates of location, morphological features of the ponds, macrophytes taxa, type and percent of coverage are specified in Supplement 1. The ponds are

Fig. 5.1 Photographs of some investigated Mediterranean ponds. A–pond DO3, Dugi otok island; B–pond DO2, Dugi otok island; C–pond RAB2, Rab island; D–pond KOR7, Korčula island. For abbreviation and detail description see Supplement 1. Photos by T. Dražina (A, B) and A. Štih (C, D)

located on islands where dominates the karst relief and geological composition is mainly formed of limestone and dolomite. High air temperatures in summer, low rainfall and continuous wind during the warmer part of the year influence the strong evapotranspiration and extreme dryness of the islands, making these ponds specific freshwater habitats, "islands within islands".

Macrophytes were hand collected and their coverage was estimated using standard Braun-Blanquet method [28]. Zooplankton samples (if possible), were taken from three different locations in each pond: open water, along the macrophytes edge towards open water and within macrophytic stands. At each study site, triplicates were taken by filtering 10 L of water through a 26 μm mesh net and fixed with a 4% formalin solution. In the laboratory, zooplankton comprising of rotifers, cladocerans and copepods was counted and identified to the lowest taxonomic level under the inverted microscope (Opton-Axiovert 35) at 400 $\times$ and 1000 $\times$ magnification. We classified all identified zooplankton species according to frequency level into four groups: constant ($\geq$ 50%), accessory, (21–49%), accidental (10–20%) and sporadic (< 10%). Macroinvertebrates were sampled using a 25 $\times$ 25 Surber sampler, preserved in 70% ethanol and subsequently isolated by systematic groups. Only predatory macroinvertebrates were taken under observation. The fish were sampled with the net and only their presence/absence was recorded.

Temperature, oxygen, pH, conductivity and transparency were measured in situ. Water samples were taken for laboratory analysis of total phosphorus (TP), total nitrogen (TN), chlorides (Cl^-) and chlorophyll *a* concentration. Salinity was considered through conductivity and chloride concentrations.

Data were analysed in Statistica 13 and Canoco 5 software.

5.3.2 Abiotic Factors, Macrophytes and Faunal Assemblage

Physico-chemical variables measured in the studied ponds are shown in Table 5.1. Ponds did not differ significantly in measures parameters (Kruskal–Wallis H test, $p > 0.05$). Water temperatures were relatively similar among ponds (with a mean value of 24.1 °C and a standard deviation of 3.8), as well as pH. In general, these water bodies were well oxygenated, with low TN and TP values. Most of the ponds were shallow and transparent to the bottom. Only three ponds from Dugi otok island were deeper than 1 m, and transparency values did not match maximum depth: DO3, DO6 and DO10. Investigated ponds were freshwater habitats and only two ponds were slightly brackish: DO9 in Dugi otok island and KOR6, from Korčula island.

Altogether 15 different macrophyte taxa were identified (Supplement 1). Only three ponds were without any vegetation, while in thirteen ponds more than 50% of surface were covered with dense macrophyte coverage. Submerged (*Chara* spp. and *Potamogeton pectinatus* Linnaeus 1753) and floatant (*Potamogeton natans* Linnaeus 1753) species prevailed in density.

Regarding the zooplankton assemblage, a total of 72 taxa were found in the studied ponds, being rotifers the most dominant group, with 51 different species, followed with 12 cladoceran and 9 copepod taxa (Supplement 2). In average 15 ± 5 taxa were found per one pond. Littoral rotifer species from genera *Lecane* (7 species) and *Lepadella* (7 species) showed highest diversity in these habitats. In density, rotifers were also the dominant group in 14 of the investigated ponds (Fig. 5.2). Seven species contributed to 92% of total rotifer density: *Anuraeopsis fissa* (Gosse, 1851),

Table 5.1 Average values (± standard deviation) of measured abiotic parameters in ponds of investigated islands (Adriatic Sea, Croatia)

	Dugi otok	Korcula	Rab
Temperature (°C)	22.5 ± 3.91	26.76 ± 4.41	20.90 ± 2.12
Oxygen (mg L^{-1})	12.54 ± 5.37	10.07 ± 4.41	11.07 ± 2.88
pH	9.45 ± 1.16	8.29 ± 0.88	7.74 ± 0.63
Conductivity(μS cm^{-3})	475.47 ± 745.61	1148.38 ± 2373.79	327.50 ± 106.77
TN (mg L^{-1})	1.66 ± 1.03	4.18 ± 1.04	2.24 ± 2.43
TP (mg L^{-1})	0.09 ± 1.03	0.13 ± 0.17	1.32 ± 1.75
Chlorophyl a (μg 1^{-1})	7.52 ± 6.06	20.17 ± 3.20	26.64 ± 20.09
AFDM (g L^{-1})	0.14 ± 0.07	0.14 ± 0.18	0.28 ± 0.20
Cl^- (mg L^{-1})	129.45 ± 91.49	364.09 ± 915.85	61.90 ± 2.23

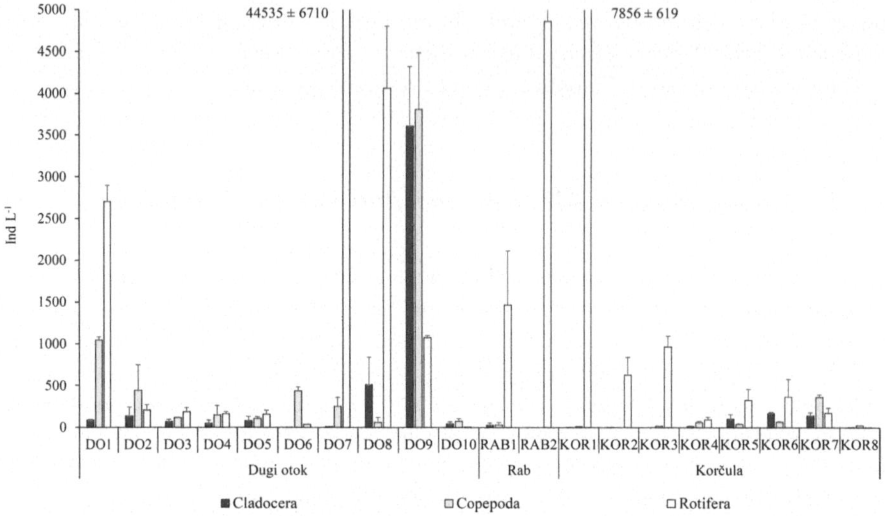

Fig. 5.2 Mean density (± standard deviation) of zooplankton in ponds of three investigates island (Adriatic Sea, Croatia)

Keratella testudo (Ehrenberg, 1832), *Lecane bulla* (Gosse, 1851), *Lecane closterocerca* (Schmarda, 1859), *Lecane hamata* (Stokes, 1896), *Polyarthra dolichoptera* Idelson, 1925 and *Trichocerca pusilla* (Jennings, 1903). Both, littoral species (all *Lecane* species) and planktonic species (*A. fissa*, *K. testudo*, *P. dolichoptera*, *T. pusilla*) were constant pond inhabitants. In other six ponds, the zooplankton was dominated by copepods, mostly with their larval stages (copepodites + nauplii). Both calanoid and cyclopoid copepods were present (Supplement 2). Most common copepod species in these habitats were *Eucyclops serrulatus* (Fischer, 1851), with frequency of 50%. Cladocerans were the less abundant and did not dominate the zooplankton assemble in any pond. Small sizes cladoceran species (*Alona rectangula* Sars, 1861, *Chydorus sphaericus* (O. F. Muller, 1776) and *Moina brachiata* (Jurine, 1820)) were more common in ponds, with frequency of occurrence from 35 to 45% (accessory species, Supplement 2). Rotifers dominated in zooplankton density (Figs. 5.2, 5.3), while cladocerans and copepods dominated in biomass (Fig. 5.4). Maximum density was achieved in ponds DO7 and KOR1, where species *A. fissa* extremely high density values of 41,383 Ind L^{-1} and 3394 Ind L^{-1}, receptivity (Fig. 5.2). This species is euryvalent, indicating eutrophic or even hypertrophic condition in these ponds. In other ponds, zooplankton shoved smaller density values. Mean values for rotifer were 3495 ± 988 Ind L^{-1}, for copepods 365 ± 85 Ind L^{-1} and for cladocerans 254 ± 98 Ind L^{-1}.

Macrozoobenthos were also diverse in these habitats, even if we only consider higher taxonomical taxa. Strictly freshwater arthropod taxa were Hydrachnidia, Amphypoda (represent with species *Gammarus aequicauda* (Martynov, 1931), found only in pond on Korčula island), Ephemeroptera (with one species, *Cloeon dipterum*

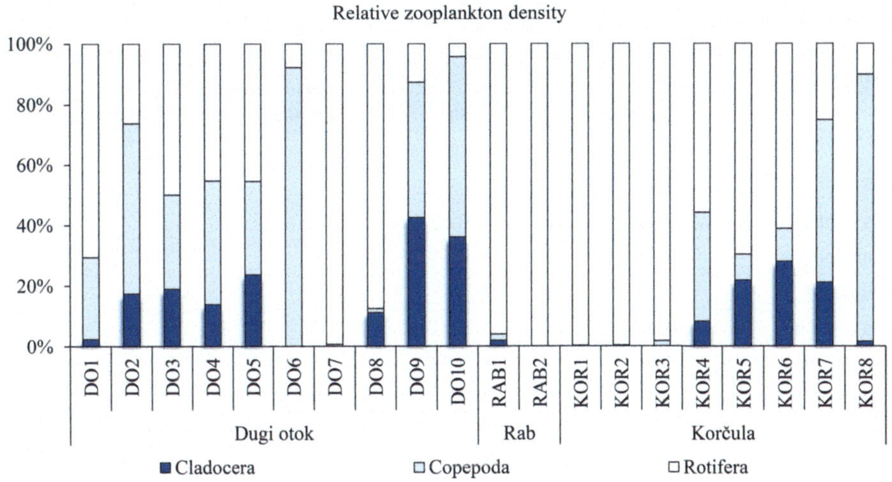

Fig. 5.3 Relative zooplankton density in ponds of three investigates island (Adriatic Sea, Croatia)

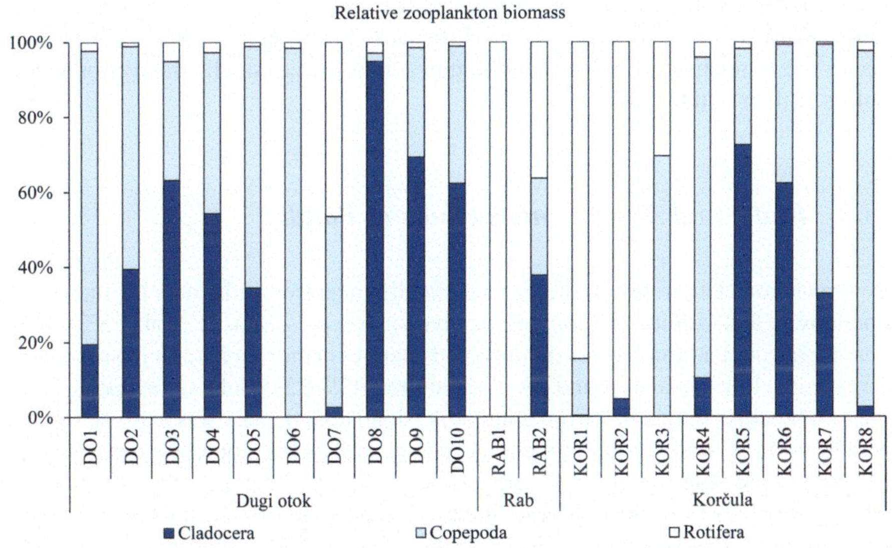

Fig. 5.4 Relative zooplankton biomass in ponds of three investigates island (Adriatic Sea, Croatia)

(Linnaeus, 1761), found in several pods of Dugi otok island [17]), and Odonata [16, 17, 23]. Also, some other insect orders were recorder with several freshwater families: Diptera (Ceratopogonidae, Chaoboridae, Chironomide, Culicidae, Syrphidae and Stratiomyidae), Coleoptera (Dytiscidae, Elmidae, Hydraenidae and Hydrophilidae) and Hemiptera (Corixidae, Notonectidae, Mesoveliidae and Pleidae). Most

Table 5.2 Results of Spearman Correlations ($p < 0.05$) test among number of zooplankton species (S), Shannon's diversity index (H '), zooplankton density, zooplankton biomass and analysed environmental factors

	Macroph	Fish	Chl a	Cond
S	0.37			
H '	0.43			−0.37
Cla (Ind/L)	0.44	−0.55		
Cop (Ind/L)	0.33	−0.68		
Rot (Ind/L)		0.34		
Cla (μg/L)	0.43	−0.51		
Cop (μg/L)	0.29	−0.58		
Rot (μg/L)		0.35	0.726	

abundant was dipteran family Chironomide, which were constant inhabitants, present in 16 ponds.

Fish were most important predatory species in the investigated ponds. Two species have been recorded: *Gambusia holbrooki* Girard, 1859 (ponds DO1, RAB1, RAB2, KOR1, KOR3) and *Carassius auratus* (Linnaeus, 1758), in pond KOR2. Both fish species are nonnative, introduced by human activities, i.e. small islands ponds are fishless in their natural state.

5.3.3 Biotic and Abiotic Interactions in Ponds

Spearman correlation test indicate statistically negative influence of increased conductivity and salinity on zooplankton diversity (expressed as H'; Table 5.2). This decrease in zooplankton species richness and changes in the species composition with increasing salinity is also found by other authors [29–32]. Only some species can adapt: in brackish pond DO9 from Dugi otok island high density was achived by two species, rotifer *L. bulla* and cladoceran *A. rectangular*, while copepods were present only by nauplii stages (Fig. 5.3). Increased salinity also has negative affected on macrozoobenthos [33]. For example, in brackish pond KOR8 dominated amphipod *G. aequicauda*. This species is widely distributed along the Mediterranean and Black Sea coasts. It penetrates brackish waters and it is abundant in shallow waters between macroalgae [34]. In general, increased conductivity in ponds leads to the dominance of euryhaline and euryvalent species, while freshwater species are eliminated.

Other statistically significant factors that shaped the composition, diversity and density of zooplankton were biotic. Spearman correlation test (Table 5.2) together with Monte Carlo permutation test (Table 5.3) indicate a positive effect of macrophyte stands, while the presence and predation of fish had a negative effect on the density and biomass of crustaceans. The results of the canonical-correlation analysis (CCA) of zooplankton density versus 9 environmental variables (listed in description of

Table 5.3 Results of Monte Carlo permutation test (p values and % of explained variance) of relationships among zooplankton density and measured environmental variables (temperature, oxygen, conductivity, total nitrogen, total phosphorous, Chl *a*, AFDM (ash free dry mass, i.e. amount of organic matter), fish and macrophyte coverage)

	%	p
Macroph	19.4	0.002
Fish	16.2	0.004
Chl a	14.2	0.01

Table 5.3) from Mediterranean ponds are shown in Fig. 5.5. Macrophytes explained most of variations in zooplankton, followed by fish presence. This indicates strong "top-down" regulation of zooplankton in this small water bodies. Chl *a* concentration, as measure for phytoplankton density, was also important biotic factor, and it has positive influence only on rotifers. (Tables 5.2, 5.3).

Macrophytes in general act as ecosystem engineers–they modify the abiotic environment (water column in this particular case), making freshwater littoral zone more heterogeneous habitats [27, 35, 36]. They promote higher dissolved oxygen concentrations and diverse food resources, i.e. periphyton assemblage, composed of detritus, bacteria, algae, fungi, protozoan and minute metazoan [37]. This is also the case in our investigated sites – littoral rotifers and cladocerans were abundant among macrophyte stands. These stands are important for macrozoobenthos too, as numerous species are grazing upon periphyton [38]. Macrophytes also provide refuge for invertebrates, primary from fish predation, although this statement is not a general rule [39]. Other positive effects of macrophytes on freshwater ecosystems is reducing nutrient concentrations and sediment resuspension [40]. Thus, macrophytes stands

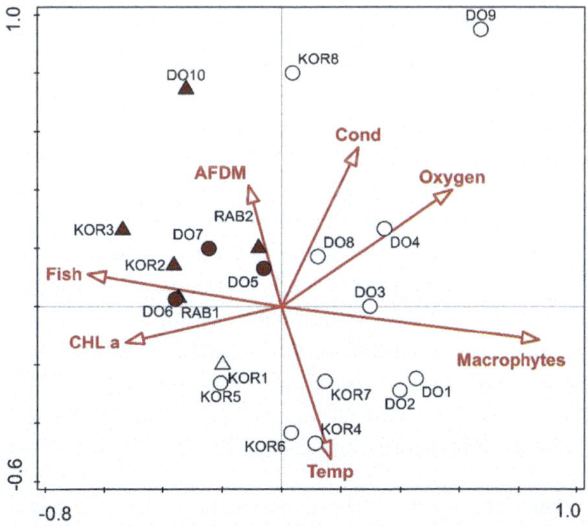

Fig. 5.5 Canonical-correlation analysis (CCA) presenting the influence of the environmental factors on zooplankton density. Empty symbols–dense macrophyte cover; Full symbols–ponds without/or with sparse macrophytes; Circles–fishless ponds; Triangles–ponds with fish

are important drivers for clear water state maintenance, both for shallow lakes and ponds (Fig. 5.5).

Fish presence in ponds have the strongest effect on zooplankton, and both non-native fish species (*G. holbrooki* and *C. auratus*) almost entirely eliminated Cladocera and Copepoda from this small water bodies (Fig. 5.2). Fish have a major structuring impact on the zooplankton assemblages in lakes, that may cascade to the lower trophic levels [40]. This is more pronounced in shallow lakes, where the absence of zooplankton, due to the high fish predation, allows the proliferation of phytoplankton. This leads to reduced water transparency, and lake can shift from vegetation dominated clear water state and in a turbid phytoplankton dominated state in which submerged plants are largely absent [39, 40]. *G. holbrooki* is small fish species, abundant in the vegetation, implying that zooplankton cannot use submerged macrophytes as a good refuge to the same extent as in the freshwater lakes dominated by larger planktivorous fish [32, 33]. *G. holbrooki* is among the 100 worst invasive species in the world, according to The International Union for Conservation of Nature [41]. The presence of *G. holbrooki* negatively affects both invertebrate [42] and amphibian fauna [43] representing a serious threat for conservation of native communities. So, introduced alien fish are widespread threat to all freshwater environments, including small ponds.

5.4 Conclusion

Investigated Mediterranean temporary ponds were small and shallow water bodies, but with diverse and constant zooplankton and macroinvertebrate assemblages. Most of the ponds have dense macrophyte stands, and these plants have positive effect on abundance and biomass of Cladocera and Copepoda. Rotifers, on the other hand, showed a dual respond to macrophytes: littoral, (semi)planktonic species preferred macrophyte stands while true, (eu)planktonic species were extremely abundant in ponds with low macrophyte cover or in open water. Fish presence strongly reduced zooplankton density and diversity, and especially negatively affected Cladocera and Copepoda. From abiotic factors, increased salinity was most important factor and it negatively affected species diversity.

Habitat heterogeneity is one of the most important driver in formation of "healthy" ponds and in slowing down inevitable succession. Fishless ponds with mixture of submerged macrophyte stands and open water support higher invertebrate diversity. Thus, one of the key elements in future conservation and management of small isolated ponds should be maintaining of mosaic habitats. On the Adriatic islands these small, shallow ponds are often the only freshwater habitats. It is necessary to implement their continuous research and to undertake appropriate conservation and restoration measures, in order to prevent their terrestrification.

Acknowledgements Authors are grateful to Ana Štih and other member of Association Hyla for collecting part of the samples, as part of the project Mediterranean Island Wetlands Project

(MedIsWet). We are also very grateful to Antun Alegro and Tomislav Hudina, for the identification of macrophyte taxa.

Appendix

Supplement 1. Coordinates, morphometry and some biotic features of the studied pond from Adriatic islands (Croatia)

Pond	Coordinates		Lenght (m)	Width (m)	Max depth (m)	Fish	Macrophytes	Coverage (%)	Type
DO1	43°55'01"	15°10'04"	15.7	11.2	1.6	-	*Potamogeton pusillus, Ranunculus trichophyllus*	100	sub
DO2	43°56'03"	15°08'55"	10.5	9	0.715	-	*Chara connivens*	95	sub
DO3	43°56'12"	15°07'03"	15.7	16.8	3.2	-	*Potamogeton trichoides, Thypa* sp.	80	sub, em
DO4	43°56'12"	15°07'02"	7.4	5.5	1.5	-	*Chara globularis, Potamogeton pusillus*	80	sub
DO5	43°56'13"	15°06'59"	12.3	5.4	1.27	-	*Drepanocladus aduncus, Thypa* sp.	50	sub, em
DO6	43°56'13"	15°07'01"	10.8	10.6	1.2	-	-		
DO7	44°06'41"	14°55'46"	12.88	11.04	0.7	-	-		
DO8	44°08'22"	14°50'59"	8.8	8.06	0.37	-	*Chara globularis*	50	sub
DO9	44°08'50"	14°49'58"	10.54	6.8	0.22	-	*Chara contraria, Potamogeton pectinatus*	100	sub
DO10	43°56'38"	15°06'59"	17.3	15.6	2.78	+	*Ranunculus trichophyllus, Typha* sp.	25	sub, em
RAB1	44°48'11"	14°44'46"	20	10	1	+	*Nymphaea alba*	40	flo
RAB2	44°47'12"	14°41'28"	14	7	1	+	*Potamogeton natans, Potamogeton pectinatus, Zannichellia palustris*	75	flo, sub
KOR1	42°56'16"	16°53'43"	20	12	1	+	*Chara* sp., *Potamogeton natans*	75	flo, sub
KOR2	42°56'43"	16°46'50"	25	20	1	+	*Potamogeton natans*	10	flo
KOR3	42°57'33"	16°57'29"	9	9	1	+	-		
KOR4	42°55'47"	16°52'54"	15	15	2	-	*Potamogeton natans, Ranunculus aquatilis*	75	flo
KOR5	42°55'51"	16°52'53"	15	15	1.8	-	*Potamogeton pectinatus, Zannichellia palustris*	25	sub
KOR6	42°58'37"	16°41'11"	10	10	0.6	-	*Chara* sp., *Eleocharis palustris, Potamogeton natans, Ranunculus aquatilis*	50	em, flo, sub
KOR7	42°55'55"	17°05'13"	15	15	0.8	-	*Chara* sp., *Eleocharis palustris*	50	em, sub
KOR8	42°56'14"	17°09'09"	15	15	1.2	-	*Chara* sp., *Scirpus maritimus, Typha angustifolia, Zannichellia palustris*	75	em, sub

Supplement 2. List of zooplankton taxa from three investigates island (Adriatic Sea, Croatia), with overall frequency of occurrence

	Dugi otok	Rab	Korčula	F
Rotifera				
Anuraeopsis fissa (Gosse, 1851)	*	*	*	45%
Asplanchna herricki Guerne, 1888	*			5%
Asplanchna sp.	*			5%
Bdelloidea	*	*	*	55%
Brachionus angularis Gosse, 1851	*			10%
Brachionus calyciflorus Pallas, 1766			*	5%
Brachionus quadridentatus Hermann, 1783		*		5%
Brachionus urceolaris Müller, 1773	*	*	*	35%
Cephalodella auriculata (Müller, 1773)	*			5%
Cephalodella catellina (Müller, 1786)	*	*	*	40%
Cephalodella forficata (Ehrenberg, 1832)	*		*	15%
Cephalodella gibba (Ehrenberg, 1830)	*	*	*	15%
Cephalodella sp. 1	*	*		10%
Cephalodella sp. 2	*		*	10%
Cephalodella sp. 3	*		*	10%
Colurella adriatica Ehrenberg, 1831	*			5%
Colurella obtusa (Gosse, 1886)	*	*	*	60%
Colurella uncinata (Müller, 1773)	*	*	*	40%
Epiphanes macroura (Barrois & Daday, 1894)			*	5%
Epiphanes senta (Müller, 1773)	*			5%
Euchlanis dilatata Ehrenberg, 1830	*		*	25%
Filinia brachiata (Rousselet, 1901)	*	*	*	20%
Hexarthra fennica (Levander, 1892)	*		*	10%
Hexarthra mira (Hudson, 1871)	*			5%
Keratella quadrata (Müller, 1786)	*			10%
Keratella testudo (Ehrenberg, 1832)	*			10%
Lecane bulla (Gosse, 1851)	*	*	*	65%
Lecane closterocerca (Schmarda, 1859)	*	*	*	65%
Lecane flexilis (Gosse, 1886)	*			5%
Lecane furcata (Murray, 1913)		*	*	25%
Lecane hamata (Stokes, 1896)	*	*	*	35%
Lecane inermis (Bryce, 1892)	*	*		15%
Lecane ludwigii (Eckstein, 1883)	*		*	20%
Lecane luna (Müller, 1776)	*		*	20%
Lecane lunaris (Ehrenberg, 1832)	*		*	15%
Lepadella acuminata (Ehrenberg, 1834	*			5%
Lepadella ovalis (Müller, 1786)	*	*	*	60%
Lepadella patella (Müller, 1773)	*	*	*	70%
Lepadella rhomboides (Gosse, 1886)		*		10%
Lepadella triptera (Ehrenberg, 1830)			*	10%
Mytilina brevispina (Ehrenberg, 1830)	*		*	30%
Plationus patulus (Müller, 1786)	*		*	10%
Polyarthra dolichoptera Idelson, 1925	*		*	50%
Polyarthra vulgaris Carlin, 1943		*		5%
Squatinella lamellaris f. *mutica* (Ehrenberg, 1832)	*			5%
Synchaeta pectinata Ehrenberg, 1832	*			10%
Testudinella patina (Hermann, 1783)	*	*		15%
Trichocerca bidens (Lucks, 1912)			*	10%
Trichocerca cavia (Gosse, 1886)	*			20%
Trichocerca elongata (Gosse, 1886)	*			5%
Trichocerca pusilla (Jennings, 1903)	*	*	*	45%
Cladocera				
Alona quadrangularis (O. F. Muller, 1776)	*		*	15%
Alona rectangula Sars, 1861	*	*	*	35%

Alonella excisa (Fischer, 1854)	*			5%
Biapertura affinis (Leydig, 1860)			*	5%
Ceriodaphnia quadrangula (O.F. Müller, 1785)	*			10%
Ceriodaphnia reticulata (Jurine, 1820)	*		*	25%
Chydorus sphaericus (O. F. Muller, 1776)	*	*		45%
Daphnia longispina (O.F. Müller, 1776)	*			5%
Dunhevedia crassa King, 1853	*			5%
Moina brachiata (Jurine, 1820)	*			5%
Moina micrura Kurz, 1875			*	25%
Simocephalus vetulus (O.F. Müller, 1776)	*		*	45%
Copepoda				
Arctodiaptomus sp.			*	5%
Cyclops vicinus Uljanin, 1875	*			5%
Diacyclops sp.	*			15%
Eucyclops serrulatus (Fischer, 1851)	*	*	*	50%
Eudiaptomus sp. 1	*		*	10%
Eudiaptomus sp. 2			*	10%
Megacyclops viridis (Jurine, 1820)	*		*	10%
Neolovenula alluaudi (Guerne & Richard, 1890)			*	10%
Tropocyclops prasinus (Fischer, 1860)	*		*	45%
Copepodites	*	*	*	75%
Nauplii	*	*	*	75%

References

1. Céréghino R, Biggs J, Oertli B, Declerck S (2008) The ecology of European ponds: defining the characteristics of a neglected freshwater habitat. Hydrobiologia 597:1–6
2. Meester De L, Declerck S, Stoks R, Louette G, Meutter Van de F, Bie De T, Michels E, Brendonck L (2005) Ponds and pools as model systems in conservation biology, ecology and evolutionary biology. Aquatic Conserv: Mar Freshw Ecosyst 15:715–725
3. Sahuquillo M, Miracle MR (2013) The role of historic and climatic factors in the distribution of crustacean communities in Iberian Mediterranean ponds. Freshw Biol 58:1251–1266
4. Scheffer M, van Geest GJ, Zimmer K, Jeppesen E, Søndergaard M, Butler MG, Hanson MA, Declerck S, De Meester L (2006) Small habitat size and isolation can promote species richness: second-order effects on biodiversity in shallow lakes and ponds. Oikos 112:227–231
5. Council of the European Communities (1992) Council directive 92 / 43 / EEC of 21. May 1992 on the conservation of natural habitats and of wild fauna and flora. Off J Eur Communities 35:7–50
6. Previšić A, Walton C, Kučinić M, Mitrikeski PT, Kerovec M (2009) Pleistocene divergence of Dinaric Drusus endemics (Trichoptera, Limnephilidae) in multiple microrefugia within the Balkan Peninsula. Mol Ecol 18:634–647
7. Deharveng L, Gibert J, Culver DC (2012) Diversity patterns in Europe. In: White WB, Culver DC (eds) Encyclopedia of Caves. Academic Press, Chennai, pp 219–228
8. Bonacci O (2017) Preliminary analysis of the decrease in water level of Vrana Lake on the small carbonate island of Cres (Dinaric karst, Croatia). In: Parise M, Gabrovsek F, Kaufmann G, Ravbar N (eds) Advances in Karst research: theory, fieldwork and applications, vol 466. Geological Society, London, Special Publications, pp 307–317
9. Bukvić-Ternjej I, Kerovec M, Mihaljević Z, Tavčar V, Mrakovčić M, Mustafić P (2001) Copepod communities in karstic mediterranean lakes along the eastern Adriatic coast. Hydrobiologia 453(454):325–333
10. Kerovec M, Kerovec M, Brigić A (2016) Croatian freshwater oligochaetes: species diversity, distribution and relationship to surrounding countries. Zootaxa 4193:73–101

11. Mihaljević Z, Kerovec M, Ternjej I, Popijač A (2004) Long-term changes in the macroinvertebrate community structure of a shallow Mediterranean lake. Ekologia (Bratislava) 23:421–429
12. Ternjej I, Stanković I, Mihaljević Z, Furač L, Želježić D, Kopjar N (2009) Alkaline comet assay as a potential tool in the assessment of DNA integrity in freshwater zooplankton affected by pollutants from water treatment facility. Water Air Soil Pollut 204:299–314
13. Ternjej I, Mihaljević Z, Stanković I, Kerovec M, Sipos L, Želježić D, Kopjar N (2010) Estimation of DNA integrity in blood cells of Eastern Mosquitofish (Gambusia holbrooki) inhabiting an aluminium-polluted water environment: an alkaline comet assay study. Arch Environ Contam Toxicol 204:299–314
14. Habdija I, Primc B, Špoljar M, Sertić Perić M (2011) Ecological determinants of rotifer vertical distribution in a coastal karst lake (Vrana Lake, Cres Island, Croatia). Biologia 66:130–137
15. Gligora M, Plenković-Moraj A, Ternjej I (2003) Seasonal distribution and morphological changes of Ceratium hirundinella in two mediterranean shallow lakes. Hydrobiologia 506(509):213–220
16. Grgić I (2019) The dragonfly fauna of Dugi otok island. In: Hocenski K, Mišerić I (eds) A collection of scientific studies of the educational and research project "Insula Tilagus 2017". Biology Students Association–BIUS, Zagreb, pp 246–252 (in Croatian with English summary)
17. Polović L, Dražina T, Špoljar M (2019) Inventory of macrozoobenthos and zooplankton in ponds of Dugi otok island, and research of macrophytes impact and physicochemical parameters of water, as main assemblage architects. In: Hocenski K, Mišerić I (eds) A collection of scientific studies of the educational and research project "Insula Tilagus 2017". Biology Students Association–BIUS, Zagreb, pp 118–128 (in Croatian with English summary)
18. Bućan D, Špoljar M, Dražina T, Fiorentin C, Alegro A, Zrinščak I, Landeka N, Štih A (2019) Benthic invertebrates, fish and zooplankton coupling in freshwater Mediterranean ponds. In: Ivković M, Stanković I, Matoničkin Kepčija R, Gračan R (eds) 3rd Symposium of freshwater biology book of abstract, Zagreb, 2019
19. Dražina T, Špoljar M, Fiorentin C, Polović L, Bućan D, Hudina T (2019) Zooplankton in Mediterranean ponds–implications for conservation measures. In: Sertić Perić M, Miliša M, Gračan R, Ivković M, Buj I, Mičetić Stanković V (eds) Symposium for European freshwater sciences 11 Abstract book, Zagreb, 2019
20. Dražina T, Špoljar M, Kahriman K, Cvetnić M, Štih A (2018) Zooplankton in small fishless Adriatic ponds (Mediterranean Sea). In: 8th European pond conservation network workshop, abstract book. Torroella de Montgrí, Spain, 21–25 May 2018
21. Franković M, Bogdanović T (2011) Vretenca (Insecta: Odonata) i njihova staništa otoka Mljeta. In: Benović A, Durbešić P (eds) Zbornik radova simpozija Dani Branimira Gušića (Proceedings of the Symposium Branimir Gušić Days), Mljet, 2010
22. Temunović M, Šeric Jelaska L (2009) First record of diving beetle Cybister tripunctatus africanus Laporte, 1835 in the Croatian Hydradephagan fauna. In: Proceedings of 21 Symposium Internationale Entomofaunisticum Europae Centralis. Communications and abstracts. Ceške Budejovice, Czech Republic, 2009
23. Belančić A, Bogdanović T, Franković M, Ljuština M, Mihoković N, Vitas B (2008) Red data book of dragonflies of croatia. State Institute for Nature Protection, Republic of Croatia, Zagreb, Ministry of Culture
24. Trilar T, Bedjanic M (1999) Contribution to the knowledge of the dragonfly fauna of Lastovo island, Dalmatia, southern Croatia. Exuviae 6:1–6
25. Kučinić M, Ćukušić A, Plavec H, Landeka M, Plantak M, Vuković M, Bukvić V, Franjević M, Žalac S, Lukač G (2019) Caddisfly fauna characteristics (Insecta, Trichoptera) of four Adriatic islands with a note on DNA barcoding. Nat Croat 28:403–413
26. Wallace RL, Snell TW, Ricci C, Nogrady T (2006) Rotifera, Biology, ecology and systematics. In: Dumont HJF (ed) Guides to the identification of the microinvertebrates of the continental waters of the world, vol 1, 2nd edn. Kenobi Productions, Ghent
27. Wetzel RG (2001) Limnology lake and reservoir ecosystems. Academic Press, San Diego

28. Braun-Blanquet J (1932) Plant sociology: the study of plant communities. McGraw-Hill, New York
29. Moss B (1994) Brackish and freshwater shallow lakes different systems or variations on the same theme? Hydrobiologia 275:1–14
30. Boix D, Gascón S, Sala J, Badosa A, Brucet S, López-Flores R, Martinoy M, Gifre J, Quintana XD (2008) Patterns of composition and species richness of crustaceans and aquatic insects along environmental gradients in Mediterranean water bodies. Hydrobiologia 597:53–69
31. Brucet S, Boix D, Gascón SJ, Quintana XD, Badosa A, Søndergaard M, Lauridsen TL, Jeppesen E (2009) Species richness of crustacean zooplankton and trophic structure of brackish lagoons in contrasting climate zones: north temperate Denmark and Mediterranean Catalonia (Spain). Ecography 32:692–702
32. Jensen E, Brucet S, Meerhoff M, Nathansen L, Jeppesen E (2010) Community structure and diel migration of zooplankton in shallow brackish lakes: role of salinity and predators. Hydrobiologia 646:215–229
33. Brucet S, Boix D, Nathansen LW, Quintana XD, Jensen E, Balayla D, Meerhoff M, Jeppesen E (2012) Effects of temperature, salinity and fish in structuring the macroinvertebrate community in shallow lakes: implications for effects of climate change. PLoS ONE 7(2):e30877
34. Rađa B, Rađa T (2012) The first record of Gammarus aequicauda (Martynov, 1931) (Amphipoda, Gammaridea) in the estuary of the river Jadro (Adriatic Sea, Croatia). Crustaceana 85:987–991
35. Špoljar M, Dražina T, Šargač J, Kralj Borojević K, Žutinić P (2012) Submerged macrophytes as a habitat for zooplankton development in two reservoirs of a flowthrough system (Papuk Nature Park, Croatia). Ann Limnol Int J Limnol 48:161–175
36. Špoljar M, Dražina T, Lajtner J, Duić Sertić M, Radanović I, Wallace R, Matulić D, Tomljanović T (2018) Zooplankton assemblage in four temperate shallow waterbodies in association with habitat heterogeneity and alternative states. Limnologica 71:51–61
37. Špoljar M, Zhan C, Dražina T, Zhao G, Lajtner J, Radonić G (2017) Development of submerged macrophyte and epiphyton in a flow-through system: assessment and modelling predictions in interconnected reservoirs. Ecol Indic 75:145–154
38. Jones JI, Sayer CD (2003) Does the fish-invertebrate-periphyton cascade precipitate plant loss in shallow lakes? Ecology 84:2155–2167
39. Jeppesen E, Søndergaard M, Søndergaard M, Christoffersen K, (Eds) (1998) The structuring role of submerged macrophytes in lakes. Ecological Studies, vol 131, p 427
40. Scheffer M (2004) Ecology of shallow lakes: population and community biology series, vol 22, p 378
41. Lowe S, Browne M, Boudjelas S, De Poorter M (2000) 100 of the world's worst invasive alien species: a selection from the global invasive species database. Auckland, New Zealand: The International Union for Conservation of Nature Invasive Species Specialist Group (ISSG)
42. Margaritora FG, Ferrara O, Vagaggini D (2001) Predatory impact of the mosquitofish (Gambusia holbrooki Girard) on zooplanktonic populations in a pond at Tenuta di Castelporziano (Rome, Central Italy). J Limnol 60:189–193
43. Vannini A, Bruni G, Ricciardi G, Platania L, Mori E, Tricarico E (2017) Gambusia holbrooki, the 'tadpolefish': the impact of its predatory behaviour on four protected species of European amphibians. Aquat Conserv 28:476–484

Chapter 6
Riparian Springs—Challenges from a Neglected Habitat

Vladimir Pešić, Dejan Dmitrović, and Ana Savić

Abstract Riparian springs bordering streams, rivers and lakes are under strong influence of flooding events, and therefore more sensitive to the ongoing climatic changes. However this type of habitats is still commonly neglected in limnological literature. This chapter presents modern knowledge on these habitats gained in recent years by researching the riparian springs in the Dinaric karst area and opens some important questions about their ecology and conservation. The current concept of riparian springs seems to be too broad, and we have redefined it as springs that are flooded by adjacent lotic/lentic reaches at least part of the year. The existence of "flood" and "spring" aquatic phases affects the biodiversity and ecosystem functioning of riparian springs, but also development of appropriate management strategies and policies for the protection of these valuable habitats.

Keywords Springs · Riparian · Flood · Macroinvertebrate · Generalists · Specialists · Climate changes

6.1 Introduction

Springs are classically defined as spatially restricted inland freshwater ecosystems which occur in places where groundwater reaches the surface [1]. In most cases, they are defined as small water bodies (of a few square meters), characterized by relative

V. Pešić (✉)
Faculty of Sciences, Džordža Vašingtona bb, University of Montenegro, 81000 Podgorica, Montenegro

D. Dmitrović
Faculty of Natural Sciences and Mathematics, Department of Biology and Department of Ecology and Environment Protection, University of Banja Luka, Mladena Stojanovića 2, 78000 Banja Luka, Republic of Srpska, Bosnia and Herzegovina
e-mail: dejan.dmitrovic@pmf.unibl.org

A. Savić
Department of Biology and Ecology, Faculty of Sciences and Mathematics, University of Niš, Višegradska 33, 18000 Niš, Serbia
e-mail: anka@pmf.ni.ac.rs

V. Pešić et al. (eds.), *Small Water Bodies of the Western Balkans*, Springer Water,
https://doi.org/10.1007/978-3-030-86478-1_6

stability of environmental parameters, primarily water temperature [2, 3]. This is only partially true, as size and relatively constant environmental conditions of springs may vary due to differences in their hydromorphological properties. Rheocrene springs (which emerge into one or more stream channels with a fast flowing water) are mainly small and cold stenothermic habitats, while limnocrenes (which emerge as one or more lentic pools), may be significantly larger in size and show greater annual water temperature amplitudes characteristic of small standing water bodies. The amplitude of variation of environmental parameters may often get quite impressive: for example, small rheocrenes in the Dinaric region may experience both diurnal and annual temperature variability from zero to up to 15 °C [4].

As shown by numerous taxonomic and faunistic studies, springs significantly contribute to maintenance of aquatic biodiversity, especially in areas with just a few surface water bodies, such as the Dinaric Karst [5–15]. Surface waters in large karstic areas of the Western Balkans are rather scarce as most of water supply disappears into complex underground structures, and karstic springs are often the only surface water bodies that occur in such areas [16]. Heterogeneous systems such as the springs in the Dinaric karst often support extremely rich and diverse aquatic biota which is in research focus of a large number of studies [5–16]. The composition and biotic interactions of spring assemblages are mainly driven by substrate composition and anthropogenic impacts. A large number of studies emphasized the importance of environmental parameters such as physicochemical factors, discharge, substrate composition [17–20], or hydromorphological modification [21] as important determinants of the composition of macroinvertebrate assemblages in springs of Dinaric karst.

A significant part of the local biota inhabiting the Dinaric karst springs is represented by endemic species [6, 10, 16]. The proportion of endemic fauna in karstic springs depends on their geographical isolation and differs between different types of springs as well as among the faunistic groups that inhabit them. Available data on water mites, which are one of the most specialized invertebrate group inhabiting springs [11, 22, 23] have shown that helocrenes (springs that emerge in a diffuse fashion in wet meadows) and to a lesser extent rheocrenic springs host the highest number of endemics, while the lowest number is present in limnocrenes.

A special type of spring that has long been neglected by limnologists (as well as by crenobiologists) are the so-called "riparian" springs that appear along adjacent lotic (streams and rivers) or lentic (lakes, ponds) water bodies. Lately these springs are often defined as a riparian springs, as they are usually located in the riparian belt of flowing or standing adjacent water bodies [20, 24]. Nevertheless, the latter definition seems to be too broad and often unusable in practice. A large number of springs located along the nearby lotic/lentic reaches become flooded during a certain period of the year, and consequently they are directly exposed to the influence of the adjacent water body. On the other hand, some springs may be located in the riparian belt, but at a higher altitude than the adjacent lotic/lentic reaches, and thus avoid being flooded, while in reality do not differ from "non-riparian" springs located in their near or far surroundings. In this chapter, the term riparian springs refers to those springs that get flooded by adjacent lotic/lentic reaches in at least one part of the year. Poor understanding of the concept of riparian springs is likely one of the factors that prevented quantification of their extent within the spring network, resulting in this type of habitat generally remaining forgotten in the biomonitoring programs committed to the EU Water Framework Directive implementation [25].

Riparian springs are very common in karstic environment and represent a significant part of the spring network at the local and regional scale. Of the 50 springs situated along 12 km of the Cvrcka River basin in Bosnia and Herzegovina, fourteen of them were flooded (Fig. 6.1). Riparian springs are also common on floodplains where lateral connectivity with nearby water systems is permanently present; for example, most of the karstic springs that appear around Skadar Lake are periodically flooded. As emphasized by von Fumetti et al. [20], the number of flooded springs is likely to increase in the future as a result of increasing frequency of flood events caused by the ongoing climate changes.

Most of the existing studies on riparian springs in the Dinaric Karst have concentrated on the study of springs in the Cvrcka River basin in the north-western part of Bosnia and Herzegovina [7, 18–21, 24]. Along the lower part of the Zeta River in Central Montenegro there are more than 20 springs that are flooded on a regular basis. Available studies that include data on riparian springs of Dinaric karst have been concentrated either on particular faunistic groups such as gastropods [17, 19], chironomids [7], Ephemeroptera, Plecoptera and Trichoptera (EPT) [18] or on entire macroinvertebrate assemblages (see 20, 21, 24).

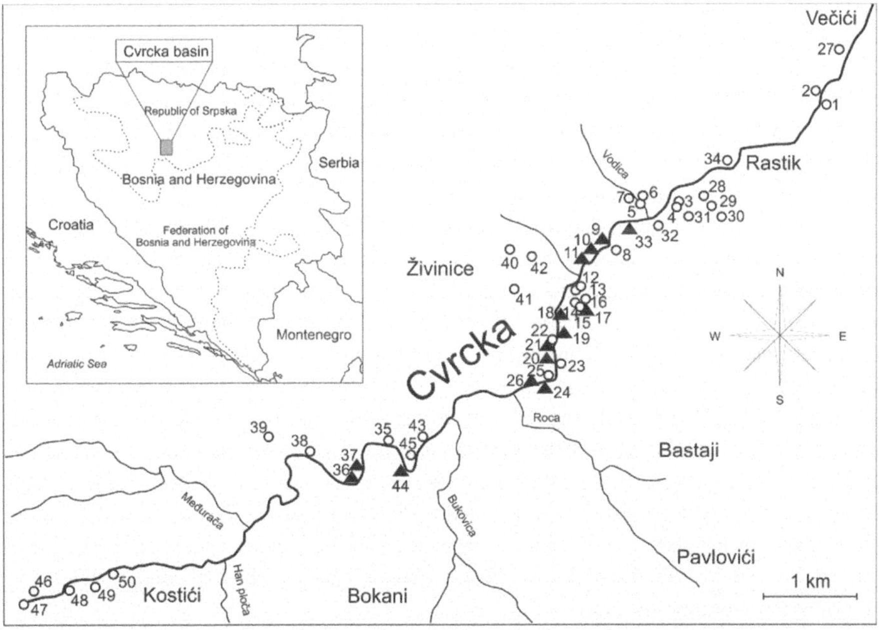

Fig. 6.1 Map of the Cvrcka river basin. The numbers correspond to numbers assigned to individual springs as listed in Pešić et al. [26]. Flooded springs are marked by a black triangle

With the exception of recent studies on riparian springs along the Krapiel river in Poland [27–29], this habitat type has been largely ignored by the ongoing biodiversity and ecology studies of small water bodies in Europe. The comprehensive eco-faunistic studies of riparian springs at a broader scale are still missing. Moreover, riparian springs become mainly forgotten in projects aimed at assessing river integrity, regardless of their potentially great significance in the context of river alteration. In this study we compiled recent knowledge about this type of habitat based on data obtained from the investigation of riparian springs in the Dinaric Karst region.

6.2 Faunistic Diversity of Riparian Springs is Driven by Substrate Composition

The influence of substrate composition in springs has been well-studied and many researchers [30–32] have emphasized that the type and number of different substrate types are the main discriminatory factor affecting the species composition and abundance of different faunistic groups. Von Fummeti et al. [20] have shown that distribution of macroinvertebrate communities in studied rheocrenes of the Cvrcka River basin is significantly correlated with substrate composition, while differences between studied assemblages were a result of differing substrate composition.

The procedure for estimating the representation of different types of substrates in the springs is based on the percentage of their areal coverage [30, 31]. In most of crenobiological studies in which this procedure was used (for examples see [20, 21, 24, 26, 31]), five classes of frequency were usually defined: 0-absent; 1-low level (1–25% coverage); 2-medium level (25–50% coverage); 3-strong level (51–75% coverage) and 4-continuous level (76–100% coverage). The presence of different substrate types and accompanying higher habitat heterogeneity are generally expected to increase the diversity of species [31].

Table 6.1 shows recorded presence and coverage of the studied substrate types in 35 investigated rheocrenes within the canyon of the Cvrcka river, as well as differences in their representation between flooded and non-flooded springs. The Mann–Whitney U-test has revealed significant differences in the percentage of sand cover between flooded and unflooded springs of the Cvrcka River basin. In the flooded springs, the substrate composition was dominated by coarse mineral material: rocks, gravel and sand (Fig. 6.2). On the other hand, in the substrate composition of springs that were not flooded, coarse particular organic material (e.g. leaf litter, dead wood) were more prominent. Moreover, filamentous algae and spring-fed herbaceous macrophytes were more dominant in springs that were not flooded. Von Fumetti et al. [20] stated that the presence of macrophytes may be viewed as a visual indicator of undisturbed, non-flooded springs, as most of springs with a lot of macrophytes were not flooded.

The structure and substrate composition in a riparian spring is likely to depend on the relief of the catchment. Riparian springs in hilly terrain such as the Cvrcka

Table 6.1 Results of the Mann–Whitney U-test for the substrate composition for comparing flooded and unflooded rheocrenes in the Cvrcka river basin

Substrate type	Cvrcka River basin		p-value
	Unflooded	Flooded	
Stones	2.48 ± 1.21	2.93 ± 0.83	0.325
Gravel	1.38 ± 0.59	1.57 ± 0.65	0.359
Sand	1.29 ± 0.78	2.21 ± 0.80	**0.004**
Calcareous sinter	1.95 ± 1.69	1.93 ± 1.07	0.934
Clay	0.67 ± 1.16	0.43 ± 0.94	0.702
Anoxic mud	0.29 ± 0.90	0.00 ± 0.00	0.495
Detritus	1.05 ± 0.87	1.07 ± 0.62	0.855
Leaf litter	2.05 ± 0.87	1.71 ± 0.73	0.325
Dead wood	1.19 ± 0.75	1.00 ± 0.39	0.516
Roots	0.38 ± 0.59	0.36 ± 0.5	0.987
Moss	1.38 ± 1.07	1.43 ± 1.02	0.855
Macrophytes	0.76 ± 1.14	0.50 ± 0.76	0.727
Algae	0.10 ± 0.44	0.00 ± 0.00	0.829

canyon, dominated by erosional habitats, are characterized by steep gradients and predominant direction of movement of both biota and materials is downstream. As a consequence, after a rainfall event the sediments and particulate organic matter in such springs would be readily transported downstream. During the "spring" phase, however, the accumulation of organic material in these springs may be significant and sampling in that period does not necessarily show differences between flooded and unflooded springs as is the case with the studied rheocrenes of the Cvrcka river basin characterized by diverse microstructures related to surrounding forestland cover. Riparian springs in lower relief landscapes located in the floodplain zone of adjacent lentic reaches may become inundated with flooded water and would likely have lower velocities, favoring deposition of finer organic material in the benthos (Fig. 6.3).

Groundwater discharge is another important factor that affects deposition of material and nutrients in riparian springs. Groundwater levels in springs may rise, often abruptly, although with a lag to surface flows. In the case of a rapid flow in the spring itself, finer organic material may quickly be flushed downstream. Rheocrenic springs with a lower discharge (and consequently a slow flows) or a limnocrene might capture large quantities of leaf litter and other organic material.

The presence of different types of microhabitats in riparian springs affects the occurrence, species composition and abundance of different faunistic groups. Different groups of macroinvertebrates show preference for certain types of microhabitats. For example, Savić et al. [18] found that Ephemeroptera show higher abundance in flooded than in unflooded rheocrenes of the Cvrcka river basin. The later study has shown that the number of EPT (Ephemeroptera, Plecoptera, Trichoptera) species in the studied rheocrenes along the Cvrcka River positively correlates with areal coverage of sand and stones, respectively, and negatively with the areal coverage of algae [18].

Fig. 6.2 Photographs of flooded rheocrenes in the Cvrcka river basin. Spring numbers correspond to the numbers listed in Appendix 4 of Pešić et al. [26]. Photos by D. Dmitrović

Most of the mayfly and caddisfly species recorded in riparian rheocrenes in the Cvrcka river basin prefer micro- and mesolithal. Interestingly, some species were more frequent in flooded springs. Five (out of eight) springs in which *Sericostoma sp.* was found and four (out of the seven) springs that hosted recently described Drusinae species *Drusus crenophylax* were flooded. The latter species shows a preference to the coarse mineral substrate (epilithon) using the latter microhabitat as a feeding ground (functional feeding group–FFG: an epilithic grazer [33]).

The significance of microhabitat substrates, however, is difficult to determine without analyzing the distribution of feeding groups, between different types of springs and within the assemblages of particular faunistic groups or the whole community. The FFG abundance analysis of EPT assemblages of 35 rheocrenes of the Cvrcka river basin has shown that each of four feeding groups (i.e., predators, scrapers, collector-gatherers, and shredders) was in average more abundant in the flooded springs (Table 6.2). However, a statistically significant difference between flooded and unflooded springs was found only in the average number of scrapers ($p = 0.042$).

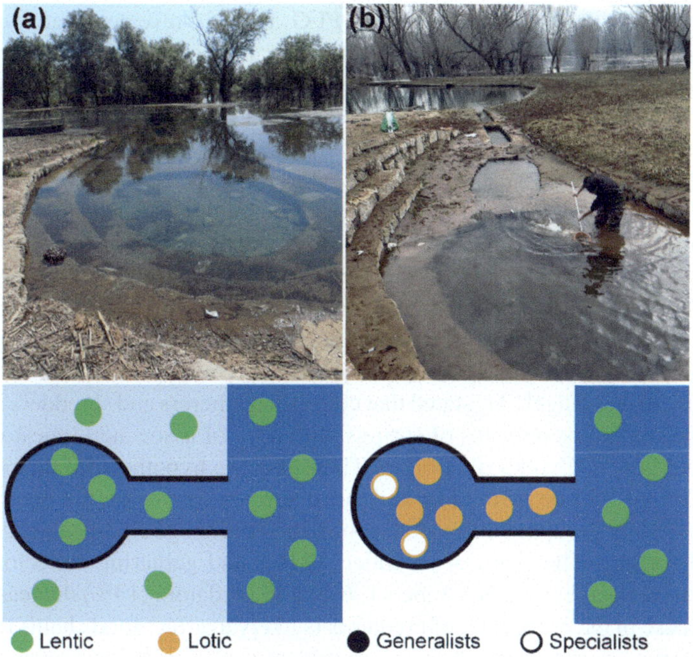

Fig. 6.3 The spring at Vitoja located on the northwestern shore of Lake Skadar, Montenegro, during (**a**) "flood", and (**b**) "spring" aquatic phases. Below: a plane view of the occurrence of lentic and lotic generalist and specialist species in spring habitat. Photos by V. Pešić

Table 6.2 Results of the Mann–Whitney U-test for the functional feeding groups used for comparison of EPT and the entire macroinvertebrate assemblage, in flooded and unflooded rheocrenes in the Cvrcka river basin, respectively

FFG	Spring	EPT		Macroinvertebrates	
		Mean ± SD	p-value	Mean ± SD	p-value
Shredders (Sh)	unflooded	1.055 ± 2.31	0.56	11.86 ± 9.99	0.516
	flooded	1.5388 ± 2.50		14.36 ± 10.51	
Collector-gatherers (Co-Ga)	unflooded	1.89 ± 4.89	0.068	7.67 ± 8.87	0.293
	flooded	3.84 ± 4.89		9.64 ± 9.20	
Scrapers (Sc)	unflooded	1.55 ± 2.7	0.042	13.38 ± 15.37	0.145
	flooded	2.92 ± 2.43		20.93 ± 16.91	
Predators (Pr)	unflooded	1.55 ± 2.50	0.89	5.00 ± 4.99	0.096
	flooded	1.92 ± 2.75		8.36 ± 6.42	

At the level of the entire community, FFG analysis revealed that each of the four feeding groups had higher values of average number of individuals in flooded springs of the Cvrcka river basin; however, no statistically significant difference was found in the distribution of feeding groups between flooded and unflooded springs. The flooded springs in this study were predominated by scrapers (39%), followed by shredders (27%). Predators (16%) and gathering-collectors (18%) were present in a lower share. The obtained results show that in-spring periphyton production is important in flooded springs dominated by coarse inorganic substrates. *Elmis aenea* was the most dominant grazer species in the assemblages of our study. On the other hand, allochthonous leaf-litter material entering the spring represents an important source of energy for shredders (dominated by *Gammarus fossarum)*, the most abundant organisms in riparian springs of the Cvrcka river basin [20, 21].

Von Fumetti and Nagel [34] stated that collector-gatherers and shredders are likely to dominate in the assemblages of springs with a lot of associated limestones such as karstic springs in Cvrcka river basin. Therefore, we hypothesized that shredders and collectors would be favored in unflooded springs. However, the percentage of functional feeding groups in unflooded springs did not change significantly compared to flooded springs. The share of shredders (31%) and gathering-collectors (20%) increases, while percentage of scrapers (36%) and predators (13%) decreases.

An increase in the proportion of predators is likely to affect food chains in flooded springs. While scrapers, collectors and shredders use microhabitats as a source of food intake or the substrate itself is a food source, predators such as leeches and water mites use these microhabitats as their hunting areas. Some of the predators, such as water mites, may have different feeding strategies as larvae (parasitic) and as adults (predators) [35]. In most cases, predators and their prey share similar substrate preferences. The slightly higher prevalence in flooded springs recorded in our study is probably due to the fact that predators are likely to be more capable of colonizing flooded springs, either during the "flood" phase or later during the "spring" phase via upwards migration. During the "flood" phase, many predators, including fish, extend into riparian springs along large lowland rivers or standing bodies.

6.3 Is There a Longitudinal Pattern in Environmental Variables and the Community in Riparian Springs?

In the 1960s, Ilies and Botosaneanu first proposed dividing the springs into two zones: the springhead (eucrenal) and the springbrook (hypocrenal) [36]. The boundary between these two zones was primarily based on thermal criteria. Erman and Erman [37] believed that a temperature difference of 2 °C is required for the differentiation of eucrenon and hypocrenon. Later on, von Fumetti et al. [38] claimed that a sufficient temperature difference at which the communities of these two zones may be distinguished might even be reached at 1 °C.

Nevertheless, as some recent studies have shown, the boundaries between spring zones are not always clear, especially in the case of riparian springs [24]. The latter spring type is generally characterized by short springbrooks, often entering the main river after only a few meters. Pešić et al. [24] studied riparian springs along the Cvrcka River and found no significant differences in macroinvertebrate assemblages between springhead and springbrook. The latter authors state that the riparian springs with a short outflow "likely do not exhibit true spring zonation but may show a quasi-zonation" meaning that that assemblage shifts along the eucrenal-hypocrenal gradient was caused by transition [24]. They speculated that the main factors that blurred zonation in riparian springs of Dinaric karst are length of spring outflow and distance from the mainstream, and possibly also the input of nutrients in the period when the springs are flooded during high spates [24].

In the above-mentioned study on riparian springs along the Cvrcka River mainstream the temperature variations between eucrenal and hypocrenal did not exceed 1 °C during the year [24]. The length of the springbrook of the studied springs was less than 20 m, likely not allowing temperature difference to exceed 1 °C. Von Fumetti et al. [38], who studied the zonation of springs in the Swiss Alps, found that the point at which the water temperature of springbrook differed from the temperature at the springhead more than 1 °C varied from 3.5 m up to over 40 m. On the other hand, Di Sabatino et al. [39] found no distinct longitudinal separation between the eucrenal and hypocrenal sections along the 45-m reach of Vera Spring in Central Italy. In their study, the investigated spring reach was characterized by almost constant values of abiotic parameters. Stable environmental variables along even a 100-m spring–springbrook gradient were also found by Carroll and Thorp [40] in springs of Missouri, USA.

The short springbrook is characteristic of small karstic springs located along the river corridors in hilly terrain, as is the case with Cvrcka canyon. In the lowland areas, however, as a result of a larger floodplane zone, springs which are also located at a greater distance from the mainstream may be inundated by the flood water, resulting in their comparatively longer springbrook and possibly greater environmental variability. Analysis of ten flooded springs in the lowland Zetsko-Bjelopavlicka valley in Central Montenegro did not reveal any significant difference between eucrenal and hypocrenal sections in the substrate composition and selected physico-chemical parameters (i.e., temperature, pH, concentrations of oxygen, ammonium, nitrates and phosphates); the only recorded significant difference (Mann-Whittney p value: 0.041) was for the flow velocity, which was significantly higher in springbrook (0.26 $\pm$ 0.24 m/s) than in springhead (0.05 $\pm$ 0.04 m/s).

Seasonal changes of abiotic and biotic parameters are expected to affect longitudinal changes along the eucrenal–hypocrenal ecotone. Pešić et al. [24] found no statistically significant seasonal differences in substrate composition and/or physico-chemical parameters between the eucrenal and hypocrenal sections of three investigated springs in the Cvrcka river basin. In their study, macroinvertebrate assemblages of studied spring sections were more uniform in summer than in other seasons, possibly due to the lower discharge and higher temperature common in small springs of Dinaric karst [24]. The lack of clear longitudinal changes of abiotic variables along the eucrenal-hypocrenal gradient may suggest that other factors, such as changes in autochthonous (periphyton) and allochthonous resources (leaf-litter input), may be more important in structuring the seasonal response of crenic assemblages [39].

Some authors (e.g., [40, 41]) have emphasized that eucrenal–hypocrenal differences in crenic assemblages do not necessarily occur at the level of the entire community, but may be expressed at the level of some particular faunistic groups. Plociennik et al. [7] have shown that eucrenon and hypocrenon sections of the springs in Cvrcka valley have clearly distinct assemblages of chironomid larvae. Some species, such as *Prodiamesa olivacea,* preferred the springhead, while *Paraphaenocladius* sp. shows preference for the springbrook [7]. The latter study reveals that chironomid assemblages of the eucrenal were more distinct and better defined than the hypocrenon communities, likely due to the greater environmental uniformity of eucrenal sites, despite the lack of recorded environmental differences between the investigated spring sections [7].

In a study conducted by Pešić et al. [11] no differences in the assemblages of spring-dwelling water mites were found along the eucrenal-hypocrenal gradient in 14 karst springs located in the Mediterranean part of Montenegro. Most of these springs were located along the River Zeta mainstream valley and they were partially flooded by the river [11]. In six of these flooded springs, Karouzas et al. [42] collected seven caddisflies species. The average number of caddisflies species was higher in eucrenal, whereas the average number of specimens was higher in hypocrenal; however, no significant differences in the number of species and abundance between the studied spring sections were recorded.

Significantly higher occurrences of insect larvae (mainly Trichoptera, Plecoptera and Ephemeroptera) downstream from the springhead have been demonstrated by many authors [35, 39, 43]. Pešić et al. [24] found that the total number of mayfly larvae in hypocrenal of the springs in the Cvrcka river basin was more than three times higher than in the eucrenal section of the latter springs. The larvae of dragonfly *Cordulegaster bidentatus* were also more abundant in the hypocrenal than in the eucrenal [24]. Nevertheless, in flooded springs it may be expected that the ecotonal shift towards increasing share of insect larvae along the eucrenal-hypocrenal gradient may be less pronounced, as a consequence of a short spatial gradient of these habitats and a dominance of generalist rhitrobiontic species in their crenic assemblages. The percentage of insects at the eucrenal site of the flooded spring SPR-1 in the Cvrcka river basin (see Pešić et al. [24] for a description of a spring) was 59% of the total individuals sampled at the springhead and did not change significantly at the downstream hypocrenal site (62%).

6.4 Transition Between Flood and Spring Phases May Enhance Biodiversity of Riparian Springs

As shown by the few studies on riparian springs of Dinaric karst [18, 20, 21, 24], the diversity of the macroinvertebrate community of flooded springs on the local scale is at the level of unflooded springs or even greater. Transition between "flood" and "spring" aquatic phases results in greater variation in environmental conditions along both spatial and temporal gradient, in comparison to unflooded springs that in turn may enhance the biodiversity of these habitats. Regional-scale diversity (γ-diversity) of flooded rheocrenic springs, especially those inundated by flood waters from the adjacent lentic reaches, may be greater than of the unflooded rheocrenes, when an annual cycle is considered. During the flood phase, the generalist lentic colonists arriving from the nearby lentic reaches predominated in the spring assemblage. Once the flooding ceases, the lentic colonists disappear and the crenic assemblage become dominated by the generalist lotic colonists and to a lesser extent by specialist taxa (Fig. 6.3).

Overall, 63 taxa belonging to 40 families were collected in 14 flooded rheocrenes along the 12 km mainstream of Cvrcka river (list of taxa is available in [26]). The average Shannon diversity of flooded springs was higher (1.82 ± 0.45) than of unflooded springs (1.57 ± 0.50) (Table. 6.3). Under the assumption of intermediate disturbance hypothesis, an increase of species diversity in disturbed systems such as flooded springs should be expected [20]. However, a Mann–Whitney U-test did not reveal a significant difference in the Shannon–Wiener diversity between flooded and unflooded springs in our study.

The results of the additional six macroinvertebrate community metrics (i.e., BMWP Score, ASPT, 1-GOLD index, EPT, PTH, and IBE) are given in Table 6.3.

Table 6.3 Results of the Mann–Whitney U-test for the seven selected macroinvertebrate community metrics for comparing flooded and non-flooded rheocrenes in the Cvrcka river basin

Macroinvertebrate Metrics	Unflooded springs	Flooded springs	p-value
Biological Monitoring Working Party Index (BMWP Score)	31.67 ± 8.91	42.36 ± 20.41	**0.052**
Average Score Per Taxon (ASPT)	5.247 ± 0.87	5.468 ± 0.93	0.495
Percentage of individuals from Gastropoda + Oligochaeta + Diptera Index (1-GOLD index)	0.71 ± 0.22	0.73 ± 0.15	0.855
Shannon–Wiener Diversity Index	1.57 ± 0.50	1.82 ± 0.45	0.263
Ephemeroptera, Plecoptera and Trichoptera Richness index (EPT)	1.76 ± 1.14	2.79 ± 1.58	**0.048**
Index of presence/absence of Plecoptera, Trichoptera and Hydrachnidia (PTH)	1.43 ± 1.03	1.36 ± 1.15	0.752
Indice Biotico Esteso (IBE)	3.89 ± 1.67	4.46 ± 0.95	0.293

Only two of examined metrics, i.e., EPT ($p = 0.048$) and BMWP ($p = 0.052$) metrics were found to have a statistically significant difference.

SIMPER analysis was used to identify the taxa responsible for the greatest differences between flooded and unflooded springs. The analysis has shown that the common species such as amphipod *Gammarus* cf. *fossarum* and grazer *Elmis aenea* (adults) significantly contribute to structure of the assemblages both in flooded and unflooded springs. Hydrobiid snail *Bythinella schmidti* and subterranean amphipod *Niphargus* sp. play a significant role in structuring the assemblages of unflooded springs, while *Dugesia gonocephala* and *Ancylus recurvus* were characteristic representatives of assemblages of flooded springs. The SIMPER test has shown that the assemblages of flooded and unflooded springs are quite different (average dissimilarity = 79.27). Table 6.4 shows that the assemblages of flooded springs are better defined and have a higher internal similarity (average similarity: 25.78) than the assemblages of unflooded springs.

Table 6.4 Results of SIMPER analysis for assemblages of flooded and unflooded spring groups of the Cvrcka river basin

Group 1–unflooded springs average similarity: 20.25 Group 2–flooded springs average similarity: 25.78			Group 1 and 2 average dissimilarity = 79.27		
Species	Av. Abund	Av. Sim	Sim/SD	Contrib %	Cum %
Group 1					
Gammarus cf. *fossarum*	10.67	13.61	0.85	67.22	67.22
Elmis aenea adults	3.57	1.80	0.38	8.89	76.11
Bythinella schmidti	2.43	1.16	0.42	5.72	81.83
Niphargus sp.	1.14	0.66	0.32	3.26	85.09
Belgrandiella bozidarcurcici	2.67	0.56	0.21	2.77	87.85
Hydraena sp. adults	0.52	0.38	0.20	1.90	89.75
Micropsectra type A larvae	1.48	0.24	0.20	1.17	90.92
Group 2					
Gammarus cf. *fossarum*	12.86	13.86	1.68	53.75	53.75
Elmis aenea adults	9.07	3.73	0.42	14.46	68.21
Dugesia gonocephala	2.86	1.67	0.49	6.47	74.68
Ancylus recurvus	3.21	1.13	0.46	4.39	79.07
Ecdyonurus sp.	1.00	0.77	0.48	2.99	82.06
Baetis rhodani	1.36	0.48	0.36	1.88	83.94
Electrogena sp.	0.93	0.48	0.39	1.88	85.82
Eiseniella tetraedra	0.79	0.45	0.29	1.74	87.56
Bythinella schmidti	2.50	0.43	0.17	1.68	89.24
Lumbriculus variegatus	0.71	0.37	0.39	1.45	90.69

Colonization of riparian springs occurs by different pathways. Active dispersers can reach the spring either via the active dispersive stage and/or by migrating upwards via springbrook into the spring [20]. In the latter case, at least two conditions must be met, 1) there must be a connection between the river and the spring, and 2) species that are capable for successful upwards migration must be able to survive both in river and in spring (euryoecious species) [20]. Von Fumetti et al. [20] showed that in the case of springs with a short springbrook that flows directly into the adjacent lotic reaches, upwards migration may be the main pathway for their colonization. As a consequence, the macroinvertebrate assemblages of flooded springs have a higher share of rhithrobiontic taxa, which also occur in the adjacent river [20]. Von Fumetti et al. [20] investigated 35 rheocrene springs in the Cvrcka river basin and found that the percentage of generalist species in flooded springs was higher (87%) in comparison to unflooded springs (75%). The latter study revealed that flooding significantly influences the assemblage composition of riparian springs, but this impact is not linked with distance of eucrenal to the nearby mainstream [20]. Approximately about two-thirds of all taxa found in rheocrenes along the Cvrcka river mainstream belong to insects with a flying, dispersive adult stage, indicating that colonization occurs via air dispersal along the watercourse corridor [20]. For taxa with fully aquatic life cycles such as two hydrobiid snails abundant in the latter study, crenobiontic *Belgrandiella bozidarcurcici*, an endemic of Cvrcka river basin, and *Bithynella schmidti* which inhabits also the epirithral zone of the latter river, dispersal via the groundwater passage was assumed [20].

The duration of the flood period and the type of the nearby water body also influences the assemblage composition of riparian springs, modifying the differences between the assemblages of springs and the adjacent lotic/lentic reaches; for example, as duration of the "flood" phase increases, differences decrease, while with increasing spatial isolation of a spring, differences between communities become more pronounced. Transitions between "flood" and "spring" phases necessitate repeated recolonization of riparian springs from adjacent lotic/lentic reaches, neighboring riparian springs, as well as more distant refuges. It is important to emphasize that changes in the diversity and composition of the riparian spring community cannot be interpreted only as a response to differences in the hydrological cycle (reflected in the existence of "flood" and "spring" phases) and variation of local abiotic factors, but is also strongly structured by the meta-community dynamics that influence dispersal processes.

In addition, spring slopes likely represent constraints affecting colonization by filtering out the species that cannot enter the spring. Pešić et al. [11] found the highest number of water mites in springs that flow directly into the river, without slope, indicating that these conditions are likely to facilitate easier migrations of many rhithral species into springs. Other characteristics of riparian habitat such as bank-top height and type of vegetation might also affect the rate and extent of dispersal, but more extensive research is needed in order to assess the importance of these factors for ecological controls of dispersal paths. It is worth mentioning that the importance of potential upward migration depends on the degree of coupling or connectivity between the springs and nearby lotic reaches; it is possible that during the "spring" phase the connection between the springbrook and the mainstream is interrupted, not allowing upward migration.

Generalist species are generally expected to invade disturbed habitats such as flooded springs more easily, and sometimes include some exotic species especially in springs located along larger lentic reaches, such as lowland rivers (for example the records of invasive Japanese fish louse *Argulus japonicus* in the springbrook of riparian springs along the Zeta River in Central Montenegro [44]).

Another consequence of flooding is reduction in number of specialist taxa (crenobionts) in the flooded springs. Von Fumetti et al. [20] found that the share of specialist taxa in the macroinvertebrate assemblages of the flooded springs along the Cvrcka river mainstream was lower than in the assemblages of unflooded springs (flooded: 13%, non-flooded: 25%). This conclusion was supported by research conducted by Pešić et al. [11] who studied water mite assemblages in 14 karstic springs located in the Mediterranean part of Montenegro. Water mites (Hydrachnidia) are among the aquatic animal groups with the highest percentage of crenobiontic taxa [22, 23]. In the above-mentioned study on water mites, Pešić et al. [11] found a negative response of the diversity of crenobiontic taxa to proximity of the nearby water body: in eight regularly flooded springs, only one specimen of crenobiontic *Atractides fonticulus* was found [11].

Information on specialist taxa in riparian springs remains mainly undocumented, both at the local and regional scale, but limited data suggest that such specialists generally do not contribute much to crenic assemblages [20]. Most crenobiont species are not able to colonize riparian springs via upwards migration [20, 21]. Besides, it is possible that some crenobiontic species are absent from flooded springs due to interspecific competition with generalist competitor [20]. Von Fumetti et al. [20] suggested that generalist species are the strongest competitors in flooded springs. Crenobiontic hydrobiid snail *Belgrandiella bozidarcurcici* occurs in high abundance in non flooded rheocrenes but in low abundance in flooded springs [20]. Interestingly, this species does not occur in the flooded springs inhabited by another hydrobiid snail *Bithynella schmidti*. A similar pattern has been observed for crenobiontic *Anagastina zetaevalis*, an endemic hydrobiid snail that inhabits numerous springs in the Skadar Lake basin [6, 10], where it may be found together with the hydrobiid snail *Radomaniola curta* which is a generalist and occurs in many streams and rivers, including the Zeta River. During our research, *Anagastina zetaevalis* was found only in unflooded springs along the Zeta River whereas it was absent in flooded springs where *Radomaniola curta* dominates.

Nevertheless, occurrence of crenobiont taxa in flooded springs is possible if there is no stronger generalist competitor at the site (Fig. 6.4). This is possible in riparian springs that are flooded by standing waters or downstream sections of lowland rivers and their floodplains. Downstream, the absence of a stronger competitor is often conditioned by decrease of rheophilic taxa as conditions become lentic in the most downstream reaches of the basin. *Radomaniola elongata*, an endemic hydrobiid snail, is known only from a single site, a spring on the island of Vranjina, at the place where left branch of river Morača enters the Skadar Lake. The seasonal dynamics of the macroinvertebrate assemblages of the spring on the island of Vranjina were studied by

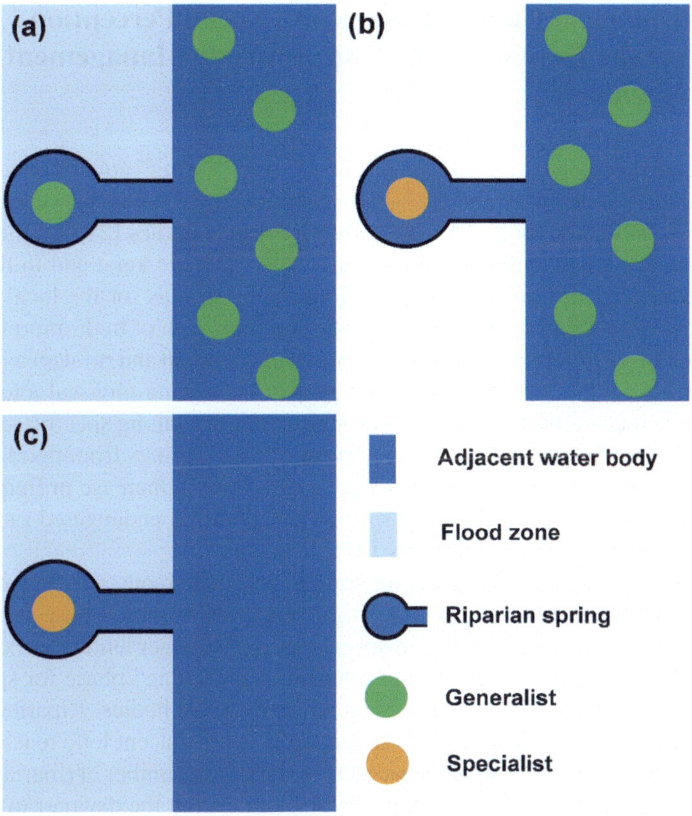

Fig. 6.4 Occurrence of specialist taxa in flooded springs. Three possible scenarios are shown: (**a**) crenobiontic species is suppressed due to interspecific competition with a stronger generalist competitor inhabiting nearby lotic reaches; this scenario is also present in the "flood" phase of riparian springs located along the lentic reaches; (**b**) the second scenario allows occurrence of specialist species in flooded springs when absence of a stronger competitor from adjacent lentic reach is conditioned by environmental factors during their "spring" phase; (**c**) third scenario allows occurrence of specialist species in flooded springs during their "flood" phase due to absence of a possible competitor generalist in the adjacent lentic reach

Šundić and Pešić [45]. For most of the year, the spring is connected with the river/lake waters. During the summer period (August and/or September) communication with the river/lake is interrupted. The highest number of *Radomaniola* snails was recorded when river/lake water level was high, when the spring was flooded, which coincides with the period of most intensive development of moss growth on the wall of the captured eucrenal section [45]. In September, when the water level in the spring drops sharply (below the level of moss growth), *Radomaniola* snails disappear [45]. These findings underline the importance of exploring biotic interactions that may enable the survival of crenobiont species in riparian springs.

6.5 Hydrological Characteristics and Social Perception Affect the Development of Appropriate Management Strategies

Flooding and the existence of different "flood" and "spring" aquatic phases, have an effect on the biodiversity and ecosystem functioning of flooded springs, but also affect their social perception. As emphasized by recent studies [21, 46] the human-induced changes are evident at many springs in the Dinaric karst and in that sense the riparian springs are no an exception. Tapping of springs for the local drinking water supply or just for aesthetic reasons is a common type of hydromorphological modification of springs bordering streams and rivers in urban and rural areas [21, 46].

Pešić and co-authors [21] stressed that projects involving physical alteration of banks such as their re-profiling (i.e. construction of levees along specific sections of rivers) may be implemented in order to prevent riparian springs from flooding. Such interventions should be an option only in places where an increase in frequency of flood events is expected to endanger springs inhabited by endangered or endemic species so their protection is necessary [21].

Riparian systems, including riparian springs and their biota, perform important ecosystem functions, often supporting high biodiversity which may include top predators (for example fish species) in springs along the larger lentic reaches. Moreover, these habitats may provide refuges during their "spring" phase for specialists who would be surpassed by generalists in adjacent water bodies. Riparian springs may significantly contribute to the discharge regimes of adjacent lotic reaches, especially during the times of drought. Besides, there is a large number of riparian springs in Dinaric karst that often cease to flow, most often during the dry part of the year. Such systems represent an infrequent source of organic matter for adjacent lotic/lentic reaches, but their dynamics as well as the fauna-habitat relationship are still barely studied.

6.6 Conclusion

Our main objective was to compile the most important scientific knowledge gained in recent years by researching riparian springs in the Dinaric karst area. We found that the concept of riparian springs was too broad, so we redefined it as springs under strong impact of an adjacent water body, in practice meaning that it was regularly or irregularly flooded. We provided new information about most important environmental variables in structuring macroinvertebrate assemblages of these springs as well as on the mechanisms of their colonization. Our survey confirmed the importance of these springs in maintaining freshwater biodiversity, especially in light of ongoing climatic changes and the expected increase of flooding events which will surely lead to higher frequency of occurrence of these habitats in future.

Special emphasis is placed on the concept of spring specialists and their maintenance in flooded springs, with the aim of making protection of these habitats more manageable and pragmatic.

Acknowledgements This study was partially supported in the framework of the "DNA/Eco" Project financed by Montenegrin Ministry of Science.

References

1. Kresic N (2010) Types and classification of springs. In: Kresic N, Stevanovic Z (eds) Groundwater hydrology of springs. Oxford: Elsevier, pp 31–86
2. Odum HT (1971) Fundamentals of Ecology. Saunders, Philadelphia
3. Gooch JL, Glazier DS (1991) Temporal and spatial patterns in Mid-Appalachian springs. In: Williams DD, Danks HV (eds) Arthropods of springs, with particular reference to Canada. Memoirs of the Entomological Society of Canada, vol 155, pp 29–49
4. Berlajolli V, Płóciennik M, Antczak-Orlewska O, Pešić V (2019) The optimal time for sampling macroinvertebrates and its implications for diversity indexing in rheocrenes case study from the Prokletije Mountains. Knowl Manag Aquat Ecosyst 420:6
5. Ivković M, Miliša M, Baranov V, Mihaljević Z (2015) Environmental drivers of biotic traits and phenology patterns of Diptera assemblages in karst springs: the role of canopy uncovered. Limnologica 54:44–57
6. Pešić V, Glöer P (2013) A new freshwater snail genus (Hydrobiidae, Gastropoda) from Montenegro, with a discussion on gastropod diversity and endemism in Skadar Lake. ZooKeys 281:69–90
7. Płóciennik M, Dmitrović D, Pešić V, Gadawski P (2016) Ecological patterns of Chironomidae assemblages in Dynaric karst springs. Knowl Manag Aquat Ecosyst 417:11
8. Gligorović B, Savić A, Protić Lj, Pešić V (2016) Ecological patterns of water bugs (Hemiptera: Heteroptera) assemblages in karst springs: a case study in central Montenegro. Oceanol Hydrobiol St 45:554–563
9. Pešić V, Gligorović B, Savić A, Buczyński P (2017) Ecological patterns of Odonata assemblages in karst springs in central Montenegro. Knowl Manag Aquat Ecosyst 418:3
10. Pešić V, Gadawski P, Gligorović B et al (2018) The diversity of the Zoobenthos communities of the Lake Skadar/Shkodra basin. In: Pešić V, Karaman GS, Kostianoy AG (eds) The Skadar/Shkodra lake environment. Springer, Berlin, Heidelberg, pp 255–293
11. Pešić V, Savić A, Jabłonska A et al (2019) Environmental factors affecting water mite assemblages along eucrenon-hypocrenon gradients in Mediterranean karstic springs. Exp Appl Acarol 77:471–486
12. Vilenica M, Mičetić-Stanković V, Sartori M et al (2017) Environmental factors affecting mayfly assemblages in tufa-depositing habitats of the Dinaric Karst. Knowl Manag Aquat Ecosyst 418:14
13. Kamberović J, Plenković-Moraj A, Borojević KK et al (2019) Algal assemblages in springs of different lithologies (ophiolites vs. limestone) of the Konjuh Mountain (Bosnia and Herzegovina). Acta Bot Croat 78:66–81
14. Marinković N, Karadžić B, Pešić V, Gligorović B, Grosser C, Paunović M, Nikolić V, Raković M (2019) Faunistic patterns and diversity components of leech assemblages in karst springs of Montenegro. Knowl Manag Aquat Ecosyst 420:26
15. Pozojević I, Pešić V, Goldschmidt T, Gottstein S (2020) Crenal habitats: sources of water mite (Acari: Hydrachnidia). Diversity 12:316. https://doi.org/10.3390/d12090316

16. Pešić V, Grabowski M, Hadžiablahović S et al (2019) The biodiversity and biogeographical characteristics of the River Basins of Montenegro. In: Pešić V, Paunović M, Kostianoy A (eds) The Rivers of Montenegro. The Handbook of Environmental Chemistry, vol 93. Springer, Cham, pp 157–200
17. Dmitrović D, Savić D, Pešić V (2016) Discharge, substrate type and temperature as factors affecting gastropod assemblages in springs in northwestern Bosnia and Herzegovina. Arch Biol Sci Belgrade 68:613–621
18. Savić A, Dmitrović D, Pešić V (2017) Ephemeroptera, Plecoptera and Trichoptera assemblages of karst springs in relation to environmental factors: a case study in central Bosnia and Hercegovina. Turk J Zool 41:119–129
19. Savić A, Dmitrović D, Glöer P, Pešić V (2020) Assessing environmental response of gastropod species in karst springs: what species response curves say us about niche characteristic and extinction risk? Biodivers Conserv 29:695–708
20. von Fumetti S, Dmitrović D, Pešić V (2017) The influence of flooding and river connectivity on macroinvertebrate assemblages in rheocrene springs along a third-order river. Fund Appl Limnol 190(3):251–263
21. Pešić V, Dmitrović D, Savić A et al (2019) Application of macroinvertebrate multimetrics as a measure of the impact of anthropogenic modification of spring habitats. Aquat Conserv 29:341–352
22. Gerecke R, Martin P, Gledhill T (2018) Water mites (Acari: Parasitengona: Hydrachnidia) as inhabitants of groundwater-influenced habitats–considerations following an update of Limnofauna Europaea. Limnologica 69:81–93
23. Zawal A, Pešić V (2018) The diversity of water mite assemblages (Acari: Parasitengona: Hydrachnidia) of Lake Skadar/Shkodra and its catchment area. In: Pešić V, Karaman GS, Kostianoy AG (eds) The Skadar/Shkodra lake environment. Springer, Cham, pp 311–323
24. Pešić V, Dmitrović D, Savić A, von Fumetti S (2016) Studies on eucrenal-hypocrenal zonation of springs along the river mainstream: a case study of a karst canyon in Bosnia and Herzegovina. Biologia 71:809–817
25. Council of the European Communities (2000) Directive 2000/60/EC of the European Parliament and the council of 23rd October 2000 establishing a framework for community action in the field of water policy. Off J Eur Communities L327:1–72
26. Pešić V, Dmitrović D, Savić A et al (2018) Data on substrate composition, species list and values of selected macroinvertebrate community metrics of 50 investigated springs of the Cvrcka river basin (Bosnia and Herzegovina). Mendeley Data, v5. https://doi.org/10.17632/ck5t89pwmm.5
27. Pakulnicka J, Buczyński P, Dąbkowski P et al (2016) Aquatic beetles (Coleoptera) in springs of a small lowland river: habitat factors vs landscape factors. Knowl Manag Aquat Ecosyst 417:1–13
28. Zawal A, Stryjecki R, Buczyńska E et al (2018) Water mites (Acari, Hydrachnidia) of riparian springs in a small lowland river valley: what are the key factors for species distribution? PeerJ 6:e4797
29. Szlauer-Łukaszewska A, Pešić V, Zawal A (2021) Environmental factors shaping assemblages of ostracods (Crustacea: Ostracoda) in springs situated in the River Krąpiel valley (NW Poland). Knowl Manag Aquat Ecosyst 422:14
30. Hahn H-J (2000) Studies on classifying of undisturbed springs in Southwestern Germany by macrobenthic communities. Limnologica 30:247–259
31. Von Fumetti S, Nagel P, Scheifhacken N, Baltes B (2006) Factors governing macrozoobenthic assemblages in perennial springs in north-western Switzerland. Hydrobiologia 568:467–475
32. Dumnicka E, Galas J, Koperski P (2007) Benthic invertebrates in karst springs? Does substratum or location define communities? Int Rev Hydrobiol 92:452–464
33. Vitecek S, Previšić A, Kučinić M et al (2015) Description of a new species of Wormaldia from Sardinia and a new Drusus species from the Western Balkans (Trichoptera, Philopotamidae, Limnephilidae). ZooKeys 496:85–103
34. Von Fumetti S, Nagel P (2012) Discharge variability and its effect on faunistic assemblages in springs. Freshwater Sci 31:647–656

35. Gerecke R, Meisch C, Stoch F et al (1998) Eucrenon-hypocrenon ecotone and spring typology in the Alps of Berchtesgaden (Upper Bavaria, Germany). A study of microcrustacea (Crustacea: Copepoda, Ostracoda) and water mites (Acari: Halacaridae, Hydrachnellae). In: Botosaneanu L (Ed) Studies in crenobiology. The biology of springs and springbrooks. Backhuys Publishers, Leiden, pp 167–182
36. Illies J, Botosaneanu L (1963) Problèmes et méthodes de la classification et de la zonation écologique des eaux courantes, considérées surtout du point de vue faunistique. Mitteilungen der Internationalen Vereinigung für Theoretische und Angewandte Limnologie 12:1–57
37. Erman NA, Erman DC (1995) Spring permanence, Trichoptera species richness and the role of drought. J Kans Entomol Soc 68:50–64
38. Von Fumetti S, Nagel P, Baltes B (2007) Where a springhead becomes a springbrook-a regional zonation of springs. Fundam Appl Limnol 169:37–48
39. Di Sabatino A, Coscieme L, Miccoli FP, Cristiano G (2021) Benthic invertebrate assemblages and leaf-litter breakdown along the eucrenal–hypocrenal ecotone of a rheocrene spring in Central Italy: are there spatial and seasonal differences? Ecohydrology e2289. https://doi.org/10.1002/eco.2289
40. Carroll TM, Thorp JH, Roach KA (2016) Autochthony in karst spring food webs. Hydrobiologia 776:173–191
41. McCabe DJ, Sycora JL (2000) Community structure of caddisfly along a temperate springbrook. Arch Hydrobiol 148:263–282
42. Karaouzas I, Zawal A, Michoński G, Pešić V (2019) Contribution to the knowledge of the caddisfly fauna of Montenegro - New data and records from the karstic springs of Lake Skadar basin. Ecol Montenegrina 22:34–39
43. Crema S, Ferrarese U, Golo D et al (1996) Ricerche sulla fauna bentonica ed interstiziale di ambienti sorgentizi in area alpina e prealpina [A research on benthonic and interstitial fauna in Alpine and pre-Alpine springs]. Report Centro Ecologia Alpina 8:1–104
44. Pešić V (2020) First record of the Japanese fish louse (Argulus japonicus) in Montenegro (Crustacea: Branchiura). Ecol Montenegrina 38:141–143
45. Šundić M, Pešić V (2007) Seasonal changes in the abundance of benthic assemblages in the spring on Vranjina island (Skadar Lake National Park). Glas Republ Zavoda Zašt Prirode–Prirodnjačkog Muzeja 29–30:125–130
46. Dedić A, Gerhardt A, Kelly MG et al (2020) Innovative methods and approaches for WFD: ideas to fill knowledge gaps in science and policy. Water Solut 3:30–42

Chapter 7
Ecological Characteristics and Specifics of Spring Habitats in Bosnia and Herzegovina

Svjetlana Stanić-Koštroman, Jasmina Kamberović, Dejan Dmitrović, Anita Dedić, Dragan Škobić, Anđelka Lasić, Marija Gligora Udovič, and Nevenko Herceg

Abstract Due to its geographical position, relief features, specific geological past and great heterogeneity of space in a relatively small area, Bosnia and Herzegovina has specific hydro-morphological features, with numerous spring habitats and unique aquatic biodiversity, mainly composed by stenothermic and relict taxa. Spring habitats, as well as the flora and fauna that inhabit them are considered particularly vulnerable to climate change and anthropogenic pressures, due to their isolation and fragmentation. According to forecasts, these ecosystems will be exposed to multiple pressures caused by changes in water regime, rising water temperatures and a continuous decline in water quality, which will lead to the disappearance of more sensitive crenobiont species and a reduction in biodiversity. Therefore, in this chapter we present the ecological characteristics and specifics of spring habitats in Bosnia and

S. Stanić-Koštroman (✉) · A. Dedić · D. Škobić · A. Lasić · N. Herceg
Faculty of Science and Education, University of Mostar, Rodoč bb, 88 000 Mostar, Bosnia and Herzegovina
e-mail: svjetlana.stanic.kostroman@fpmoz.sum.ba

A. Dedić
e-mail: anita.dedic@fpmoz.sum.ba

D. Škobić
e-mail: dragan.skobic@fpmoz.sum.ba

N. Herceg
e-mail: nevenko.herceg@fpmoz.sum.ba

J. Kamberović
Faculty of Natural Sciences and Mathematics, University of Tuzla, Urfeta Vejzagića4, 75 000 Tuzla, Bosnia and Herzegovina
e-mail: jasmina.kamberovic@untz.ba

D. Dmitrović
Faculty of Natural Sciences and Mathematics, University of Banja Luka, Mladena Stojanovića 2, 78000 Banja Luka, Bosnia and Herzegovina
e-mail: dejan.dmitrovic@pmf.unibl.org

M. Gligora Udovič
Faculty of Science, University of Zagreb, Rooseveltov trg 6, 10 000 Zagreb, Croatia
e-mail: marija.gligora.udovic@biol.pmf.hr

V. Pešić et al. (eds.), *Small Water Bodies of the Western Balkans*, Springer Water,
https://doi.org/10.1007/978-3-030-86478-1_7

Herzegovina, with a preliminary list of species that inhabit them and special focus on endemic, sensitive and rare species, which are potentially threatened by predicted changes in habitat quality.

Keywords Spring habitats · Endemic species · Anthropogenic pressures · Climate change · Water Framework Directive · Natura 2000 · Bosnia and Herzegovina

7.1 Introduction

The water resources of Bosnia and Herzegovina belong to two main basins: the Black Sea basin (38,719 km^2) and the Adriatic Sea basin (12,410 km^2). The Black Sea basin covers 76% of the territory of Bosnia and Herzegovina, while the remaining part belongs to the Adriatic basin [1]. The relative annual availability of water resources per capita ranks Bosnia and Herzegovina among the countries of "average water availability" between 5,000 and 10,000 m^3/capita. Rivers in Bosnia and Herzegovina are characterized by high gradients and relatively high flow rate [2].

Fig. 7.1 Springs: **a** Konjuh Mountain (photo by J. Kamberović); **b** Vilenjska vrela (photo by D. Dmitrović); **c** Klokun (photo by A. Dedić); **d** Mandića vrilo, the Lištica River (photo by S. Stanić-Koštroman)

In addition, large rivers and streams of Bosnia and Herzegovina are also rich in natural springs (Fig. 7.1) that represent specific unique habitats, with stable physico-chemical factors of water and the composition of biocenoses that differ from all other aquatic habitats [3]. Generally, springs are inhabited by rare species, most of which are endangered, which is why they are recognized as important centers for biodiversity conservation [3, 4]. In recent years, there has been significant interest in researching the springs area of Bosnia and Herzegovina [5, 6].

7.2 Physical and Chemical Characteristics of the Spring Waters in Bosnia and Herzegovina

The largest part of the terrain in Bosnia and Herzegovina was formed during the Mesozoic Era, and these rocks are predominantly carbonate type and are present as limestones and dolomites [7]. Carbonate sediment is under the constant influence of water circulation and other influences, forming a special type of relief called karst. Karst makes up more than two thirds of the territory of Bosnia and Herzegovina [8, 9]. Karst springs contain interesting features as results of hydrological, physical and chemical characteristics that form them. As a special type of landscape, karst is characterized by the appearance of karst springs as places where groundwater comes to the surface of the Earth creating a visible flow. Also, an interesting character of karst springs are their heterogeneous origin, which significantly affects the distribution and number of species, the interaction between species and the trophic structure of biological communities [10]. According to Steinmann [11] and Thieneman [12], local topological conditions and hydromorphological properties, such as flow velocity and turbulence, determine three type of spring classification: (1) rheocrenic springs with fast flowing or falling water; (2) helocrenic springs with diffuse or laminar flowing water; and (3) limnocrenic springs with a still water pool [13]. Di Sabatino et al. [14] consider rheocrene springs to be most common in South Central Europe, helocrenes in Scandinavia, and limnocrenes in the karst area.

The springs are mostly small, but complex and species-rich systems, with a mosaic structure and a high degree of individuality [15]. They are characterized by physico-chemical stability. Almost all springs have a low water temperature (8–12 °C) with small annual fluctuations, high concentrations of oxygen (8–12 mg/dm^3) and high concentrations of carbon dioxide (up to 60 mg/dm^3) [16–20]. Karst springs have high alkalinity, which is mainly due to carbonates, and high carbonate hardness [16, 21]. Carbonate sedimentary rocks in natural waters are the main source of alkalinity [3, 17].

Ammonia, nitrites and nitrates are inorganic nitrogen compounds present in natural waters. Ammonia is formed by the decomposition of nitrogen-containing organic compounds [22]. Since they are readily oxidized they present in trace amounts. Nitrates in groundwater are considered important indicators of anthropogenic impact [23]. They are present in low concentrations (0.4–8.0 mg/dm^3) [3, 17].

Sulphates are usually present with concentrations of a few tenths of mg/dm^3 while hundreds of mg/dm^3 can be found in springs fed by aquifers that also include evaporites or geological formations which contain gypsum [17]. Phosphorus is also present in low concentrations (0.005–0.050 mg/dm^3) [16, 24] but in central Bosnia (the Fojnica River and the Lašva River) the concentration of phosphorus may be higher depending on structure of sediment [25]. Underground waters in siliceous aquifers usually hold 3–10 mg/dm^3 and in carbonate aquifers about 1 mg/dm^3 SiO_2 [17]. A common characteristic of karstic springs, whether permanent or temporary, is the strong dependence of discharge on precipitation. As a consequence, the ratio between minimum and maximum discharge is great (1:60, or more). For example, the minimum and maximum discharges registered for the spring of the Bunica River were 0.72 and 207 m^3/s, respectively [26]. All values of physico-chemical parameters during major rainfall may be increased.

7.3 Algae and Cyanobacteria in Spring Habitats

Studies on algae in Bosnia and Herzegovina date back to the late nineteenth century [27], and were initially focused on the exploration of river and lake ecosystems while springs were omitted. The first detailed studies of algae in springs of Bosnia and Herzegovina were carried out by Blagojević [16, 28-30], who focused on diatoms, chlorophycaea, xantophyceae and cyanobacteria in the crenon area of the karstic springs Mošćanica (Sarajevo) and Radobolja (Mostar). The main research topic in these studies was the dynamics of periphytic colonization on artificial (Plexiglas plates) and natural substrates. Studies on algae and cyanobacteria in the springs of Bosnia and Herzegovina have recently intensified and so far, the focus have been on karstic springs [3, 31–37], small rheocrenic and rheohelocrenic springs of the Vranica Mountain [38, 39], and springs of the Konjuh Mountain [18–20]. Most of the research has been focused on diatoms, while cyanobacteria and other benthic algae were less studied (Table 7.1).

Since the Dinarides are mostly composed of carbonates, most algal studies refer to karstic springs. A study conducted in 2009 in three karstic springs in Herzegovina (Ljubovija, Lištica and Klokun) reveals a high diversity of diatoms [31, 37], whilst research on cyanobacteria assemblages in other three karstic springs (Bunica, Buna and Radobolja) [32, 37] confirmed that spring ecosystems are heterogeneous habitats in terms of species composition. This study pointed out that indicators of oligosaprobial zone were predominant among cyanobacteria with genera *Phormidium* and *Plectonema* with the highest number of taxa (11 and 4, respectively). A similar, but lower number of taxa has been found at springs of the Bosna River [33], revealing *Odontidium mesodon* (Ehrenberg) Kützing and *Melosira arenaria* Moor as the most abundant. In the karstic springs of the Studenčica River in western Herzegovina (Mlin

Table 7.1 The list of studies conducted on cyanobacterial and algal communities in spring ecosystems in Bosnia and Herzegovina. Studies are listed chronologically

Springs	Results
Karstic springs Mošćanica and Radobolja [28]	57 algal taxa
Karstic springs Mošćanica and Radobolja [16]	116 algal taxa
Karstic springs Mošćanica and Radobolja [29]	26 taxa of Cyanobacteria
Karstic springs Mošćanica and Radobolja [30]	35 taxa of Chlorophyta
Spring ecosystems of the Bosna River: Spring Stojčevac, Spring 1, 2, 4 of the Bosna River, Spring Borim 1 and Borim 2 [33]	174 algal taxa
Wet habitats (eight locations) developed around springs in subalpine belt of the Vranica Mountain: Glavica 1, Glavica 2, Jezero 3, Zavol 4, Točilo 5, Suhoperka 6, Potok 7 and Podovi 8 [38]	221 diatom taxa
Karstic springs in Herzegovina: Ljubovija, Lištica, Klokun [31, 37]	112 diatom taxa
Karstic springs in Herzegovina: Bunica, Buna, Radobolja [32, 37]	42 taxa of Cyanobacteria
Wet habitat near springs Čavljak on the Ozren Mountain [34]	139 diatom taxa
Karstic springs Kajtaz and Mlin of the Studenčica River in the Western Herzegovina [35]	43 algal taxa
Nine karstic springs in Herzegovina: Buna, Bunica, Mlin, Kajtaz, Vrioštica, Nezdravica, Modro oko, Studena and Klokun [30]	113 diatom taxa
Spring of the Bunica River, Herzegovina [36]	104 diatom taxa
20 springs of the Konjuh Mountain: Studešnica, Krabašnica, Krabanja, Gluha Bukovica, Tarevo, Tuholj, Zapaučki potok, Bebroštica, Muška voda, Katranica, Miljevica, Kesovača 1, Kesovača 2, Zlaća, Vukotići 1, Vukotići, Salihovo korito, Borovnica 1, Borovnica 2, Podgornica [18, 20]	234 algal taxa, 187 diatom taxa and 34 cyanobacteria
Five springs and five streams of the Konjuh Mountain: Tarevo, Miljevica, Gluha Bukovica, Katranica and Kesovača [38]	45 diatom taxa in springs and 17 taxa in streams
Six springs and streams on the Vranica Mountain [39]	45 diatom taxa, new location for *Hydrurus foetidus* (Villars)

and Kajtaz) the most represented genera were: *Navicula* (4 taxa), *Phormidium* (4), *Batrachospermum* (4), *Cocconeis* (3 taxa), *Planothidium* (3) and *Gomphonema* (3).

In addition to the above mentioned studies, nine karstic springs in the southern part of Bosnia and Herzegovina were investigated by Dedić [30] from the aspect of dynamics of colonization of periphytic diatoms on artificial (glass slide) substrates.

Similar studies were presented in a paper by Dedić et al. [3] on the spring of the Bunica River. Results in both studies showed a large number of widespread species a consequence of anthropogenic influences and morphological changes. Generally, abundant taxa in carbonate springs in Bosnia and Herzegovina are *Achnanthidium minutissimum* (Kützing) Czarnecki, *Achnanthidium pyrenaicum* (Hustedt) H. Kobayasi, *Odontidium mesodon* (Ehrenberg) Kützing, *Nitzschia fonticola* (Grunow) Grunow, *Amphora pediculus* (Kützing) Grunow ex A. Schmidt, and *Meridion circulare* (Greville) C. Agardh [19, 36], which is in concordance with well-known research into alpine springs [15]. Moreover, a study on springs in Bosnia and the continental Dinarides on the Konjuh Mountain has been carried out by Kamberović et al. [19, 40] with focus on the influence of geological substrate (ophiolites or carbonates) and spring types (rheocrenes and rheohelocrenes) on the algal composition. Since the phycological studies focusing on ophiolite springs (silicate rocks) are very rare, Kamberović et al. [19] singled out the following taxa as having a preference for ophiolites: *Amphora lange-bertalotii* var. *tenuis* Levkov et Metzeltin, *Cymbella hantzschiana* Krammer, *C. parva* (W. Smith) Kirchner, *Encyonopsis cesatii* (Rabenhorst) Krammer, *E. krammeri* Reichardt, *Gomphonema parvulum* (Kützing) Kützing, *G. tergestinum* (Grunow) Fricke, and *Navicula leistikowii* Lange-Bertalot. The above-mentioned study implied a higher similarity in algal composition in ophiolites vs. carbonate springs than ophiolites vs. other siliceous springs, and suggested that algal assemblages in springs emerging from ophiolites and springs emerging from other types of siliceous substrata should be separately analyzed in the future. Analyzing the influence of seasonality and different microhabitats (epibryon, epilithon, epipelon) on the algal composition in the springs of the Konjuh Mountain, the highest algal diversity was found in the spring season and epibryon samples [18]. Actually, algal assemblages in spring ecosystems were most influenced by the spring types. It has been noted that small rheohelocrenic and helocrenic springs provide habitats for many taxa and have great potential for species conservation [17–19]. According to the Red List for Algae [41], a total of 35 taxa belonging to the categories of endangered, rare and vulnerable algae species were found in small rheohelocrenic and rheocrenic springs of the Konjuh Mountain [19]. A total of 9 taxa are endangered: *Achnanthidium caledonicum* (Lange-Bertalot) Lange-Bertalot, *Achnanthidium rosenstockii* (Lange-Bertalot) Lange-Bertalot, *Brachysira vitrea* (Grunow) R. Ross, *Encyonema hebridicum* Grunow ex Cleve, *Eucocconeis flexella* (Kützing) Meister, *Gomphonema procerum* E. Reichardt & Lange-Bertalot, *Planothidium joursacense* (Héribaud) Lange-Bertalot, *Epithemia parallela* (Grunow) Ruck & Nakov and *Achnanthidium pusillum* (Grunow) Czarnecki. Unfortunately, most of the studied springs are affected by anthropogenic influences and morphological changes [19]. However, the study on comparative analysis of epilithic diatom assemblages of springs and streams in the Konjuh Mountain [40] showed higher species diversity in springs compared to downstream sites. This indicates the purpose of conserving these rare and extremely endangered ecosystems, which can serve as refugia for endangered species. Uncontrolled management and capture of mountain springs and alteration of spring habitats may lead to a reduction in diversity of rare taxa in the future. Particular attention should be paid to the conservation of small rheohelocrenic springs, which are mostly ignored in management and conservation plans, although they have great potential for species conservation [18, 20]. The finding of new species for the algal flora of Bosnia and Herzegovina in spring ecosystems supports this statement. The following new taxa for the algal flora of Bosnia were found in rheohelocrenic springs at the Ozren Mountain [34]: *Aulacoseira* cf. *alpigena* (Grunow) Krammer, *Decussata hexagona* (Torka) Lange - Bert., *Diploneis* cf. *fontium* E. Reichardt & Lange - Bert., *Stauroneis kriegeri* R. M. Patrick and *Pinnularia brandeliformis* Krammer, whilst *Cymbella tridentina* Lange-Bertalot, M. Cantonati & A. Scalfi and *Achnanthidium dolomiticum* M. Cantonati & H. Lange-Bertalot were found for the first time in Bosnia and Herzegovina in small rheocrenic ophiolite springs of the Konjuh Mountain [18, 20].

Rheohelocrenic and helocrenic springs on the Vranica Mountain were also the subject of algological research in Bosnia and Herzegovina. The most dominant genera in these springs were *Eunotia*, *Navicula* sensu lato and *Cymbella* sensu lato [38]. Recently, Mašić et al. studied the diversity of diatoms in oligotrophic habitats (springs and streams) of the Vranica Mountain and the distribution of the rare golden algae *Hydrurus foetidus* (Villars) Trevisan [39]. *Hydrurus foetidus*, which is an indicator of clean water and good ecological status, was identified at six sites in the springs and streams of the Vranica Mountain, followed by 48 diatom taxa.

Another aspect of algological research was conducted in 2019. Trophic relationships between diatoms and the endemic caddisfly (Trichoptera; Insecta) species *Drusus ramae* Marinković-Gospodnetić, 1971 were studied at the spring of the Lištica River [42]. The latter study reveals that the samples from different substrate types showed a great diversity of diatom taxa as well as their different dominance considering the substrates.

7.4 Plant Communities of Springs

In the springs of the Neretva and Cetina basins, mosses are the most abundant both by number of taxa and by coverage: *Cinclidotus aquaticus* (Hedw.) Bruch & Schimp., *Cinclidotus riparius* (Host ex Brid.) Arn., *Fontinalis antipyretica* Hedw., *Cratoneuron filicinum* (Hedw. Ripycepius) Spruce (Hedw.) Dix., *Marchantia polymorpha* L., *Eucladium verticillatum* (With.) Brush set Shrimp. and *Conocephalum conicum* (L.) Dumort. The most common taxa are: *Mentha aquatica* L., *Agrostis stolonifera* L., *Equisetum arvense* L., *Adiantum capillus-veneris* L., *Berula erecta* (Hudson) Coville, *Epilobium hirsutum* L. and *Solanum dulcamara* L., *Eupatorium cannabinum* L., *Carex pendula* Huds [43].

At the springs of the Tihaljina River and the Vrioštica River As. *Eucladio-Adiantetum capilli-veneris* Br.-Bl. ex Horvatić 1934 develops the association of southern maidenhair fern of the rheocrenic type. The spring is surrounded by moist and shaded steep cliffs. The floral composition is dominated by *Adiantum capillus-veneris*, and of the mosses *Fontinalis antipyretica*, *Eucladium verticillarum* and *Eucladium angustifolium*.

The association of the river water-crowfoot and the cut-leaved water parsnip (As. *Ranunculo fluitanti-Sietum erecto-submersi* (Roll 1939) Th. Müller) occurs in the sprigs of the rivers Lištica, Trebižat and Bistrica at a depth of up to 0.5 m. The floral composition is dominated by *Berula erecta*, *Ranunculus fluitans* and *Nuphar luteum*, while *Sparaganium erectum*, *Myosotis scorpioides*, *Mentha aquatica*, *Erysimum repandum*, *Alisma plantago aquatica* and others often appear from the companions [43].

Associations of mosses are quite common in rheocrenes fast flowing springs and belong to the class *Platyhypnidio—Fontinalietea antypireticae* Philippi 1956. Among them in the springs of the Neretva and Cetina basins most commonly occur: *Oxyrrhynchietum rusciformis* Kaiser ex Hübschmann 1953, *Leptodictyo riparii-Fissidentetum crassipedis* 1953, *Fontinaletum antipyreticae* Kaiser 1926 [44].

Plant communities in the spring habitats of the Konjuh Mountain in karstic and fast flowing springs belong to the previously mentioned class *Platyhipnidio-Fontinalietea antipyretica*e Philippi 1956 and alliance *Platyhypnidion rusciformis* Philippi 1958, whilst small rheocrene and rheohelocrene spring on ophiolites are inhabited with vegetation of the class *Montio-Cardaminetea* Br.-Bl. & R. Tx. ex Klika & Hadač. 1944 em. Zechmeister 1993, and two alliances *Cratoneurion commutati* Koch 1928 and *Caricion remotae* Kästner 1941. Common moss species in small springs emerged from ophiolites are *Cratoneuron filicinum* (Hedw.) Spruce, *Bryum pseudotriquetrum* (Hedw.) Gaertn. et al., *Fissidens adianthoides* Hedw., *Rhizomnium punctatum* (Hedw.) T. J. Kop., *Brachythecium rivulare* Schimp. and *Conocephalum conicum* (L.) Corda. Karstic spring on the Konjuh Mountain with variable discharge are inhabited predominantly by *Platyhypnidium riparioides* (Hedw.) (Fig. 7.2).

Fig. 7.2 Mosses: **a** *Platyhypnidium riparioides* (Hedw.) Dixon in the carbonate karstic spring Krabašnica and **b** *Cratoneuron filicinum* (Hedw.) Spruce in the ophiolite spring on the Konjuh Mountain (photos by J. Kamberović)

7.5 Endemic and Rare Crenobiontic Fauna

Springs in Bosnia and Herzegovina are known as habitats with numerous endemic macroinvertebrate taxa [45–57]. However, the exact number of true crenobiontic species and subspecies endemic to Bosnia and Herzegovina used to be unknown. In that sense a list of true crenobiontic macroinvertebrates endemic to Bosnia and Herzegovina was formed for the first time in this study (Table 7.2). This preliminary list was formed on the basis of data from the available literature. All taxa in the list were identified as rare according to the criteria listed by Işik [58].

This list includes 28 species and subspecies of rare and endemic crenobiontic macroinvertebrates from the following groups: Gastropoda (8), Hirudinea (2) and Insecta (18) (Fig. 7.3). Eucrenal and hypocrenal zones are inhabited by 16 taxa, where each zone has 6 taxa exclusively present in that particular zone. The greatest number of rare and crenobiontic taxa is endemic for Bosnia only (21), two taxa are distributed in Bosnia and Herzegovina as a whole (*Sarajana apfelbecki apfelbecki* and *S. a. travnicensis*), while five taxa are endemic exclusively to Herzegovina. In comparison to other macroinvertebrate groups, rare and endemic insects are the group with the greatest species richness. The springs in Bosnia are inhabited by 15 species, springs in Herzegovina by 3 species, while representatives that would characteristic for spring zones throughout Bosnia and Herzegovina were not determined. The species-richest group of insects was of the order Diptera (13 species, family Psychodidae), followed by representatives of Trichoptera (4 species, genus *Drusus*). Endemic and rare insects from the order Plecoptera were represented with just a single crenobiontic species (*Leuctra aptera*). This species is known only from three springbrooks situated in the southeastern part of Bosnia [55, 63]. While there were five rare and endemic crenobiontic species and subspecies of gastropods recorded in Bosnia, only a single species (*Travunijana vruljakensis*) was recorded in Herzegovina. Surprisingly, this single species is a true crenobiont [45], while all other described species in its genus are subterranean [71]. Hirudinea were represented by just two rare and crenobiontic species. One of these was recorded at a handful of localities in Bosnia (*Dina sketi* [48, 61]), while the other is known from a single locality in Herzegovina (*Piscicola hadzii* [62]). In addition to the true crenobiontic taxa of macroinvertebrates endemic to Bosnia and Herzegovina, some taxa may be found in the springs only occasionally, such as some subterranean species of gastropods [46, 60, 71–73] or amphipods [49, 74], while others, such as some insects, may also inhabit the rhithral zone [54, 56].

7.6 Discussion

Springs are extremely valuable ecosystems in terms of biodiversity [75–77]. They are characterized by diverse communities of plants and animals and in many cases the biota shows a high degree of endemism. Cantonati et al. [15] suggest that springs may play an important role as refuges for aquatic flora and fauna, while Botosaneanu [78]

Table 7.2 List of the rare crenobiontic macroinvertebrates endemic to Bosnia and Herzegovina; Distribution: B—only in Bosnia, H—only in Herzegovina, B&H—in Bosnia and Herzegovina; Habitat: euc—eucrenal zone (springhead), hyc—hypocrenal zone (springbrook)

Taxa	Distribution	Habitat	References
GASTROPODA			
Belgrandiella bozidarcurcici Glöer & Pešić, 2014	B	euc, hyc	[57, 59]
Belgrandiella dabriana Radoman, 1975	B	euc	[60]
Bythinella samecana Clessin, 1911	B	euc, hyc	[60]
Graziana vrbasensis Radoman 1975	B	euc, hyc	[60]
Sarajana apfelbecki apfelbecki (Brancsik, 1988)	B&H	euc, hyc	[60]
Sarajana apfelbecki erythropama (Schütt, 1959)	B	euc	[60]
Sarajana apfelbecki travnicensis Radoman, 1975	B&H	euc, hyc	[60]
Travunijana vruljakensis Grego & Glöer, 2019	H	euc, hyc	[45]
HIRUDINEA			
Dina sketi Grosser & Pešić, 2014	B	euc, hyc	[48, 61]
Piscicola hadzii Sket, 1985	H	euc, hyc	[62]
PLECOPTERA			
Leuctra aptera Kaćanski & Zwick, 1970	B	hyc	[55, 63]
DIPTERA			
Jungiella jadarica Krek, 1979	B	hyc	[54]
Mormia villosa Krek, 1972	B	hyc	[54, 64]
Pericoma (Vaillantella) antennata Krek, 1983	H	euc, hyc	[54]
Pericoma (Leptopericoma) ljubiniensis Krek, 1967	B	euc, hyc	[54, 64]
Satchelliella jasnae (Krek, 1990)	B	hyc	[54]
Satchelliella sanae Krek, 1990	B	hyc, erh	[54]
Sycorax trifida Krek, 1970	B	euc, hyc	[54]
Telmatoscopus orbiculatus (Krek, 1971)	B	hyc	[54, 64]
Threticus optabilis Krek, 1970	B	euc	[54, 64]
Ulomyia (Sijarićija) erinacea (Krek, 1970)	B	euc	[54, 64]
Ulomyia spinosa Krek, 1972	B	euc	[54, 64]
Vagmania ramulosa Krek, 1972	B	euc, hyc	[54, 64]
Vaillantia ramae (Krek, 1977)	H	hyc	[54]

(continued)

Table 7.2 (continued)

Taxa	Distribution	Habitat	References
TRICHOPTERA			
Drusus crenophylax Graf &Vitecek, 2015	B	euc	[57, 65]
Drusus radovanovici Marinković-Gospodnetić, 1971	B	euc, hyc	[52, 56, 66, 67]
Drusus ramae Marinković-Gospodnetić, 1971	H	euc, hyc	[52, 53, 56, 66, 68]
Drusus septentrionis Marinković-Gospodnetić 1976	B	euc, hyc	[52, 56, 69, 70]

Fig. 7.3 Endemic crenobionts of Bosnia and Herzegovina: **a** *Dina sketi* (photo by D. Dmitrović) and **b** *Drusus ramae* (photo by S. Stanić-Koštroman)

suggests that springs provide refuges in geological time, providing stable habitats for relict species. Given their distinct physico-chemical and biological characteristics, springs provide ideal settings to studying the effects of environmental stability on biocenoses [79–81].

The specific geological history and relief, hydro-morphological and geological diversity of Bosnia and Herzegovina, has caused a great diversity of spring habitats in a relatively small area, with a high degree of biodiversity and numerous endemic, rare and sensitive species [82]. Based on all conducted cyanobacterial and algological studies a great diversity of taxa have been represented, as well as the presence of rare, and sensitive species, especially in the oligotrophic habitats of the Vranica Mountain, and the springs of the Konjuh Mountain. Studies on the vegetation of spring ecosystems are extremely sporadic and rare in Bosnia and Herzegovina [18]. Since these ecosystems are often inhabited by relics of glacial flora, more detailed research is needed for a detailed syntaxonomic review in the future. Dealing with invertebrate fauna, spring habitats in Bosnia and Herzegovina are important centers of endemism. For example, the karst rheocrene spring Vilenjska vrela, located in the catchment of the Cvrcka River (NW Bosnia and Herzegovina), is inhabited by three species of endemic aquatic invertebrates: *Belgrandiella bozidarcurcici* [59], *Dina sketi* [61] and *Drusus crenophylax* [65]. The first two species also inhabit the springs of the catchment of the Crna Rijeka River in the northwestern part of Bosnia and Herzegovina [47, 48], while the last species is a micro-endemic restricted to the watershed of the Cvrcka River [65]. The area of the Cvrcka River basin is also inhabited by some other endemic species, *Symphyandra hofmannii* Pant. (1881) [83] and *Rana graeca* Boulenger, 1891 [84]. The first species, known as Hoffmann's Ringed Bellflower, is an Illyrian endemic plant species, and the second, the Greek frog, is endemic to the Balkan Peninsula. Representatives of these species also inhabit the immediate vicinity of the Vilenjska vrela spring [57].

Because of their features, primarily isolation and fragmentation, spring habitats and the organisms that inhabit them, are particularly vulnerable to environmental changes caused by various anthropogenic pressures [82]. One of the main stressors in the spring habitats of Bosnia and Herzegovina is the change of hydro-morphological features due to catchment for water supply purposes or their immersion by the construction of dams and reservoirs. A striking example this is the immersion of the springs of the Rama River (Buk and Krupić) after the formation of the hydro-accumulation of Rama in the 1970s. These springs were habitats of the endemic species *Drusus ramae*, whose populations disappeared after immersion, and the only known sites of this species to date are the two springs of the Lištica River [53]. Water pollution is also a significant stressor in spring ecosystems, especially pollution caused by the leaching of nutrients from the catchment area. Increased concentrations of nutrients are especially present in the springs of the karst area, which is characterized by high permeability of the substrate and low ability of autopurification [82].

In addition to hydro-morphological changes, as well as various chemical and biological pollutants, the springs are considered habitats particularly sensitive to climate change [85]. The predicted global warming will lead to further fragmentation, and even the disappearance of certain aquatic habitats, and the organisms that live in them will face numerous, multiple pressures [82, 86]. According to Hershkovitz et al. [87], stenoendemic and cold stenothermic species with specific physiological requirements will be particularly vulnerable to the predicted climate change, and primarily organisms inhabiting the crenal and epirhithral habitats. The spring habitats of Bosnia and Herzegovina are inhabited by several endemic species that could be endangered by global warming, especially species of the genus *Drusus* from the *bosnicus* group [53, 85]. All eight species from this group are (micro)-endemic, among which six species (*D. bosnicus* Klapálek, 1899, *D. crenophylax, D. medianus* Marinković-Gospodnetić, 1976, *D. radovanovici, D. ramae, D. septentrionis*) are endemic to Bosnia and Herzegovina, while *D. klapaleki* Marinković-Gospodnetić, 1971 and *D. vespertinus* Marinković-Gospodnetić, 1976 are also known from spring habitats in the border area with Montenegro and Croatia, respectively. The range of these species is limited to a very small number of source habitats within the same mountain range, primarily in the eucrenal and hypocrenal zones. According to the "climate change vulnerability score" – CCVS, which includes six autecological traits that are known to be associated with vulnerability to climate change: endemism, micro-endemism, temperature preference, altitudinal preference, stream zonation preference, and life history [87, 88], the endemic species of the genus *Drusus* from the *bosnicus* group are estimated as highly vulnerable (CCVS $\geq$ 4) [82, 85].

Despite their high biodiversity and sensitivity to anthropogenic pressures, initiatives aimed at preserving the spring ecosystems are rare [82, 89]. Spring ecosystems are not adequately covered by EU environmental legislation: The Water Framework Directive (Council Directive 2000/60/EC), or The Habitats Directive (Council Directive 92/43/EEC). Therefore, spring habitats and aquatic ecosystems in general, are less represented within the Natura 2000 ecological network [87, 90]. The only spring type that is under protection as a habitat type are Petrifying springs with tufa formation (*Cratoneurion*). This type of source is extremely sensitive to temperature increase and nutrient influx, and is recognized as a priority habitat type in Annex 2 of the Habitats Directive (Natura code 7220).

In order to preserve species diversity of rare and endemic taxa in spring ecosystems which are typical by their high heterogeneity, systematic research and conservation of springs as a type of habitats is required in the future.

Acknowledgements This publication was partly supported by the Ministry of Education and Science of Federation Bosnia and Herzegovina (*The explorations of biocenoses and ecological integrity of the Dinaric streams and rivers*) and Ministry of Scientific and Technological Development, Higher Education and Information Society of Republic of Srpska (*Project No 19.032/961-104/19*).

References

1. Dakić E (2020) The impact of climate change on water resources in Bosnia and Herzegovina. Electron Collection Papers Faculty Civil Eng 10(19):30–38
2. Hadžić E, Imamović A (2019) Water resources in Bosnia and Herzegovina. In: Negm AM, Romanescu G, Zelenakova M (eds) Water resources management in Balkan Countries. Springer Water, Springer Nature Switzerland AG, Switzerland, pp 177–210
3. Dedić A, Plenković-Moraj A, Borojević- Kralj K, Hafner D (2015) The first report on periphytic diatoms on artificial and natural substrate in the karstic spring Bunica Bosnia and Herzegovina. Acta Bot Croat 74(2):393–406
4. Cantonati M, Spitale D (2009) The role of environmental variables in structuring epiphytic and epilithic diatom assemblages in springs and streams of the Dolomiti Bellunesi National Park (south-eastern Alps). Fundam Appl Limnol 174(2):117–133
5. Mašić E (2020) Bibliography of phycological research in Bosnia and Herzegovina (1849–2019). Phytologia Balcanica 26(3):437–443
6. Pešić V, Dmitrović D, Savić A, von Fumetti S (2016) Studies on eucrenal-hypocrenal zonation of springs along the river mainstream: A case study of a karst canyon in Bosnia and Herzegovina. Biologia 71:809–817
7. Slišković I (2014) Vode u kršu slivova Neretve i Cetine. Hrvatski geološki institut, Zavod za hidrogeologiju i inženjersku geologiju, Zagreb
8. Jovanović R, Avdagić I (1981) Some matters concerning the development and hydrogeological function of Karst polje. Naš krš 6(10–11):55–78
9. Lučić I (2012) Krš Bosne i Hercegovine: Malo znamo al je znano. In: Zukanović A (eds) Krš i pećine Bosne i Hercegovine. Centar za krš i speleologiju, Sarajevo, Bosnia and Herzegovina, pp 8–15
10. Levin SA (1976) Population dynamic models in heterogeneous environments. Annu Rev Ecol Syst 7:287–310

11. Steinmann P (1915) Praktikum der Süßwasserbiologie. Tail 1: Die Organismen des fließenden Wassers. Borntraeger, Berlin
12. Thienemann A (1924) Hydrobiologische Untersuchungen an Quellen. Arch Hydrobiol 14:151–189
13. Reiss M, Chifflard P (2017) An opinion on spring habitats within the earth's critical zone in headwater regions. Water 9(9):645
14. Di Sabatino A, Gerecke R, D'Alfonso S, Cicolani B (1997) Prime considerazioni sulla biodiversità delle sorgenti italiane: la taxocenosi ad acari acquatici (Acari, Actinedida, Hydrachnidia). Atti Società Italiana di Ecologia 18:171–174
15. Cantonati M, Gerecke R, Bertuzzi E (2006) Springs of the Alps–sensitive ecosystems to environmental change: from biodiversity assessments to long-term studies. Hydrobiologia 562:59–96
16. Blagojević S (1974) Razvoj perifitona u otvorenim uređajima vodovoda na krškim vrelima. Prirodoslovno-matematički fakultet, Sveučilište u Zagrebu, Zagreb, Doktorska disertacija (PhD-Thesis)
17. Cantonati M, Bertuzzi E, Spitale D (2007) The spring habitat: biota and sampling methods. Museo Tridentino di Scienze Naturali, Trento. Monografie del Museo Tridentino di Scienze Naturali, 4
18. Kamberović J (2015) Struktura i sezonska dinamika kreonskih zajednica algi i vodenog bilja na području planine Konjuh. Univerzitet u Tuzli, Tuzla, Doktorska disertacija (PhD-Thesis)
19. Kamberović J, Plenković-Moraj A, Kralj Borojević K, Gligora Udovič M, Žutinić P, Hafner D, Cantonati M (2019) Algal assemblages in springs of different lithologies (ophiolites vs. limestone) of the Konjuh Mountain (Bosnia and Herzegovina). Acta Bot Croat 78(1):66–81
20. Kamberović J (2020) Ekosistemi izvora planine Konjuh. OFF-SET Štamparija Tuzla.
21. Štambuk Giljanović N (1998) Neretva. Dubrovnik 9(4):311–372
22. Fenessy MS, Cronk JK (1997) The effectiveness and restoration potential of riparian ecotones for the management of nonpoint source pollution, particularly nitrate. Crit Rev Environ Sci Technol 27(4):285–317
23. Bratovcic A, Petrinic I (2020) Quality assessment and health safety of natural spring water. Technol Acta 13(1):33–40
24. Werum M (2001) Die Kieselalgen Gesellschaften in Quellen: Abhängigkeit von Geologie und anthropogener Beeinfl ussung in Hessen (Bundesrepublik Deutschland). Hessische Landesamt für Umwelt und Geologie, Wiesbaden
25. Škobić D, Šljivić A (2015) Usporedba fizikalno-kemijskih svojstava krških rijeka i utjecaj onečišćenja. In: 2nd Scientific and expert conference with international participation "Karst water management and protection" 15-16 Oct 2015. Mostar, Bosnia and Herzegovina, p 65
26. Milanović PT (2006) Karst istočne Hercegovine i Dubrovačkog priobalja. Asocijacija speleoloških organizacija Srbije, Srbija
27. Kummer A, Sendtner O (1849) Enumerationplantarum in itinere Sedtneriano in Bosnia lectorumcum, definitionibus novarum specearum etadumbrationibus obscurum varietatumque. Flora 32(1):1–10
28. Blagojević S (1966) Prilog poznavanju algi kraških izvorišta u Bosni i Hercegovini. I. Chrysophyceae, Xanthophyceae, Bacillariophyceae. God Biol Inst Univ Sarajevo 29:5–22
29. Blagojević S (1976) Prilog poznavanju cijanoficeja (Cyanophyceae) krških izvorišta. Acta Bot Croat 35:207–215
30. Blagojević S (1979) Prilog poznavanju algi kraških izvorišta u Bosna i Hercegovini. II Kloroficeje (Chlorophyceae). Acta Bot Croat 38:125–132
31. Hafner D (2009) Dijatomeje krških izvora u Hercegovini. Radovi Hrvatskog Društva za Znanost i Umjetnost XI:209-220
32. Hafner D, Jasprica N, Lončar A (2010) Cyanobacteria of the karstic springs in Herzegovina (Bosnia and Herzegovina). Zbornik radova Trećeg hrvatskog botaničkog kongresa. Hrvatsko botaničko društvo, Murter-Zagreb, p 80
33. Trožić-Borovac S, Hafner D (2004) Fitobentos i zoobentos hidroekosistema šireg područja Vrela Bosne u ocjeni kvaliteta vode. Voda i mi 37:18–26

34. Lončar A, Kapetanović T, Hafner D (2011) Knjiga radova. Dijatomeje vlažnih staništa planine Ozren, Bosna i Hercegovina. Međunarodni naučni skup „Zaštita prirode u XXI vijeku" Žabljak, Crna Gora, pp 88–97
35. Dedić A, Hafner D, Jasprica N, Šakić D, Grizelj Z (2014) Algal composition of two karstic springs in western Herzegovina (Bosnia and Herzegovina). Glas Zem Muz BiH Sarajevo 34:41–50
36. Dedić A (2015) Dinamika naseljavanja perifitonskih dijatomeja u krškim izvorima Bosne i Hercegovine. Prirodoslovno-matematički fakultet, Sveučilište u Zagrebu, Zagreb, Doktorska disertacija (PhD-Thesis)
37. Hafner D, Dedić A (2020) Dijatomeje slijeva Jadranskog mora u Bosni i Hercegovini/Diatoms of the Adriatic Sea basin in Bosnia and Herzegovina. Pressum, Mostar
38. Kapetanović T, Hafner D (2007) Diatoms of wet habitats in the subalpine belt of Mt. Vranica (Bosnia and Herzegovina). In: Kusber WH, Jahn R (eds), Proceedings of the 1st Central European Diatom Meeting 2007 Berlin-Dahlem. BGBM Press, Berlin, pp 73–78
39. Mašić E, Barudanović S, Žero S, Ramić E, Macanović A, Boškailo A, Fejzić S (2019) Diversity of diatoms in freshwater oligotrophic habitat types on Vranica Mountain (Bosnia and Herzegovina). Eur J Phycol 54 (128):1–109
40. Kamberović J, Kišić A, Hafner D, Plenković-Moraj A (2016) Comparative analysis of epilithic diatom assemblages of springs and streams in the Konjuh Mountain (Bosnia and Herzegovina). Works Fac Forest Univ Sarajevo 2:54–67
41. Lange-Bertalot H, Steindorf A (1996) Rote Liste der limnischen Kieselalgen (Bacillariophyceae) Deutschlands. Schr Vegetationsk 28:633–677
42. Dedić A, Stanić-Koštroman S, Đolo S, Lasić A, Škobić D (2019) Preliminary study of trophic relation between diatoms and endemic species *Drusus ramae* Marinković-Gospodnetić (1970) (Insecta: Trichoptera) at the Lištica spring, Bosnia and Herzegovina. In: The proceedings of the ISEM8 2–5 Oct 2019, Budva, Montenegro, p 77
43. Lasić A, Jasprica N (2016) Vegetation diversity of the two Dinaric karstic rivers in Bosnia and Herzegovina. Biologia 71(7):777–792
44. Weekes L, Kącki Z, FitzPatrick Ú, Kelly F, Matson R, Kelly-Quinn M (2018) An Irish national vegetation classification system for aquatic river macrophytes. Appl Veg Sci 21(2):322–340
45. Grego J, Glöer P (2019) A New hydrobiid (Mollusca, Gastropoda, Hydrobiidae) genus of Bosnia and Hercegovina. Ecol Monten 21:86–89
46. Dmitrović D, Savić A, Pešić V (2016) Discharge, substrate type and temperature as factors affecting gastropod assemblages in springs in northwestern Bosnia and Herzegovina. Arch Biol Sci 68(3):613–621
47. Savić A, Dmitrović D, Glöer P, Pešić V (2020) Assessing environmental response of gastropod species in karst springs: what species response curves say us about niche characteristic and extinction risk? Biodivers Conserv 29:695–708
48. Dmitrović D, Pešić V (2020) An updated checklist of leeches (Annelida: Hirudinea) from Bosnia and Herzegovina. Ecol Monten 29:10–19
49. Karaman GS (2010) The current approach to the fauna Amphipoda (Crustaceae) in Bosnia and Herzegovina (contribution to the knowledge of Amphipoda 250). In: Berberović LJ, Redžić S (eds) Simpozijum-Panel „Darvin danas" povodom 200-godišnjice rođenja i 150-godišnjice "Postanka vrsta", Zbornik radova, November 2009. Special editions, vol 129, Odjeljenje prirodnih i matematičkih nauka, vol 17. Akademija nauka i umjetnosti Bosne i Hercegovine, Sarajevo, pp 17–28
50. Trožić-Borovac S, Bašić L, Hafner D, Mušović A (2012) Endemic amphipod *Niphargus ilidzensis* Shäfferna 1922, (Crustacea: Amphipoda) within the protected area of canton Sarajevo. Work Fac For 2:73–87
51. Čepić V, Marinković-Gospodnetić M (1978) Zoobentos rijeke Toplice. God Biol Inst Univ Sarajevo 31:5–32
52. Kučinić M, Previšić A, Mihoci I, Ćuk R, Delić A, Žganec K, Cerjanec D, Vučković I (2014) The first finding of the *Drusus bosnicus* group (Insecta, Trichoptera, Limnephilidae) in Croatia with some notes on diversity, taxonomy, distribution and ecology of genus *Drusus* in Croatia and in Dinaric karst of the Balkan Peninsula. Nat Croat 23(2):365–377

53. Stanić-Koštroman S (2009) Faunističke, ekološke i biogeografske značajke tulara (Insecta, Trichoptera) Bosne i Hercegovine. Prirodoslovno-matematički fakultet, Sveučilište u Zagrebu, Zagreb, Doktorska disertacija (PhD-Thesis)
54. Krek S (1999) Psychodidae (Diptera, Insecta) Balkanskog poluotoka. Ministarstvo obrazovanja, nauke, kulture i sporta, Studentska štamparija Univerziteta Sarajevo, Sarajevo
55. Kaćanski D (1987) O naselju endemičnih Plecoptera (Insecta) u tekućim vodama Bosne i Hercegovine. In: Vuković T (ed) Naučni skup – Zaštita endema u živom svijetu Jugoslavije, May 1986. Posebna izdanja, knjiga 83, Odjeljenje prirodnih i matematičkih nauka, knjiga 14. Akademija nauka i umjetnosti Bosne i Hercegovine, Sarajevo, pp 33–38
56. Graf W, Murphy J, Dahl J, Zamora-Muñoz C, López-Rodríguez MJ (2008) Distribution and ecological preferences of european freshwater organisms, vol 1, Trichoptera. Pensoft, Sofia–Moscow
57. Dmitrović D (2017) Makrozoobentos odabranih krenona sliva rijeke Cvrcke. Prirodno-matematički fakultet Univerziteta u Banjoj Luci, Doktorska disertacija (PhD-Thesis)
58. Işik K (2011) Rare and endemic species: why are they prone to extinction? Turk J Bot 35(4):411–417
59. Glöer P, Pešić V (2014) *Belgrandiella bozidarcurcici* n. sp., a new species from Bosnia and Herzegovina (Gastropoda: Hydrobiidae). Arch Biol Sci 66(2):461–464
60. Radoman P (1983) Hydrobioidea a superfamily of Prosobranchia (Gastropoda) I Sistematics. In: Pantić V (ed) Monographs, vol 547, Department of Sciences, no 57. Serbian Academy of Sciences and Arts, Beograd, p 256
61. Grosser C, Pešić V, Dmitrović D (2014) *Dina sketi* n. sp., a new erpobdellid leech (Hirudinida: Erpobdellidae) from Bosnia and Herzegovina. Zootaxa 3793(3):393–397
62. Sket B (1985) *Piscicola hadzii* sp. n. (Piscicolidae, Hirudinea), a probably endemic species of leeches from Hercegovina, Yugoslavia. Biol Vestn 33(2):89–94
63. Kaćanski D, Zwick P (1970) Neue und wenig bekannte Plecopteren aus Jugoslawien. Mitt Schweiz Entomol Ges 43(1):1–16
64. Krek S (1973) Ekološka klasifikacija i cenotički odnosi Psychodidae u tekućicama jugo-istočne Bosne. God Biol Inst Univ Sarajevo 26:57–95
65. Vitecek S, Previšić A, Kučinić M, Bálint M, Keresztes L, Waringer J, Pauls SU, Malicky H, Graf W (2015) Description of a new species of *Wormaldia* from Sardinia and a new *Drusus* species from the Western Balkans (Trichoptera, Philopotamidae, Limnephilidae). Zookeys 496:85–103
66. Marinković-Gospodnetić M (1970) Descriptions of some species of Trichoptera from Yugoslavia. God Biol Inst Univ Sarajevo 23:77–84
67. Kučinić M, Previšić A, Graf W, Šerić Jelaska L, Stanić-Koštroman S, Waringer J (2011) Larval description, genetic and ecological features of *Drusus radovanovici radovanovici* Marinković-Gospodnetić, 1971 (Trichoptera, Limnephilidae) with some phylogenetic and taxonomic data on the *bosnicus* group in the Balkan Peninsula. Dtsch Entomol Z 58(1):135–153
68. Kučinić M, Previšić A, Stanić-Koštroman S, Franjević M, Šerić Jelaska L, Delić A, Posilović H (2010) Description of the larvae of *Drusus ramae* Marinković-Gospodnetić and *Drusus medianus* Marinković-Gospodnetić (Trichoptera: Limnephilidae) with some genetic, distributional, ecological, faunal and conservation notes. Zootaxa 2484(1):1–24
69. Marinković-Gospodnetić M (1976) The differentiation of *Drusus* species of the group *bosnicus*. In: Malicky H (ed) Proceedings of the 1th International symposium on Trichoptera. Junk, pp 77–85
70. Kučinić M, Previšić A, Gottstein S, Hrašovec B, Stanić-Koštroman S, Pernek M, Delić A (2008) Description of the larvae of *Drusus radovanovici septentrionis* Marinkovic-Gospodnetic, 1976 and *Drusus croaticus* Marinkovic-Gospodnetic, 1971 (Trichoptera: Limnephilidae) from Bosnia and Herzegovina, and Croatia. Zootaxa 1783:1–17
71. Grego J (2020) Revision of the stygobiont gastropod genera *Plagigeyeria* Tomlin, 1930 and *Travunijana* Grego & Glöer, 2019 (Mollusca; Gastropoda; Moitessieriidae and Hydrobiidae) in Hercegovina and adjacent regions. Eur J Taxon 691:1–56
72. Boeters HD, Glöer P, Pešić V (2013) Some new freshwater gastropods from southern Europe (Mollusca: Gastropoda: Truncatelloidea). Folia Malacol 21(4):225–235

73. Glöer P, Grego J (2015) New subterranean freshwater Molluscs from Bosnia and Hercegovina (Mollusca: Hydrobiidae). Ecol Monten 2(4):307–314
74. Karaman GS (2014) Further studies of genus *Niphargus* Schiödte, 1849 (fam. Niphargidae) from western Balkan peninsula (Contribution to the knowledge of the Amphipoda 274). Agricult Forest 60(1):203–220
75. Williams DD, Danks HV, Smith IM, Ring RA, Cannings RA (1990) Freshwater springs: a national heritage. Entomol Soc Can Suppl Bull 22:9
76. Williams DD, Danks HV (1991) Arthropods of springs: Introduction. Mem Entomol Soc Can 155:3–5
77. Botosaneanu L (1998) Studies in crenobiology: the Biology of springs and springbrooks. Backhuys Publishers, Leiden
78. Botosaneanu L (1995) Springs as refugia for geographic relicts. Crunoecia 4:5–9
79. Barquín J, Death RG (2004) Patterns of invertebrate diversity in streams and freshwater springs in Northern Spain. Archiv für Hydrobiologie 161:329–349
80. Barquín J, Death RG (2006) Spatial patterns of macroinvertebrate diversity in New Zealand springbrooks and streams. J N Am Benthol Soc 25:768-786
81. Barquín J, Scarsbrook M (2008) Management and conservation strategies for coldwater springs. Aquat Conserv Mar Freshwat Ecosyst 18(5):580–591
82. Herceg N, Stanić-Koštroman S, Šiljeg M (2018) Čovjek i okoliš / The man and the environment. Sveučilište Sjever, Hrvatska akademija za znanost i umjetnost Bosne i Hercegovine, Synopsis. Koprivnica-Mostar-Zagreb
83. Škondrić S, Šumatić N, Perić R, Filipović S, Dmitrović D, Šukalo G (2013) New record of endemic *Symphyandra hofmannii* Pant. (1881) (Campanulaceae) in Cvrcka gorge (Bosnia and Herzegovina). In: Govedar Z, Dukić V (eds) International Scientific Conference – "Forestry science and practice for the purpose of sustainable development of forestry", 20 years of the Faculty of Forestry in Banja Luka, Proceedings, November 2012. University of Banja Luka, Faculty of Forestry, Banja Luka, pp 197–200
84. Šukalo G, Dmitrović D, Filipović S, Kovačević M, Đorđević S, Lj T (2015) New findings of the Greek Frog, *Rana graeca* Boulenger, 1891 (Anura: Ranidae) in the north-western Bosnia and Herzegovina. Ecol Monten 2(2):74–77
85. Stanić-Koštroman S, Kučinić M, Herceg N, Čordaš D, Durbešić P (2015) Caddisflies (Insecta, Trichoptera) as indicators of climate changes and ecological status in karst water ecosystems. In: 2nd Scientific and expert conference with international participation "Karst Water Management and Protection" 15-16 Oct 2015. Mostar, Bosnia and Herzegovina
86. Woodward G, Perkins DM, Brow LE (2010) Climate change and freshwater ecosystems: impacts across multiple levels of organization. Phil Trans R Soc B Biol Sci 365:2093–2106
87. Hershkovitz Y, Dahm V, Lorenz AW, Hering D (2015) A multi-trait approach for the identification and protection of European freshwater species that are potentially vulnerable to the impacts of climate change. Ecol Ind 50:150–160
88. Schmidt-Kloiber A, Hering D (2015) www.freshwaterecology.info - an online tool that unifies, standardises and codifies more than 20,000 European freshwater organisms and their ecological preferences. Ecol Ind 53:271–282
89. Hotzy R (2007) Spring protection in Bavaria. In: Cantonati M, Bertuzzi E, Spitale D (eds) The spring habitat: biota and sampling methods. Monografie del, Trento, Italy, pp 317–326
90. Cardoso P (2012) Habitats Directive species lists: urgent need of revision. Insect Conserv Divers 5(2):169–174

Chapter 8
Algae in Shallow and Small Water Bodies of Serbia: A Frame for Species and Habitat Protection

Ivana Trbojević and Dragana Predojević

Abstract The shallow aquatic ecosystems of Serbia are treasuries of algal biodiversity, unfairly neglected in both scientific studies and legislation concerning species and habitat protection. Underestimating the algal diversity in these ecosystems derives primarily from poor knowledge and/or interest in algal taxonomy and phylogeny, as well as ignorance of their role in the maintenance of ecosystem equilibrium and in the bioindication of water quality. The issue of conserving algae (i.e. their habitats) is challenging due to many aspects, including the problematic biogeography concept (particularly for microalgae), low taxonomic resolution of available data, and undersampling. Still, progress in the conservation of algae is noticeable worldwide. Although macroalgae are mainly recognized as endangered and protected species, they are still overlooked in conservation management in Serbia. Simultaneously, the data on microalgal diversity in shallow and small water bodies are scarce and sporadic. There are no long term monitoring programs towards recognizing the remarkable algal diversity characteristic for these habitats. This chapter offers an overview of the biodiversity of algae—both microscopic and macroscopic in shallow and small water bodies of Serbia, along with the frame and guidelines for protecting algae and their habitats.

Keywords Microalgae · Macroalgae · Diversity · Protection · Conservation

8.1 Introduction

Shallow lakes and ponds are a prevalent type of freshwater ecosystems worldwide [1], but limnological and algological studies have been traditionally focused on deep stratified lakes [2]. In recent decades, freshwater scientist's interest has started to broaden due to recognition of the significant role of shallow freshwater ecosystems

I. Trbojević (✉) · D. Predojević
Faculty of Biology, University of Belgrade, Studentski trg 16, 11000 Belgrade, Serbia
e-mail: itrbojevic@bio.bg.ac.rs

D. Predojević
e-mail: d.predojevic@bio.bg.ac.rs

V. Pešić et al. (eds.), *Small Water Bodies of the Western Balkans*, Springer Water,
https://doi.org/10.1007/978-3-030-86478-1_8

in the vital processes of nutrient cycling, carbon sequestration, removal of different pollutants, mitigation of the adverse effects of climate change, and providing various and essential ecosystem functions and services. This is why these ecosystems are referred to as "the kidneys of our planet" [3] as well as "biological supermarkets" [4]. Bearing in mind that they are very sensitive and vulnerable, the significance of shallow aquatic ecosystems for biodiversity and conservation have gained importance as scientific data has accumulated [2, 5]. However, compared to the other topics in small water habitats [6, 7], algological studies still seem to straggle behind.

Small water bodies (SWB) are increasingly represented in freshwater science, particularly attracting attention to the importance of these ecosystems for biodiversity and ecosystem services [8]. Although there is growing evidence for the importance of these ecosystems (as well as a gap in knowledge about them), the term SWB is still considered ambiguous. Lacking a universal/legal definition, we support the proposal of Biggs et al. [8] for SWB characterization that refers to ponds and small lakes, small streams including headwaters, ditches, and springs. However, we suggest also other SWB types should be covered in the SWB definitions, such as wet meadows and flood zones of small rivers.

In this chapter, in analyzing SWB and reviewing algological studies in Serbia, we have considered a wider ecological spectrum including both shallow and small water bodies (SSWB). We decided to refer to this wider concept due to various reasons, including the lack of precise definitions and data for waterbody classification in literature (such as the total lack of locality descriptions, no GPS coordinates, and imprecise descriptions). We found it appropriate to treat shallow and small water bodies as one group, regardless of the surface area, which we were not able to determine in a vast number of cases. Our aim was to cover all ecosystems truly and potentially qualifying for the SWB category.

The competition of primary producers in SSWB selected in this overview is extraordinarily strong and complex. Bearing in mind that the littoral zone is wider than the pelagic zone, macrophytic vegetation and benthic communities commonly prevailed. In addition, methaphyton as a specific assemblage of densely packed relatively large and filamentous algal representatives is developed among macrophytic vegetation (Fig. 8.1). The phytoplankton community is composed of many tychoplanktonic elements (benthic, epiphytic, and methaphytic), and it is usually less abundant, except in the situations when algal blooms appeared (sometimes as Harmful Algal Blooms—HABs, if toxic cyanobacteria form a bloom). According to the theory of alternative stable states, SSWB can be found in two alternative equilibrium states along the nutrient concentration gradient. The first state is clear water state, with well-developed macrophytic vegetation, and low phytoplankton biomass, while the second—turbid state implies densely developed phytoplankton and the absence or only slight development of macrophytic vegetation [9]. The clear state is preferable since it supports a higher biodiversity of both planktonic algae and macrophytic vegetation [9].

Considering that the waterbodies corresponding to the frames of SSWB are mainly situated in the Serbian Northern Province Vojvodina, the majority of relevant algological studies took place in this area. Thus, in the following review, we are going

Fig. 8.1 Methaphyton community (labeled with yellow arrows) characteristic for SSWB. Photo by D. Predojević

to cover the regions of Vojvodina in detail, including the Mačva and the Belgrade region herein (which geographically represents lowlands), and Central and Southern Serbia will be treated as one region (Fig. 8.2). We are going to present the overview and analyze the diversity and the distribution of algae in SSWB. We stress out that the following review of algological studies is the selection made according to the estimated relevance to the chapter subject, and not a comprehensive review of all algological studies which have ever been conducted in Serbia.

8.2 Diversity of Algae in Shallow/Small Water Bodies of Serbia

8.2.1 Microalgae

Microalgae are an extremely heterogeneous group of primarily photoautotrophic organisms belonging to different kingdoms. They may have a single-celled, colonial, or multicellular morphological organization of the talus, but what they have in common is that they are invisible to the naked eye. They can be members of plankton, the most frequently researched community in SSWB, but may also be found in periphyton and benthos, communities that, as a rule, dominate in this type of waterbody, but are far less researched. Additionally, we emphasize that microalgae in this

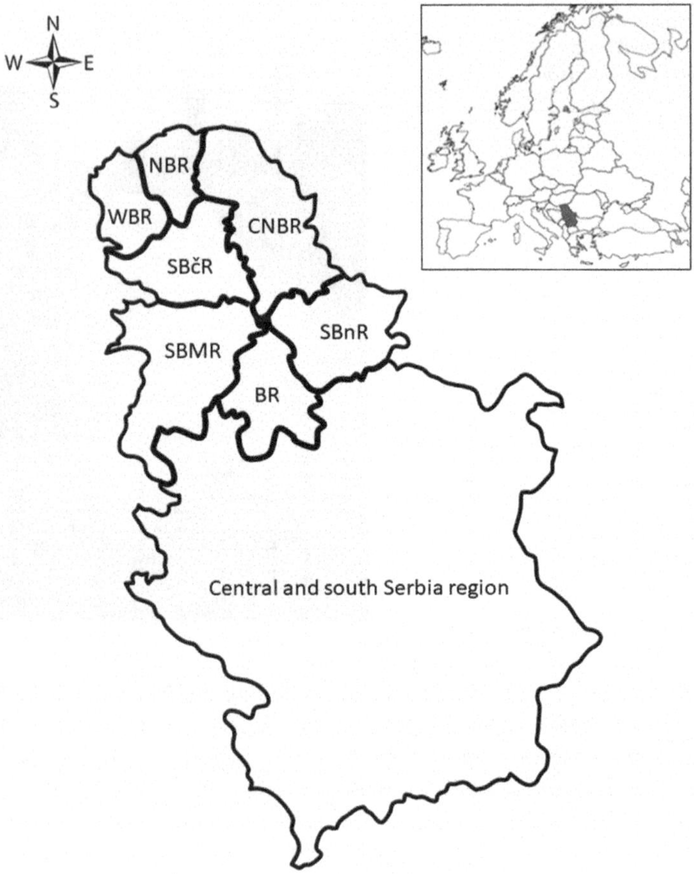

Fig. 8.2 The map of Serbia, showing the position in Europe and the regions covered with the review of algological studies in this chapter, Sect. 8.2. NBR—North Bačka region, WBR—West Bačka region, SBčR—South Bačka region, SBMR—Srem and Mačva region, BR—Belgrade region, SBnR—South Banat region, CNBR—Central and north Banat region

review will be considered representatives of divisions according to the classification proposed by Reynolds [10].

8.2.1.1 Microalgae in Vojvodina

In the **North Bačka region** (Fig. 8.2), the lakes Palić, Ludaš, Krvavo, and Zobnatica accumulation have long been the subject of phytoplankton research, especially Lake Palić, where surveys date back to 1781 [11]. Only the latter water body is not under the protection regime. Lakes Palić and Ludaš are aeolian lakes that represent probably the most investigated shallow aquatic ecosystems in Serbia. Although these lakes cannot be considered small water bodies, they found a place in this diversity review due to their shallowness and long-term algological exploration. Long-term research has left a lot of data behind. However, the phytoplankton studies lasted for a different period of time, covering a different number of seasons. Hence, the number of identified taxa, in that case, is not a convenient indicator of diversity that can be compared. On the other hand, many published papers deal with combined investigation results of some or all mentioned lakes, even some outside this region [12, 13]. All these obstacles make it challenging to present the diversity of these four water bodies. It can be observed that Lake Ludaš has the greatest taxa richness among them (e.g. 190 species were recorded from 1970 to 1981 [14, 15]); in general, green algae dominate in all ecosystems, followed by cyanobacteria, diatoms, and euglenoid. Sometimes cyanobacteria take over, and this usually happens during the summer months. Thus, from 1981 to 1990, in the Zobnatica accumulation, only 40 phytoplankton taxa were detected, and cyanobacteria were the dominant group, with the most numerous and blooming *Aphanizomenon flos-aquae* Ralfs ex Bornet & Flahault and *Anabaena spiroides* Klebahn [16]. In Lake Palić, the usual richness is about 100 taxa, with a maximum of 180 taxa recorded after the revitalisation process when the bloom of *Microcystis aeruginosa* (Kützing) Kützing and *A. flos-aquae* were recorded at the same time [17]. The mentioned cyanobacteria were dominant in this lake until 2012 when *Raphidiopsis raciborskii* (Woloszynska) Aguilera, Berrendero Gómez, Kastovsky, Echenique & Salerno became the dominant summer species that lead to blooming [18]. Bloom of *R. raciborskii* was also noted in 2014, when another invasive cyanobacterium *Sphaerospermopsis aphanizomenoides* (Forti) Zapomelová, Jezberová, Hrouzek, Hisem, Reháková & Komárková appeared as an accompanying species, along with some other potentially toxic and blooming cyanobacteria [18]. The same situation happened in Kravavo Lake that year [18]. Cyanobacteria in Lake Ludaš were present and dominant, especially in the summer months, ever since the first algological research in 1970. *M. aeruginosa* was responsible for the first detected cyanobacterial bloom [14]. The complete list of the cyanobacteria that appeared and bloomed from 1970 to 2015 is shown in Tokodi et al. [19]. In the research during 2011 and 2012, *Limnothrix redekei* (Goor) Meffert, *Pseudanabaena limnetica* (Lemmermann) Komárek, *Planktothrix agardhii* (Gomont) Anagnostidis & Komárek, and *Microcystis* spp. were dominant [19], while in the period from 2013 to 2017, cyanobacterial species that were also frequently found were *Microcystis delicatissima* (West & G.S.West) Starmach, *P. agardhii*, *Oscillatoria putrida* Schmidle, *Planktolyngbya limnetica* (Lemmermann) Komárková-Legnerová & Cronberg, and *A. spiroides* [20]. Furthermore, an investigation in 2018 revealed the dominance of *L. redekei* and *P. limnetica* with also numerous *M. aeruginosa* and *Microcystis wesenbergii* (Komárek) Komárek ex Komárek populations as well. *R. raciborskii* was noticed in autumn, and this cyanobacterium was firstly detected in this lake in 2013 [20].

Horgoš peat bog is another specific shallow water body situated in this region. This is a unique lowland peat bog, at 75 m a.s.l, apparently under strong anthropogenic pressure, as in the peat bog's nearby even a fishpond could be created, due to long term peat exploitation [21]. The lack of coordinates in the literature sources [22] prevented precise locating of this habitat since many similar locations are situated in the area of the Subotičko-Horgoška sands. Šovran et al. [21] report the results of the first algological survey of this specific habitat, providing the list of planktonic desmid algae. In total, they report 30 taxa from 5 genera—*Closterium*, *Cosmarium*, *Cosmocladium*, *Gonatozygon* and *Staurastrum*, mainly meso and eutrophic, related to the neutral to alkaline habitats. Among others, two taxa are found only in this habitat in Serbia—*Cosmarium zonatum* Lundell and *Cosmocladium saxonicum* De Bary.

The only known phytoplankton studies in **the West Bačka region** (Fig. 8.2) were those conducted as part of the diversity investigation of SNR "Gornje Podunavlje" (from 1996 to 1998 at 14 localities without stating their exact locations [23]), which preceded the proclamation of this reserve. Since the dynamics of the research were also different, only some general dynamic patterns of this community can be drawn. A total number of 377 taxa were recorded at the end of that period, with green algae being the most diverse, followed by diatoms. The most diverse genera among green algae were *Scenedesmus* (23 taxa), *Cosmarium* (20), *Staurastrum* (18) and *Closterium* (12), and *Navicula* (14), *Cymbella* (12), *Gomphonema*, and *Nitzschia* (both with 8 registered taxa) among diatoms. The qualitative phytoplankton composition varied depending on the location and sampling time, but some general trends could be established. Summer assemblages were the most diverse, with a large number of green algae and increased cyanobacteria taxa compared with spring and winter assemblages, when the total and the taxa richness of those groups were drastically decreased [23]. As a conclusion to the mentioned phytoplankton survey, this community is designated as characteristic for lowland aquatic ecosystems.

The phytoplankton research in this area was repeated in 2016 when 5 ponds (Semenjača, Šarkanj, Široki Rit, Ribolov, and Sakajtaš) in the Monoštorski Marsh were investigated from May to July [24]. Sakajtaš is the only pond in the flood zone of the Danube River, and it is in the third protection zone, while others are in the second, and located behind the embankment. Additionally, three ponds were exposed to the revitalisation process (excavation and removal of sludge and partial removal of macrophytic vegetation)—Semenjača in 2011, Šarkanj in 2015, while in Široki rit, the revitalisation process began at the same time as the beginning of the phytoplankton research. High taxa richness was recorded during only three months but at almost three times fewer localities than in the previous investigation (350 taxa in total) [24]. The dynamic of phytoplankton was similar to the previously recorded data for the spring and summer community, with a high number of green algae (157 taxa) and cyanobacteria (55 taxa). Also, a significant richness of euglenoid representatives (66 taxa) was detected, and, within this group of mixotrophs, the most diverse genus *Euglena* (30 taxa) was found [24]. The results of a three-month phytoplankton study showed that the qualitative and quantitative composition of ponds were the most dependent on the origin of the water supplying them. Thus, the Semenjača and Šarkanj, supplied with water from the Tisza River, were more similar than others, supplied with water from the Danube River [25]. Generally, non-revitalised ponds showed higher taxa richness than revitalised ones, but Ribolov, as the most protected pond, that is located deep in the forest, had the highest taxa richness (224 detected taxa). These results confirm that natural, well-protected ecosystems have the potential to be a hot spot of phytoplankton diversity.

Although there are many small and shallow aquatic ecosystems in the area of **South Bačka** (Fig. 8.2), phytoplankton research is known only for the Jegrička River that is a part of NP “Jegrička” proclaimed in 2003. This lowland river was the longest autochthonous watercourse of the Vojvodina Province until the middle of the last century when it was turned into a channel by digging the riverbed and joining to the DTD hydro system. Since then, it has been divided into three parts, with the central one retaining the previous characteristics the most, while the third, the part closest to the confluence, was turned into a carp fishpond [26]. The first phytoplankton research of this aquatic ecosystem, more precisely of the fishpond, was conducted in 1959 and 1960 [27]. As a result, 170 algal taxa were recorded, but the authors did not list all the species present. They only pointed out the differences between the examined years. During 1959, *Euglena lepocincloides* Drezepolski and the green alga *Closterium acerosum* Ehrenberg ex Ralfs appeared to be abundant, while during 1960, the condition changed and the bloom of *A. spiroides*, *Cylindrospermum stagnale* Bornet & Flahault, and *Oscillatoria* spp. appeared. After that, in the monograph on this ecosystem from 1996, all previous research were enumerated, and a list of 56 most common taxa was given: Chlorophyta (23 taxa), Euglenophyta (12), Bacillariophyta (12 taxa), Cyanobacteria (7), and Chrysophyta (2) [28]. Additionally, the same authors stated that the number of taxa varied depending on the locality, season, and year. In the twentieth century, phytoplankton studies were conducted during the autumn of 2006, in the central part of the Jegrička, when a total of 68 taxa were detected, with diatoms being the most diverse, followed by green, and euglenoid algae. The most numerous taxa were *Romeria leopoliensis* (Raciborski) Koczwara, *Stephanodiscus hantzschii* Grunow, *Ulnaria ulna* (Nitzsch) Compère, *Nitzschia sigmoidea* (Nitzsch) W. Smith, *Cyclotella* spp., *Synura uvella* Ehrenberg, and *Lepocinclis acus* (O. F. Müller) B. Marin & Melkonian [26].

The Zasavica swamp, as a part of the Special Nature Reserve of the same name proclaimed in 1997, is located in the northeast of **the Mačva region** (Fig. 8.2). Zasavica is specific because, during the extensive research on biodiversity that preceded the proclamation of this nature reserve, microalgae were also analyzed, which is a rare practice in nature protection in Serbia. Emphasis was placed on phytoplankton studies, so samplings were performed once a month for a year (July 1995–June 1996) at 9 localities along the Zasavica. A total of 234 taxa from 8 divisions (Cyanobacteria, Bacillariophyta, Chlorophyta, Cryptophyta, Chrysophyta, Euglenophyta, Dinophyta, and Xanthophyta) were registered [29]. Green algae and diatoms were dominant in most samples in both taxa richness and abundance [29]. Only at some localities, usually during the summer months, the abundance and richness of representatives of the Cryptophyta and Euglenophyta were very significant, which is a good indication of organic pollution [29]. After the proclamation, SNR “Zasavica” became an interesting area for biodiversity research, so more detailed algological studies continued [30–32], and new species for this reserve, but also for Serbia, were recorded [32–35]. For example, during the detailed investigation in 2012 and 2013, a total of 442 microalgae taxa were registered in the phytoplankton and metaphyton communities. It is essential to note the significant richness of representatives of the three divisions (Chrysophyta, Cyanobacteria, and Euglenophyta), although the most significant taxa richness was also in green algae and diatoms (143 and 109 taxa, respectively). Within Chrysophyta, there were 28 recorded taxa [32], of which even 16 were new to the algal flora of Serbia (e.g. *Mallomonas splendens* (G.S.West) Playfair, *Uroglena skujae* Matvienko (Fig. 8.3a, b)) [33]. Within the well-researched euglenoid algae in Serbia [36], 89 representatives of this group with 9 new taxa for the Serbian flora were recorded in the Zasavica (e. g. *Lepocinclis acicularis* Francè, *Phacus monilatus* (A.Stokes) Lemmerman (Fig. 8.3c, d)) [34], while within the Cyanobacteria among a total of 50 detected taxa, 12 of them were registered for the first time in Serbia [35].

Referring to the available literature at that time, 13 potentially toxic cyanobacterial taxa that can cause Harmful Algal Blooms (HABs) were recorded [32], and three of them are also potentially invasive algae, according to Kaštovský et al. [37]. Those are *Raphidiopsis mediterranea*, *R. raciborskii*, and *S. aphanizomenoides* (Fig. 8.4), species that increased their distribution in Serbia in the last two decades, especially in SSWB [20, 35, 38].

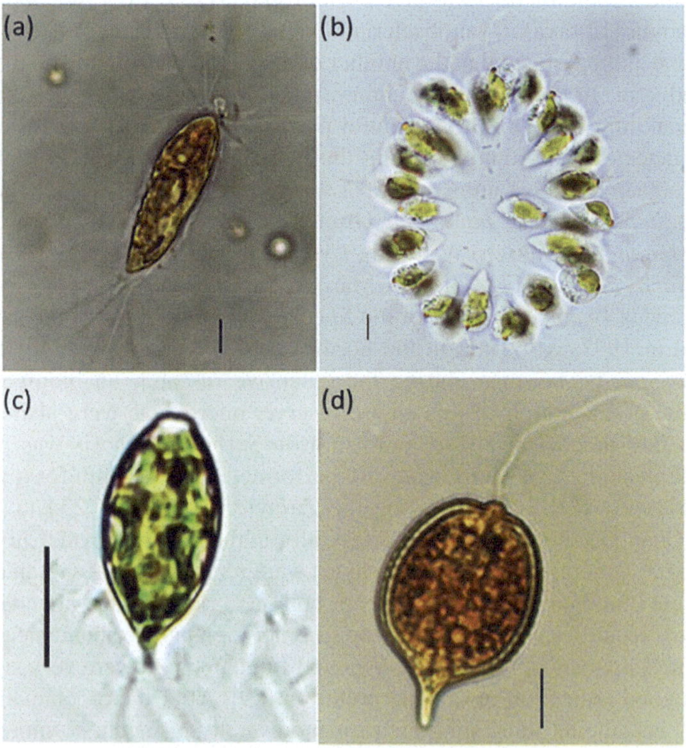

Fig. 8.3 Some new taxa for Serbian algal flora recorded in the Zasavica in 2013: **a** *Mallomonas splendens*, **b** *Uroglena skujae*, **c** *Lepocinclis acicularis* and **d** *Phacus monilatus*. Scale bar 10 μm

Fig. 8.4 Potentially toxic and invasive cyanobacteria: **a** *Raphidiopsis mediterranea*, **b** *Raphidiopsis raciborskii*, **c** *Sphaerospermopsis aphanizomenoides*. Scale bar 10 μm

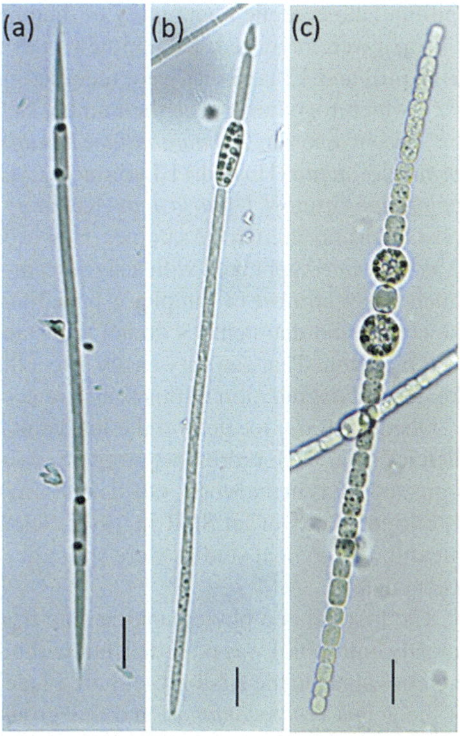

Additional research of phytoplankton and metaphyton of this SSWB was conducted during the spring and summer months of 2018 and 2019, when additional 112 new taxa for the algal flora of the Zasavica were recorded, mostly from euglenoid and green algae (40 and 35, respectively) [39]. It is important to stress that this re-analysis did not detect the presence of the three above-mentioned potentially invasive and toxic cyanobacteria, which were observed in 2013 in an extremely small number only in metaphyton [39]. That may indicate that this ecosystem with high taxa richness potentially withstood the impact of invasive planktonic algae. Including the taxa recorded during 1995 and 1996 [29], whose presence was not confirmed by late analyses [32, 33–35, 39], the total number of microalgae taxa in phytoplankton and metaphyton Zasavica detected so far is 605.

The phytobenthos in this ecosystem was examined from the aspect of diatoms because they are successfully used for water quality assessment by diatom indices. In addition to phytoplankton and metaphyton (2012–2013), the diversity of epilithic diatoms was observed, and 134 taxa were registered [32]. In the vast number of samples, *Amphora pediculus* (Kützing) Grunow, and *Planothidium frequentissimum* Lange-Bertalota appeared as dominant or subdominant taxa [32]. *Diadesmis confervacea* Kützing, known as an invasive diatom, was recorded in most samples at that time but in very low numbers. Because pebbles are a less common substrate in this ecosystem, in 2018 and 2019 the studies of diatoms were expanded, so in addition to epilithic, the epipelic community of diatoms was also examined. On that occasion, 131 taxa were recorded, with 55 new ones compared to the previous research from 2012 [40] and 12 taxa that were recorded only in plankton in the previous research [32], which together makes the number of 201 diatoms observed in the benthos in the Zasavica so far. *Achnanthidium minutissimum* (Kützing) Czarnecki, *Fragilariforma mesolepta* (Hustedt) Kharitonov, *Cyclotella meneghiniana* Kützing, *Nitzschia amphibian* Grunow, *Ulnaria acus* (Kützing) Aboal, and *U. biceps* (Kützing) Compère most often appeared in the samples. However, opposite to the previous study, invasive *D. confervacea* appeared with a significant share in the community (up to 75%) in a much larger number of samples especially in the epipelon [40]. All this points us to the conclusion that benthos do not withstand the invasion of diatoms, as is the case with plankton. That can very easily affect the future decrease of diatom richness and the general degradation of this sensitive ecosystem.

Obedska Bara, located on the left bank of the Sava River in southeastern **Srem district** (Fig. 8.2), represents an old Sava's meander. This is one of the two oldest protected areas in the world. Obedska Bara is protected since 1874 [41], and it gained its current status of an SNR in 1994. Scientific studies here have a long tradition, but still, algological studies were sporadic, and data seems to lack (particularly the recent data).

The first data on phytoplankton date from 1948 [42], when richness of 151 taxa (mostly eutrophic) was recorded, but without a detailed analysis of diatoms and filamentous algae in the Krstonošić shaft. Flagellate algae dominated, especially species of the genus *Trachelomonas*, and this group of algae had the greatest abundance in summer and autumn, while diatoms were the most numerous in spring. The research conducted in 1964 and 1965, besides the Krstonošić shaft, included the Vujića shaft also [43]. The Krstonošić shaft is deeper, with free water and under a stronger influence of the Sava River in relation to the shallow Vujić shaft. A diverse community of phytoplankton was established in the Krstonošić shaft with 260 recorded taxa, predominantly mixotrophic flagellates and green algae with many desmids during the summer and with a great richness of diatoms during the winter. The shallower Vujić shaft was characterized by a greater variety of metaphytonic elements like genera *Zygnema*, *Spirogyra*, *Oedogonium*, *Mougeotia*, *Bulbochaete* [43].

Another detailed research of phytoplankton took place in the Krstonošić shaft from 1982 to 1985 [44], when a high richness of 369 algal taxa was recorded. The most diverse were diatoms, followed by greens, but flagellate algae from the divisions Euglenophyta, Dinophyta, and Chrysophyta also had an important share, while cyanobacteria did not exceed 10% of the recorded taxa [44]. Detected representatives are characteristic for heleoplankton, with many periphytic (e.g. *Cocconeis*, *Epithemia*) and tychoplanktonic (e.g. *Oscillatoria*, *Phormidium*, *Cymatopleura*) elements [44].

At the beginning of the nineties, there were hydrological changes; the connection with the Sava River was lost, the water level dropped, and the amount of mud increased in all shafts. All studies conducted during that period [13, 45, 46] gave similar results; the patterns of phytoplankton dynamics were unchanged (mixotrophic flagellates and green algae dominate in summer and diatoms in winter), but a slight decrease in taxa richness was observed. After this period, as far as we know, phytoplankton research has not been done.

The only examination of benthic diatoms concerns epiphytic representatives, but published data about their taxa richness were presented along with the data for planktonic diatoms in summary, without separating which taxa belonged to which community. On that occasion, a total number of 151 diatom taxa were recorded, and the most numerous genera were *Navicula* and *Gomphonema* [47].

The most numerous and significant phytoplankton research in **the Belgrade region** (Fig. 8.2), the most developed and the most densely populated area in Serbia, took place in the Sava Lake, one of the most popular places for sports and recreation near the city center, which is also used for water supply. It has good water quality, considering that it is the only reservoir that has been awarded the Blue Flag in terms of quality by the Foundation for Environmental Education [48]. It is in an equilibrium state with macrophytes dominating, since they are never completely removed during the bathing season. This prevents mass blooming of cyanobacteria above all. The first phytoplankton studies date back to 1975 when a diverse community was observed [49–52] with the domination of eutrophic elements [52]. The most diverse were green algae, especially during summer, followed by diatoms in winter. In addition, short-term blooms were registered (*Stephanodiscus astrea* var. *minutula* (Kützing) Grunow in winter and *Microcystis* sp. and *Aphanizomenon* sp. in late summer) [50]. One of the last phytoplankton studies was the one conducted in 2014, after heavy floods that confirmed good water quality. The sampling was done weekly, from July to September, at four depths on one point, and a total of 172 taxa were detected [38, 48]. The most diverse were green algae (98 taxa), followed by Cyanobacteria (30 taxa—with the most numerous *M. aeruginosa* and *Aphanocapsa holsatica* (Lemmermann) G.Cronberg & Komárek), and diatoms (22 taxa) [38, 48]. A detailed analysis of the cyanobacteria present in this water body was done, and since the lake was explored, a total of 49 taxa were recorded in plankton and benthos [38]. Further, it is important to stress the first finding of the rare diatom *Stauroneis balatonis* Pantocsek in Serbia occured in the periphyton community of the Sava Lake. This diatom was previously known to exist only in the lakes Ohrid and Prespa [53].

The South Banat region (Fig. 8.2) is characterized by unprotected salt SSWB in the flood zone of the river Tamiš, shallow aquatic ecosystems (up to 1 m deep), which have often been the subject of algological research due to their specific properties of water [54–59]. They hide a specific diversity of organisms, and the most studied are Velika Slatina near the Sefkerin, and Slatina and Pečena Slatina near the Baranda village. These SSWB are fossil riverbeds of the Tamiš River, horseshoe-shaped, and their salinity and water level depend on the inflow of its water and meteorological factors that occasionaly dry them up during the summer. Phytoplankton studies took place in 2003, 2004, and 2006 in those alkaline aquatic ecosystems. In the Velika Slatina, the most diverse and numerous were diatoms, followed by cyanobacteria in 2003 and at that time, the most abundant taxon was *Navicula slesvicensis* Grunow [56]. It is alkalophilic species that inhabits the water with a pH higher than 7 and a range of chloride concentration from 500 to 1000 mg/l. In 2006, the phytoplankton composition changed, and the euglenoids became the most dominant and numerous, followed by cyanobacteria. The most dominant taxa was *Trachelomonas volvocina* Ehrenberg, known as cosmopolitan species. Fifty-six taxa were recorded in each study year in the Velika Slatina pond [56]. The number of taxa recorded in the Slatina pond was 79, while in the Pečena Slatina there were 40 taxa recorded during 2003 and 2004 [55]. The most diverse were cyanobacteria, euglenoids, and green algae. Generally, phytoplankton of such unique habitats is mainly comprised of small-sized coccoid cyanobacteria and some eukaryotic algae [60]. A lot of detected taxa in those two ponds are characteristic for salty and brackish water with increased chloride concentration (e.g. cyanobacteria—*Kamptonema chlorinum* (Kützing ex Gomont) Strunecký, Komárek & J. Smarda, *Leptolyngbya fragilis* (Gomont) Anagnostidis & Komárek, euglenoids—*Euglena agilis* Carter, *E. vermicularis* Proskina-Lavrenko, *Phacus salinus* (F. E. Fritsch) E. W.Linton & Karnkowska and diatoms—*C. meneghiniana* and *Craticula halophila* (Grunow) D.G.Mann) [54, 56].

It seems that the taxa richness in salty ponds is not so impressive, but many new records for Serbian algal flora have been made here, especially Cyanobacteria, which makes these habitats extremely significant. One of these species is the potentially toxic and invasive *R. raciborskii*, a cyanobacterium already mentioned, which was detected in the Slatina pond for the first time in Serbia in 2006, but in a very small number [59]. This trichal, unbranched cyanobacterium with a characteristic tear-like terminal heterocyst is known as a tropical and subtropical species. However, in recent years, it spreads its distribution and can be found worldwide in the temperate zone. It produces cyanotoxins, primarily of hepatotoxic cylindrospermopsin, but also neurotoxic anatoxin and saxitoxin [59]. It probably did not show its invasive character in the Slatina due to the specific conditions in which it failed to adapt [59]. Unfortunately, some ecosystems in which it was found later were not as lucky, so the summer bloom was detected in them (e.g. the Ponjavica River [61, 62], the Kapetanski rit fishpond [63, 64]). Still, some managed to resist its invasion, like the mentioned Zasavica swamp [32]. This salty pond is the place w, here the first record of *Chrysosporum minus* (Kiselev) Komárek for Serbia was made in 2006, a species that is characteristic for saline waters [57]. Another interesting cyanobacterium, *Limnospira fusiformis* (Voronichin) Nowicka-Krawczyk, Mühlsteinová & Hauer, was firstly found in the Slatina and the Pečena Slatina ponds in 2003 [58]. That was the first findings for Europe, too. This tropic species, a producer of microcystins and anatoxin-a, from salty alkaline lakes, frequently forms blooms in Africa and Asia. In Slatina, it was found sporadically, but the unialgal bloom was registered in the Pečena Slatina. The Velika Slatina pond is no exception in terms of the presence of interesting cyanobacterial taxa. *Gloeotrichia natans* f. *bucharica* Kissel, a benthic species that turns into a planktic form under favorable conditions, was detected here forming a surface bloom in 2006, when the water level was increased, and salinity was reduced [56].

The diversity of benthic diatoms from salty aquatic ecosystems in the Vojvodina Province have been investigated and summarized by Vidaković et al. [60]. Among all those unique alkaline habitats, the mentioned small water bodies of Velika Slatina, Pečena Slatina, and Slatina found their place. Authors emphasized that a total of 11

Table 8.1 First findings of benthic diatoms in Serbia from salty ponds in the South Banat

Species list	Slatina	Pečena Slatina	Velika Slatina
Caloneis permagna (Bailey) Cleve	+		+
Gyrosigma macrum (W.Smith) J. W. Griffith & Henfrey			+
Haslea spicula (Hickie) Bukhtiyarova	+		
Mastogloia elliptica (C. Agardh) Cleve	+		
Nitzschia elengantula Grunow	+		
Nitzschia incognita Legler & Krasske			+
Nitzschia pellucida Grunow			+
Nitzschia vitrea var. *salinarum* Grunow	+		
Pseudofallacia tenera (Hustedt) Y. Liu, Kociolek & Q.Wang			+
Rhopalodia constricta (Brébisson) Krammer	+		
Rhopalodia gibba var. *minuta* Krammer		+	+

taxa new to the Serbian diatom flora were recorded in the period from 2003 to 2019, just in those south-located ponds (Table 8.1). They have been found mainly in the epiphytic community, and it is essential to highlight that the genus *Haslea* with *H. spicula* (Hickie) Bukhtiyarova species was registered in Serbia for the first time on that occasion. Most of the identified taxa are characteristic for brackish and/or waters with increased conductivity or are commonly found on marine coasts [60].

Another significant water body in the area of the South Banat is a small, lowland, slow-flowing Ponjavica River that is a part of a Nature Park of the same name proclaimed in 1995. Only a part from Omoljica to Banatski Brestovac villages (about 10 km of the flow) that remained autochthonous and authentic was protected. In the NP "Ponjavica", the first phytoplankton studies were conducted in September 1984 and in April 1989, and soon after that, studies were repeated in 1991, 1992, and 1993 [65, 66]. A high diversity of algae, especially greens and diatoms, was observed during all that research, while numerous *Ceratium hirundinela* (O. F. Müller) Dujardin populations and poor cyanobacterial blooms occasionally occurred. Phytoplankton investigations in early 2000 were relatively numerous (studies in 2001/2002, 2005, in October 2006, and March–November 2008) but discontinuous. What is evident at the beginning of the twenty-first century is an increase in the trophic status and a decrease in depth along with an increased amount of sludge. These conditions have led to more intense and long-lasting blooms of potentially toxic cyanobacteria during the summer, but with the retention of relatively high taxa richness, especially in green algae [61, 62, 67–70].

Karadžić [61] observed the richness of 444 algal taxa during studies in 2001/2001, 2005, and 2008 in the Ponjavica River. The most diverse divisions were Chlorophyta (167 registered taxa), Bacillariophyta (97), Cyanobacteria (76), and Euglenophyta (74), and the most diverse genera were *Scenedesmus* (41), and *Euglena* (33). The same author also pointed to the presence of 21 potentially toxic cyanobacterial taxa, three of which have an invasive character (*R. raciborskii*, *S. aphanizomenoides*, and *R. mediterranea*). In Serbia, *S. aphanizomenoides* was first recorded in the summer phytoplankton samples in the Ponjavica River in 2008 [61, 62]. Periodically, it was dominant in this ecosystem. Its detailed morphological and ecological properties were presented in Jovanović et al. [70]. As previously mentioned, *R. raciborskii* was first recorded in the Slatina, but in the Ponjavica River its bloom was recorded for the first time in Serbia in 2008 [61, 62]. It replaced the previous summer bloom of *M. aeruginosa* and *A. flos-aqua*, and frequently was followed by also the invasive cyanobacterium *R. mediterranea* [62]. The invasive halophilic diatom *Actinocyclus normanii* f. *subsalsus* (Juhlin-Dannfelt) Hustedt appeared in low abundance, and its presence is associated with an increase in eutrophication [37], so a general conclusion could be that all algological parameters indicated a high degree of eutrophication and a very poor ecological status of this important aquatic ecosystem in 2008 [62].

The Central and North Banat regions (Fig. 8.2) are characterised by numerous but sporadic and unorganized studies of SSWB that are very different, from channels, artificial reservoirs for various purposes to salt marshes and fishponds. Therefore, the algal taxa richness and history of research will be shown only for selected ecosystems in the text below.

Typical lowland aquatic ecosystems that have been exposed to various types of human activity for centuries are the Stari Begej—an old riverbed of the Begej River and Carska bara—swamp, as a part of SNR "Stari Begej-Carska Bara" [71]. One of the first phytoplankton studies of Carska Bara dates back to July 1960, when the authors listed 43 detected taxa [72]. Then, a complete list of 103 taxa recorded from 1982 to 1985 was presented [73]. In all these studies, green algae were the most diverse, followed by diatoms, cyanobacteria, and euglenoids. The study of the Carska bara and the Stari Begej (April 1991–May 1993) revealed 293 recorded taxa in those two ecosystems together. Again, green algae were the most diverse (135 taxa), then diatoms (70 taxa), euglenoid algae (41 taxa), and cyanobacteria (25 taxa). The authors did not present the entire list of recorded taxa but listed the characteristic taxa that are indicators of all saprobic levels [71]. Later phytoplankton studies in this area are unknown.

As in South Banat, this region is also characterized by the existence of numerous saline water bodies, both natural and artificial, permanent as well as ephemeral ones. Only a few of them are protected on the national level (SNR "Slano Kopovo", one of the last preserved salt marshes in Serbia; NP "Rusanda" with a salty lake and few salty habitats; SNR "Okanj Bara" also characterised by saline soil and waters). On the other hand, there are saline ecosystems that are not protected like Novo Ilje I, Novo Ilje II, and artificial lakes near Kikinda city made by the digging of clay. They are among the most investigated ones in terms of phytoplankton research.

Novo Ilje I is an artificial salty channel, while Novo Ilje II is a natural salty swampy meadow, both near Melenci village, usually drying up during the summer. Phytoplankton studies of those aquatic ecosystems took place in April 2003 and in March 2004. A total number of 163 taxa were detected in both studied ecosystems [74]. The most diverse were Euglenophyta (50 taxa) with genus *Trachelomonas* (28 taxa), followed by diatoms (41 taxa), green algae (29 taxa), and cyanobacteria (24 taxa) [74]. In the first year of exploration, the most numerous were *T. volvocina* in both ecosystems, with the addition of cyanobacterium *Johanseninema constrictum* (Szafer) Hasler, Dvorák & Poulícková in Novo Ilje II, while in the second year, green algae prevailed with the most numerous being *Chlamydomonas lapponica* Skuja and *Hindakia tetrachotoma* (Printz) C. Bock, Pröschold & Krienitz [75]. In the Novo Ilje II, *Leptolyngbya foveolarum* (Gomont) Anagnostidis & Komárek and *Tribonema regulare* Pascher were also abundant [75]. Since then, those water bodies have not been investigated.

Another unprotected ecosystem in this region is the abandoned artificial and hyposaline pond near Kikinda city locally called Kop, where the phytoplankton studies took place in July 2018 [76]. Low biomass but a relatively high abundance of cells was detected due to small-dimension algae dominance, mostly chroococcalean cyanobacteria, green algae, and diatoms [76]. The decreased surface to volume ratio seems to be a competitive advantage in osmoregulation processes in saline environments [77] and ecosystems with the light-limiting conditions caused by macrophyte domination [78, 79]. Among the total of 27 taxa detected in the Kop pond, there are a few characteristic phytoplankton taxa for inland saline or brackish water—*Oocystis submarina* Lagerheim, *Merismopedia warmingiana* (Lagerheim) Forti, and *Euglenaformis proxima* (P. A. Dangeard) M. S. Bennett & Triemer, but it should be stated that the majority of detected phytoplankton representatives can be characterised as halotolerant taxa rather than halophilic ones [76].

Of course, protected ecosystems have not been neglected in terms of phytoplankton research. The only known phytoplankton investigation of the Velika Rusanda, an alkaline soda lake protected in 2011 as a nature park, is about planktonic diatoms [80]. In this extremely saline aquatic ecosystem, diatoms in plankton were examined during 2017 and 2018, together with those in benthic communities (epipelon, epilithon, and epiphyton) and the plankton assemblage had the lowest diatom taxa richness among them. Nine recorded taxa are either cosmopolitan or characteristic for waters with moderate to high electrolyte content. It is worth mentioning that *Navicula stafordiae* Bahls, as species new for European diatom flora, was detected in all examined communities [80]. In addition to diatoms, there is only one older record, and that was the appearance of cyanobacterium *Nodularia crassa* (Woronichin) J. Komárek, M. Hübel, H. Hübel & J. Smarda [78, 80]. Additionally, in the nearby salty lake Okanj bara, which is a part of the SNR of the same name, *Pinnularia schimanskii* Krammer was detected for the first time in Serbia [60].

Benthic diatoms in this region were only investigated in salty habitats. The research was mostly sporadic, short-lived, and conducted in different seasons, except for a two-year detailed study of the Lake Velika Rusanda. During 2017 and 2018, 27 diatom taxa were recorded [79, 80]. The community attached to the reed was the most diverse with 25 taxa, and the richest in species was the genus *Nitzschia* with *N. supralitorea* Lange-Bertalot presented in all investigated communities (epipelon, epilithon, epiphyton, and plankton). Generally, studies of benthic diatoms in those specific salty habitats are significant due to the findings of new species to the Serbian flora. Thus, in the Velika Rusanda, five species were recorded as new for Serbia, namely *Craticula halopannonica* Lange-Bertalot, *Hantzschia weyprechtii* Grunow, and *Nitzschia thermaloides* Hustedt found in epipelon, epilithon, and epiphyton, *Navicymbula pusilla* (Grunow) Krammer was found in epiphyton and epilithon, and *Navicula stafordiae* in all benthic assemblages and plankton [79, 80]. Further, in the SNR "Okanj Bara", a diatom investigation revealed *Mastogloia elliptica* (C. Agardh) Cleve and *Nitzschia vitrea* var. *salinarum* Grunow in the epiphytic community, and *Sellaphora harderi* (Hustedt) J. Foets & C. E. Wetzel in phytobenthos as new species in Serbia [60]. *Achnanthidium saprophilum* (H. Kobayashi & Mayama) Round & Bukhtiyarova was recorded in the Novo Ilje I, while in the Novo Ilje II, *Pinnularia ammerensis* Kulikovskiy, Lange-Bertalot & Metzeltin was recorded for the first time in Serbia [60]. Although these habitats are characterised by a low richness of diatoms, they are important due to newly discovered ones.

8.2.1.2 Microalgae in Central and South Serbia

Although we previously postulated that the majority of algological studies in SSWB took place in Vojvodina, and gave a detailed review, here, we provide a selected overview and discuss the most important findings on the diversity of algae in SSWB in the territory of Central and South Serbia (Fig. 8.2), though the fact is that the majority of corresponding studies in this region took place in deep stratified lakes and reservoirs.

As a specific type of habitat, peat bogs and their specific planktonic desmid flora are selected to be presented here in detail. Relatively recent algological studies on desmid flora in peat bogs were conducted in Serbia, and the detailed historical review of previous surveys was provided by Fužinato [22]. Before Fužinato [22], 524 taxa of desmids were recorded in Serbia, while in her survey she detected 69 new ones. Desmid algae are exclusively freshwater algae, inhabiting commonly nutrient-poor water with acid pH. They are also the most diverse and abundant, thus being under constant threat of biodiversity decline and loss due to habitat eutrophication. This threat is estimated to be much more pronounced in the densely populated and agriculturally important lowlands, than in the pristine mountainous areas [81]. Considering the estimated threat and possible biodiversity loss, an initiative started in the UK to nominate the most important UK areas/sites for algae [81], and it resulted in over 140 designated spots vital to freshwater algal IPAs, of which 80 were important for desmids [81]. Desmids are a group of microscopic green algae, single-celled or filamentous, easily recognised by their central constriction, which divides the cell content bilaterally symmetrically; they are usually very beautiful and decorative (Fig. 8.5) [81].

The first surveys on desmid algae flora in Serbia date from XIX century and were continued through the XX century in two significant periods—in the beginning by Košanin, Katić, and Đorđević, particularly in peat bogs, and later in the period after the second world war by Milovanović [21]. The most recent detailed reports on desmid flora taxonomy and ecology in Serbia were provided by Cvijan and Laušević [82], Fužinato [22], and Šovran et al. [21].

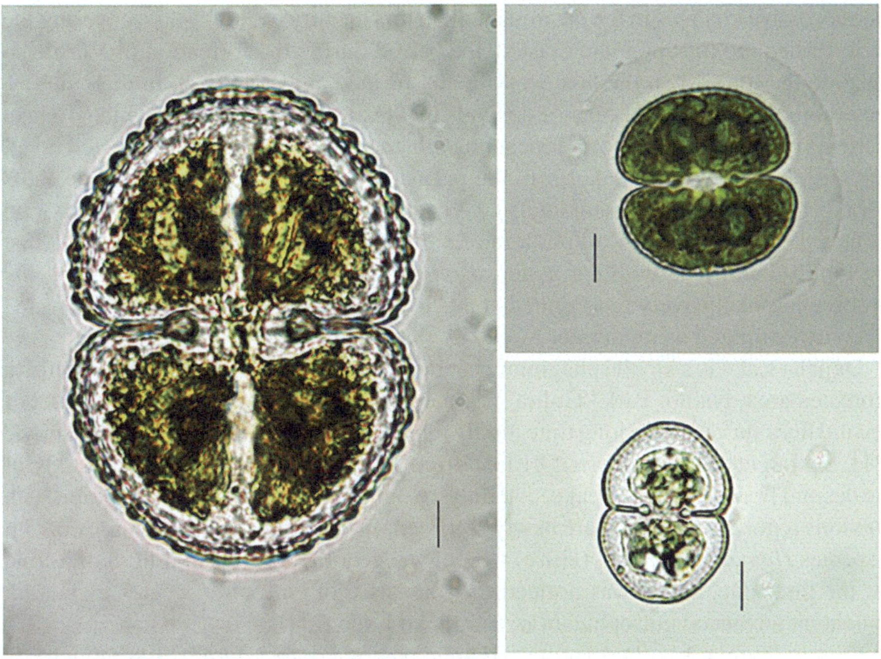

Fig. 8.5 Desmid representatives—*Cosmarium* spp. Scale bar 10 μm

Horgoš peat bog is a specific lowland peat bog situated in the North Bačka (it was reviewed in the North Bačka section), and according to Šovran et al. [21], who provided the results of the first phytoplankton study related to the desmids flora, this is a habitat to 30 taxa of desmids.

Pešter peat bog is situated in the Pešter plateau, at 1155–1161 m a.s.l., at the very southwestern point of Serbia [21, 83]. This area is characterized by its unique geological, geomorphological, hydrological, and climate features. Along with the Vlasina peat bog, which is today largely flooded after the formation of the hydro accumulation Vlasinsko Lake, this one is among the greatest peat bogs in Serbia but also the Balkans. With other peat bogs in the region, it represents the southernmost border of peat bogs areal in Europe [83]. Pešter peat bog is formed at the bottom of a disappeared lake, and today it is a protected area—Special Nature Reserve, as well as designated Ramsar site [84]. Until Šovran et al. [21], detailed algological studies were not conducted here, and these authors report 189 desmid taxa from as many as 18 genera for the first time. Many species were recorded here for the first time. Fužinato [22] highlights even 25 planktonic species of desmids characteristic for Pešter peat bog.

Vlasina peat bog was the largest in Serbia until it was anthropogenically transformed, as previously mentioned, when it was flooded and Vlasinsko Lake (protected area, Outstanding Natural Landscape "Vlasina") was made by damming the Vlasina River in 1949 [82]. In their comparative review, Cvijan and Laušević [82] presented the successive changes in the desmids flora in this locality over 80 years, covering the period when the peat bog had existed, the period during the formation of Vlasinsko Lake, to the time when the lake was already formed and stabilized. In their survey, these authors recorded 71 taxa of desmids in Vlasinsko Lake, among which 46 taxa were in common with the former peat bog (characterised by even 329 detected taxa), and only 8 taxa with the lake from the period when it was formed (when desmids flora disappeared almost completely). Cvijan & Laušević on this occasion reported 24 new desmid taxa for the algal flora of Serbia [82]. Fužinato [22] pointed out that the family Peniaceae with the genera *Penium* and species *Penium margaritaceum* Brébisson is exclusively represented at this locality in Serbia, while overall detected diversity estimated as moderate.

Dajićko Lake is a small sphagnum peat bog on the northwest side of Mt. Golija (a protected area, Nature Park "Golija"), situated at 1556 m a.s.l. It was the subject of desmid flora surveys for a long time, firstly reported by Košanin and later Milovanović [21]. Šovran et al. [21] reported 81 taxa from 19 genera in the most recent study of the desmid flora of Dajićko Lake, and only 16 out of these taxa were in common with previous reports, while 37 were newly recorded. In this study, genus *Heimansia* with 1 species *Heimansia pusilla* (Hilse) Coesel was recorded for desmid flora of Serbia for the first time, and it was noticed that the share of eutrophic species increased, indicating advanced eutrophication of Dajićko Lake [21].

Crvene Pode at Mt. Tara (protected area, National Park "Tara"), 1080 m a.s.l, is partially sphagnum and partially forest peat bog located at the plateau in the central part of Mitrovac [21]. In comparison to the latest (and first) report when Milovanović recorded 56 desmid taxa from 10 genera, Šovran et al. [21] reported 44 desmid taxa from 15 genera, and species *Hyalotheca mucosa* Ralfs was found to be specific for this habitat only in Serbia [22].

Considering phytobentic algal diversity overview, we will stay with peat bogs, i.e. the Pešter peat bog, for which Vidaković et al. [85] provided the most recent report on diatoms diversity, not only in benthos but also epiphyton and plankton. These authors reported extraordinary diatom taxa richness of the Pešter peat bog, in total 250 diatom species from 53 genera and 45 new taxa for Serbian diatom flora. As details were only provided for newly discovered taxa, we may conclude that diatoms were mainly represented in the benthic community, although present in the other two as well. Authors characterized the detected diatom flora as composed of cosmopolitan species with a large share of alpine and subalpine taxa, indicators of dystrophic, oligotrophic up to mesotrophic ecosystem state. The share of newly detected taxa for diatom flora in Serbia suggests a high specificity of diatom diversity in these types of habitat.

8.2.2 Macroalgae

The term macroalgae refers to algal species visible with the naked eye, without the need for magnificition instruments. Here we are going to consider the species which occur as macroscopic in terms of individual appearance, and not macroscopic colonies and/or cenobia. Macroscopic algae found in freshwater ecosystems in Serbia are mainly representatives of charophytes and red algae.

8.2.2.1 Macroalgae in Vojvodina

According to the available literature, the most characteristic representatives of macroalgae in this region (except three species of red macroalgae, but detected mainly in types of water bodies not belonging to the SSWB [86], and thus excluded from further review in this region) belong to charophytes, green algae of complex morphology, which resemble and grow together with the vascular aquatic macrophytes. According to the latest reviews and updates, the diversity of charophytes in Serbia consists of 24 species in total, out of which 22 species are represented in Vojvodina, and 6 species are exclusive for this lowland area of Serbia (Table 8.2).

According to the available literature [87, 88], in **the North Bačka region** (Fig. 8.2), charophytes were recorded in 9 localities—Tresetište, Majdan pond, the road to Kelebija, Kelebija, Graničar—Makova Sedmica, Kireš (Kereš) Hajdukovo, Palić, Selevenj Road, Lofej. These records were made in various types of SSWB, including a small lowland river (Kireš), various ponds, lake (Palić), and channels (most probably irrigational, not specified [88]), supporting the need for these habitats recognition and protection. The majority of localities are already covered by protected areas, except the pond near Selevenj road—which is besides Majdan pond, the only currently known habitat of *Chara intermedia* in Serbia [88]. Majdan pond, on the other hand, is the first and only currently known habitat of *Tolypella glomerata* in Serbia [88]. A great diversity of charophytes is represented in this area, although relatively few localities were re-visited on several occasions [88]. In total, nine charophyte species were recorded in this region (Table 8.2), but many of the listed findings are relatively old records [89]. Although localities Kireš River, Makova Sedmica, Tresetište, and Palić Lake were re-visited recently [88], the presence of charophytes was not confirmed. The records of *C. vulgaris* in Palić Lake and the Kereš River date back to 1989, and although this particular species is considered cosmopolitan and regularly present along the whole gradient of trophic levels [94], these records are important in terms of future plans for the revitalization of these currently highly degraded ecosystems [95, 96]. The detected loss of charophyte diversity in particular habitats in this region opens the question of general habitat degradation and biodiversity loss due to the natural succession process or various anthropogenic activities and highlights the need for active monitoring and revitalization plans of charophytes in SSWB.

In **the West Bačka region** (Fig. 8.2), the following review will focus on Special Nature Reserve (SNR) "Gornje Podunavlje", as charophyte studies mainly took place in this protected area particularly. The only record made out of this protected area is the record of *Chara canescens* in channel Vrbas-Bezdan, near Mali Stapar [97], but this record is challenged by later authors [88, 92] and will not be considered in this review. SSWB in SNR "Gornje Podunavlje" are very rich in charophyte diversity

Table 8.2 Charophytes diversity and their presence overview across the regions of Vojvodina. Data based on literature sources [87–93], IUCN categorisation of detected species according to Blaženčić [87]. In bold—species present only in Vojvodina

Species list	IUCN	NBR	WBačR	SBR	SBMBR	BR	SBanR	CNBR
1. ***Chara baueri*** A. Braun	–		+					
2. *Chara braunii* C. C. Gmel	VU		+	+	+	+		+
3. *Chara canescens* Desv. & Loisel	EW (?)							+
4. *Chara connivens* Salzm. ex A.Braun	CR						+	
5. *Chara contraria* A.Braun ex Kütz	LR	+	+		+	+	+	+
6. *Chara globularis* Thuill	VU	+	+	+	+	+	+	
7. *Chara hispida* (L.) Hartm	EN	+						
8. *Chara intermedia* A.Braun	CR	+						
9. *Chara tenuispina* A.Braun	CR	+	+	+				
10. *Chara virgata* Kütz	EN							+
11. *Chara vulgaris* L	LR	+	+	+	+	+	+	
12. *Nitella brachyteles* A.Braun *	EW (?)					+		
13. *Nitella capillaris* (Krock.) J.Groves & Bull.-Webst	DD	+	+		+			
14. ***Nitella confervacea*(Bréb.)** A. Braun ex Leonh	CR		+		+			
15. *Nitella gracilis* (Sm.) C. Agardh	CR		+	+			+	
16. *Nitella mucronata* (A. Braun) Miq	CR		+	+	+	+	+	+
17. *Nitella opaca* (Bruzelius) C. Agardh	CR	+						
18.*Nitella syncarpa* (Truill.) Chevall	CR		+	+	+	+		
19. ***Nitellopsis obtusa*(Desv. in Lois.)** J. Grov	CR						+	

(continued)

Table 8.2 (continued)

Species list	IUCN	NBR	WBačR	SBR	SBMBR	BR	SBanR	CNBR
20. ***Tolypella glomerata*** (Desv.) Leonh	CR	+						
21. ***Tolypella intricata*** (Trentepohl ex Roth) Leonhardi	CR		+	+	+			
22. ***Tolypella prolifera*** (Ziz ex A. Braun) Leonhardi	CR		+		+			

* This species presence in Europe is disputable (comment by authors)

[88, 90, 93]. A literature overview reveals even 21 localities where charophytes were detected, among which ponds are the habitats where most of the charophyte records were made and where the highest diversity was recorded [88, 90, 93]. Ruts are very specific ephemeral habitats of stoneworts in SNR "Gornje Podunavlje", pointing to still active oospores in soil diaspore banks, even where the original wetland habitats are modified. Channels as habitats in which records of charophytes were made are not reconfirmed in recent studies, and due to the strong and continued processes of habitat degradation during the twentieth century, diversity once detected here is probably lost [88]. Extraordinary and unique charophyte diversity adorns this area, particularly the Monoštorski Marsh, where up to 8 species could be recorded in one locality—Široki rit pond [93]. In the Štrbac area, also situated in the Monoštorski Marsh, a remarkable finding of *Chara baueri* was recently recorded [90]. *Chara baueri* is one of the rarest charophytes worldwide; the contemporary distribution of this species is limited to a few localities in Germany, Poland, and Russia in Europe, and Kazakhstan in Asia. Specific attention should be paid to SWB such as artificial or natural field ponds in particular, as *C. baueri* was recorded in such habitats. The threat status of *Chara baueri* has not been estimated in Serbia yet, since it was recently recorded, but this species has the potential to be declared threatened at the global level due to its very restricted distribution. In total, 13 charophyte species were recorded in this region (Table 8.2).

Findings of charophytes in **the South Bačka region** (Fig. 8.2) are all sporadic and mostly single findings, besides the ones made by Vesić [88] in the systematic research. The total diversity consists of only 8 species, but many are critically endangered or vulnerable (Table 8.2), which should surely attract attention to the protection of the known species and habitats and the need for their systematic monitoring. In total, 12 localities as habitats of charophytes were recorded in this area, mainly in Koviljsko-Petrovaradinski Marsh. When reviewing habitat types where charophyte records were made [88], various pond types dominate, and there are also recent findings in ruts on dirt roads. Particularly interesting is the finding of *C. tenuispina* in the river Čik. The river Čik is still a completely unprotected area, and along with the Jergička and the Beljanska Bara (protected areas), is a very specific and unique habitat of "flowing ponds". Today, these rivers (due to regulation and natural succession) are a complex mosaic of ponds, wet meadows, and inland salt marshes, but water in these habitats still flows very slowly [98, 99]. The record of *Chara tenuispina* in the Čik River in 2013 was the first after more than 100 years in the territory of Serbia [87]. After that, it was reconfirmed in this locality in 2014 and recorded in few others in Vojvodina [88, 93]. This is, however, rare and critically endangered species in Serbia. Thus, its habitat in the Čik River should be monitored and put under a protection program. The finding of *Chara* sp. in the Beljanska Bara is old and not precise [89], but still, it points to the presence of charophytes in this locality.

Srem and Mačva region (Fig. 8.2) are conjoined in the review since the majority of algological studies took place in the protected areas SNR „Zasavica " and SNR "Obedska Bara", which stretches in both regions. The territory of Mačva and particularly Srem is rich in SSWB, but charophyte studies are lacking, or the localities were visited only sporadically, without significant findings [88]. A detailed survey of charophytes was conducted in the Zasavica during the period 1998–2010 [100, 101] when these macroalgae were detected in even 19 localities. Charophytes were mainly detected in ponds, and particularly interesting are findings in ruts, extremely small and ephemeral habitats made by vehicle passing [88]. The diversity of charophytes detected in the Zasavica comprises 9 species, out of which the most are estimated as critically endangered in Serbia (Table 8.2). The diversity of the genus *Nitella* is emphasized. *Nitella confervacea* was detected here for the first time in Serbia. At the same time, *Tolypella intricata* and *T. prolifera* records in the Zasavica, along with the very few others in Serbia, have been the only reliable findings of these species in the entire region of central and western Balkan for the last 100 years [88]. The first literature data on charophytes records in the Obedska Bara area date from the early and mid-twentieth century [102, 103]. Afterward, this area was searched for charophytes on several occasions, but none of these surveys confirmed charophytes in the very basin of Obedska Bara, but exclusively in wetlands and particularly small (even ephemeral) water bodies in the area [88]. A review of the localities and types of habitats where charophytes were detected [87–89, 104] reveals 13 localities, mainly in Pećinci and Obrež area, and ponds—ephemeral and permanent as the most represented habitats.

The region of the capital city of Serbia—**the Belgrade region** (Fig. 8.2) is the most developed and the most densely populated area, but at the same time it is an area rich in various SSWB, especially in the left Danube bank, the so-called Foreland of the left bank of the Danube. Charophytes used to be recorded in various types of SSWB and many localities (14 in total) in this region [87–89]. Nevertheless, these records are mostly sporadic and single, prevalently dating from the twentieth century, while the recent presence of charophytes in most localities was not reconfirmed [88]. An exception is a case of Sava Lake, where in a recent study, only one previously detected species was reconfirmed—*C. contraria*, while *Tolypella intricata* and *Nitella* sp. are newly recorded [91]. Findings in Sava Lake and a nearby small lake Ada Safari [87, 91] are at least intriguing, considering high anthropogenic pressure on this recreational center. Wider area in this region, despite formerly recorded extraordinary and unique diversity and presence of even 10 charophyte species (which is, disregarding

SNR "Gornje Podunavlje" in WBR, more than in any other region reviewed here (Table 8.2)), have not been the subject of any recent algological studies.

In a recent systematic survey in **the South Banat region** (Fig. 8.2) conducted by Vesić [88], a great area of the Danube, Tamiš, and Nera rivers floodplain was searched. Unfortunately, charophytes were detected in very few localities—in two oxbow lakes near Tamiš, a brickyard pond, Danube bay Dolnice, and a pond near Nera. In 2017, Trbojević et al. [91] started detailed surveys in the Labudovo Okno wetland area, finding unique and intriguing diversity in one new locality—Dulin Pond. All the other records from this area (in total 17 localities) are old, sporadic, and single records (some of which very old, i.e. finding of *Chara vulgaris* near Vršac dates from 1902). Though these old records implicate high possibilities for detection of rich biodiversity, the majority of habitats are yet to be adequately surveyed in the South Banat region. It would be exciting to search the salt marshes near Opovo, the ponds and lakes near Bela Crkva, Kovin, Debeljača, and Vršac, as well as many agricultural ponds over the region since charophytes are common inhabitants of these SSWB [105]. The total detected diversity of charophytes in this area is relatively poor; only 7 species were recorded (Table 8.2). Still, the presence of species *Nitellopsis obtusa* and *Chara connivens* for sure makes this area specific and unique regarding charophyte diversity in Serbia. The South Banat region, more precisely water bodies in the Labudovo Okno and nearby pond in the Nera floodplain, are the only known localities where *Nitellopsis obtusa* is recorded in Serbia. The finding of the taxon *C. connivens* in the Dulin Pond (Fig. 8.6) in Labudovo Okno is peculiar, as according to the relevant literature, this is a brackish species, and the Dulin Pond definitely is not saline [91]. Genetic analyses proved specimens from the Dulin pond to be more closely related to *C. globularis* than to the other *C. connivens* specimens used in the phylogenetic analyses. However, more studies on freshwater populations of *C.*

Fig. 8.6 Specimens of *Chara connivens* (male) from the Dulin pond. Photo by V. Simić

connivens are needed to confirm consistent genetic specificities and taxonomic status of these specimens [91]. *C. connivens* in Serbia is considered critically endangered [87], and currently, the Dulin Pond is its only confirmed habitat in Serbia.

Although **the Central and North Banat region** (Fig. 8.2) is partially bordered by the Tamiš River in the south, and all along the western border by the Tisa River, this area is generally arid, especially the northern part, and characterized by unique saline SSWB of both natural and artificial origin. All charophyte records in this region (Table 8.2), except the recent findings of *C. canescens* [92], are old and have not been confirmed by Vesić [88] in the latest systematic survey of this region. Recordings were made in only four localities near Ečka, Čenta, Kikinda, and Zrenjanin, and prevalently in various types of ponds [87]. The artificial saline pond Plava Banja near Kikinda city is the second known locality where *Chara canescens* (Fig. 8.7) was detected in Serbia, and currently is the only known. *Chara canescens* is a typically brackish species inhabiting saline habitats of a wide range. It is heliophilic and tolerable towards nutrients and ion anomalies [106–108]. Populations found across Europe are mainly parthenogenetic (as well as the one in Serbia), bisexual populations are extremely rare [107, 109]. One more specific characteristic of the species is semelparity [106, 110], which might explain why Vesic [88] has not made a record when surveying the Plava Banja pond.

Fig. 8.7 Specimens of *Chara canescens* (female) from the Plava Banja pond. Photo by D. Predojević

8.2.2.2 Macroalgae in Central and South Serbia

Macroalgae diversity and distribution in the Central and South Serbia region (Fig. 8.2) will be presented in a more comprehensive overview of habitats, based on available corresponding literature sources, and here, red algae will also be considered along with charophytes.

According to Vesić [88], until 2014, 18 species of charophytes were detected in total in the territory of Serbia without Vojvodina since 1851, when we assume that surveys started. Vesić [88] noted that three species (out of 18) were found only in the localities southern of the Danube and the Sava rivers—*Chara connivens*, *C. tomentosa*, and *C. rochleanae*, while she put the presence of *Chara canescens* in Vojvodina under a question mark. Now, we have solid evidence of the presence of this taxon in Vojvodina nearby Kikinda [92]. As Trbojević et al. [91] recorded, the taxon resembling *Chara connivens*, whose status is still unclear taxonomically and phylogenetically, was found in Dulin Pond in Vojvodina. This is most probably also the case with the old (and only) record of *Chara connivens* in Serbia from the channel near Srebrno Lake in 1980 and 1984 [87]. However, this species is now common for Vojvodina and the rest of Serbia, leaving only two species characteristic for the territory south of the Danube and Sava rivers, and we are going to review these records in more detail.

There is only one and relatively old record on *Chara tomentosa* in Serbia, near Poklek Spa, Kosovo [87], but interestingly the habitat is described as mineral water in the flood zone of the stream [88]. This habitat would fully correspond to SWB, but the lack of deposited herbarium material and lack of the later findings makes this record even debatable. Still, since this species is known as typical for highly mineralized, alkaline habitats and despite being robust to large, these specimens are occasionally found in shallow water [111], which makes the habitat described near Poklek Spa fully suitable. *Chara tomentosa* is considered to be critically endangered in Serbia [87].

The other charophyte taxon we have not already discussed as it is not found in Vojvodina is *Chara rochlenae*. *C. rochleanae* is an interesting and important species in many aspects. It was first described from Montenegro (Mratinje Stream) in 1912. Afterwards, it was reported from Greece by Langangen [112], but essentially this author revised and reported an old herbarium record, from 1885 (De Heidrich collection, initially reported as *Chara gymnophylla*). These records clearly outline this taxon as endemic to the Balkans along with the ones in Bosnia and Herzegovina and in Serbia [113]. In the review of Balkan charophytes [114], *C. rochleanae* was even estimated as extinct at the global level (EX glob?), as there were no records for almost 80 years. The most recent finding was in 2010 and 2012 in a small mountain puddle, the Ponor locality, Mokra Gora Mountain in Serbia, at an altitude of 1600 m (species is described as prevalently characteristic for high altitudes) [113]. This habitat in which *C. rochleanae* was rediscovered in Mokra Gora Mountain is a typical SWB. After these records were reported by Blaženčić & Stevanović [113], new ones have never been made in Serbia, and this species is estimated as critically endangered [87]. We support and strongly express concerns [113] for the urgent need for efficient protection measures to be established so that this taxon is preserved in charophytes flora of Serbia.

When habitat types presented in Table 8.3 are reviewed, SSWB types clearly dominate, and as the only habitats which are definitely out of scopes of SWB, the Vlasinsko Lake and the Blačko Lake are recognized (for the lake near Pirot we could

Table 8.3 The list of charophyte species and localities where records were made in Central and South Serbia. Data presented according to the review provided in Vesić [88] and Blaženčić [87], IUCN categories according to Blaženčić [87]. Habitats estimated as not qualifying for SSWB in italic and underlined. In bold – species present only in the Central and South Serbia regio

Species list	IUCN	Localities
1. *Chara braunii* C. C. Gmel	VU	*Vlasinsko Lake*, Niška Spa—artificial pond near Radon hotel
2. *Chara canescens* Desv. & Loisel	EW (?)	Suva Česma saline spring, near Prokuplje
3. *Chara connivens* Salzm. ex A.Braun	CR	Channel near Srebrno Lake, Veliko Gradište
4. *Chara contraria* A.Braun ex Kütz	LR	Many localities, the most representative as SWB: ponds near Drenovac, Kragujevac, ponds near dam on Uvac Lake, peat bogs and streams on Tara mountaine, peat bog lake on Pešter, ponds near Gračanica Lake, Priština, pons on Šar mountaine, fish pond in Zvonačka spa, saline spring in Lalinačka salt marsh, Niš
5. *Chara globularis* Thuill	VU	Many localities, the most representative as SWB: channel near Srebrno Lake, Veliko Gradište, streams and channels in Vrujci spa, Kolubara river oxbows near Obrenovac, ponds near Kragujevac, ponds along Južna Morava River, fish pond near Priština
6. *Chara hispida* (L.) Hartm	EN	Ždraljica River near Kragujevac
7. *Chara intermedia* A.Braun	CR	Ponds near Negotin, Toidže village, near Semeteš, Kopaonik mountain—in spring of the Lisine stream and small ponds on the meadows
8. ***Chara rohlenae*** Vilh	CR	small mountain puddle, Ponor locality, Mokra Gora Mountainin
9. *Chara tenuispina* A.Braun	CR	Great Lake, Crepuljnik Golija mountain
10. ***Chara tomentosa*** L	CR	Flood zone of a stream, near Poklek spa
11. *Chara virgata* Kütz	EN	channel near Srebrno Lake, Veliko Gradište, Veliko Nedžinatsko Lake (Metohija) in Prokletije Mt., ponds by Drina River between Mali Zvornik and Ljubovija

(continued)

Table 8.3 (continued)

Species list	IUCN	Localities
12. *Chara vulgaris* L	LR	Many localities, the most representative as SWB: ponds near Dobroselica and Gostilje, Zlatibor Mt., channel near Srebrno Lake, Veliko Gradište, ponds near Ždraljica and Drenovac, Sušički stream, spring near Ljubić, Kragujevac, Poklek spa, ponds by Drina River between Mali Zvornik and Ljubovija, ponds in Lalinačka salt marsh, Niš, open pit mine, Kostolac
13. *Nitella brachyteles* A.Braun *	EW (?)	channel near Srebrno Lake, Veliko Gradište
14. *Nitella capillaris* (Krock.) J.Groves & Bull.-Webst	DD	*Lake by Pirot*
15. *Nitella gracilis* (Sm.) C. Agardh	CR	Stream Rgoška spa, Knjaževac, channel Vrujci spa, Ljig, *Vlasinsko Lake*
16. *Nitella mucronata* (A. Braun) Miq	CR	*Lake near Blace,* Prokuplje
17. *Nitella opaca* (Bruzelius) C. Agardh	CR	*Vlasinsko Lake*
18. *Nitella syncarpa* (Truill.) Chevall	CR	Current *Vlasinsko Lake* and former Vlasina mud and peat bog

* This species presence in Europe is disputable (comment by authors)

not decide, it is an old record by Košanin [115]). In this review, habitats for the most commonly found taxa - *C. globularis*, *C. vulgaris* and *C. contraria* were filtered and preselected (due to many findings), and only SSWB types are presented in Table 8.3, but for the other species, all available records are listed [87, 88]. Many of the records listed in Table 8.3 are very old, from the beginning of the twentieth century (and many habitats does not exist anymore), and very few are from the twenty-first century (more details in Blaženčić [87]).

In contrast to Vojvodina, the Central and South Serbia region (Fig. 8.2) geography and geology support the occurrence of a specific type of SWB—high mountain streams. Specific macroalgal flora is associated with this type of waterbody—red algae; thus, we found it noteworthy also to give a review of red algae diversity and distribution in Serbia. Red algae are considered stenovalent towards various ecological factors, thus being rare and sensitive and consequently susceptible even to the low pressures on their habitats [86]. Ten taxa of these macroalgae are recognized as protected and strictly protected species in Serbia [116, 117] (marked with * in Table 8.4), and also categorized according to the international criteria in Red Lists in few European countries [86]. Simić et al. [86] made the first steps toward constituting the first Red List of red algae in Serbia by defining threatening factors to these algae habitats and giving a preliminary Red List according to which 5 species were considered

Table 8.4 List of red algae species and localities where records were made in Central and South Serbia. Label * for protected and strictly protected species in Serbia [116, 117], underlined species found only in SWB habitats

Species list	Springs	Spa	Fountains	Fishponds/Small Tanks	Brook	Rivers
1. *Audouinella hermannii* (Roth) Duby	–	–	–	–	–	+
2. *Bangia atropurpurea* (Mertens ex Roth) C.Agardh *	–	–	–	small tank of the Gvozdačka River	–	+
3. *Batrachospermum cayennense* Montagne ex Kützing *	Crnica River (Kučajske mountains),Mlava River (Well, Žagubica)	–	–	–	–	+
4. <u>*Batrachospermum confusum*(Bory) Hassall *</u>	Banja (Petnica cave), Spring near the village Krupac	–	–	Vlasina mud	–	–
5. *Batrachospermum ectocarpum* Entwisle & Kraft *	Banja (Petnica cave), Small springs (left bank of the Gradac River)	-	–	–	–	+
6. *Batrachospermum gelatinosum* (Linnaeus) De Candolle	Banja (Petnica cave), Krupajska River (Beljanica mountains), Popovo Well (Gradac River gorge), Spring near Bela Palanka, Stanjanska River	–	–	–	–	+
7. <u>*Batrachospermum turfosum*Bory *</u>	Banja (Petnica cave)	–	–	–	–	–
8. <u>*Batrachospermum virgato-decaisneanum*Sirodot *</u>	Banja (Petnica cave)	–	–	–	–	–

(continued)

Table 8.4 (continued)

Species list	Springs	Spa	Fountains	Fishponds/Small Tanks	Brook	Rivers
9. *Batrachospermum* sp. Roth	Crni Timok (Kučajske mountains), Moravica River	–	–	–	–	+
10. *Chantransia chalybea* (Roth) Fries	Barevačka čuka (river Lepenac, NP Šara)	Niška Banja Spa (Učiteljska česma), Niška Banja Spa (The Source Suva Banja)	Belgrade (Hajdučka česma), Belgrade (Fountain in the Vodovodska streat)	Small tank of the Gvozdačka River	–	+
11. *Chantransia pygmaea* Kützing	–	–	–	–	–	+
12. *Hildenbrandia rivularis* (Liebmann) J.Agardh *	–	–	–	–	Cvetića Brook and Bioštanska Banja Brook (tributaries of the Vrutci reservoir)	+
13. *Lemanea fluviatilis* (Linnaeus) C.Agardh	–	–	–	–	Fishpond on the Resava River	+
14. *Lemanea rigida* (Sirodot) De Toni	–	–	–	–	–	+
15. *Lemanea fucina* Bory	–	–	–	–	–	+
16. *Lemanea* sp. Bory	–	–	–	–	–	+

(continued)

Table 8.4 (continued)

Species list	Springs	Spa	Fountains	Fishponds/Small Tanks	Brook	Rivers
17. *Paralemanea annulata* (Kützing) M. L.Vis & R.G.Sheath	–	–	–	–	–	+
18. *Paralemanea catenata* (Kützing) M. L.Vis & Sheath *	–	–	–	–	–	+
19. *Paralemanea torulosa* (Roth) Sheath & A. R.Sherwood *	–	–	–	–	–	+
20. *Thorea hispida* (Thore) Desvaux *	–	–	–	–	–	+

as endangered—*Thorea hispida* and *Batrachospermum cayennense* as CR (critically endangered), *Bangia atropurpurea* and *Paralemanea catenata* as EN (endangered) and *Paralemanea annulata* as VU (vulnerable), while 4 species were categorized as NT (near threatened)—*Batrachospermum gelatinosum*, *Lemanea fluviatilis*, *Chantransia chalybea*, *C. pygmaea*. For the species *Batrachospermum confusum*, *B. ectocarpum*, *B. turfosum*, *B. virgato-decaisneanum* and *Hildenbrandia rivularis*, it was established that the data was deficient (DD data deficient). The official and final red list with estimated categories for species still has not yet been published.

According to the reviewed literature [86, 118–120], the diversity of red algae in Serbia is comprised of the 20 species presented in Table 8.4, along with their detailed distribution in the SWB habitats (and only the note on presence in riverine ecosystems) [86, 119,120]. In the Rivers category in Table 8.4, possibly a few streams generally belonging to the SWB are included (for more details, see Simić [86], Blagojević et al. [119], and Mitrović et al. [120]).

It is evident that the distribution of red algae taxa is spread over various types of SWB habitats, but still, the majority of species are present in (and characteristic to) riverine habitats [86,119, 120] (Table 8.4). In this review, species characteristic to the SWB habitats, *Batrachospermum confusum*, *B. turfosum*, and *B. virgato–decaisneanum* were distinguished, and all three taxa at the same time are listed as protected and strictly protected species in Serbia [116, 117]. Simić et al. [86], in their review on the diversity and distribution of red algae in Serbia, noted that 46% of the localities inhabited by red algae were situated in protected areas.

8.3 Frame for Species and Habitat Protection

8.3.1 Protection of Algae: Problems and Efforts Made in Finding a Solution

Algal diversity conservation in the system of natural ecosystems protection has been promoted and more recognized lately [121]. With reference to Juran and Kaštovsky [122], in the review on algae diversity in SSWB in Serbia, we aimed to highlight the need for algal biodiversity protection. We propose that habitat protection should be considered synonymous for species conservation. Besides regarding biodiversity and genetic resources conservation, the importance of algae protection in the small and shallow habitats relies also on their biondication capacities in the habitat quality evaluation as well as on the basic and crucial role in ecosystem services provision.

Juran and Kaštovsky [122] provided the most recent and relevant review on overall progress made in the field of conservation and protection of algae. They are particularly reviewing the published Red Lists across Europe, focusing on microalgae that were mostly neglected and overlooked as endangered or prone to extinction, even by many phycologists [123]. The complexity and controversy of understanding, and subsequently protecting microalgae in practice is recognized at several levels, starting with challenging the ubiquitous concept for the microorganisms that "everything is everywhere", by the evidence provided by Brodie et al. [123] and Juran and Kaštovsky [122]. Biogeography i.e. distribution data deficiency is one of the obstacles in the microalgae protection (and endemism) assessment process. At the same time, distributional patterns seem to be much easier to determine for macroscopic species. This is only the first in the line of problems in the microalgae conservation process, including (not) precise determination (taxonomic resolution), undersampling, introduced and/or invasive taxa, lack of long-term monitoring at the localities and subsequently the lack of comprehensive data on taxonomy and ecology [122]. Still, Juran and Kaštovsky [122] proposed an efficient methodological approach (based on the euglenophytes as a model group) for categorizing microalgal species and making Red Lists as a starting point in practical habitat conservation.

When macroscopic algae are considered, in the majority of European countries, they are recognized in Red Lists which predominantly include representatives of brown, red and green algae (mainly charophytes) [122]. Here, the problem with the distribution determination would be expected to be diminished, but still, geographic patterns are considered to be poorly described even in macroalgae [123]. This problem is also reflected in low taxonomic resolution, due to the deficiency of data and/or unavailability of deposited specimens (herbariums) for verification, and lack of appropriate expertise to identify newly collected or deposited material. Related to this, herbarium specimens of macroalgae have to be highlighted as a valuable resource for the estimation of the species and habitat status [123]. The establishment of whether macroalgal species can be classified as endangered, whether on a local, regional or global level, is undoubtedly necessary for the development of adequate protection and conservation measures, and this is rather hard to do due to scarce or non-existing data. The treats can obviously be recognized in habitat destruction, but many other factors could also contribute, such as climate change. Joy and Boissezon [124] provided an insight into climate change effects (temperature rise and precipitation reduction during the vegetative season) on charophytes distribution patterns, identifying potential losers and winners in the new scenario—interestingly, half of the species showed potential to be winners, and those were mostly the species colonizing SWB. Still, if the climate change scenario and the other threatening factors would lead to drying up, which is likely if habitats are not included into the conservation program and active protection measures are not applied, it remains questionable how much success can be predicted for species colonizing SWB. When there is no available method to protect a habitat so that diversity could be preserved, ex-situ cultivation seems to be a powerful tool for species protection, and although this approach seems to be more suitable in terms of microalgae, it has been chiefly applied for macro species [123].

8.3.2 State of the Art in Serbia

There are no Red Lists or any other official form of threat/endangerment status assessment for microalgae species in Serbia. Although algological studies in Serbia and the Balkans in general have a long tradition, interest in the scientific community (and consequently in the general public) seems to have been declining for the last 30 years. The last review of algological studies during one whole century 1883–1983 was provided in 1986 by Blaženčić [125], and the next one, although based on a selection of relevant sources concerning SSWB, is provided here, 35 years later. In this regard, the consequences of climate change are becoming apparent in this period, as industrial and technological developments have greatly exceeded ecological and environmental awareness, interests and principles. We have to ask ourselves—have we lost significant biodiversity of algae in Serbia, and can we do something now to realize the current state, protect what we have left and potentially recover what we have lost?

It is very important to highlight the existence of Novi Sad Cyanobacterial Culture Collection (NSCCC), which currently counts about 500 strains of aquatic and terrestrial cyanobacteria isolated from natural ecosystems in Serbia [126]. This collection consists of cyanobacterial strains exclusively, and it is not registered with the World Federation for the Culture Collections. It was primarily used for productivity and biochemical analyses, for screening of antibacterial, antifungal, and cytotoxic capacities and, finally, toxicity tests [126]. The significance of this collection in terms of ex-situ conservation of both genetic and species biodiversity is, however, unquestionable, although sequencing of cultured strains still (as far as authors are aware) is yet to be done. NSCCC is the only microalgal collection in Serbia.

Regarding macroalgae collections in Serbia, in the wet collection and the herbarium of the University of Belgrade Herbarium (BEOU) until 2011 [127], there were in total 2348 labelled and named specimens of charophytes (in the wet collection 2287, and in herbarium sheets 61), originating mainly from the Balkans, but also from all over the world. This rich wet collection was constituted thanks to the devoted and systematic studies of charophytes from the 1970s onwards by prof. Jelena Blaženčić. The herbarium collection originates mainly from a much earlier period, and unfortunately, it was ruined and currently only what was left of it – 61 specimens are preserved in BEOU. The Pančić's collection of charophytes in Serbia—the first one, was completely lost as supposed by Vukojičić et al. [127] in turbulent historical events in Serbia in the twentieth century. Professor Košanin, who continued charophyte studies in Serbia after Pančić, and published Pančić's records later at the beginning of XX century [115, 128], noted that Pančić's collection was well preserved at the time. Recently intensified studies on charophytes by Vesić and Trbojević [88, 90–93] reflected in many more deposited specimens in the wet collection, but particularly in the herbarium sheets, which is also important from the aspect of potential genetic studies on collected specimens.

In comparison to the microalgae, there was progress made in Serbia recently towards the protection of species, mainly since Blaženčić [87] provided an overview of the stoneworts (Charales) of Serbia with the estimation of the threat status and the preliminary Red List of red algae was suggested by Simić et al. [86]. In the previous section of this chapter, species review is given parallel to the categories provided in these publications, and it can be clearly concluded that SSWB mainly are colonized by macroalgal species in the category of endangered. Also, national legislation in Serbia recognized 25 species of macroalgae, 15 species of charophytes and 10 species of red algae, as needing to be protected and strictly protected in Serbia [116, 117]. An assumption can be made that the main prerequisites for conservation of macroalgae officially exist. Unfortunately, the situation in the field is much different.

When management plans were reviewed, at least for the protected areas treated in the previous section of this chapter, none even note the presence of charophytes, and there are no proposed conservation measures towards their protection. This is mainly the consequence of not recognizing charophytes by protected area managers and the usual overlook of charophytes in regular macrophytes surveys (whether due to lack of expertise or not adequate monitoring plans). However, their presence in protected areas is noticeable, where they benefit from the overall conservation in the areas, but the resilience of their habitats there—mainly SSWB is questionable without particular conservation plans and direct actions towards the protection of charophyte diversity and their fragile habitats. These habitats are highly endangered, and the reasons are numerous. Some of the principal risk factors recognized are: eutrophication, the spread of invasive alien species, reed burning, illegal fishing and hunting, infrastructure and agriculture developments, ploughing of pastures etc. Charophytes have a whole spectrum of important roles in maintaining these ecosystem health and function, such as sediment stabilization, nutrients sequestration, water clarity enhancement, providing habitat for zooplankton and zoobenthos, a substrate for fish spawning and shelter for fish fry. All of these lead to restoring and/or conservation of the natural ecosystem function and subsequently related biodiversity [129]. To achieve the goal of development and in-field application of conservation plans to protect charophytes and their habitats, long-term and devoted educational work with the protected area managers and the wider public is needed. Education of the wider public on the charophytes importance for habitat conservation and revitalization is of special importance since community participation is described as the most powerful tool for environmental conservation in protected areas that have not received enough attention at the practical level in most developing countries [130].

8.3.3 Instead of Conclusion—Guidelines Proposal

The following guidelines leading towards successful protection of algae and their habitats in Serbia could be proposed, based on the review provided in the previous section and state-of-the-art in Serbia concerning the protection of algae:

1. Implementation of long term and continuous monitoring of algae, compiling comprehensive data on the taxonomic composition and distributional patterns of

phytoplankton, phytobenthos and macrophytes (prerequisite for this is support and development of experts in algal taxonomy and ecology).
2. Development of the Red Lists for at least some groups of microalgae, based on guidelines provided by Juran and Kaštovsky [122]. Updating and reviewing the existing estimation of the threat status of charophytes [87] due to new records. A preliminary Red List of red algae [86] should be finalised and made official.
3. National legislation updating related to the endangered species of algae.
4. Education of the wider public and particularly of protected area managers and local community and authorities in non-protected areas on algae biodiversity and their important roles in ecosystem conservation and providing ecosystem services.
5. Development of specific conservation management plans for algae and their habitats.
6. Monitoring of practice in species and habitat protection based on developed plans.

Guidelines proposed and an overview of algae diversity in SSWB of Serbia provided represent only the starting point for the protection of algae and their habitats in Serbia, and many intermediate steps need to be done before results in each point are visible. Authors strongly encourage the scientific public in Serbia to get more involved in species and habitat conservation issues.

Acknowledgements This work was supported by the Rufford foundation (Grant No. 25789-1), and Ministry of Education, Science and Technological Development of the Republic of Serbia (Grant No. 451-03-9/2021-14/ 200178).

References

1. Hilt S, Brothers S, Jeppesen E, Veraart AJ, Kosten S (2017) Translating regime shifts in shallow lakes into changes in ecosystem functions and services. Bioscience 67:928–936. https://doi.org/10.1093/biosci/bix106
2. Scheffer M (2004) Ecology of shallow lakes. Springer-Science & Business Media, B.V.
3. Silva JP, Phillips L, Jones W, Eldridge J, O'Hara E (2007) LIFE and Europe's wetlands: restoring a vital ecosystem. Office for Official Publications of the European Communities, Luxembourg
4. Laskar HS, Gupta S (2013) Phytoplankton community and limnology of Chatla floodplain wetland of Barak valley, Assam North-East India. Knowl Manag Aquat Ecosyst 411:06. https://doi.org/10.1051/kmae/2013073
5. Scheffer M, van Geest GJ, Zimmer K, Jeppesen E, Søndergaard M, Butler MG, Hanson MA, Declerck S, de Meester L, Persson L (2006) Small habitat size and isolation can promote species richness: second-order effects on biodiversity in shallow lakes and ponds. Oikos 112:227–231. https://doi.org/10.1111/j.0030-1299.2006.14145.x
6. Beklioğlu M, Meerhoff M, Davidson TA, Ger KA, Havens KE, Moss B (2016) Preface: shallow lakes in a fast changing world, the 8th International Shallow Lakes Conference. Hydrobiologia 778:9–11. https://doi.org/10.1007/s10750-016-2840-5

7. Milovanović V, Predojević D, Subakov Simić G, Vidaković D, Trbojević I (2021) Abandoned artificial saline pond–safe place for rare/endangered species. In: Abstracts of the 10th international shallow lakes conference, online, Feb 2021
8. Biggs J, von Fumetti S, Kelly-Quinn M (2017) The importance of small waterbodies for biodiversity and ecosystem services: implications for policy makers. Hydrobiologia 793:3–39. https://doi.org/10.1007/s10750-016-3007-0
9. Scheffer M, van Nes EH (2007) Shallow lakes theory revisited: various alternative regimes driven by climate, nutrients, depth and lake size. Hydrobiologia 584(1):455–466. https://doi.org/10.1007/s10750-007-0616-7
10. Reynolds CS (2006) The ecology of phytoplankton. Cambridge University Press, Cambridge, UK
11. Seleši Đ (2000) The water of the Palić Lake from 1781 to 1999. Public company "Palić-Ludaš", Subotica
12. Seleši Đ (1979) Contribution to the determination of the trophic level of some shallow waters in Vojvodina. In: Proceedings of the second congress of Yugoslav ecologists, Zagreb, 1979
13. Branković D, Budakov LJ, Sekulić N (1998) Phytoplankton as indicator of the saprobiological characteristics of some protected water ecosystems. Nat Prot 50:291–296
14. Seleši Đ (1981) Limnological investigations of Lake Ludoš. Vode Vojvodine 9:333–352
15. Seleši Đ (2006) The water of Ludaš Lake. Public company "Palić-Ludaš", Subotica
16. Đukić N, Pujin V, Maletin S, Gajin S, Gantar M, Petrović O, Ratajac R, Seleši Đ, Matavulj M (1991) Lentic waters eutrophication in Vojvodina part I "Zobnatica." Zbornik radova Instituta za biologiju 31:39–40
17. Seleši Đ (1982) Limnological investigations of Lake Palić. Vode Vojvodine 10:345–358
18. Jovanović J, Predojević D, Trbojević I, Popović S, Blagojević A, Karadžić V, Subakov Simić G (2015) Expansion of *Cylindrospermopsis raciborskii* in stagnant waters of northern Vojvodina. In: Proceedings of the 44th annual conference on current problems of water use and protection "Water 2015", Kopaonik (Serbia), 2–4 June 2015
19. Tokodi N, Drobac D, Meriluoto J, Lujić J, Marinović Z, Važić T, Svirčev Z et al (2018) Cyanobacterial effects in Lake Ludoš, Serbia—Is preservation of a degraded aquatic ecosystem justified? Sci Total Environ 635:1047–1062https://doi.org/10.1016/j.scitotenv.2018.04.177
20. Tokodi N, Drobac Backović D, Lujić J, Šćekić I, Simić S, Đorđević N, Svirčev Z (2020) Protected freshwater ecosystem with incessant cyanobacterial blooming awaiting a resolution. Water 12(1):129. https://doi.org/10.3390/w12010129
21. Šovran S, Jovanović V, Krizmanić J, Cvijan M (2013) Desmid flora from four peat bogs in Serbia. Arch Biol Sci 65(2):721–732. https://doi.org/10.2298/ABS1302721S
22. Fužinato S (2012) Floristic, taxonomic and ecological research of desmid algae in Serbia. Doctoral dissertation, University of Belgrade, Faculty of Biology
23. Technical report (2011) Management plan for the Special Nature Reserve "Gornje Podunavlje" in 2011–2021. Public company "Vojvodinašume", Petrovaradin and Forest Management "Sombor", Sombor
24. Jovanović I (2016) Phytoplankton of selected ponds in the Gornje Podunavlje. Master thesis, University of Belgrade, Faculty of Biology
25. Predojević D, Nikolić N, Jovanović I, Trbojević I, Popović S, Subakov Simić G (2017) Pond phytoplankton community in Special Nature Reserve "Gornje Podunavlje": first report. In: Abstracts of the SEFS 10-Symposium for European Freshwater Sciences, Olomouc (Czech Republic), 2–7 July 2017
26. Miljanović B, Jurca T, Mijić I, Pankov N, Šipoš Š (2007) Macrozoobenthos and phytoplankton in the assessment of the water quality of the Nature Park "Jegrička". In: Proceedings of the 36th annual conference of the Yugoslav water pollution control society "Water 2007", Tara (Serbia), 26–29 June 2007
27. Milovanović D, Živković A (1962) The composition and dynamics of plankton in the fishpond Jegrička in 1959 and 1960. Zbornik radova Biološkog instituta N. R. Srbije 6(2):3–30

28. Pujin V, Đukić N, Maletin S, Miljanović B, Ivanc A (1996) Hydrobionts as indicators of Jegrička water quality. In: Lazić V (ed) Monograph Jegrička. Edicija Tija voda, PČESA, Novi Sad, pp 69–78
29. Branković D, Budakov LJ, Kovačev N, Mijović D, Stanković M (1996) Study proposal for protection of natural resource "Zasavica" as a Special Nature Reserve. Institute for Nature Protection of Serbia, Belgrade, Novi Sad
30. Ratajac R, Rajković D, Stojković S, Bobić M (1998) Investigation of microflora and some groups of microfauna in the protected ecosystem Zasavica. In: Proceedings of the conference on current problems of water protection "Water protection '98", Kotor (Montenegro)
31. Cvijanović M, Miljanović B (2007) Influence of beaver dam on water quality in the river Zasavica. In: Proceedings of the scientific conference ZASAVICA 2007 with international participation, Sremska Mitrovica (Serbia)
32. Predojević D (2017) Ecological status assessment of the Zasavica River based on algological parameters. Doctoral dissertation, University of Belgrade, Faculty of Biology
33. Predojević D, Kljajić Ž, Kovačević E, Milosavljević J, Papić P, Lazić M, Subakov-Simić G (2014) Diversity of Chrysophyceae (Heterocontophyta) in the Zasavica River (Serbia). Arch Biol Sci 66(3):1195–1204. https://doi.org/10.2298/ABS1403195P
34. Predojević D, Subakov-Simić G, Kovačević E, Papić P, Ćuk M, Kljajić Ž, Lazić M (2015) Diversity of the Euglenophyta division in the Zasavica River, Serbia. Botanica Serbica 39(1):23–29
35. Predojević D, Popović S, Kljajić Ž, Subakov-Simić G, Blagojević A, Jovanović J, Lazić M (2015) Diversity of Cyanobacteria in the Zasavica river Serbia. Arch Biol Sci 67(2):355–366. https://doi.org/10.2298/ABS141023024P
36. Subakov-Simić G (2006) Floristic, taxonomic and ecological research of pigmented Euglenophyta in Serbia. Doctoral dissertation, University of Belgrade, Faculty of Biology
37. Kaštovský J, Hauer T, Mareš J, Krautová M, Bešta T, Komárek J, Janeček E (2010) A review of the alien and expansive species of freshwater cyanobacteria and algae in the Czech Republic. Biol Invasions 12(10):3599-3625. https://doi.org/10.1007/s10530-010-9754-3
38. Jovanović J (2020) The distribution of Cyanobacteria in surface waters in relation to water supply and recreation in Serbia. Doctoral dissertation, University of Belgrade, Faculty of Biology
39. Predojević D, Trbojević I, Pećić M, Jakovljević O, Subakov Simić G (2020) Phytoplankton of Zasavica after 20 years since the proclamation of the Special Nature Reserve. In: Proceedings of the 49th annual conference on current topics of water use and protection "Water 2020", Trebinje (Bosnia and Hercegovina), 19–20 Nov 2020
40. Jakovljević O, Trbojević I, Popović S, Pećić M, Subakov-Simić G, Krizmanić J, Predojević D (2020) Water quality assessment of the unique ecosystem of Zasavica (Special Nature Reserve) based on communities of desmids. In: Proceedings of the 49th annual conference on current topics of water use and protection "Water 2020", Trebinje (Bosnia and Hercegovina), 19–20 Nov 2020
41. Technical report (2010) Management plan for the Special Nature Reserve "Obedska bara" for the period 2011–2020. Public company "Srbijašume" and Forest management "Sremska Mitrovica"
42. Milovanović D (1949) Seasonal composition of algae in one shaft of Obedska bara. Glasnik Prirodnjačkog muzeja Srpske zemlje Ser. B, 1–2:127–162
43. Milovanović D, Obušković LJ (1978) Structure and typological characteristics of phytoplankton in permanent standing waters of Obedska bara. Ekologija 13(1):11–23
44. Martinović-Vitanović V (1994) Ecological study of phytoplankton in the "Obedska bara" Nature Reserve. Doctoral dissertation, University of Belgrade, Faculties of Natural Sciences and Mathematics, Faculty of Biological Sciences
45. Budakov LJ, Branković D (1995) The condition of Obedska bara with conditions for its preservation. In: Proceedings of the conference on current problems of water protection "Water protection '95", Tara (Serbia), 7–9 June 1995

46. Ratajac R, Rajković D, Stojković S, Bobić M, Čokić S (1997) The importance of individual members of the biocenosis in preserving the ecological balance of the protected ecosystem of Obedska bara. In: Proceedings of the conference on current problems of water protection "Water protection '97", Sombor (Serbia), 3–6 June 1997
47. Martinović-Vitanović V (1985) Bacillariophyceae from Obedska bara—Krstonošića okno. Biosistematika 11(2):91–98
48. Jovanović J, Trbojević I, Subakov Simić G, Popović S, Predojević D, Blagojević A, Karadžić V (2017) The effect of meteorological and chemical parameters on summer phytoplankton assemblages in an urban recreational lake. Knowl Manag Aquat Ecosyst 418:48. https://doi.org/10.1051/kmae/2017038
49. Dimitrijević LJ (1977) Dynamics of phytoplankton and ecological conditions in Sava Lake near Belgrade. Master thesis, University of Belgrade
50. Obušković LJ (1982) Bloom appearance of the species *Stephanodiscus astrea* var. *minutula* (Kg.) Grun. in Sava Lake near Belgrade. Biosistematika 5(2):127–138
51. Obušković LJ (1982) Dynamics of phytoplankton and some ecological factors as a reflection of eutrophication in Sava Lake near Belgrade. Vodopriveda 14(75–76):123–128
52. Martinović-Vitanović V, Kalafatić V (1990) Classification of some reservoirs in SR Serbia (SFR Yugoslavia) based on analysis of plankton species as indicators of trophic conditions. Arch Hydrobiol Beih 33:831–837
53. Trbojević I, Jakovljević O, Predojević D, Vidaković D, Popović S, Subakov Simić G, Krizmanić J (2018) The first finding of the genus Aneumastus D. G. Mann & A. J. Stickle and distribution of the rare taxon Stauronei balatonis Pantocsek on the territory of Serbia. In: Abstracts of the second congress of biologists of Serbia, basic and applied research, teaching methods, Kladovo (Serbia), 25–30 Sept 2018
54. Subakov–Simić G, Plemić N, Karadžić V, Cvijan M, Krizmanić J (2004) Qualitative and quantitative composition of the Slatina near Opovo. In: Proceedings of the 33th annual conference of the Yugoslav water pollution control society "Water 2004", Borsko jezero (Serbia), 2004
55. Subakov–Simić G, Karadžić V, Fužinato S, Fodora A, Cvijan M (2006) Algological research of the Slatina and the Pečena Slatina near Opovo (Vojvodina). In: Proceedings of the 35th annual conference of the Yugoslav water pollution control society "Water 2006", Zlatibor (Serbia), 2006
56. Subakov–Simić G, Karadžić V, Krizmanić J (2007) Phytoplankton production of the Velika Slatina near Opovo. In: Proceedings of the 36th annual conference of the Yugoslav water pollution control society "Water 2007", Tara (Serbia), 2007
57. Cvijan M, Krizmanić J (2009) *Anabaena bergii* Ostenf. [f. *minor* (Kisselev) Kossinsk.] (Cyanoprokaryota)—The first record in Serbia, its taxonomic status, and that of the genus *Anabaena* Bory ex Born. & Flah. Arch Biol Sci 61(4):883–890. https://doi.org/10.2298/ABS0904883C
58. Fužinato S, Fodora A, Subakov–Simić G (2010) *Arthrospira fusiformis* (Voronichin) Komárek et Lund (Cyanoprokaryota)—a new species for Europe. Algol Stud 134:17–24. https://doi.org/10.1127/1864-1318/2010/0134-0017
59. Cvijan M, Fužinato S (2011) The first finding of *Cylindrospermopsis raciborskii* (Woloszinska) Seenayya et Subba Raju, 1972 (Cyanoprokaryota) in Serbia. Arch Biol Sci 63(2):507–510. https://doi.org/10.2298/ABS1102507C
60. Vidaković D, Ćirić M, Krizmanić J (2020) First finding of a genus *Haslea* Simonsen in Serbia and new diatom taxa for the country's flora in extreme and unique habitats in the Vojvodina Province. Botanica Serbica 44(1):3–9. https://doi.org/10.2298/BOTSERB2001003V
61. Karadžić V (2011) Eutrophication and its consequences on the example of the river Ponjavica (municipality of Pancevo). Doctoral dissertation, University of Belgrade, Faculty of Biology
62. Karadžić V, Subakov-Simić G, Natić D, Ržaničanin A, Ćirić M, Gačić Z (2013) Changes in the phytoplankton community and dominance of *Cylindrospermopsis raciborskii* (Wolosz.) Subba Raju in a temperate lowland river (Ponjavica, Serbia). Hydrobiologia 711:43–60. https://doi.org/10.1007/s10750-013-1460-6

63. Ćirić M, Marković Z, Dulić Z, Subakov-Simić G (2010) First report of cyanobacterium *Cylindrospermopsis raciborskii* from carp ponds in Serbia. In: Abstracts of the 8th international conference on toxic cyanobacteria (ICTC8), Istanbul (Turkey), 2010
64. Ćirić M (2013) Influence of different types of supplementary food in semi—intensive production of juvenile carp (*Cyprinus carpio*) on the structure and dynamics of the pond ecosystem. Doctoral dissertation, University of Belgrade, Faculty of Agriculture
65. Obušković LJ (1991) Phytoplankton and saprobiological characteristics as an indicator of accelerated eutrophication of the Ponjavica River (South Banat). In: Proceedings of the conference on current problems of water protection "Water protection '91", Neum, 1991
66. Đordan S, Obušković LJ (1995) Ponjavica watercourse nature park. In: Proceedings of the international conference on "Industrial Water", Pančevo (Serbia), 1995
67. Karadžić V, Subakov-Simić G (2002) High production of phytoplankton in the Ponjavica River (South Banat) in the winter period. Int Assoc Danube Res 34:153–161
68. Karadžić V, Subakov-Simić G, Krizmanić J (2007) Diversity of phytoplankton in Ponjavica River (Serbia). In: Abstracts of the 3th congress of ecologists of the Republic of Macedonia with internation participation
69. Trifunov S, Planojević I, Jurca T, Nemeš K, Teodorović I (2007) Influence of hydromorphological changes on the ecological status of "Ponjavica" Nature Park. In: Proceedings of the conference on current problems of water protection "Water protection '07", Tara (Serbia)
70. Jovanović J, Karadžić V, Predojević D, Blagojević A, Popović S, Trbojević I, Subakv-Simić G (2016) Morphological and ecological characteristics of potentially toxic invasive cyanobacterium *Sphaerospermopsis aphanizomenoides* (Forti) Zapomelová, Jezberová, Hrouzek, Hisem, Reháková & Komárková (Nostocales, Cyanobacteria) in Serbia. Rev Bras Bot 39(1):225–237. https://doi.org/10.1007/s40415-015-0223-1
71. Branković D, Budakov LJ (1995) Phytoplankton as indicator of the "Stari Begej—Carska bara" Special Nature Reserve waters quality. Prot Nat 46–47:145–150
72. Gigov A, Djerfi B (1960) Previous announcement about the plant cover of Carska bara near Zrenjanin and its history. Zaštita prirode 9:64–71
73. Pujin V, Ratajac R, Đukić N, Svirčev Z, Kilibarda P (1987) Saisonmassige Variationen der Zusammensetzung des Planktons und der Bodenbesiedlung in der Carska bara (Jugoslawien) [Seasonal variations in the composition of plankton and soil colonization in Carska bara (Yugoslavia)]. Tisca 22:83–91
74. Fužinato S, Fodora A, Subakov-Simić G, Krizmanić J (2007) Algological research of the Novo Ilje I and the Novo Ilje II near Melenci (Vojvodina, Serbia). In: Proceedings of the 36th annual conference of the Yugoslav water pollution control society "WATER 2007", Tara (Serbia), 26–29 June 2007
75. Fužinato S, Fodora A, Karadžić V, Subakov-Simić G, Krizmanić J, Andrejić J (2008) Phytoplankton production of the salt puddles Novo Ilje I and Novo Ilje II near Melenci (Vojvodina, Serbia). In: Proceedings of the 37th annual conference of the Yugoslav water pollution control society "WATER 2008", Mataruška Banja (Serbia), 3–6 June 2008
76. Predojević D, Trbojević I, Pećić M, Subakov-Simić G (2018) Phytoplankton community structure in artificial salty lake near Kikinda city. Water Res Manag 8(4):3–9
77. Waiser MJ, Robarts RD (2009) Saline Inland Waters. In: Likens GE (ed) Encyclopedia of Inland Waters, vol 2. Elsevier, Oxford, pp 634–644
78. Boros E, Ecsedi Z, Olàh J (2013) Ecology and management of soda pans in the Carpathian Basin. Hortobágy Environmental Association, Balmazújváros
79. Földi A, Ács É, Grigorszky I, Ector L, Wetzel CE, Várbíró G, Kiss KT, Dobosy P, Trabert Z, Borsodi AK, Duleba M (2018) Unexpected consequences of bombing. Community level response of epiphytic diatoms to environmental stress in a saline bomb crater pond area. PLoS One 13(10): e0205343. https://doi.org/10.1371/journal.pone.0205343
80. Vidaković D, Krizmanić J, Dojčinović BP, Pantelić A, Gavrilović B, Živanović M, Novaković B, Ćirić M (2019) Alkaline soda Lake Velika Rusanda (Serbia): the first insight into diatom diversity of this extreme saline lake. Extremophiles 23(3):347–357. https://doi.org/10.1007/s00792-019-01088-6

81. Brodie J, Jonh DM, Tittley I, Holmes MJ, Williamson DB (2007) Important plant areas for algae. A provisional review of sites and areas of importance for algae in the United Kingdom. Plantlife International, Scotland
82. Cvijan M, Laušević R (1991) Desmids of Vlasinsko Lake—from Peat Bog to Lake. Arch Protistenkd 139:21–37. https://doi.org/10.1016/S0003-9365(11)80004-7
83. Lazarević P (2014) Floristic characteristics of the Peštersko polje area in the Pešter plateau (SW Serbia). In: Mijović D (ed) Nature conservation 64/1. Institute for Nature Protection of Serbia, Belgrade (Serbia), pp 11–20
84. Technical report (2019) Management plan of the Special Nature Reserve "Peštersko polje" for the period 2020–2029. Turistička organizacija, Sjenica
85. Vidaković D, Krizmanić J, Šovran S, Cvijan M (2016) Diatoms from a peat bog on the Pešter plateau (Southwestern Serbia): new records for diatom flora of Serbia. Arch Biol Sci 68(1):107–116. https://doi.org/10.2298/ABS150512132V
86. Simić S, Pantović N, Vasiljević B (2010) Factors threatening the habitats of rare species of Rhodophyta in Serbia. In: Abstracts of the BALWOIS, Ohrid (Republic of Macedonia), 25–29 May 2010
87. Blaženčić J (2014) Overview of the stoneworts (Charales) of Serbia with the estimation of the threat status. Botanica Serbica 38(1):121–130
88. Vesić A (2016) The ecological study of Charophytes (Charophyceae) of standing and slow running waters of Vojvodina. Doctoral dissertation, University of Belgrade, Faculty of Belgrade
89. Blaženčić J, Blaženčić Ž, Ljaljević M (1995) Contribution to the knowledge of the distribution and ecology of stoneworts (Charophyta) in Vojvodina (Serbia, Yugoslavia). Ekologija 30(1–2):1–18
90. Trbojević I, Milovanović V, Subakov Simić G (2020) The Discovery of the Rare *Chara baueri* (Charales, Charophyceae) in Serbia. Plants 9:1606. https://doi.org/10.3390/plants9111606
91. Trbojević I, Marković A, Blaženčić J, Subakov Simić G, Nowak P, Ballot A, Schneider S (2020) Genetic and morphological variation in *Chara contraria* and a taxon morphologically resembling *Chara connivens*. Botany Lett 167(2):187–200. https://doi.org/10.1080/23818107.2019.1703808
92. Trbojević I, Subakov Simić G, Blaženčić J, Predojević D (2019) Rediscovery of *Chara canescens* Loiseleur in Serbia. Botanica Serbica 43(1):97–102. https://doi.org/10.2298/BOTSERB1901097T
93. Trbojević SI, Predojević DD, Šinžar-Sekulić BJ, Nikolić VN, Jovanović MI, Subakov-Simić VG (2019) Charophytes of Gornje Podunavlje ponds: revitalization process aspect Matica Srpska. J Nat Sci 136:123–131. https://doi.org/10.2298/ZMSPN1936123T
94. Lambert-Servien E, Clemenceau G, Gabory O, Douillard E, Haury J (2006) Stoneworts (Characeae) and associated macrophyte species as indicators of water quality and human activities in the Pays-de-la-Loire region, France. Hydrobiologia 570:107–115. https://doi.org/10.1007/s10750-006-0169-1
95. Technical report (2014) Palić Nature Park management plan for the period 2014–2023. Public company "Palić-Ludaš", Subotica
96. Technical report (2019) Management plan of the Special Nature Reserve "Ludaš Lake" for the period 2020–2029. Public company "Palić-Ludaš", Subotica, Serbia
97. Stojanović S, Butorac B, Vučković M, Stanković ŽS, Žderić M, Kilibarda P, Radak LJ (1994) Flora of the Vrbas-Bezdan channel. Prirodno-Matematički fakultet, Novi Sad
98. Technical report (2011) Nature Park "Beljanska bara", proposal for protection as a protected area of III category. Pokrajinski zavod za zaštitu prirode, Novi Sad
99. Technical report (2012) Nature Park "Jegrička", proposal for protection as a protected area of II category. Pokrajinski zavod za zaštitu prirode, Novi Sad
100. Blaženčić J, Stanković M (2008) *Tolypella intricata* (Trent. Ex Roth.) Leonhardi (Characeae) a new species in the flora of Serbia. Zaštita prirode 60(1–2):215–221
101. Vesić A, Blaženčić J, Stanković M (2011) Charophytes (Charophyta) In the Zasavica special nature reserve. Arch Biol Sci 63(3):883–888. https://doi.org/10.2298/ABS1103883V

102. Filarszky N (1931) Adtok Horvát-Szlavonország és a Balkán néhány más országának Chara-vegetációjához / Beitrage zur Kenntnis der Charenvegetation Kroatien-Slavonien und einiger Lander der Balkan-Halbinsel [Addition to the *Chara* vegetation of Croatia-Slavonia and some other countries in the Balkans]. Ungarische Botanische Blatter 1(12):81–99
103. Slavnić Ž (1956) Aquatic and marsh vegetation of Vojvodina. Zbornik Matice Srpske 10:5–72
104. Blaženčić J, Cvijan M, Laušević R (1995) Diversity of freshwater algae of Yugoslavia with an overview of species of international importance. In: Stevanović V, Vasić V (eds) Biodiversity of Yugoslavia with overview of species of international importance. Ekolibri, Biološki fakultet, Beograd, pp 159–172
105. Alderton E, Sayer CD, Davies R, Lambert SJ, Axmacher JC (2017) Buried alive: Aquatic plants survive in 'ghost ponds' under agricultural fields. Biol Conserv 212:105–110. https://doi.org/10.1016/j.biocon.2017.06.004
106. Blindow I, Schubert H (2004) *Chara canescens* Desv. et Loisel. in Loisel. 1810. In: Schubert H, Blindow I (eds) Charophytes of the Baltic Sea. Koeltz, Königstein, pp 70–81
107. Küster A, Schaible R, Schubert H (2004) Light acclimation of photosynthesis in three charophyte species. Aquat Bot 79:111–124. https://doi.org/10.1016/j.aquabot.2004.01.010
108. Schubert H, Telesh I, Nikinmaa M, Skarlato S (2017) Physiological adaptations. In: Snoeijs-Leijonmalm P, Schubert H, Radziejewska T (eds) Biological oceanography of the Baltic Sea. Springer, Dordrecht, pp 255–278
109. Schaible R, Bergmann I, Schubert H (2011) Genetic structure of sympatric sexually and parthenogenetically reproducing population of *Chara canescens* (Charophyta). ISRN Ecol 2011:501838, 13. https://doi.org/10.5402/2011/501838
110. Schaible R, Gerloff-Elias A, Colchero F, Schubert H (2012) Two parthenogenetic populations of *Chara canescens* differ in their capacity to acclimate to irradiance and salinity. Oecologia 168:343–353. https://doi.org/10.1007/s00442-011-2113-y
111. Urbaniak J, Gabka M (2014) Polish Charophytes. An illustrated guide to Identification. Uniwersytet Przyrodniczy We Wroclaviu. Wroclav
112. Langangen A (2005–2006) *Chara rohlenae* Vilhelm (Charales) found in Greece. Allionia 40:31–33
113. Blaženčić J, Stevanović B (2018) Rediscovery of the endemic species Chara rohlenae Vilh 1912 (Characeae)-believed extinct-on the Balkan Peninsula. Botanica Serbica 42(1):109-115. https://doi.org/10.5281/zenodo.1173564
114. Blaženčić J, Stevanović B, Blaženčić Ž, Stevanović V (2006) Red data list of charophytes in the Balkans. Biodivers Conserv 15:3445–3457. https://doi.org/10.1007/s10531-005-2008-5
115. Košanin N (1907) Our Characeae, a contribution to the knowledge of the algae flora of Serbia. Nastavnik 18:264–267
116. Official Gazette RS (2010) Rulebook of the proclamation and protection of strictly protected and protected wild species of plants, animals and fungi, number 5/2010-46, 32/2016-59, 98/2016-97
117. Official Gazette RS (2011) Rulebook of the proclamation and protection of strictly protected and protected wild species of plants, animals and fungi, number 47/2011-134, 32/2016-59, 98/2016-97
118. Simić BS, Đorđević BN (2017) Morphology, distribution and ecology of the freshwater red algae *Paralemanea* (Batrachospermaceae, Batrachospermales, Rhodophyta) in Serbia. Arch Biol Sci 69(1):167–174. https://doi.org/10.2298/ABS160211093S
119. Blagojević A, Subakov Simić G, Blaženčić J, Ilić M, Petrović J, Kostić D, Marjanović P (2017) First record of *Paralemanea torulosa* (Roth) Sheath & A.R. Sherwood and new findings of *Lemanea fluviatilis* (Linnaeus) C. Ag. and *Hildenbrandia rivularis* (Liebmann) J. Agardh (Rhodophyta) in Serbia. Botanica Serbica 41(1):55–63. https://doi.org/10.5281/zenodo.454096
120. Mitrović AB, Đorđević NB, Simić SB (2021) A review of research on the *Lemanea* genus in Serbia. Oceanol Hydrobiol St 50:47–59. https://doi.org/10.2478/oandhs-2021-0006
121. Komulaynen S (2010) Algae protection, conservation areas and the red data book of the Republic of Karelia. Oceanol Hydrobiol St 39(2):147–152. https://doi.org/10.2478/v10009-010-0028-2

122. Juráň J, Kaštovský J (2019) The procedure of compiling the Red List of microscopic algae of the Czech Republic. Biodivers Conserv 28:2499–2529. https://doi.org/10.1007/s10531-019-01792-x
123. Brodie J, Andersen RA, Kawachi M, Millar AJK (2009) Endangered algal species and how to protect them. Phycologia 48(5):423–438. https://doi.org/10.2216/09-21.1
124. Joye DA, Rey-Boissezon A (2015) Will charophyte species increase or decrease their distribution in a changing climate? Aquat Bot 120:73–83. https://doi.org/10.1016/j.aquabot.2014.05.003
125. Blaženčić J (1986) Review of the development of algology in Serbia from 1883 to 1983. Glasnik Instituta za Botaniku i Botaničke Bašte Univerziteta u Beogradu 20:99–108
126. Tokodi N (2016) Toxic Cyanobacteria from territories of the Republic of Serbia. Doctoral dissertation, University of Novi Sad, Faculty of Natural Sciences and Mathematics, Department of Biology and Ecology
127. Vukojičić S, Lakušić D, Jovanović S, Marin PD, Tomović G, Sabovljević M, Šinžar-Sekulić J, Veljić M, Cvijan M, Blaženčić J, Stevanović V (2011) University of Belgrade Herbarium–treasury of data and challenges for future research-On the occasion of the 150th anniversary of University of Belgrade Herbarium (1860–2010). Botanica Serbica 35(2):163–178
128. Košanin N (1907) Characeen Serbiens [Characeae of Serbia]. Österreische Botanische Zeitschrift 57:280–282
129. Schneider SC, García A, Martín-Closas C, Chivas AR (2015) The role of charophytes (Charales) in past and present environments: an overview. Aquat Bot 120:2–6. https://doi.org/10.1016/j.aquabot.2014.10.001
130. Zhang Y, Xiao X, Cao R, Zheng C, Guo Y, Gong W, Wei Z (2020) How important is community participation to eco-environmental conservation in protected areas? from the perspective of predicting locals' pro-environmental behaviours. Sci Total Environ 739:139889 https://doi.org/10.1016/j.scitotenv.2020.139889

Chapter 9
Springs and Headwater Streams in Serbia: The Hidden Diversity and Ecology of Aquatic Invertebrates

Ivana Živić, Katarina Stojanović, and Zoran Marković

Abstract As places where groundwater comes to the surface of the earth, springs are of relatively uniform abiotic properties. Springs are habitats of numerous organisms dominated by macrozoobenthos. Different geological structures, different altitudes, terrain aspect, capacities and plant communities surrounding them initiate a wide range of conditions for developing benthic communities. Such a collage of uniformity, but also diversity, make springs special and specific, often the habitat of endemic and relict species. Research on aquatic invertebrates, an essential segment of springs and biocenoses of headwater streams, is critical, both from the point of view of better knowledge of aquatic ecosystems and preserving the diversity of spring communities. Upper reaches of streams are increasingly exposed to the effects of anthropogenic influence. In addition to the frequent uncontrolled capture, these fragile ecosystems have become endangered by the construction of small hydropower plants in recent years. With the construction of small hydropower plants, hilly mountain streams, as well as their banks, are being turned into landfills of dirt and gravel, with a small amount of water preventing the survival of plants and animals that inhabit these streams and inland ecosystems surrounding them.

Keywords Macrozoobenthos · Diversity · Bioindicators · Fragile ecosystems · Anthropogenic impact

---

I. Živić (✉) · K. Stojanović
Faculty of Biology, University of Belgrade, Studentski Trg 16, 11000 Belgrade, Serbia
e-mail: ivanas@bio.bg.ac.rs

K. Stojanović
e-mail: k.bjelanovic@bio.bg.ac.rs

Z. Marković
Faculty of Agriculture, University of Belgrade, Nemanjina 6, 11080 Belgrade, Serbia
e-mail: zoranmm@agrif.bg.ac.rs

V. Pešić et al. (eds.), *Small Water Bodies of the Western Balkans*, Springer Water,
https://doi.org/10.1007/978-3-030-86478-1_9

9.1 Introduction

Although it takes up only 2.1% of the European mainland, Serbia is characterized by exceptional species, genetic, and ecosystem diversity, conditioned primarily by the geographical position and influence of the continental and Mediterranean climate [1], but also by its geomorphological and geological diversity. A significant share of Serbia's biodiversity is related to aquatic ecosystems. There are almost all types of inland waters in Serbia, of which running waters with a drainage area of about 88.360 km^2 are the most important [2].

How interesting the area of the Serbian waters is, from the biodiversity point of view, is best understood by looking at the aquatic communities belonging to separate river basins. Illies [3] lists 5 ecoregions for the territory of Serbia (Fig. 9.1): Dinaric Western Balkans (ER 5), Hellenic Western Balkans (ER 6), Eastern Balkans (ER 7), the Carpathians (ER 10), and Pannonian Lowland (ER 11). In general, the territory of Serbia can be divided into two major geographical and orographic regions: the Pannonian Plain—with relatively homogeneous environmental factors that characterize this climate, and the hilly-mountainous area south of the Sava and Danube, which actually is the part of Serbia belonging to the Balkan Peninsula [4]. The Pannonian area (ER 11) consists of plains and their rivers that belong to the Danube and Sava basins. According to the revised boundaries of ecoregions based on macroinvertebrate communities' composition, this ecoregion also includes the lower reaches of the rivers Drina, Kolubara, Mlava, and Velika Morava [5]. On the other hand, the basins of the upper reaches of these rivers in Serbia are more in line with ER 5 in relation to the composition and diversity of macroinvertebrate communities [6, 7].

In the context of complex ecological factors, the hilly-mountainous area of Serbia is particularly interesting. This area is characterized by a great number of springs, brooks and streams with specific: substrate types, surrounding vegetations, fauna and microclimatic conditions. The remaining ecoregions (ER 5, ER6, ER7 and ER 10, Fig. 9.1) are located within this area whose streams form a dense river network and which predominantly belong to the Black Sea (complete ER 5 and ER 10, as well as most of ER 7), but also the Adriatic (ER 6) and the Aegean basins (ER 7e, Fig. 9.1). As a border between ER 5 and ER 7, Illies [3] defined basins of Great and South Morava rivers. Paunovic et al. [5] suggested the shift of the border between ER 5 and ER 7 to the east, presented by the western border of the Timok river basin. The change of the border between these two ecoregions is justified by the fact that the basin of the Timok river is characterized by macrozoobenthos communities, which can be separated from the communities of other basins in terms of their composition and diversity [7, 8].

Due to the incredible diversity (the number of determined species of macroinvertebrates is usually higher than 100 in small watercourses of Serbia, [9]), systematic research of macroinvertebrates of springs, brooks and streams in Serbia began almost a century ago [10, 11] and have been ongoing with varying intensity to date [12–21]. The earliest research on macrozoobenthos dealt with springs and brooks. The first work was about spring and brook planarians [10], then about the larvae and nymphs

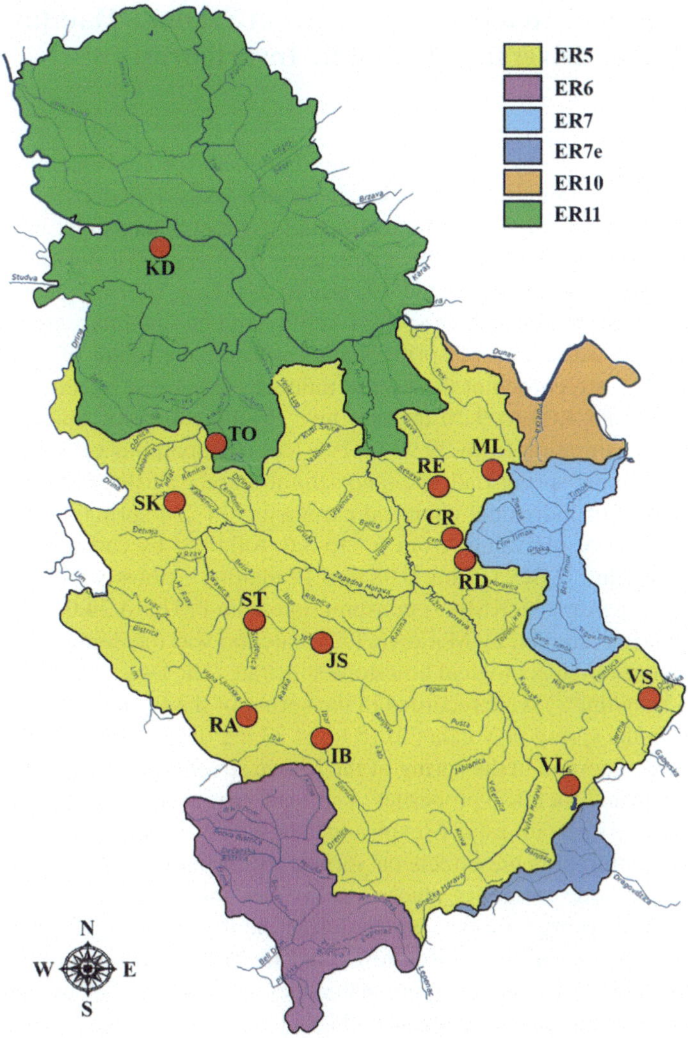

Fig. 9.1 Ecoregions of Serbia according to Illies (modified by Paunović et al. [5]) with researched localities. KD—Kudoški potok brook, TO—Toplica stream, SK—Skrapež stream, ST—Studenica stream, JS—Jošanica stream, RA—Raška stream, IB—Ibar stream, VL—Vlasina stream, VS—Visočica stream, RD—Radovanska reka stream, CR—Crnica stream, RE—Resava stream, ML—Mlava stream

of the torrential trichoptera *Thremma* sp. from brooks and springs of Western and Southern Serbia [11], and there is also a record from 1882 of the species *Ancylus fluviatilis* from the spring of the Zlotska river and the spring of the Mlava near Žagubica (a collection of freshwater snails by Lazar Dokić, [22]).

9.2 Springs and Headwater Streams in Serbia—Habitat of Hidden Diversity of Aquatic Invertebrates

Springs, places where groundwater naturally springs to the surface, are the border area of two biotopes: the biotope of organisms that inhabit groundwater flows and the biotope of water communities of brooks, rivers, and lakes. Individually, springs are characterized by relatively uniform conditions for the organisms that inhabit them. Spring water is often characterized by a deficiency of oxygen. It is not uncommon for spring waters to have a higher presence of carbon dioxide, as well as other undesirable gases [23]. During the research of springs in hilly and mountainous areas of Serbia [23], temperature variations of spring waters were in the range of 7–16 °C. An oxygen deficit was recorded at the largest number of 234 researched springs [23], which ranged from 4.09 to 9.20 mg/dm dm^{-3} in the source of Sušica (followed by low oxygen saturation from 39.94 to 79.94%) to maximum values from 9.8 to 12.6 mg/dm^{-3} in the spring Lađevac (and saturation range from 99.69 to 135.5%). Measured values of individual chemical parameters in 234 springs in Serbia [23], varied in the case of pH values from 6.1 to 8.6, followed by carbon dioxide from 2.2 to 29.04 mg/dm^{-3}, BOD_5 from 0.08 to 6.19 mg/dm^{-3} (mostly the spring values were up to 3.0 mg/dm^{-3}), $KMnO_4$ consumption from 0.073 to 24.02 mg/dm^{-3}. In most of the springs studied by Marković [23], calcium was the dominant ion, which is a consequence of the geological limestone substrate. The lowest values of chloride (1.88 to 2.55 mg/dm^{-3}) were recorded in the spring of the Pečinska river, while the highest was in the spring of Vojala (20.15 to 24.22 mg/dm^{-3}), and the highest value of sulfates was recorded in the spring of the Marec stream (88.23 mg/dm^{-3}).

The physical and chemical properties of springs, terrain configuration, spring type (Fig. 9.2a, b, c, d), terrain aspect (exposure), the effect of plants and animals, and above all, the anthropogenic effects on springs differ between them. These differences creating a wide variety of abiotic parameters, that is, the whole spectrum of characteristics of spring biotopes, which determine the diversity of qualitative and quantitative properties of spring biocenoses, and, the dominant community of springs – benthocenoses [23]. It is not uncommon that they are habitats of many stenothermic species that are often relics or endemics [24–27].

Springs (the krenal zone, [28]) represent the initial parts of streams, brooks and rivers, but also a natural supply of water to lakes. The next sections belong to rhithral (the zone of mountain brooks—salmonid region, divided into epirhithral, hyporhithral, metarhithral zones) and potamal zone (zone of rivers, divided into epipotamal, hypopotama and metapotamal zones) where it is not easy to notice a sharp border between them (Fig. 9.2e, f, g, h). However, the impression is that they are clearly separated from the downstream part of the flow. That "border" between the spring and brook benthic fauna is at different distances from the place where the water comes to the surface—from usually a few meters to a few tens of meters. Setting the "borderline" and the transition point between spring benthocenosis and brook benthocenosis is based on the analysis of the diversity of macrozoobenthos

Fig. 9.2 Springs and streams of Serbia: **a** Spring of Kaluđerska Vrela; **b** Spring of Crnica stream; **c** Spring of Raška stream; **d** Spring of Mlava stream; **e** Ljuboviđa stream; **f** Studenica stream; **g** Jerma stream; **h** Rogačica stream. Photos by K. Stojnović

communities. In the case of a comparative analysis of macrozoobenthos communities of springs and downstream communities, the similarity expressed through the Jaccard index generally ranges from 10 to 30%, which clearly confirms that the spring community is something specific in relation to communities of further river sections [23]. The explanation of the more uniform composition of macrozoobenthos communities of springs compared to macrozoobenthos communities of downstream sections can be found in the uniformity of ecological factors of springs. Comparing any abiotic or biotic factor and analyzing it along the river course will result in a much smaller amplitude of variation (whether it is about quantifiable factors or those with descriptive characteristics) at springs compared to the more "colorful" mosaic of environmental factors in streams.

9.3 Macrozoobenthos Communities of Springs and Headwater Streams

Lower variety of ecological factors in springs (primarily water temperature and the amount of dissolved oxygen) causes a lower diversity of macrozoobenthos. The presence of 10 taxa (identified at the genus or species level) from 8 groups of macroinvertebrates was recorded in five Fruška Gora springs [23]—Turbellaria, Hirudinea, Amphipoda, Mollusca, Plecoptera, Diptera, Trichoptera. Eleven taxa were found in the spings of three Fruška Gora brooks—Borkovački, Jelenački and Kudoški [29].

In the ecological study of macrozoobenthos of 234 springs of Serbia [23], 253 taxa from 23 groups of macrozoobenthos organisms were identified. The most diverse insect orders were Trichoptera (43 taxa), Ephemeroptera (37), Plecoptera (26) and Diptera (44 taxa), and from non-insect group Oligochaeta, with 30 idetified taxa. *Dugesia gonocephala* (Turebllaria), *Ancylus fluviatilis* and *Radix peregra* (Mollusca), *Erpobdella octocullata* (Hirudinea), *Gammarus balcanicus* (Gammaridae), *Elmis aenea* (Coleoptera), *Sericostoma personatum*, *Lithax niger*, *Goera pilosa* and *Polycentropus irroratus* (Trichoptera) were found in the largest number of localities.

The composition of macrozoobenthos communities at the springs of Radovanska reka and the Crnica streams (Figs. 9.1 and 9.2b) is rather similar [30]. In addition to the species *Ancylus fluviatilis* and *Gammarus balcanicus* that can form dense populations, the presence of the following species of oligochaetes was also recorded in the springs of both streams: *Bythonomus lemani*, *Haplotaxis gordioides*, *and Stylodrilus heringianus*. The community composition of dipteran at the spring of the Radovanska reka and the Crnica streams comprises the following species: *Brillia bifida*, *Diamesa insignipes*, *Atherix marginata*, *A. ibis*, *Dicranota bimaculata,* and *Hexatoma bicolor*. EPT communities of the springs of these two rivers are similar, but specific taxa for each river have also been recorded. For example, in the case of Trichoptera, the species *Thremma anomalum* and *Ecclisopteryx dalecarlica* were noted in the Radovanska reka stream, while the species *Goera pilosa, Lithax niger,* and *Polycentropus flavomaculatus* were found in the Crnica stream [30].

Areas of upper reaches (epiritral and metaritral) with high flow velocity, rocky substrate, good aeration, shallows and uniform annual temperature are characterized by communities of macrozoobenthos of greater diversity than in springs. These areas in Serbia, which mostly include the South and West Morava basins, are centers of diversity of many groups of aquatic insects (Diptera, Ephemeroptera, Plecoptera, Trichoptera) and can support genetically isolated species, thus contributing to the overall biodiversity of the region [31]. EPT taxa are the most diverse in our brooks and rivers. Today, the Serbian Trichoptera fauna has 227 species from 59 genera and 17 families [31]. Mayflies (Ephemeroptera) are represented by 85 species with 31 genera from 12 families [17], and Plecoptera by 90 species, 17 genera from 7 families [32]. At the same time, the species within the EPT group belong to the sensitive groups of aquatic insects. Among them, the most sensitive to changes in abiotic environmental conditions are the species of the order Plecoptera. Many species of freshwater invertebrates in Serbia are endemic, rare, and at the same time, they are endangered and are on the international and national lists under some degree of protection. Out of the total number of EPT taxa in Serbia, 29 species are protected at the national level [33]. Moreover, stone and noble crayfish are among the most endangered aquatic invertebrates, and that is the reason why they were in the monitoring process for several years. In addition to the EPT group, the Odonata order showed greater diversity too. There are 67 dragonfly species in the fauna of Serbia [34].

As for the non-insect groups, our fauna has Hirudinea, Turbellaria, Mollusca, Crustacea, Acari (Hydracarina). The most diverse ones are Mollusca, Hirudinea and Hydracarina. There are about 65 and 21 species of freshwater snails and bivalves, respectively in Serbia's fauna (unpublished data). Typical species of upper and middle ritral communities from Mollusca phylum are stenoendemics snails such as *Bythinella nonveilleri* and *B. pesterica* [35], as well as endemics *Bythinella serborientalis*, *Iglica (Raphica) illyrica,* and *Belgrandiella bunarbasa.* There are 39 species of leeches recorded from 16 genera and 5 families [36], with only some of them inhabiting springs and upper reaches: *Dina prokletijaca* [37], *Dina lineata dinarica* [38], *Hirudo verbena* [29, 39]. Planaria *Crenobia alpina*, *Polycelis felina*, and *Planaria montenegrina* [7] were often found. Crayfish are represented by species of the genus *Gammarus* (*G. balcanicus*, *G. fossarum*, *G. dulensis*), stone crayfish (*Austropotamobius torrentium*), and noble crayfish (*Astacus astacus*). About 20 species of water mites can be found in spring regions and upper reaches of Serbian rivers [40, 41].

9.4 Springs and Headwater Streams—Diversity Refugia

Along with the Iberian and Apennine peninsulas, the Balkan Peninsula is one of the three most important refugial areas in Europe, with the addition of the area south of the Caucasus Mountains [42, 43]. The reason for the high biodiversity in the Balkans lies in the fact that many species found refugium within specific isolated habitats [44] during the retreat of the organisms to the south in the Pleistocene glaciation period. The example of certain groups of aquatic insects showed that species that inhabit isolated habitats also have a minimal possibility of dispersion [45]. From the metacommunity point of view, it is interesting to analyse whether, above all, the spatial processes or environmental gradient shape the composition of these isolated aquatic insect's communities [46].

Rare and endemic taxa of aquatic insects in Serbia were mostly found in isolated habitats (most often mountain springs) and/or are often with disjunct areal [23–25, 30, 47–49]. A group of aquatic insects of mountain springs and brooks from the genus *Drusus* Stephens, 1837 (Trichoptera, Limnephilidae, Drusinae) is a convenient example. Within this genus, a large number of endemic species of the Balkans have been described. The reason for such high species diversity of the mentioned genus lies in its populations' isolation and the uneven geographical distribution of species [50]. This is the reason why the genus *Drusus* was a model organism for studying the process of speciation and diversification in Trichoptera [27, 50–53]. Two stenoendemic species of this genus are particularly illustrative in the fauna of Serbia—*Drusus serbicus* Marinković-Gospodnetić, 1971 (Fig. 9.3) and *Drusus zivici* Kučinić, Previšić, Stojanović & Vitecek, 2017. *D. serbicus* is a stenoendemic of southwestern Serbia. The species is originally described on adult individuals collected near the Zlošnica brook, Akmačići village, Nova Varoš area [51]. After almost half a century, a larva of this species was described, and its identity was confirmed by molecular methods in association with adult individuals [26]. Larvae and also adults were collected in the Ilinac spring (Golija Mountain) which is at a distance of about 60 km in a straight line from a type locality. The species hadn't been found again in the Zlošnica brook and its tributaries. The assumption is that the species *D. serbicus* disappeared from the locality from where it had been originally described due to the strong anthropogenic influence. Recently described *D. zivici* is a stenoendemic species of the Stara Planina Mountain in Serbia [27]. It is a typical rithrobiont, registered in the Midžor massif, at altitudes of about 1500–1900 m, in the springs of the Tovarnička, Rekička and Javorska streams, as well as in the spring of Kaluđerska Vrela [27].

The mentioned species of the genus *Drusus* are also excellent examples of the fauna of springs and headwater streams being endangered and subdued to increasing anthropogenic pressure. Based on the fact that *D. serbicus* has not been re-registered at the type locality as well as that the type locality of *D. zivici* is located near a well-known tourist destination in Serbia (Stara Planina ski resort), the survival of these species is at risk. It is necessary to emphasize that both species are found within protected natural areas. In addition to *D. serbicus,* which has the status of a protected species in Serbia [33], the species *D. zivici* is in the process of obtaining a status of a protected species at the national level.

Fig. 9.3 The larva of *Drusus serbicus* Marinković-Gospodnetić, 1971. Photo by K. Stojnović

9.5 Specific Communities of Macrozoobenthos of Thermal Springs and Brooks

Changes in the composition of benthic communities of aquatic ecosystems that are recipients of thermal brooks could indicate the effects and mechanisms of thermal pollution, free from the influence of other types of anthropogenic pollution [54]. The chosen model was a hydrological system of a typical hilly-mountain stream (Toplica stream, Fig. 9.1) and its tributary—a thermal brook (in Banja Vrujci). In this model, the effect of all abiotic factors (except temperature) on macrozoobenthos community is minimized [54]. It was shown that the bentocenoses of the thermal brook are characterized by less diversity and larger biomass (which is primarily a consequence of the high number of Gastropods) compared to the bentocenoses of cold and eurythermal water.

The basic ecological factor driving changes in diversity and biomass along the thermal brook stream is the gradient of mean annual temperatures—not their variability. With a decrease in mean annual temperature, there is an increase in diversity and a reduction in biomass [54]. Also, a specific community of macrozoobenthos was recorded in the thermal brook, which shows a small degree of similarity with the community of eurythermal and cold waters. The warm brook community of macrozoobenthos is dominated by non-insect groups Gastropoda, Gammaridae, and Turbellaria. At the same time, in the cold waters of the Toplica stream, the larvae of insect groups have the greatest diversity, which dominate with Gammaridae in number, except the locality 20 m downstream from the mouth of the thermal brook, where non-insect groups dominate in the total macrozoobenthos with 59.3% [54].

In another model system, there was a research of the effects of geothermal water inflow (from Banja Vrdnik) on the composition of the benthic macroinvertebrate community in Kudoški potok brook (located on the edge of the Pannonian plain in a temperate zone, Fig. 9.1 [55]). Due to the inflow of hot water from the spa into Veliki potok brook, the average annual water temperature increases by as much as 10.5 °C [55]. This temperature gradient was used to investigate the influence of water temperature on the macroinvertebrates' community structure. Of the 11 biotic parameters used to quantify the effect of the increase in average temperature (t_{av}), species richness, characterized by a significant decrease with thermal gradient, proved to be the most sensitive parameter, while the total abundance was the least sensitive and increased along with the t_{av}. The increase in the relative abundance of Gastropoda and the decrease in number of species of Ephemeroptera, Plecoptera, and Trichoptera orders were the earliest response of taxonomic groups to the increase in $t_{av,}$ which resulted in the potamonization of the macrozoobenthos community. The community shifted from the hyporhithral to the metapotamal community at the highest temperature gradient point [55].

These studies confirmed the general stance that Gastropoda are the most numerous inhabitants of thermal brooks [54, 56, 57], while the increase in abundance of the Chironomidae family was due to a significant increase in relative abundance of the tolerant tribe Chironomini. On the other hand, relative abundance decreased within the more sensitive subfamily Orthocladinae, which was also observed in the geothermal brooks in Yellowstone [58]. Due to this group's great diversity, the conclusion was that the presence of the family Chironomidae is not a proper indicator of the effects of rising water temperature. One must consider taxonomic categories of subfamilies and tribes when this group of organisms is in question [59].

In contrast to the thermal brook in Banja Vrujci, where they were the dominant group, the absence of Gammaridae in localities downstream from the inflow of geothermal waters in Kudoški potok brook is most likely due to significantly higher temperature fluctuations in Kudoški potok brook. This indicates that Gammaridae can survive and be very competitive at high water temperatures as long as their annual variation is small [55]. It is important to emphasize that the changes in the community structure did not follow the t_{av} linearly—there was a disproportionately large change in their structure at $t_{av} = 20$ °C [59]. One of the conclusions of this paper was that such a thermal brook model system could be used to predict the effects of global warming on macroinvertebrate communities in running waters.

9.6 Suborganismal Responses as an Endpoint in Biomonitoring of Headwater Streams

By applying biomarkers, biomonitoring of aquatic organisms takes on a new dimension of physiological (functional) monitoring. Molecular biomarkers (oxidative stress enzymes, acetylcholinesterases, changes in the DNA molecule) point to the presence of stressors and changes in environmental factors (oxygen, pH, and temperature) even when their concentrations are low or on the verge of detection [60].

In recent years in Serbia, there have been analyses of the influence of trout farms on the macrozoobenthos communities of headwater streams by examining the oxidative stress enzymes response [61–65]. This response is the first line of defense of the organism against pollutants from the environment [30]. There is special emphasis on research into the effects of trout farm discharges on components of the antioxidant defense system (superoxide dismutase—SOD, catalase—CAT, glutathione peroxidase—GPx, reductase—GR and S transferase—GST enzyme activities, and total glutathione concentration) in the larvae of *Dinocras megacephala* (Plecoptera), *Ephemera danica* (Ephemeroptera), *Ecdyonurs venosus* (Ephemeroptera), and *Gammarus dulensis* (Amphipoda) [61–64].

The effects of trout farms were investigated on the streams of Raška (Southern Serbia, Fig. 9.1), Crnica (Eastern Serbia, Fig. 9.1), and Skrapež (Western Serbia, Fig. 9.1). The impact of fish farm wastewater on the chemical parameters of the water of those streams was not significant, but had been expected, as it was assumed that molecular biomarkers should show a high sensitivity to pollution. The effects of trout farm discharges on water chemistry decreased from Raška, through Skrapež to Crnica, where it was negligible [61–63]. The effects of farms were primarily reflected in a decrease in the concentration of dissolved oxygen and an increase in the concentration of ammonium ions and total phosphorus, which indicates dominant organic pollution [61–63].

The influence of farm effluents on SOD and CAT activity in *D. megacephala* larvae was examined in the Raška stream (Fig. 9.1), and it was shown that there is a significant increase in the activity of both enzymes after an inflow of such water, which indicates induction of oxidative stress [61]. The effects on CAT were more pronounced and could be explained by an increase in NH_4^+ concentration levels and decreased in dissolved oxygen concentration. In contrast, the relationship between the increase in SOD activity could not be related to the measured chemical parameters of water.

In bioindicator species *Ephemera danica* and *Ecdyonurus venosus* (Ephemeroptera), the activity of SOD, CAT, and GPx was measured, as well as the concentration of GSH in the Skrapež and Crnica streams (Fig. 9.1) [62, 64, 65]. Characteristically, CAT activity could not be registered, SOD activity was not sensitive to farm effects, while the most sensitive biomarker was GPx activity which was characterized by a statistically significant increase, except for *E. venosus* in the Crnica stream. If we take into account that waste water did not influence increase in GPx activity in *E. venosus* larvae in Crnica [64], and that in the case of *E. danica* it was 6 times lower than in Skrapež [65], we can conclude that this biomarker shows significant sensitivity to the intensity of organic pollution, as well as that *E. danica* is a more sensitive bioindicator of organic pollution compared to *E. venosus* [62, 64, 65].

However, *G. dulensis* turned to be the most sensitive bioindicator species in these studies [63]. Unlike Ephmeroptera larvae, specimens of *G. dulensis* were significantly more pronounced and reflected in a statistically significant decrease in the activity of SOD and CAT and a significant increase in the activity of GST and GR. Also, there was no significant change in GPx activity in *G. dulensis*. The decrease in SOD activity could be correlated with an increase in total phosphates and a reduction in dissolved oxygen concentration, while an increase in GST activity could be in relation with an increase in BOD5 and TOC, and a decrease in CAT activity with an increase in toxic metals (Cr, Ni and Cd) concentrations in the sediment [63]. If the effects of farms on CAT and SOD activities are compared between *D. megacephala* larvae and *G. dulensis* individuals, it can be seen that the effects are opposite, which can be partly explained by different pollution intensities (higher in Raška stream, [61]) and different stressor types (toxic metals in Crnica stream, [63]), but the specifics of bioindicator species must undoubtedly be taken into account.

The presented results clearly indicate that the components of the antioxidant defense system are biomarkers that are sensitive to pollution originating from trout farms, but that the direction and level of this sensitivity depend not only on the intensity and type of stressor but also on the selected bioindicator species, so there should be no generalizations and comparisons made.

9.7 The Effects of Trout Farms on Macrozoobenthos Communities

One of the potential factors disrupting the ecosystems of the upper reaches of mountain streams are trout farms. The first indication that the water flow has been enriched with nutrients is the disappearance or decline in abundance of sensitive taxa such as the larvae of the insect orders Plecoptera, Ephemeroptera, and Trichoptera. Different types of community metrics [66] can be applied to obtain information about the influence of trout aquaculture on the recipient streams. Many of the standard types of biotic indices used in our studies have been shown to be very effective in assessing the status of the headwater streams under the infuence of trout farms [30, 67, 68]. We showed that biotic indices developed in the United States [66] could quantify the effects of stressors at an equal level as the indices in standard use. It was shown that a modified biotic index—MBI [69] quantified the impact of fish farm on the Trešnjica stream in Serbia very successfully [67, 68]. Indices that include the number of organisms proved to be even better for monitoring the impact of trout farm discharges. The family biotic index—FBI [70] was among the three most sensitive of the 46 biotic parameters used to quantify the impact of 9 trout farm effluents on receiving watercourses [30], along with the MASPTar index [71]. The FBI index clearly indicated that the water quality on the stream Raška (Fig. 9.1), where a trout farm with a smaller production capacity is located, changes at the localities below the farm outlet [30]. The water in the reach downstream from the trout farm is significantly to quite significantly polluted in relation to the water quality at the control locality (control locality—FBI = 4.20 ± 0.10; downstream locality—FBI = 5.20 ± 0.20). Also, an index indicating a change in the relative abundance of EPT [72] resulted as sensitive [30, 67]. Thus, for example, in the Radovanska stream (Fig. 9.1), it was registered that the decrease of the relative number of EPT taxa was statistically significant in the locality immediately downstream from the trout farm (EPT = 4.34 ± 1.29) in relation to the control locality (EPT = 24.63 ± 6.97). In contrast, an increase in the number of pollution-tolerant indicator groups, such as the Oligochaeta and Chironomidae communities, can be expected at localities downstream of the farm outlet.

From the diversity point of view, several studies [67, 73, 74] report a decline in the alpha diversity of macrozoobenthos communities downstream the outlet. However, our research indicates that this component of diversity was not sensitive to the influences of trout farms [30]. Observed in relation to the communities of Chironomidae and Oligochaeta, their diversity has a significant contribution to the overall diversity of macroinvertebrate communities in the localities below the farms. On the other hand, the overall diversity in localities further from the farm outlet may still remain high, but in favor of EPT taxa that take up a larger share in those localities due to the recovery of their communities [30].

Changes in the trophic structure of macrozoobenthos communities (quantified through indices based on functional feeding groups [75]) under the influence of trout farm discharges follow the concept along the longitudinal river gradient, indicating trout farms causing a potamological effect on the trophic structure of macrozoobenthos communities downstream. Our research [30, 67] shows a statistically significant increase in the total participation of collectors (% COLL) and detritivores (% DET) at the localities downstream from the trout farm, which is a clear sign of recipient eutrophication. Thus, for example, following discharge water on the Crnica stream [30], the share of total collectors and detritivores in downstream localities, in relation to the control ones, doubles (for example, control locality % DET $= 11.29 \pm 2.29$; downstream locality % DET $= 23.61 \pm 2.67$). In terms of trophic, a decrease in the total amount of shredders (% SHR) was observed too in the same stream (control locality – % SHR $= 1.99 \pm 0.74$; downstream locality % SHR $= 0.69 \pm 0.24$, [30]).

To obtain more precise information about trout farm pollution, the finest taxonomic resolution should be implemented, identifying taxa on the species level. This is especially important for chironomids since they present one of the main components of macroinvertebrate communities and also shows a wide spectre of ecological preferences. Following this concept, the study of Milošević et al. [13] indicated how the trout farm pollution had an impact on chironomids community. Results of the study suggest, according to Categorical principal components analyses, that diversity metrics (total number of taxa, abundance, Shannon–Wiener index), FFG metrics (the total amount of grazers and gatherer collectors) and the proportion of Tanypodinae showed to be promising metrics as a bioassessment tool in the research of trout farm pollution on headwater streams.

9.8 The Effects of Spring Capture on Macrozoobenthos Communities

There are almost no springs whose uniqueness, beauty, or drinking water quality have not attracted human attention. There are fewer and fewer which have remained unspoiled in their original natural beauty, while more and more are entirely changed and adapted to human needs. The only springs without noticeable anthropogenic influence are the ones beyond the reach of man. Those are difficult to access and away from roads and human settlements or springs located in areas under a certain degree of protection.

The least significant changes at springs occur when building smaller barriers, which direct (the source of the Valja Zoni river), slow down (the fountainhead of the Banja river), or speed up the flow (the source of the Sušica river). The change in water velocity initiates the redistribution of macrozoobenthos organisms according to the newly created conditions. Namely, by slowing down the flow, water-borne particles create sediments, thus creating living conditions for argyloreophilic or peloreophilic organisms, such as organisms from the Oligochaeta group and some Molluscs. In contrast, accelerating the flow prevents the survival of organisms without special adaptations to fast flows or passive protection in fast water currents (sheltering behind rocks, moss…) [23].

The changes in the physiognomy of springs for livestock feeding needs are slightly more significant, creating smaller stagnant water reservoirs. Such changes significantly change the biotope, as well as the composition of benthocenoses. In addition to the biotope changes, the contribution of livestock coming to the watering-place, i.e., organic pollution, creates preconditions for the habitat of Oligochaeta, Hirudinea, Chironomidae, and larvae of other Diptera tolerant to organic pollution, while pollution sensitive organisms (such as Plecoptera or Tricladida) disappear [23].

Building fountains (Obudojevica spring, the Matica česma spring, the Hajdučka voda spring, the source of the Pocibrava brook; the source of the Marecova river; the source of Pocibrava brook) or setting up other ways of utilizing drinking water from springs can lead to drastic changes in environmental conditions and communities of organisms that inhabit such biotopes. Since these small water reservoirs are often cleaned by users of deposited sludge, sand, or gravel, the living conditions for organisms get changed completely, bringing the change over to original communities of organisms [23].

When capturing water, there is a complete change of a spring when all of the water is drained primarily for water supply needs [23, 76]. Concrete catchment objects completely change the habitat of aquatic organisms and thus the living world itself, which is often accompanied by complete destruction.

9.9 The Influence of Small Hydropower Plants on Macrozoobenthos Communities

In recent years, hilly and mountainous streams in Serbia and the biotic component of these ecosystems have been threatened by the construction of small hydropower plants (SHPPs) [77]. In the Balkans, Serbia ranks third (after Albania and Bosnia and Herzegovina) in terms of the threat posed to hilly and mountain streams caused by the construction of SHPPs. So far, over 90 SHPPs have been built in Serbia (primarily in the territory south of the Sava and the Danube), with as many as 850 planned, many of which are in protected natural areas [31, 77]. The most significant consequences caused by SHPPs in Serbia were recorded on the Ibar river (southwestern Serbia, Fig. 9.1) and in the upper parts of its tributaries (Jošanička, Samokovska, and Studenica streams, Fig. 9.1). Also, the streams on Stara Planina (eastern Serbia) are endangered: Visočica (Fig. 9.1), Crnovrška reka, Rakitska reka, as well as the rivers Vlasina (southeastern Serbia, Fig. 9.1), Resava (eastern Serbia, Fig. 9.1), and Mlava (eastern Serbia, Fig. 9.1) [77].

In addition to the changes that affect the shape of the riverbed, hydrobionts suffer tremendous consequences due to SHPPs. Changes in water temperature and flow make aquatic insects particularly sensitive to the impact of SHPPs, which is reflected in their development, growth, fecundity, and emergence [78]. Moreover, on each of the above-mentioned watercourses and their tributaries, at least one taxon from the EPT group was found, which is on the list of strictly protected or protected species of the Republic of Serbia. As many as six protected species of Trichoptera were recorded on the Vlasina stream: *Silo nigricornis, Lithax niger, Helicopsyche bacescui, Potamophylax cingulatus, Hydropsyche fulvipes, Wormaldia subnigra* [31]. Also, the survival of protected species of Plecoptera is questionable in the Vrla stream: *Dinocras megacephala, Amphinemura sulcicollis, Taeniopteryx nebulosa, Perlodes microcephalus*, all of which were recorded in previous studies [30, 41]. It is expected that there will be a declining trend in the number of populations of sensitive taxa of aquatic insects in degraded watercourses, but also in the time of their disappearance from the habitat. A typical example is a cold stenothermic Trihoptera species *Drusus discolor*, which was not registered in the Samokovska stream (a tributary of the Ibar river) on the locality where earlier research had pointed to its presence in isolated populations [79]. In addition to insects, the population of stone crayfish (*A. torrentium*)—a strictly protected species in our country and the world (Habitats Directive, IUCN)—is in danger of declining and extinction in parts of the watercourse where individual SHPPs have been built [31]. Having all this in mind, future research should provide an answer to the degree of habitat disturbance and how changes caused by the influence of SHPPs, primarily in temperatures and the volume of the water flow, led to the extinction of endangered macroinvertebrate taxa in the upper and middle parts of Serbian mountain streams.

9.10 Proposed Measures for the Preservation of Springs and Headwater Streams in Serbia

At first glance, one could say that human needs for the use of water from springs and upper river reaches, as well as the need to preserve biotopes and communities of springs and headwater streams, are entirely opposed. However, designing and building water intake structures integrated into the natural appearance of watercourses, as well as considering the amount of water being taken, would preserve the living world and the appearance of the natural environment, and thus create the possibility of rational and sustainable use of water from these streams.

Bearing in mind that the capture of water sources is often unplanned, spontaneous, and the construction of water intake structures on hilly and mountain streams is insufficiently professional, even without real willingness to preserve streams and keep them sustainable, both in terms of fitting into the natural environment and preserving communities of living beings, it is necessary to put these water flows under a special form of supervision and protection. For capturing springs, as well as the construction of water intake structures, but also all other facilities on small streams (above all SHPPs), it is necessary to pass certain legal acts firstly which would define conditions that have to be met before water capture, as well as supervision procedures so that those conditions could be implemented. Such legal acts would define the ways and possibilities of building water intake structures harmonized with the preservation of the natural environment, which would, in turn, preserve the biotope and the community of aquatic organisms. The best way is a naturalized approach that involves fitting into the natural environment, with constant care for the biological minimum of water. This approach reconciles the need for water use with the need to preserve the natural beauty and living conditions and the watercourse community itself.

Parts of the upper reaches of hilly and mountainous rivers, especially springs, are refuges for many species, especially aquatic insects. Among them, there are certainly cryptic species in Serbia as well, waiting to be discovered and described. These species, which are in most cases ecological indicators, are potentially endangered by trout farms, primarily if they are not managed in the right way [30]. Trout farms located along the upper reaches of streams, discharging uneaten food and fish feces into them, are a potential danger for disturbing the natural habitats of sensitive indicator groups of macrozoobenthos. The intensity and the effects of fish farm on the recipient stream depend on several factors, related to the production capacity of a farm, the amount of fish, the existence of a pond water treatment system, the type of fish feed, but also the stream ability of self-purification [30]. It is necessary to conduct research on as many farms as possible and define legal measures that will properly and fully protect sensitive ecosystems downstream of trout farms. This is particularly important in developing countries where trout farming is increasing, while legislation is incomplete or non-existent [30, 67]. Measures to protect sensitive habitats include a range of different activities, from proper capacity assessment for a given recipient to technical and technological solutions in fish production. Simultaneously, the entire process would have to be accompanied by continuous environmental monitoring [30].

Despite the primary goal that should be pursued during the design of SHPPs is preserving natural resources and protecting biodiversity. Previous examples on hilly and mountain rivers in Serbia have been driven solely by more significant profits, which has caused devastating consequences on the environment. In that sense, it is imperative to determine the degree of endangerment of local populations of the sensitive taxa, which are usually under the direct influence of SHPPs, to implement protection measures to preserve their habitats and the possibility of survival. The best way to preserve this part of biodiversity would be to ban the construction of SHPPs throughout Serbia, which have been proven to be inefficient and harmful in various aspects, and to compensate for the negligible amount of energy that can be generated by shifting to other forms of energy production from renewable sources (solar, aeolian, geothermal, etc.). Defining legal acts that would regulate the use of water resources when it comes to SHPPs, is the first step towards the preservation of these fragile ecosystems [31].

9.11 Conclusion

Water is the source of life, necessary not only for man but also for all living organisms. By endangering aquatic ecosystems, disrupting the hydrological cycle, and polluting all forms of water on the planet, humans are indirectly challenging the survival of other organisms. The survival of humankind and the biosphere depends on the solution of the puzzle – how to meet the increasing needs for freshwater (which takes up only 2.5% of the hydrosphere), which is being degraded more and more at the same time. The solution to the puzzle should commence laying down particular laws and regulations, with clearly defined conditions that must be met before water is taken for the needs of water supply or electricity. Drafting and passing legal regulations is the first step towards preserving sensitive aquatic ecosystems such as springs and headwater streams in mountainous regions of Serbia, followed by education of the local population about the need to protect areas and species that inhabit them.

Moreover, the problem of adaptation of organisms to pollutants is increasingly noticeable, that is, the adaptation of species to new environmental conditions, which changes the threshold of species' sensitivity as well as the intensity of stressors. This clearly indicates the necessity for a continuous revision of biological methods for assessing the quality of aquatic ecosystems and constant biomonitoring to preserve the water resources of the Republic of Serbia. Surely, the progress in this field depends on the financial support of the state, which is a precondition for the introduction of new, more sensitive, and efficient methods in standard monitoring such as modified biotic indices for Serbia, multimetric indices, and a multivariate approach. In order to reach this goal, it is necessary to form and standardize a database on inland water ecosystems. The initial steps towards this goal and successful use of macroinvertebrates in biomonitoring require extremely good knowledge of the reference state. At the moment, this implies additional faunal, taxonomic, biogeographical, and watercourse typification research as well as defining reference localities, which is, among other things, the first step towards the implementation of the EU Water Framework Directive [80]. As the first reference localities were defined during 2020 [7], all conditions were met for their additional research and a subsequent significant expansion of the network of localities that would serve as standard biomonitoring sampling localities in Serbia. In this way, even the smallest anthropogenic influences at springs and headwater streams would be detected, allowing for timely prevention of the destruction of the most sensitive aquatic ecosystems.

While preserving clean drinking water resources and the organisms that make up specific communities of those ecosystems, we also preserve our own future [23]. It is very important to start the revitalization of springs that have already been captured, as well as already built water intake structures on upper reaches, so as to initiate revival and harmonization with the environment as soon as possible.

References

1. Amidžić L, Bartula M, Cvetković D (2014) The state biodiversity of Serbia. Nat Area J 34(2):222–226
2. Simić S, Simić V (2012) Inland water ecology (Hydrobiology I). University of Belgrade – Faculty of Biology, University of Kragujevac, Faculty of Science [in Serbian]

3. Illies J (1978) Limnofauna Europaea: Eine Zusammenstellung aller die europäishen Binnengewässer bewohnenden mehrzelligen Tierarten mit Angaben über ihre Verbreitung und Ökologie (Zweite, überarbeitete und ergänzte). Gustav Fischer Verlag, Stuttgart, Germany
4. Panjković B, Puzović S (2019) Biodiversity in Serbia. In: Pullaiah T (ed) Global biodiversity, vol 2. selected countries in Europe. Apple Academic Press Inc., Oakville, Canada, pp 305–324
5. Paunović M, Tubić B, Kračun M, Marković V, Simić V, Zorić K, Atanacković A (2012) Ecoregions delineation for the territory of Serbia. Water Resour Manag 2(1):65–74
6. Paunović M (2007) Composition of macro-invertebrate communities as indicator of running waters types in Serbia. Ph.D. thesis. Faculty of Biology, University of Belgrade [in Serbian]
7. Živić I (2020) Research monitoring of biological, physico-chemical and hydromorphological parameters in order to define reference sites in the Republic of Serbia, Ministry of Agriculture, Forestry and Water Management of the Republic of Serbia, Republic Directorate for Water, Belgrade [in Serbian]
8. Simić V (1993) Saprobiologic evaluation of the Svrljiški and Trgoviški Timok on the basis of the composition of macrozoobenthos. MSc thesis, Faculty of Biology, University of Belgrade [in Serbian]
9. Živić I, Ostojić A, Miljanović B, Marković Z (2018) Macroinvertebrates of Serbian running waters and their bioindicator significance in water quality assessment. Proceedings "Ecological and economic significance of the fauna of Serbia", scientific conferences SANU, Book CLXXI, Department of Chemical and Biological Sciences, Book 12, pp 199–229
10. Stanković S (1924) Studies on spring and stream planarians of the Balkan Peninsula. Book CXIII, Glas SKA, Belgrade [in Serbian]
11. Stanković S (1928) On a new larva and nymph of Trichoptera, *Thremma* sp. Sci Soc Bull Skopje 1:121–128 [in Serbian]
12. Milošević Đ, Simić V, Stojković M, Živić I (2012) Chironomid faunal composition represented by taxonomic distinctness index reveals environmental change in a lotic system over three decades. Hydrobiologia 683(1):69–82
13. Dj M, Stojanović K, Djurdjević A, Marković Z, Stojković Piperac M, Živić M, Živić I (2018) The response of chironomid taxonomy- and functional trait-based metrics to fish farm effluent pollution in lotic systems. Environ Pollut 242:1058–1066
14. Đuknić J, Jovanović MV, Popović N, Živić I, Raković M, Čerba D, Paunović M (2019) Phylogeography of *Simulium* subgenus *Wilhelmia* (Diptera: Simuliidae)—insights from Balkan Populations. J Med Entomol 56(4):967–978
15. Đuknić J, Jovanović MV, Čanak Atlagić J, Andjus S, Paunović M, Popović N, Živić I (2020) *Simulium reptans* (Linnaeus, 1758) and *Simulium reptantoides* Carlsson, 1962 from the Balkan Peninsula. ZooKeys 922:141–155
16. Novaković B, Teofilova T, Pandakov P, Živić I (2020) New distributional records of rare riffle beetles (Coleoptera: Elmidae) from the Balkan Peninsula. Arch Biol Sci 72(1):129–135
17. Petrović A, Milošević D, Paunović M, Simic S, Đorđevic N, Stojkovic M, Simić VM (2014) New data on the distribution and ecology of the mayfly larvae (Insecta: Ephemeroptera) of Serbia (central part of the Balkan Peninsula). Turk J Zool 38:195–209. https://doi.org/10.3906/zoo-1304-2
18. Savić A, Ranđelović V, Đorđević M, Pešić V (2016) Assemblages of freshwater snails (Mollusca: Gastropoda) from the Nišava River, Serbia: Ecological factors defining their structure and spatial distribution. Acta Zool Bulg 68(2):235–242
19. Stojanović K, Petrović A, Kučinić M, Božanić M, Vesović N, Stojanović D, Živić I (2018) *Odontocerum hellenicum* Malicky, 1972 (Trichoptera: Odontoceridae) as a host of *Agriotypus armatus* Curtis, 1832 (Hymenoptera: Ichneumonidae)—the first finding for Serbia. In: Abstracts of the 13th Croatian biological congress with international participation, Poreč, Croatia, 19–23 Sept 2018
20. Živić I, Stojanović K (2018) Trichoptera fauna of Serbia - current situation and perspectives. In: The book of abstracts, Second Congress of Biologists of Serbia, Kladovo, Serbia, 25–30 Sept 2018 [in Serbian]

21. Živić I, Stojanović K, Ćetković A (2019) *Palingenia longicauda* (Ephemeroptera: Palingenidae) - proposal for inclusion of the species in Annex II and IV of the Habitats Directive. In: Proceedings of plenary papers and summaries of the Symposium of Entomologists of Serbia with international participation, Niš, Serbia, Niš, Serbia, 25–29 Sept 2019 [in Serbian]
22. Filipović D (1954) Investigation of the living world of running waters of Serbia, I Contribution to the knowledge of the communities of the mountain stream Katušnica (Western Serbia). SANU- Institute of Ecology and Biogeography. Proceedings, Book 5, No 8, pp 117–133 [in Serbian]
23. Marković Z (1998) Springs in mountainous regions of Serbia: Ecological study of the macrozoobenthos. University of Belgrade, Belgrade, Faculty of Biology [in Serbian]
24. Živić I, Marković Z, Simić V, Kučinić M (2009) New records of *Helicopsyche bacescui* (Trichoptera, Helicopsychidae) from the Balkan Peninsula with notes on its habitat. Acta Zool Acad Sci Hung 55(1):77–87
25. Živić I, Bjelanović K, Simić V, Živić M, Žikić V, Marković Z (2013) New records of *Thremma anomalum* (Trichoptera: Uenoidae) from southeastern Europe with notes on its ecology. Entomol News 123(3):206–219
26. Waringer J, Graf W, Bálint M, Kučinić M, Pauls SU, Previšić A, Keresztes L, Ibrahimi H, Živić I, Bjelanović K, Krpač V, Vitecek S (2015) Larval morphology and phylogenetic position of *Drusus balcanicus*, *D. botosaneanui*, *D. serbicus* and *D. tenellus* (Trichoptera: Limnephilidae: Drusinae). Eur J Entomol 112 (2): 344–361
27. Vitecek S, Kučinić M, Previšić A, Živić I, Stojanović K, Keresztes L, Bálint M, Hoppeler F, Waringer J, Graf W, Pauls S (2017) Integrative taxonomy by molecular species delimitation: multi-locus data corroborate a new species of Balkan Drusinae micro-endemics. BMC Evol Biol 17:129. https://doi.org/10.1186/s12862-017-0972-5
28. Moog O (2002) Fauna Aquatica Austriaca. 2nd edition. Wassewirtschaftskataster, Bundesministerium für Land-und Forstwirtschaft, Umwelt und Wasserwirtschaft. Wien, Austria: Austrian Federal Ministry of Agriculture Forestry Environment & Water Management, Wasserwirtschaftskataster Vienna
29. Živić I, Marković Z, Brajković M (2008) Macrozoobenthos of some springs and brooks in the Fruška Gora Mountain. In: Šimić S (ed) Invertebrates (Invertebrata) of the Fruška Gora Mountain. Matica Srpska, Department of Natural Sciences, Novi Sad, Serbia, pp 19–55
30. Stojanović K (2017) Influence of trout farms on macrozoobenthos communities with special emphasis on larvae of the genus Baetis (Ephemeroptera, Insecta). Ph.D. thesis. Faculty of Biology, University of Belgrade [in Serbian]
31. Stojanović K, Živić I (2019) Small hydropower plants as a potential threat to protected taxa of aquatic insects. In: Proceedings of plenary papers and summaries of the Symposium of Entomologists of Serbia with international participation, Niš, Serbia, 25–29 Sept 2019 [in Serbian]
32. Petrović A, Simić V, Dj M, Paunović M, Sivec I (2014) Diversity and distributional patterns of stoneflies (Plecoptera: Insecta) in the aquatic ecosystems of Serbia (Central Balkan Peninsula). Acta Zool Bulg 66(4):517–526
33. Official Gazette of the Republic of Serbia, No. 98/16 (2016) Book of Regulations on the Proclamation and Protection of Strictly Protected and Protected Wild Species of Plants, Animals, and Fungi [in Serbian]
34. Jović M (2013) A proposal of Serbian names for dragonfly species (Insecta: Odonata) of the Balkan Peninsula, with the checklist of Odonata of Serbia. Acta Entomol Serb 18(1/2):1–10
35. Glöer P (2008) Three new hydrobioid species from Serbia (Mollusca, Gastropoda, Hydrobiidae). In: Pavićević D, Perreau M (eds) Advances in the studies of the fauna of the Balkan Peninsula. Papers dedicated to the memory of Guido Nonveiller. Monograph 22,. Institute for Nature Conservation of Serbia, Serbia, pp 349–356
36. Živić I, Cvetković A, Božanić M, Radojević A, Stojanović K (2017) Checklist of freshwater leeches (Hirudinea) in Serbia. Water Resour Manag 7(3):35–41
37. Grosser C, Pešić V, Berlajolli V, Gligorović B (2016) Glossiphonia balcanica n. sp. and Dina prokletijaca n. sp. (Hirudinida: Glossiphoniidae, Erpobdellidae) - two new leeches from Montenegro and Kosovo. Ecol Mont 8:17–26

38. Grosser C, Pešić V, Lazarević P (2014) A checklist of the leeches (Annelida: Hirudinida) of Serbia, with new records. Fauna Balkana 3:71–86
39. Živić I, Radosavljević T, Stojanović K, Petrović A (2015) The first molecular characterization of the genus Hirudo on the territory of Serbia: estimation of endangerment. Aquat Ecol 49(1):81–90
40. Pešić V (2003) New records of water mites (Acari, Actinedida) from Serbia with 18 species new for Serbian fauna. Natura Montenegrina 1:77–88
41. Živić I (2005) Faunistical and ecological study of macrozoobenthos in the rivers of the Southern Morava River basin with an emphasis on taxonomy of Trichoptera larvae (Insecta). Ph.D. thesis. Faculty of Biology, University of Belgrade [in Serbian]
42. Hewitt GM (1999) Post-glacial re-colonization of European biota. Biol J Linn Soc Lond 68:87–112
43. Schmitt T (2007) Molecular biogeography of Europe: Pleistocene cycles and postglacial trends. Front Zool 4(1):11. https://doi.org/10.1186/1742-9994-4-11
44. Hewitt G (2000) The genetic legacy of the Quaternary ice ages. Nature 405:907–913
45. Geismar J, Haase P, Nowak C, Sauer J, Pauls SU (2015) Local population genetic structure of the montane caddisfly Drusus discolor is driven by overland dispersal and spatial scaling. Freshw Biol 60:209–221
46. Heino J, Mykrä (2008) control of stream insect assemblges: roles of spatial configuration and local environmental factors. Ecol Entomol 33:614–622
47. Petrović A, Simić V, Paunović M, Stojanović B (2006) A new records of *Epeorus yougoslavicus* (Šamal, 1935) [Ephemeroptera] in Serbia and Montenegro. Biotechnol Equip 20:67–71
48. Oláh J, Beshkov S, Chvojka P, Ciubuc C, Coppa G, Ibrahimi H, Kovács T, Mey W, Oláh, Jr J (2017) Revision of Drusinae subfamily (Trichoptera, Limnephilidae): divergence by paraproct and paramere: speciation in isolation by integration. Opusc Zool Budapest 48 (Supplementum 1), 3–228
49. Oláh J, Vinçon G, Kerimova T, Kovács T, Manko P (2020) On the Trichoptera of the Caucasus with western and eastern relatives. Opusc Zool Budapest 51 (Supplementum 3), 03–174.
50. Kučinić M, Previšić A, Graf W, Šeric Jelaska L, Stanić-Koštroman S, Waringer J (2011) Larval description, genetic and ecological features of *Drusus radovanovici radovanovici* Marinković-Gospodnetić, 1971 (Trichoptera: Limnephilidae) with some phylogenetic and taxonomic data on the *bosnicus* group in the Balkan Peninsula. Deut Entomol Z 58:136–153
51. Marinković-Gospodnetić M (1971) The species of the genus *Drusus* in Yugoslavia. Yearb Inst Biol Univ Sarajevo 24:105–109
52. Pauls SU, Lumbsch HT, Haase P (2006) Phylogeography of the montane caddisfly *Drusus discolor*: Evidence for multiple refugia and periglacial survival. Mol Ecol 15:2153–2169
53. Previšić A, Walton C, Kučinić M, Mitrikeski PT, Kerovec M (2009) Pleistocene divergence of Dinaric Drusus endemics (Trichoptera, Limnephilidae) in multiple microrefugia within the Balkan peninsula. Mol Ecol 18:634–647
54. Živić I, Marković Z, Brajković M (2006) Influence of the temperature regime on the composition of the macrozoobenthos community in a thermal brook in Serbia. Biologia 61(2):179–191
55. Živić I, Živić M, Dj M, Bjelanović K, Stanojlović S, Daljević R, Marković Z (2013) The effects of geothermal water inflow on longitudinal changes in benthic macroinvertebrate community composition of a temperate stream. J Therm Biol 38(5):255–263
56. Duggan IC, Boothroyd IKG, Speirs DA (2007) Factors affecting the distribution of stream macroinvertebrates in geothermal areas: taupo Volcanic Zone, New Zealand. Hydrobiologia 592:235–247
57. Woodward G, Dybkjaer JB, Olafsson JS, Gislason GM, Hannesdottir ER, Friberg N, (2010) Sentinel systems on the razor's edge: effects of warming on Arctic geothermal stream ecosystems. Glob Chang Biol 16:1979–1991
58. Clements WH, Arnold JL, Koel TM, Daley R, Jean C (2011) Responses of benthic macroinvertebrate communities to natural geothermal discharges in Yellowstone National Park, USA. Aquatic Ecol 45:137–149

59. Živić I, Živić M, Bjelanović K, Dj M, Stanojlović S, Daljević R, Marković Z (2014) Global warming effects on benthic macroinvertebrates: a model case study from a small geothermal stream. Hydrobiologia 732:147–159
60. Sanz A, Trenzado CE, López-Rodríguez MJ, Furné M, Tierno de Figueroa JM (2010) Study of antioxidant defense in four species of Perloidea (Insecta, Plecoptera). Zool Sci 27(12):952–958. https://doi.org/10.2108/zsj.27.952
61. Mirčić D, Stojanović K, Živić I, Todorović D, Stojanović D, Dolićanin Z, Perić-Mataruga V (2016) Effects of the trout farm on *Dinocras megacephala* larvae. Environ Toxicol Chem 35(7):1775–1782
62. Božanić M, Todorović D, Živić M, Perić– Mataruga V, Marković Z, Živić I, (2018) Influence of a trout farm on antioxidant defense in larvae of Ephemera danica (Insecta: Ephemeroptera). Knowl Manag Aquat Ecosyst 419:47. https://doi.org/10.1051/kmae/2018036
63. Vranković J, Živić M, Radojević A, Perić-Mataruga V, Todorović D, Marković Z, Živić I (2018) Evaluation of oxidative stress biomarkers in the freshwater gammarid *Gammarus dulensis* exposed to trout farm outputs. Ecotoxicol Environ Saf 163:84–95
64. Radojević A, Mirčić D, Živić M, Perić-Mataruga V, Božanić M, Stojanović K, Lukičić J, Živić I (2019) Influence of trout farm effluents on selected oxidative stress biomarkers in larvae of *Ecdyonurus venosus* (Ephemeroptera, Heptageniidae). Arch Biol Sci 71(2):225–233
65. Božanić M (2019) Monitoring the effects of trout farms on the antioxidative enzyme activity level in Ephemera danica larvae (Insecta, Ephemeroptera). Ph.D. thesis. University of Belgrade, Faculty of Biology [in Serbian]
66. Barbour MT, Gerritsen J, Snyder B, Stribling J (1999) Rapid bioassessment protocols for use in streams and wadeable rivers: periphyton, benthic macroinvertebrates and fish, second edition. United States Environmental Protection Agency, Washington D.C.
67. Živić I, Marković Z, Filipović-Rojka Z, Živić M (2009) Influence of a trout farm on water quality and macrozoobenthos communities of the receiving stream (Trešnjica River, Serbia). Int Rev Hydrobiol 94:673–687
68. Živić I, Marković Z, Živić M (2009) Influence of a trout farm on macrozoobenthos communities of the Trešnjica River Serbia. Arch Biol Sci 61:483–492
69. Plafkin JL, Barbour MT, Porter KD, Gross SK, Hughes RM (1989) Rapid Bioassessment Protocols for use in Streams and Rivers: Benthic Macroinvertebrates and Fish. EPA 440/4–89/001. U.S. Environmental Protection Agency, 8 chapters, Appendices A-D
70. Hilsenhoff WL (1988) Rapid field assessment of organic pollution with a family-level biotic index. J North Am Benthol Soc 7(1):65–68. https://doi.org/10.2307/1467832
71. Paisley M, Trigg D, Walley W (2014) Revision of the Biological Monitoring Working Party (BMWP) score system: derivation of present-only and abundance-related scores from field data. River Res Appl 30(7):887–904. https://doi.org/10.1002/rra.2686
72. MACS (1996) Mid-Atlantic Coastal Streams Workgroup (MACS): Standard operating procedures and technical basis: Macroinvertebrate collection and habitat assessment for low-gradient nontidal streams. Delaware Department of Natural Resources and Environmental Conservation, Dover, DE
73. Camargo JA (1992) Structural and trophic alterations in macrobenthic communities downstream from a fish farm outlet. Hydrobiologia 242(1):41–49
74. Selong JH, Helfrich LA (1998) Impacts of trout culture effluent on water quality and biotic communities in Virginia headwater streams. Progress Fish Cult 60(4):247–262
75. Merritt R, Cummins K (1996) Trophic relations of macroinvertebrates. In: Hauer FR, Lamberti GA (eds) Methods in Stream Ecology. Academic Press Inc., San Diego, USA, pp 453–474
76. Pešić V, Dmitrović D, Savić A, Miloševi DJ, Zawal A, Vukašinović-Pešić V, von Fumetti S (2019) Application of macroinvertebrate multimetrics as a measure of the impact of anthropogenic modification of spring habitats. Aquat Conserv 29:341–352
77. Ristić R, Malušević I, Polovina S, Milčanović V, Radić B (2020) Small hydropower plants: insignificant energy benefit and immeasurable environmental damage. In: Proceedings "Impact of small hydropower plants on the environment", scientific conferences of SANU, Book CLXXXVII, Department of Chemical and Biological Sciences; Book 17, pp 107–133

78. Kennedy TA, Muehlbauer JD, Yackulic CB, Lylte DA, Miller SW, Dibble KL, Kortenhoeven EW, Metcalfe AN, Baxter CV (2016) Flow managment for Hydropower extirpates aquatic insects, undermining river food webs. Bioscience 66:561–575
79. Petrović A (2014) Possibilities of using a database in the strategy of conserving the biodiversity of inland water macroinvertebrates at a national level. Ph.D. thesis. University of Kragujevac–Faculty of Science [in Serbian]
80. WFD (2000) Water Framework Directive-Directive 2000/60/EC of the European Parliament and of the Council Establishing a Framework for Community Action in the Field of Water Policy

Chapter 10
Springs of Southeastern Serbia with a Focus on the Vlasina Plateau: Different Types of Challenges for the Macroinvertebrate Community

Ana Savić, Miodrag Đorđević, Milan Đorđević, Vladimir Ranđelović, Dejan Dmitrović, and Vladimir Pešić

Abstract The ecology of springs of the southeastern part of Serbia is poorly studied. The present chapter is planned as the first step towards the inventory and characterization of springs in this region. In this chapter we present the results of the investigation of macroinvertebrate communities of two groups of springs, in the valley of Niš and the Vlasina plateau, i.e. its surroundings. The investigated springs are distributed on a small spatial scale, but they are not from the same geological underground. The first group of springs is dominated by representatives of non-insect groups, while insects predominate in the springs of the Vlasina plateau. FFG analysis revealed scrappers as dominant organisms in the valley of Niš, while shredders predominated in the springs of the Vlasina region. The decisive drivers of macroinvertebrate

A. Savić (✉) · V. Ranđelović
Department of Biology and Ecology, Faculty of Sciences and Mathematics, University of Niš, Višegradska 33, 18000 Niš, Serbia
e-mail: anka@pmf.ni.ac.rs

V. Ranđelović
e-mail: vladar@pmf.ni.ac.rs

M. Đorđević
Department of Mathematics, Faculty of Sciences and Mathematics, University of Niš, Višegradska 33, 18000 Niš, Serbia
e-mail: dmiodrag@pmf.ni.ac.rs

M. Đorđević
Department of Geography, Faculty of Sciences and Mathematics, University of Niš, Višegradska 33, 18000 Niš, Serbia
e-mail: milan.djordjevic@pmf.ni.ac.rs

D. Dmitrović
Faculty of Natural Sciences and Mathematics, Department of Biology and Department of Ecology and Environment Protection, University of Banja Luka, Mladena Stojanovića 2, 78000 Banja Luka, Bosnia and Herzegovina
e-mail: dejan.dmitrovic@pmf.unibl.org

V. Pešić
Faculty of Sciences, University of Montenegro, Džordža Vašingtona b.b, 81000 Podgorica, Montenegro

V. Pešić et al. (eds.), *Small Water Bodies of the Western Balkans*, Springer Water,
https://doi.org/10.1007/978-3-030-86478-1_10

community variability include conductivity, altitude and pH of water. Differences in environmental variables between the weakly alkaline springs of the Niš valley and the weakly to moderately acidic springs of the Vlasina plateau have an impact on macroinvertebrate communities. In thermal springs, which are numerous in the study area, water temperature, but also biotic factors, such as the presence of the invasive species *Melanoides tuberculata*, affected the structure of crenic assemblages. The threats to springs in the study area range from water being diverted for water supply systems of nearby settlements, use of thermal water for therapeutic purposes and surface habitat modification, to anthropogenic groundwater depletion and pollution, as well as changes caused by the ongoing climate change. In order to improve the understanding and best practices for monitoring, management and conservation of springs, it is necessary to study various aspects of these ecosystems.

Keywords Springs · Macroinvertebrates · Southeastern Serbia

10.1 Introduction

> Because of the items that satisfy his fleeting greed, he destroys large plants that protect the soil everywhere, quickly leading to the infertility of the soil he inhabits and causing **springs to dry up**, removing animals that relied on this nature for their food and resulting in large areas of the once very fertile earth that were largely inhabited in every respect, being now barren, infertile, uninhabitable, deserted. One could say that he is destined, after making the earth uninhabitable, to destroy himself.
>
> Jean Baptiste Lamarck (Zoological Philosophy, 1809).

The first step in studying various types of aquatic ecosystems is mapping their occurrence, followed by their adequate typology classification [1]. This chapter aims to provide an overview of diversity and general environmental characteristics of the spring ecosystems of Southeastern Serbia (Fig. 10.1a). The number of publications on spring types and springs in Serbia in general is very limited, and data on biota inhabiting these springs is particularly scarce. The most comprehensive one was published by Marković more than 30 years ago [2]. The oldest publications refer to taxonomy of some specific animal groups in springs (e.g., planarians [3]), and since then only a few studies devoted to some macroinvertebrate groups in these ecosystems have been published [2, 4–6]. Regardless of the enormous importance of these ecosystems, there is a 30-year gap in research of these habitats in Serbia. In spite of continuous efforts to call attention to the importance of spring ecosystems, both from the aspect of their contribution to biodiversity [7] and the aspect of them being sources of drinking water, these ecosystems are yet to be awarded with an appropriate level of attention in Serbia.

In the last several decades, conservation activities have mostly been focused on extensive ecosystems, especially those that harbor large portions of the Earth's biodiversity [8, 9]. On the other side of the spatial gradient, there are some conservation efforts focused on Small Natural Features in contrast to the large-scale conservation

Fig. 10.1 Map of the investigated localities: **a** springs at the Niš valley-left; springs at the Vlasina plateau and its vicinity-right **b** geology of the area

[9]. Hunter et al. [9] have stated that Small Natural Features (SNFs) are analogous to keystone species because they have an ecological importance that is disproportionate to their size; sometimes because they provide resources that constrain key populations or processes that affect a much larger area; sometimes because they support unusually high diversity, abundance, or productivity. Springs are spatially restricted ecotones between surface water and groundwater that are highly complex in their environmental conditions, and have a high diversity of invertebrates [10, 11]. In some regions, springs contain many endemic species, contributing to almost one third of regional freshwater biodiversity [7]. From the conservation point of view, springs fit (in terms of dimension and structural complexity) the definition of small natural features (SNF) [9] and deserve more attention by conservationists, especially in

certain parts of the Balkan Peninsula such as Serbia. In order to preserve and underline the importance of these habitats, which are often among the most vulnerable aquatic habitats, it is necessary to get to know them from several aspects.

Macroinvertebrate communities are very important for the functioning of spring ecosystems, partly because they represent a fundamental link in the food web between the organic matter resources (algae, detritus, leaf litter) and predators (fish, amphibians) and partly due to their diversity and ubiquity. The key factors in structuring the macroinvertebrate community of a spring include climatic conditions, historical parameters, mineralogical composition of spring water, dynamics of temperature changes and discharge, composition and structure of substrate, altitude and biotic interaction [12]. The importance of these factors varies between biogeographic regions and the type of spring [12]. The goals of this chapter are based on the above issues: to determine the key environmental parameters defining the macroinvertebrate community in different groups of springs in Southeastern Serbia, to determine the key macroinvertebrate taxa in these communities, and to determine the form of their functional structure.

10.2 Southeastern Serbia—A Forgotten Chest of Crenobiology

Although not occupying a large territory, Serbia is characterized by a high level of genetic, species and ecosystem diversity [13]. The main factors responsible for this diversity include the geographic position on the dividing line between continental and Mediterranean influences, complex relief and the fact that part of its territory served as a glacial refuge during the last Ice Age [13]. There are two main well-defined orographic units in Serbia: the Pannonian basin and the hill-mountainous area, interconnected by a series of transitions [2]. Due to a hill-mountainous character, most of the territory of Serbia may be considered rich in springs [14].

The Vlasina plateau is situated in the hill-mountainous region of southeastern Serbia, close to the Bulgarian border (Fig. 10.1a-right side). The plateau is surrounded by mountains Vardenik, Gramada and Veliki Čemernik, with Besna Kobila, Dukat and Ruj in their vicinity, representing northwestern branches of the Rhodope mountain range. This area is characterized by harsh continental climate, partly altered after the construction of the dam which formed the artificial accumulation lake Vlasina [15].

The mountain massifs around the plateau possess numerous springs, many of them scattered around the plateau itself, either near or far from the lake. Peatland vegetation is almost always developed around the springs [15]. In most peat bogs in the Vlasina area, the water is slightly to moderately acidic [16]. The discharge of the springs in the crystalline schist of this area is relatively low, mostly below 0.1 ls^{-1} and only rarely reaching 0.5 ls^{-1} [17]. Some of the springs dry up in the dry period of the year, but most of them are permanent due to high precipitation in this region (above 850 mm/year; [17]). In addition to their permanent/seasonal status, the springs in this area also differ in habitat types (according to Glazier [10]): rheocrene, small limnocrene and helocrene (Fig. 10.2). On the other hand, among the springs situated next to the lake shore at the Vlasina plateau, some are flooded by the lake for most of the year, while others never get flooded at all. Regardless of the specific features of the springs at the Vlasina plateau, this region was not included in the comprehensive study given by Marković [2] on the springs of the hilly and mountainous parts of Serbia. Therefore, this chapter will contribute to our knowledge of the exceptional diversity of springs in the region of the Vlasina plateau (Fig. 10.2).

The Nišava River Basin is highly diverse and unique, both geographically and hydrologically, due to its special position at the junction of the Serbian-Macedonian

Fig. 10.2 Different types of springs (diversity of habitats) at the Vlasina plateau and surroundings: **a**, **b** small rheocrenes springs; **c**, **d**, **e** small helocrenes springs located in the riparian zone immediately next to the lake

Fig. 10.3 Different types of springs (diversity of habitats) in the valley of Niš: **a**, **b**, **c** rheocrenes springs; **d** helocrenes spring

Mass and the Carpatho-Balkan Mountains, an extremely complex geological structure and tectonic, and due to diverse climatic conditions [18]. The average annual air temperature is 8.25 °C. The average annual rainfall in the basin is 731 mm [19]. The mountains are of different ages, composed of various igneous, sedimentary and metamorphic rocks. Various types of springs were formed in these specific geological and tectonic conditions (Fig. 10.3). Some of them show considerable discharge, even reaching 30 ls^{-1} [18]. The springs with the highest discharge are most often used to supply water to the city of Niš (Fig. 10.1) and they are mostly captured. The pH reaction of most water bodies of this area is weakly alkaline [20]. Thermal springs have also been recorded in this area. Our research was concentrated on the lower part of this basin, on the springs which were not captured for water supply purposes, located on the rims of the Niš valley.

10.3 Springs in the Niš Valley Versus Springs at the Vlasina Plateau and Threats

The studies on the macroinvertebrate assemblages of the springs of the investigated areas, the Vlasina plateau and the valley of Niš, were conducted in 2017 and 2018. These two study areas are characterized by different geological histories (Fig. 10.1b). Geology is often considered as one of the most important environmental factors, producing variability in the composition of macroinvertebrate communities, which has an impact on the bio-assessment output in temperate water bodies [21, 22]. Geological impacts on macroinvertebrate communities are associated with changes in water chemistry, food resources, and flow [23]. In our study, the studied springs of the Vlasina area were divided into three subgroups: the first subgroup consisted of springs located on the hills surrounding the plateau; the second subgroup consisted of springs located in the immediate vicinity of the lake, which are not flooded by the adjacent lake; and the third subgroup consisted of springs located in the riparian zone next to the lake but regularly flooded by the lake water. The flooded springs had the lowest average altitude (1207.5 m above sea level), followed by the second subgroup (1277.5 m above sea level). The springs of the first subgroup were situated at the highest altitudes (1563.6 m above sea level) and mostly were of rheocrenic type. The first subgroup was characterized by the lowest average temperature (8.03 °C), while the springs of the third subgroup exhibited the highest average temperature (16.7 °C). The pH values were mostly similar in all three subgroups of springs and ranged between 6.7 and 6.86. The average values of conductivity were relatively low, with the lowest values in the springs of the second subgroup and the highest values in the springs of the third subgroup. The average values of oxygen concentration and saturation were highest in the springs of the first subgroup (9.2 mg/l and 92.3%, respectively). In regard to all the above-mentioned environmental parameters, statistically significant differences were not found between the three subgroups of springs.

On the other hand, our study, using Mann–Whitney test, showed a statistically significant differences in altitude ($p = 0.017$), pH values ($p = 0.017$) and conductivity ($p = 0.016$) between the springs of the Vlasina plateau and the springs of the valley of Niš. As expected, the average altitude of springs in the Vlasina area was higher (1380.1 ± 174 m) than those from the valley of Niš (388.33 ± 145.1 m). The average concentration and saturation of oxygen are also higher in the springs of the Vlasina (8.04mg/l ± 1.8; 81.5% ± 16.4; respectively) in comparison to the springs of the valley of Niš (5.98 ± 1.29 mg/l; 57.8% ± 29.9). On the other hand, the values of conductivity (50.8 S/cm ± 18.4) and pH (6.8 ± 0.08) were significantly lower in the Vlasina group than in the Niš group (620.6 Scm ± 134.8; 7.43 ± 0.01, respectively). The slightly acidic reaction of water of the Vlasina plateau springs may be partly caused by the fact that this plateau is situated in the zone of brown forest soil (dystric cambisol) that forms on quartz-silicate substrates in humid areas of hilly and mountainous regions. The pH reaction of this soil is quite acidic (pH = 4.5–4.7) [24]. All other soil types at this plateau, due to silicate substrate and high levels of precipitation leading to an increased leaching of bases, belong to the group of acidic soils [25].

The most extensive threat to springs is related to their small size. It is also worth to mention that the small size of springs may make them more vulnerable to degradation or even complete destruction [9]. In many parts of the Balkan Peninsula springs are poorly documented, rarely included in comprehensive studies and often excluded from legislative framework. Notwithstanding the requirements of the European Water Framework Directive (WFD) in achieving at least ‘good’ ecological status in all surface waters, sources are not included in ecological status assessment tools [11]. The small size of springs, on the other hand, can have advantages when it comes to their protection and restoration: inexpensive, more cost-effective options and an easier restoration process. In our study, both groups of the investigated springs were exposed to similar threats (Fig. 10.5). The springs in the valley of Niš were exposed to a high level of anthropogenic impact: the cold springs are used as a source of drinking water, while the thermal springs are used for therapeutic purposes. Many of them were hydromorphologically modified. As recently shown by some studies, hydromorphological modification of springs leads to a decrease of diversity in a macroinvertebrate community [26]. Recently, some springs at the Vlasina plateau were tapped in 2005 for the purpose of the local water factory. Since then, from 2005 until today, water use has doubled, which on the one hand has had economic benefits for the local community, but it has certainly affected springs and their biota, although detailed data on the potential impact are not yet available.

10.4 Faunistic Composition of Springs Communities and Their Main Drivers

Benthic macroinvertebrate communities of springs consist of species that differ in their preference for this environment, and also in biological and ecological traits [27]. Representatives of the following groups have been recorded in the springs of Southeastern Serbia: Gastropoda, Bivalvia, Hirudinea, Isopoda, Ephemeroptera, Plecoptera, Trichoptera, Hemiptera and Diptera. Representatives of all the above groups were recorded in the springs of the Vlasina plateau, while there were no records of representatives of Ephemeroptera and Hemiptera in the springs of the Niš valley. The lowest species diversity with just three identified species was detected in the spring Vidrište in the Niš valley, where the highest water temperature was recorded. Such low species diversity matches the data published by Marković [2]. In studies on the springs Dušničko Vrelo and Dušnička Česma, both close to the city of Niš and with a much larger habitat size, the latter author (Marković [2]) recorded just five species in each of these springs. The greatest diversity of species was recorded in the springs at the Vlasina plateau which belong to the group of springs which are located in the immediate vicinity of the lake and are not flooded by the lake. Considering the abundance of the collected faunistic groups, gastropods dominated the springs of the Niš valley (Fig. 10.4a), while in the springs of the Vlasina area the most abundant group was Diptera (Fig. 10.4b). The average value of the Shannon Index of Diversity was slightly higher in the springs of the Niš valley (H' = 1.21) than in the springs of the Vlasina plateau (H' = 1.02).

SIMPER (Similarity Percentages) analysis was used to determine the key taxa by comparing the two groups of the investigated springs. The analysis showed that the hydrobiid snails had the greatest contribution with respect to the structure of the macroinvertebrate assemblages in the springs of the Niš valley. In the springs of the Vlasina plateau, representatives of Chironomidae and Limnephilidae played the main role in structuring the macroinvertebrate community. The high dissimilarity of these two groups of springs (99.13%) is mostly caused by the latter macroinvertebrate groups (Limnephilidae—19.64%; Chironomidae—14.84%).

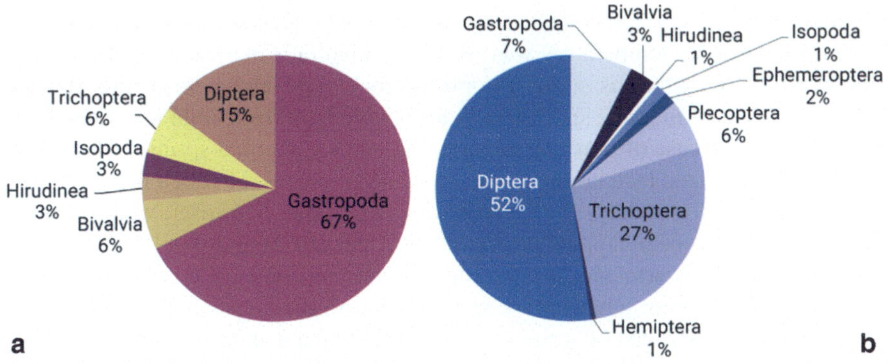

Fig. 10.4 Representation of individuals of certain groups in: **a** the springs of the Nis valley; **b** the springs in the Vlasina region

Fig. 10.5 Threats to autochthonous macroinvertebrate communities in southeastern Serbia: **a** *Melanoides tuberculate*-invasive species; **b** capturing activity; **c** waste material

It is interesting to mention that the spring Vidrište in the Niš valley, which had the lowest species diversity and the highest water temperature, was the site where the invasive species *Melanoides tuberculata* (O. F. Müller, 1774) was found. This species is native to northern Africa and southern Asia, but it has been accidentally introduced in many other tropical and subtropical areas worldwide, as well as in thermal springs in the temperate zone [28]. So far, its occurrence has been noted in Italy [32], Germany, France and Austria [30]. More recently, this species was also recorded at two localities in Bulgaria [33] and Poland [34]. In Serbia, so far it has been recorded at a single locality [30], in the stream Topla voda, which collects water from several thermal springs, but without information what the water temperature was at the site where the snail was collected. In our study, *M. tuberculatus* was recorded in a spring whose water temperature was 18.3 °C, which falls in the temperature range from 18 to 32 °C, which is stated as optimal for the development of this species [31]. The impact of invasive species such as *M. tuberculatus* should be particularly emphasized in the case of sensitive ecosystems, such as thermal springs. The valley of Niš is characterized by a large number of thermal springs, of which the springs in Niška Banja and Topilo have the highest water discharge. However, water in most of these springs has been tapped, which makes access to them difficult.

SIMPER analysis was also performed in order to compare macroinvertebrate communities in different subgroups of springs at the Vlasina plateau. Within the three already defined subgroups of springs, we identified taxa responsible for structuring the crenic communities. Our results revealed that Chironomidae and Limnephilidae structure the assemblages of the first subgroup of springs significantly; the bivalve *Sphaerium corneum* plays a significant role in structuring the assemblages of the second subgroup, while representatives of Chironomidae are characteristic of the assemblages of the third subgroup of spring.

According to the results of CCA (Canonical Correspondence Analysis) analysis for the group of springs from the Vlasina plateau, the key factors for structuring the macroinvertebrate community are conductivity (−0.776) and altitude (0.687), while for the group of springs at the Niš valley, the key factors are conductivity (−0.970) and pH value (0.896). Conductivity is a shared key factor and may be considered a measure of pollution level. Medupin [38] determined that the relationship between conductivity and benthic macroinvertebrate communities is related to contamination sources such as urban runoff, sewage outfalls and effluent from point sources. High conductivity levels are associated with high salt concentrations in more urbanized sections, and that may impair water quality following the increased nutrient enrichment [38]. The group of springs at the Niš valley has much higher average levels of conductivity than the springs at the Vlasina area.

It is also important to note that pH value is a much more important factor (according to CCA) in the springs of the Niš valley than at the Vlasina plateau. Some studies have emphasized that water chemistry does not have such an important role in the structure of the macroinvertebrate community of springs [36] while in some studies it is recognized as the most important parameter in determining faunistic variability [37]. In the studies on 50 springs in Bosnia and Herzegovina, pH was one of the key factors in defining the assemblage of EPT [37]. Glazier and Gooch [39] and Glazier [40] reported that alkalinity and pH had an important influence on spring community compositions. Similarly, Williams et al. [41] and Krouplaove et al. [27] found that water chemistry parameters, matching different levels of urbanization, were important determinants of spring communities. We have already stated that springs in the Niš valley are weakly alkaline, while springs at the Vlasina area have acidic character. Biodiversity in acidic freshwater systems is not necessarily lower in places where natural acidity has been present over extended periods. Actually, communities in naturally acidic fresh waters may have as high diversity as that in neutral ecosystems of the same geographical region, and adaptation to low pH has been proposed [42, 43]. In our study, the macroinvertebrate community in the weakly acidic springs of the Vlasina area shows only slightly lower diversity average when compared to the weakly alkaline springs at the Niš valley. In the springs of the Vlasina plateau, one of the main drivers is the acidic character of water. For example, the Ephemeroptera in the springs of Vlasina plateau (which are in the vicinity of the lake, but which are not flooded) were represented by just two species: *Baetis vernus,* considered to be an "acid-benefiting" species [27], whose density may even increase with increasing acidity [35], and the "slightly acid-sensitive" species *Baetis niger,* which tolerates weakly acidic conditions [27]. It may be concluded that the difference in the macroinvertebrate assemblages might be driven by different environmental parameters, such as water temperature (thermal springs), but biotic factors such as the presence of invasive species might also be significant.

10.5 Functional Composition of Springs Communities

Fumetti et al. [36] determine that despite the existence of different systems of spring classifications (e.g., according to habitat size, hydrological regime, variation of persistence flow, discharge, geologic structure, water chemistry, habitat type etc.), there is still no accepted spring typology based on their faunistic characteristics. One such attempt to classify springs according to faunistic characteristics was based on using functional feeding groups (FFG) [36] but so far this approach has been applied in a small number of studies. In this study, we applied this approach to springs in Southeastern Serbia.

NMDS (Non-metric Multidimensional scaling) analysis of our study (Fig. 10.6) showed two clearly distinct types of springs in relation to FFG composition. The first group includes three springs from the Niš valley, while the second group consists of springs at the Vlasina plateau.

The most important functional group in all the springs of the first group (which may be characterized as mostly lotic springs) are scrapers (70%). Fumetti et al. [36] mention scrapers as the dominant feeding group in lotic springs with prevalent inorganic material, matching the results of our study. The characteristic taxon in this group were hydrobiid snails. In the second group of springs, shredders predominated (Fig. 10.7). Shredders are, usually, a dominant group in springs with a high percentage of coarse organic material, which is the case with most springs of the Vlasina plateau. Among the studied localities, localities 7 and 8 (Fig. 10.7) have shown the greatest representation of filter feeders (collectors), with the characteristic species *Sphaerium corneum.* These two springs were characterized by a high percentage of leaf litter, and, on the other hand, a high percentage of sand.

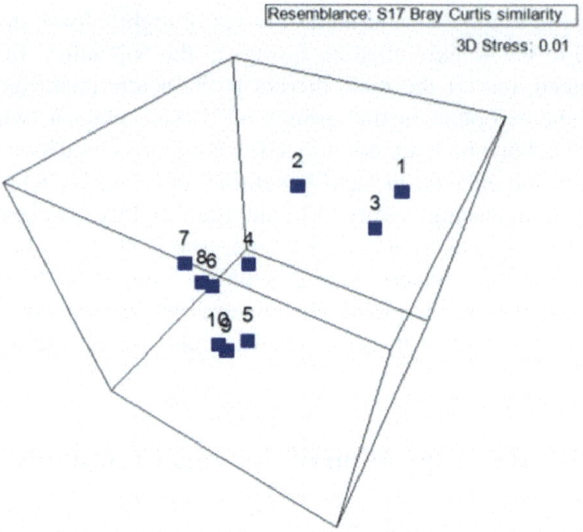

Fig. 10.6 Non-metric multidimensional scaling (NMDS) of the 10 investigated springs in southeastern Serbia: slightly separation of types of springs according to FFG

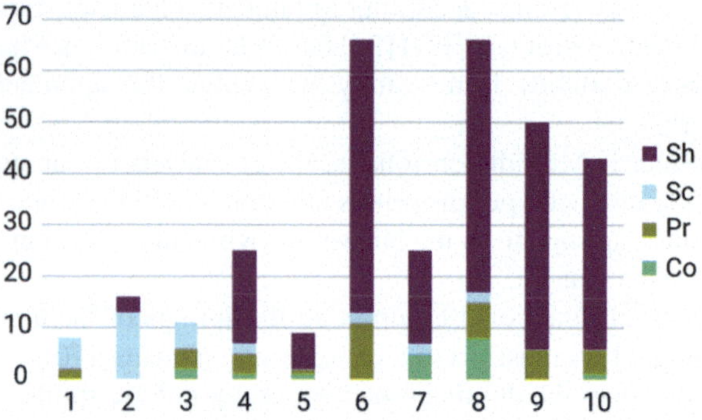

Fig. 10.7 Functional feeding group composition of the macroinvertebrate community in the investigated springs. Sh-shreders; Sc-scrapers; Pr-predators; Co-collectors

10.6 Conclusion

This chapter is designed as the first step towards a more comprehensive study that would include the inventory and characterization of springs in Southeastern Serbia. Studies on a small number of springs in a geographically very limited area have shown a high diversity of these ecosystems and an impact of a number of environmental parameters such as: temperature, pH values, habitat types in structuring the response of macroinvertebrate communities. This study showed that changes in macroinvertebrate assemblages of the springs in the Niš valley and in the springs of the Vlasina plateau were driven by different abiotic but also biotic parameters. This influence is visible at the level of the whole community, with different faunal groups inhabiting springs often exhibiting different responses. We hope that these types of studies will be helpful for the suggestion and implementation of conservation measures that will protect springs—these small natural features with great ecological roles. Future research needs to be intensified, both in the direction of more comprehensive documentation of their diversity and in understanding the mechanisms that affect the response of crenic assemblages, especially in the light of the ongoing climate changes and the increasing use of resources for water supply.

References

1. Pešić V, Pavićević A, Savić A, Hadžiablahović S (2019) The intermittent rivers of south Montenegro: ecology and biomonitoring. In: Pešić V, Paunović M, Kostianov AG (eds) The rivers of Montenegro. Springer, Cham, pp 231–252
2. Marković Z (1998) Izvori brdsko-planinskih područja Srbije. Biološki fakultet Univerziteta u Beogradu, Beograd, Ekološka studija makrozoobentosa
3. Stanković S (1924) Studija na izvorskim i potočnim planarijama Balkanskog poluostrva. Glas Srp Kralj Akad 50:39–89
4. Pljakić M (1952) Prilog poznavanju rasprostranjenja *Gammarus (Rivulogammarus) pulex fossarum* u Srbiji. Arch Biol Sci 4(1–2):81–88
5. Filipović D (1967) Struktura populacija dveju vrsta Amphipoda u Lisinskom potoku na Kopaoniku. Arch Biol Sci 19(1–2):57–74
6. Janković M (1983) Proučavanje Chironomidae (Diptera) sliva Velikog Peka. Zbornik radova o fauni SR Srbije 2:55–146
7. Pešić V, Gligorović B, Savić A, Buczynski P (2017) Ecological patterns of Odonata assemblages in karst springs in central Montenegro. Knowl Manag Aquat Ec 418(3):1–20
8. Brooks TM, Mittermeier RA, da Fonseca GAB, Gerlach J, Hoffmann M, Lamoreux JF, Mittermeier CG, Pilgrim JD, Rodrigues ASL (2006) Global biodiversity conservation priorities. Science 313:58–61
9. Hunter Jr ML, Acuna V, Bauer DM, Bell KP, Calhoun AJK, Felipe-Lucia MR, Fitzsimons JA, Gonzalez E, Kinnison M, Lindenmayer D, Lundquist CJ, Medellin RA, Nelson EJ, Poschlod P (2017) Conserving small natural features with large ecological roles: An introduction and definition. Biol Conserv 211:88-95
10. Glazier DS (2009) Springs. In: Likens GE (ed) Encyclopedia of inland waters. Academic Press Elsevier, Oxford, pp 734–755
11. Pešić V, Dmitrović D, Savić A, Milošević Đ, Zawal A, Vukašinović-Pešić V, von Fumetti S (2019) Application of macroinvertebrate multimetrics as a measure of the impact of anthropogenic modification of spring habitats. Aquatic Conserv Mar Freshw Ecosyst 1–12
12. Radkova V, Syrovatka V, Bojkoa J, Schenkova J, Kroupalova V, Horsak M (2014) The importance of species replacement and richness differences in small-scale diversity patterns of aquatic macroinvertebrates in spring fens. Limnologica 47:52–61
13. Amidžić L, Bartula M, Cvetković D (2014) The State of Biodiversity in Serbia. Nat Area J 34(2):222–226
14. Petrović J (1974) Krš istočne Srbije. Srpsko geografsko društvo, Beograd
15. Ranđelović VN, Zlatković BK (2010) Flora i vegetacija Vlasinske visoravni. Univerzitet u Nišu, Niš, Prirodno-matematički fakultet
16. Stamenković M, Cvijan M, Fužinato S (2008) A checklist of desmids (Conjugatophyceae, Chlorophyta) of Serbia. I. Introduction and elongate baculiform taxa. Cryptogamie Algol 29(4):325–347

17. Dragišić V (1997) Geološko-hidrološke karakteristike sliva Vlasinskog jezera. In: Blaženčić J (ed) Vlasinsko jezero. Biološki fakultet Univerziteta u Beogradu, Beograd, Hidrobiološka studija, pp 25–34
18. Đokić M (2015) The Nišava River – Potamology Study. Dissertation, Faculty of Science and Mathematics, University of Niš
19. Manojlovic S, Antic M, Sibinovic M, Dragicevic S, Novkovic I (2017) Soil erosion response to demographic and land use changes in the Nišava river basin Serbia. Fresenius Environ Bull 26(12):7547–7560
20. Stanković SS (2009) Banje Srbije. Zavod za udžbenike, Beograd
21. Munné A, Prat N (2011) Effects of Mediterranean climate annual variability on stream biological quality assessment using macroinvertebrate communities. Ecol Indic 11:651–662
22. Neff MR, Jackson DA (2011) Effects of broad-scale geological changes on patterns in macroinvertebrate assemblages. J N Am Benthol Soc 30:459–473
23. Carrie R, Dobson M, Barlow J (2015) The influence of geology and season on macroinvertebrates in Belizean streams: implications for tropical bioassessment. Freshw Sci 34(2):648–662
24. Stanković S, Laušević R (1997) Vlasinsko jezero. In: Blaženčić J (ed) Vlasinsko jezero. Biološki fakultet Univerziteta u Beogradu, Beograd p, Hidrobiološka studija, pp 1–23
25. Ranđelović N (1978) Fitocenološko-ekološke karakteristike brdskih travnjaka jugoistčne Srbije. Doktorska disertacija, Prirodoslovno-matematički fakultet, Zagreb
26. Dmitrović D, Savić A, Pešić V (2016) Discharge, substrate type and a temperature as a factor affecting the gastropod assemblages in springs in North-Western Bosnia and Herzegovina. Arch Biol Sci 68(3):613–621
27. Kroupalova V, Bojkova J, Scenkova J, Paril P, Horsak M (2011) Small-scale distribution of aquatic macroinvertebrates in two spring fens with different groundwater chemistry. Int Rev Hydrobiol 96(3):235–256
28. Glöer P (2015) Süßwassermollusken - Ein Bestimmungsschlüssel für die Muscheln und Schnecken im Süßwasser der Bundesrepublik Deutchland. Deutscher Jugendbund für Naturbeobachtung, Göttingen
29. Appleton CC, Forbes AT, Demetriades NT (2009) The occurrence, bionomics and potential impacts of the invasive freshwater snail *Tarebia granifera* (Lamarck, 1822) (Gastropoda: Thiaridae) in South Africa. Zool Med Leiden 83(4):525–536
30. Milenković M, Gligorijević V (2012) *Melanoides tuberculata* (Müller, 1774), a new species for the territory of Serbia (Gastropoda: Thiaridae). Biol Nyssana 3(1):43–45
31. Mitchell A, Brandt TM (2005) Temperature Tolerance of Red-Rim Melania *Melanoides tuberculatus*, an Exotic Aquatic Snail Established in the United States. T Am Fish Soc 134(1):126–131
32. Cianfanelli S, Lori E, Bodon M (2007) Non-indigenous freshwater molluscs and their distribution in Italy. In: Gherardi F (ed) Biological invaders in inland waters: Profiles, distribution, and threats. Springer, Dordrecht, pp 103–121
33. Gashtarov V, Georgiev D (2016) First record of introduction of the tropical snail *Melanoides tuberculata* (O. F. Müller, 1774) in Bulgaria (Gastropoda: Thiaridae). Ecol mont 5:26–27
34. Maciaszek R, Sosnowski W, Wilk S (2019) Tropical snail *Melanoides tuberculata* Müller, 1774 (Thiaridae) found in thermally polluted canal in Central Poland. World Sci News 122:249–254
35. Tixier G, Felten V, Guerold F (2009) Life cycle strategies of *Baetis* species (Ephemeroptera, Baetidae) in acidified streams and implications for recovery. Fund Appl Limnol 174:227–243
36. Von Fumetti S, Nagel P (2011) A first approach to a faunistic crenontypology based on functional feeding groups. J Limnol 70:147–154
37. Savić A, Dmitrović D, Pešić V (2017) Ephemeroptera, Plecoptera and Trichoptera assemblages of karst springs in relation to environmental factors: a case study in central Bosnia and Hercegovina. Turk J Zool 41:119–129
38. Medupin C (2019) Distribution of benthic macroinvertebrate communities and assessment of water quality in a small UK river catchment. SN Appl Sci 1:544
39. Glazier GS, Gooch JL (1987) Macroinvertebrate assemblages in Pennsylvania (U.S.A.) springs. Hydrobiologia 150:33–43

40. Glazier GS (1991) The fauna of North American temperate cold springs: patterns and hypotheses. Freshwater Biol 26:527–542
41. Williams DD, Williams NE, Cao Y (1997) Spatial differences in macroinvertebrate community structure in springs in southeastern Ontario in relation to their chemical and physical environments. Can J Zool 75:1404–1414
42. Collier KJ, Ball OJ, Graesser AK, Main MR, Winterbourn MJ (1990) Do organic and anthropogenic acidity have similar effects on aquatic fauna? Oikos 59:33–38
43. Dangles O, Malmqvist B, Laudon H (2004) Naturally acid freshwater ecosystems are diverse and functional: evidence from boreal streams. Oikos 104:149–155

Chapter 11
Gastropods in Small Water Bodies of the Western Balkans—Endangerments and Threats

Maja Raković, Jelena Tomović, Nataša Popović, Vladimir Pešić, Dejan Dmitrović, Valentina Slavevska Stamenković, Jelena Hinić, Natasha Stefanovska, Jasna Lajtner, and Momir Paunović

Abstract The Balkan Peninsula is a vital part of Europe's biodiversity in respect to freshwater gastropods and many of recorded taxa are stenobionts of narrow distribution. This chapter presents review of the freshwater gastropod fauna distribution in springs, small mountain streams (first order streams), and groundwaters of the Western Balkan and points to the main threats that influence the decline of diversity of this important group of aquatic organisms. Due to the complexity of the topic, small

M. Raković (✉) · J. Tomović · N. Popović · M. Paunović
Institute for Biological Research "Siniša Stanković"–National Institute of the Republic of Serbia, University of Belgrade, Bulevar despota Stefana 142, 11060 Belgrade, Serbia
e-mail: rakovic.maja@ibiss.bg.ac.rs

J. Tomović
e-mail: jelena.tomovic@ibiss.bg.ac.rs

N. Popović
e-mail: natasa.popovic@ibiss.bg.ac.rs

M. Paunović
e-mail: mpaunovi@ibiss.bg.ac.rs

V. Pešić
Faculty of Sciences, University of Montenegro, Džordža Vašingtona b.b, 81000 Podgorica, Montenegro

D. Dmitrović
Faculty of Natural Sciences and Mathematics, Department of Biology and Department of Ecology and Environment Protection, University of Banja Luka, Mladena Stojanovića 2, 78000 Banja Luka, Bosnia and Herzegovina
e-mail: dejan.dmitrovic@pmf.unibl.org

V. S. Stamenković · J. Hinić
Faculty of Natural Sciences and Mathematics, Department of Invertebrates, Institute of Biology, "Ss. Cyril and Methodius" University, 1000 Skopje, Macedonia

N. Stefanovska
Macedonian Museum of Natural History, Blvd. Ilinden 86, 1000 Skopje, Republic of Macedonia

J. Lajtner
Faculty of Science, University of Zagreb, Rooseveltov trg 6, 10000 Zagreb, Croatia
e-mail: jasna.lajtner@biol.pmf.hr

V. Pešić et al. (eds.), *Small Water Bodies of the Western Balkans*, Springer Water,
https://doi.org/10.1007/978-3-030-86478-1_11

standing waters have not been considered here. The most significant threats to the diversity of freshwater gastropods in small streams are habitat degradation (physical destruction of habitats due to mini hydropower plants construction, hydro-technical works related to flood protection, faulty forest management excavation of material), and water abstraction. Knowledge of a considerable number of Balkan gastropod species, underlines the importance for the long-term conservation of biodiversity in this part of Europe.

Keywords Gastropods · Diversity · Hydrobiids · New species · Indicator species · Anthropogenic pressures

11.1 Introduction

Streams and rivers get their starts at springs, snowmelt or even lakes, and then travel all the way to their mouths. The characteristics of a river or stream change during the journey from the source to the mouth [1]. The temperature is cooler at the rheocrene spring than it is at the mouth, the water is also clearer, has higher oxygen levels and substrate is with large stones, gravel and coarse sand typical for the rhithron [2]. Towards the middle part of the stream/river, the width and finer particles in the substrate increases. The size and texture of rivers substrates influence invertebrate abundance and species richness [3]. Macrozoobenthos communities respond to different combinations of velocity, depth, and substrate roughness [4]. It is known that springs and small river basins can harbor high regional diversity due to complex geological history and diverse habitats on local scales [5, 6]. Especially the Balkan freshwater fauna has long been recognized as highly diverse with remarkable degrees of endemism [6].

Many studies show that the Balkan Peninsula was a refuge of biodiversity during the Pleistocene [7]. It is therefore considered to be a hotspot of European biodiversity [6, 8] with its own distinctive fauna and endemic species that are concentrated in exceptionally small areas with small population sizes [8]. For some taxonomic groups and geographical regions, diversity may have been overestimated, but for most groups such as molluscs and areas such as the Dinaric region, diversity is underestimated, therefore the number of taxa is increasing annually, as more research is carried out [6, 8].

The world's freshwater gastropod fauna faces unprecedented threats from habitat loss and degradation [9]. The most significant pressures to the diversity of gastropods in small streams and springs are habitat degradation and water abstraction. The dams are a represent one of the major anthropogenic disturbances of freshwater and nutrient cycles globally [10]. Anthropogenic pressures often trigger the cascade of effects that impact a wide range of freshwater communities.

Despite the fact that the Balkan region has some of Europe's most pristine rivers and is a global hotspot of biodiversity, the region is the target of one of the most ambitious hydropower expansion plans in the world [11]. On the other hand, the climate change models for Southeastern Europe predict large reductions in hydropower potential due to reduced precipitation; for example; by 2070, a −35% drop for Bosnia, and −25% for Croatia [12, 13]. Exploiting aquatic resources in the Dinaric Alps region, whether above or beneath the earth's surface and often results in unforeseen consequences to both nearby and distant habitats. These impacts range from water level loss or gain to lakes, springs or other rivers, unintentional redirection of flows, etc. [14, 15].

Studying the diversity and ecology, with emphasizes to endangered and endemic species may substantially help to their risk assessment. Therefore, the aim of this study is to review the distribution of freshwater gastropod fauna of small water bodies of the Western Balkan and to point main threats that influence the decline of diversity.

Since the term "small water bodies" is colloquial, loosely used expression, we need to elaborate how it was addressed in this paper. There are no universal definitions for small waters or small water bodies. In general, small waters bodies include low order streams, springs, ditches, flushes, small lakes and ponds. Here we addressed springs, small mountain streams (first order streams), and groundwaters of the Western Balkan (Fig. 11.1). Such water bodies could be found in variety of landscapes and consequently are significantly different in their characteristics, both, abiotic and biotic.

The common feature for all small water bodies is that they are neglected in general water management policies and, consequently protection practice. Thus, in the context of water management practice in Europe, running waters with catchment area smaller than 10 km^2, and standing waters with surface area smaller than 0.5 km^2, in majority of the cases, are not characterized as significant for river basin management practice, since surface waters include a large number of very small waters for which the administrative burden for the management of these waters may be enormous [16].

11.2 Diversity of Freshwater Gastropods Fauna in Small Water Bodies of the Western Balkans

Freshwater gastropods represent only approximately 5% of overall gastropod diversity [17]. In Europe the freshwater gastropods represent about 94% (808 species) of the total number of freshwater mollusk species, dividing into two groups, the Prosobranchs (with the largest family Hydrobiidae) and the Pulmonates (containing the families Lymnaeidae, Acroloxidae, Planorbidae etc.), which contain a higher proportion of the widespread, more cosmopolitan species [18]. Strong et al. [9] described global patterns of freshwater gastropod species diversity as it relates to the zoogeographic regions of the world. The greatest diversity was observed in Palearctic with 1408–1711 species, followed by the Nearctic with 585 species and the Oriental, Neotropical, and Australasian with similar ranges of 509–606, 440–533, and 490–514, respectively.

Fig. 11.1 Habitat appearance of small water bodies (the springs and small mountain streams of Montenegro: **a** Gusinje, Skakavica river; **b** Šavnik, Nevidio canyon and **c** Bajlovića sige, waterfall in the Tara canyon) and Croatia: **d** Pećina, cave and brook in the Vransko jezero Nature Park; **e** Žrnovnica river). Photos by M. Raković (Figs. **a**–**c**), I. Lajtner (Fig. **d**) and I. Maguire (Fig. **e**)

Strong et al. [9] identified 25 gastropod species diversity hotspots, categorized by primary habitat. Mountainous regions in southern France and Spain; the Southern Alps and Balkans region are stood out in Europe in category springs and groundwater. Among ancient oligotrophic lakes in our area it stands out Lake Ohrid and its basin, and in category large rivers and their first and second order tributaries it has been distinguished the River Zrmanja in Croatia [19]. Protection of these hotspots will ensure the long- term survival and viability of hundreds of endemic species [19].

Freshwater snails fauna of the Western Balkan belongs to three drainage systems: Black Sea, Aegean Sea and Adriatic Sea drainage system.

Even if the areas of springs, small mountain streams (low order streams), and groundwaters of the Western Balkan countries are unevenly explored, they are characterized the great diversity is seen in the freshwater gastropods with 251 species and subspecies belonging in 16 families. The most numerous taxa are in the family Hydrobiidae presented by 153 species and subspecies, followed by families: Moitessieriidae (31 species), Bythinellidae (15), Bithyniidae (11), Planorbidae (12 species), while the rest of families were less diversified (Table 11.1).

Freshwater gastropods indicated zoogeographically great diversity and authenticity of investigated area, through the relatively high number of endemic species. There are 11 local endemics on the territory of Serbia (*Bythinella drimica alba* Radoman, 1976, *B. istoka* Glöer & Pešić, 2014, *B. nonveilleri* Glöer, 2008, *B. pesterica* Glöer, 2008, *B. serborientalis* Radoman, 1978, *Belgrandiella serbica* Glöer, 2008, *Plagigeyeria minuta* Bole & Velkovrh, 1987, *P. piroti* Bole & Velkovrh, 1987, *Saxurinator schlickumi* Schutt, 1960, *Terranigra kosovica* Radoman 1978, *Iglica illyrica* Schütt, 1975). The most of endemic species are representatives of Hydrobiidae, and inhabiting subterranean or spring habitats and have restricted distribution ranges in South Serbia (predominantly Kosovo and Metohija), or Southeast and Southwest areas of Serbia (Stara Planina, Pešter).

Globaly, with at least 900 valid species [20], Prosobranchs gastropods from Hydrobiidae family are one of the largest gastropod families. Out of 586 described Hidrobiidae species in Europe, 570 are considered endemic (97%) [18].

Several commonly accepted subfamilies and genera were excluded from the Hydrobiidae and raised to family level [21]. According to current knowledge, families Amnicolidae, Bithyniidae Bythinellidae, the Moitessieriidae as well as the Tateidae used to include in the Hydrobiidae family. Nowdays is often used the term "hydrobioid" coined by Davis [22] for all hydrobiid like taxa.

Besides Hydrobiidae, in terms of biodiversity on Europen level, family Moitessieriidae with 54 species (all endemic) is the second most important family [18]. Among Pulmonates on European scope, the greatest diversity is observed into representatives of the family Planorbidae (22 endemic out of 42 species).

With the progress of molecular systematics it is expected revealing cryptic species lineages and greater species diversity, especially into Hydrobiidae, Bithyniidae, Lymnaeidae and Planorbidae [23]. Accordingly, the overall Balkan's malacofauna diversity seems will be even higher. According to Strong et al. [9], the current gastropods diversity may represent only 25% of their real diversity.

11.3 The Newly Described Freshwater Gastropod Species of the Western Balkans—Last Two Decades

The number of described and newly described species of hydrobiids makes them the most diverse group in small water bodies, springs and groundwater. The centre of distribution of the European Hydrobiidae is the Balkan Peninsula [24]. Among numerous studies covering this region, Pavle Radoman was the leading malacologist, published numerous papers about the hydrobiids, describing a number of new genera and species from the Balkan Peninsula [25–32]. More than 25 years of his research is

Table 11.1 List of gastropod species in small water bodies from Western Balkans

Family Melanopsidae	*Bythinella marici* Glöer & Pešić, 2014
Microcolpia daudebartii acicularis (Férussac, 1823)	*Bythinella magna* Radoman, 1976
Family: Amphimelaniidae	*Bythinella melovskii* Glöer & S-Stamenković, 2015
Holandriana holandrii (C. Pfeiffer, 1828)	*Bythinella nonveilleri* Glöer, 2008
Family: Lithoglyphidae	*Bythinella opaca* (M. von Gallenstein, 1848)
Dabriana bosniaca Radoman 1974	*Bythinella pesterica* Glöer, 2008
Lithoglyphus apertus (Küster, 1852)	*Bythinella serborientalis* Radoman, 1978
Lithoglyphus naticoides (C. Pfeiffer, 1828)	*Bythinella taraensis* Glöer & Pešić, 2010
Lithoglyphus fuscus (C. Pfeiffer, 1828)	Family: Bithyniidae
Family: Bythinellidae	*Bithynia cettinensis* Clessin, 1885
Bythinella austriaca (Frauenfeld, 1857)	*Bithynia hambergerae* Reischütz, N.R. & P. L. Reischütz, 2008
Bythinella dispersa Radoman, 1973	*Bithynia leachii* (Sheppard, 1823)
Bythinella drimica alba Radoman, 1976	*Bithynia majewskyi* Frauenfeld, 1862
Bythinella istoka Glöer & Pešić, 2014	*Bithynia montenegrina* (Wohlberedt, 1901)
Bythinella golemoensis Glöer & Mrkvicka, 2015	*Bithynia mostarensis* Möllendorf, 1873
Bythinella kapelana Radoman, 1976	*Bithynia radomani* Glöer & Pešić, 2007
Bythinella luteola Radoman, 1973	*Bithynia skadarskii* Glöer & Pešić, 2007
Bithynia tentaculata (Linnaeus, 1758)	*Ecrobia ventrosa* (Montagu, 1803)
Bithynia zeta Glöer & Pešić, 2007	*Ecrobia ventrosa* (Montagu, 1803)
Pseudobithynia kirka Glöer, A. & Wilke, 2007	*Ecrobia vitrea* (Risso, 1826)
Family: Hydrobiidae	*Graziana kuesteri* (Boeters, 1970)

(continued)

Table 11.1 (continued)

Adriohydrobia gagatinella (Küster, 1852)	*Graziana lacheineri adriolitoralis* Radoman, 1975
Anagastina gluhodolica (Radoman, 1973)	*Graziana lacheineri glinensis* Radoman, 1975
Anagastina matjasici (Bole, 1961)	*Graziana papukensis* Radoman, 1975
Anagastina scutarica (Radoman, 1973)	*Graziana slavonica* Radoman, 1975
Anagastina vidrovani (Radoman, 1973)	*Graziana vrbasensis* Radoman 1975
Anagastina zetaevallis (Radoman, 1973)	*Grossuana euxina* (A.J. Wagner, 1928)
Antibaria notata (Frauenfeld, 1865)	*Grossuana euxina macedonica* Radoman 1973
Belgrandia torifera Schütt, 1961	*Grossuana maceradica* Boeters, Glöer & Stamenković, 2017
Belgrandiella bozidarcurcici Glöer & Pešić, 2014	*Grossuana serbica scupica* Radoman, 1973
Belgrandiella bumasta Schütt, 1960	*Hadziella anti* Schütt, 1960
Belgrandiella croatica (Hirc, 1881)	*Hadziella krkae* Bole, 1992
Belgrandiella dabriana Radoman 1975	*Hadziella rudnicae* Bole, 1992
Belgrandiella driniana (Radoman 1975)	*Hadziella sketi* Bole, 1961
Belgrandiella erythropoma (Schütt 1959)	*Hadziella thermalis* Bole, 1992
Belgrandiella fontinalis (F. J. Schmidt, 1847)	*Hauffenia media* Bole, 1961
Belgrandiella koprivnensis Radoman,1975	*Hauffenia tovunica* Radoman 1978
Belgrandiella krupensis Radoman, 1973	*Hauffenia wagneri* (Kuščer, 1928)
Belgrandiella kusceri (A. J. Wagner, 1914)	*Horatia novoselensis Radoman, 1966*
Belgrandiella pageti Schütt, 1970	*Horatia klecakiana* Bourguignat, 1887
Belgrandiella serbica Glöer, 2008	*Horatia knorri* Schütt, 1961
Belgrandiella travnicensis (Radoman 1975)	*Horatia ozimeci* Grego & Falniowski, 2021

(continued)

Table 11.1 (continued)

Belgrandiella zermanica Radoman, 1973	*Horatia stygorumina* Grego & Rysiewska, 2021
Bracenica plana (Bole, 1961)	*Hydrobia acuta* (Draparnaud, 1805)
Bracenica spiridoni Radoman, 1973	*Hydrobia cattaroensis* (Westerlund, 1886)
Bracenica vitojaensis Glöer, G., E. & Fehér, 2015	*Hydrobia declinata* (Frauenfeld, 1863)
Cilgia dalmatica (Schütt, 1961)	*Islamia bosniaca* Radoman, 1973
Chilopyrgula sturanyi (Brusina, 1896)	*Islamia dmitroviciana* Boeters, Glöer & Pešić, 2013
Dalmatella sketi Velkovrh, 1970	*Islamia latina* Radoman, 1973
Dalmatinella fluviatilis Radoman, 1973	*Islamia montenegrina* Glöer, Grego, Erőss & Fehér, 2015
Docleiana tabanensis (Boeters, Glöer & Pešić, 2014)	*Islamia pusilla* (Piersanti, 1951)
Ecrobia spalatiana (Radoman, 1973)	*Islamia valvataeformis* (Möllendorff 1873)
I*slamia steffeki* Glöer and Grego, 2015	*Plagigeyeria pseudocostellina* Grego, 2020
Islamia zermanica Radoman, 1973	*Plagigeyeria reischuetzorum* Grego, 2020
Istriana mirnae Velkovrh, 1971	*Plagigeyeria vriosticaensis* Grego, 2020
Iverakia hausdorfi Glöer & Pešić, 2014	*Plagigeyeria zetaprotogona pageti* Schütt, 1961
Karucia sublacustrina Glöer & Pešić, 2013	*Plagigeyeria zetaprotogona zetadidyma* Schütt, 1960
Kerkia briani Rysiewska & Osikowski, 2020	*Plagigeyeria zetatridyma* Schütt, 1960
Kerkia jadertina (Kuščer 1933)	*Pseudamnicola conovula* (Frauenfeld, 1863)
Kerkia kareli Beran, Bodon & Cianfanelli, 2014	*Pseudamnicola orsinii* (Küster, 1852)
Litthabitella chilodia (Westerlund, 1886)	*Pseudohoratia ochridana* (Polinski, 1929)
Montenegrospeum bogici (Pešić & Glöer, 2012)	*Pyrgula annulata* (Linnaeus, 1767)
Montenegrospeum sketi Grego & Glöer, 2018	*Pyrgohydrobia jablanicensis* Radoman, 1955
Narentiana albida Radoman, 1973	*Pyrgohydrobia sanctinaumi* Radoman, 1955

(continued)

Table 11.1 (continued)

Narentiana vjetrenicae Radoman 1973	*Radomaniola bosniaca* (Radoman, 1973)
Ohridohoratia polinskii (Radoman, 1960)	*Radomaniola curta anagastica* (Radoman, 1973)
Ohridohoratia pygmaea (Westerlund, 1902)	*Radomaniola curta curta* (Küster, 1853)
Ochridopyrgula macedonica macedonica (Brusina, 1896)	*Radomaniola curta germari* (Frauenfeld, 1863)
Plagigeyeria angyaldorkae Grego, 2020	*Radomaniola curta kicavica (Radoman, 1973)*
Plagigeyeria erossi Grego, 2020	*Radomaniola curta mostarensis* (Radoman 1973)
Plagigeyeria feheri Grego & Glöer, 2019	*Radomaniola curta narentana* (Radoman 1973)
Plagigeyeria gladilini Kuščer, 1937	*Radomaniola curta pivensis* (Radoman, 1973)
Plagigeyeria inflata (A. J. Wagner, 1928)	*Radomaniola elongata* (Radoman, 1973)
Plagigeyeria jakabi Grego, 2020	*Radomaniola lacustris* (Radoman, 1983)
Plagigeyeria jalzici Cindrić & Slapnik, 2019	*Radomaniola montana* (Radoman, 1973)
Plagigeyeria konjicensis Grego, 2020	*Sadleriana cavernosa* Radoman, 1978
Plagigeyeria lewarnei Grego, 2020	*Sadleriana fluminensis* (Küster, 1853)
Plagigeyeria listicaensis Grego, 2020	*Sadleriana sadleriana* (Frauenfeld, 1863)
Plagigeyeria lukai Glöer & Pešić, 2014	*Sadleriana schmidtii* (Menke, 1849)
Plagigeyeria ljutaensis Grego, 2020	*Sadleriana supercarinata* (Schütt, 1969)
Plagigeyeria minuta Bole & Velkovrh, 1987	*Sarajana apfelbecki* (Brancsik 1988)
Plagigeyeria montenigrina Bole, 1961	*Saxurinator brandti* Schütt, 1968
Plagigeyeria mostarensis Kuščer, 1933	*Saxurinator hadzii* (Bole, 1961)
Plagigeyeria olsavskyi Grego, 2020	*Saxurinator labiatus* (Schütt, 1963)
Plagigeyeria ozimeci Grego, 2020	*Saxurinator montenegrinus* (Schütt, 1959)
Plagigeyeria piroti Bole & Velkovrh, 1987	*Saxurinator orthodoxus* Schütt, 1960

(continued)

Table 11.1 (continued)

Plagigeyeria plagiostoma (A. J. Wagner, 1914)	*Saxurinator schlickumi* Schutt, 1960
Saxurinator sketi (Bole 1960)	*Lanzaia skradinensis* Bole, 1992
Strugia ohridana Radoman 1973	*Lanzaia vjetrenicae* Kuščer, 1933
Stygobium hercegnoviensis Grego & Glöer, 2019	*Lanzaia vjetrenicae latecostata* Schütt, 1968
Tanousia zrmanjae (Brusina, 1866)	*Paladilhiopsis absoloni* (A.J. Wagner, 1914)
Terranigra kosovica Radoman 1978	*Paladilhiopsis blihensis* (Glöer & Grego, 2015)
Travunijana angelovi (Schütt, 1972)	*Paladilhiopsis cattaroensis* Grego & Glöer, 2019
Travunijana edlaueri (Schütt, 1961)	*Paladilhiopsis insularis* Cindrić & Slapnik, 2019
Travunijana gloeri Grego, 2020	*Paladilhiopsis maroskoi* (Glöer & Grego, 2015)
Travunijana djokovici Grego & Pešić. 2020	*Paladilhiopsis pretneri* Bole & Velkovrh, 1987
Travunijana klemmi (Schütt, 1961)	*Paladilhiopsis robiciana illustris* (Schütt, 1970)
Travunijana nitida (Schütt, 1963)	*Paladilhiopsis serbica* (Pavlovic, 1913)
Travunijana ovalis (Kuščer, 1933)	*Paladilhiopsis solida* Kuščer, 1933
Travunijana robusta robusta (Schütt, 1959)	*Paladilhiopsis tarae* Bole & Velkovrh, 1987
Travunijana tribunicae (Schütt, 1963)	*Paladilhiopsis turrita* (Kuščer, 1933)
Travunijana vruljakensis Grego & Glöer, 2019	Family: Emmericiidae
Vinodolia fiumana Radoman, 1973	*Emmericia expansilabris* Bourguignat, 1880
Zeteana ljiljanae Glöer & Pešić, 2014	*Emmericia narentana* Bourguignat, 1881
Family Moitessieriidae	*Emmericia patula* (Brumati, 1838)
Bosnidilhia vitojaensis Grego & Glöer, 2019	*Emmericia ventricosa* Brusina, 1870
Bosnidilhia vreloana Boeters, Glöer & Pešić, 2013	Family: Thiaridae
Bythiospeum demattiai Glöer & Pešić, 2013	*Melanoides tuberculata* (O.F. Muller, 1774)

(continued)

Table 11.1 (continued)

Bythiospeum hrustovoense Glöer and Grego, 2015	Family:Valvatidae
Bythiospeum petroedei Glöer and Grego, 2015	*Borysthenia naticina* (Menke, 1845)
Bythiospeum plivense Glöer and Grego, 2015	*Valvata cristata* Müller, 1774
Iglica bagliviaeformis Schütt, 1970	*Valvata montenegrina* (Glöer & Pešić, 2007)
Iglica elongata Kuščer, 1933	*Valvata piscinalis* (Müller, 1774)
Iglica gracilis (Clessin, 1882)	*Valvata stenotrema* Polinski, 1929
Iglica illyrica Schütt, 1975	Family: Lymnaeidae
Iglica langhofferi A. J. Wagner, 1928	*Radix balthica* (Linnæus, 1758)
Lanzaia bosnica Bole 1970	*Radix labiata* (Rossmässler, 1835)
Lanzaia edlaueri Schütt, 1961	*Radix relicta* Polinski, 1929
Lanzaia kotlusae Bole, 1992	*Radix skutaris* Glöer & Pešić, 2007
Lanzaia matejkoi Grego & Glöer, 2019	*Stagnicola fuscus* (C. Pfeiffer, 1821)
Lanzaia pesici Glöer, Grego, Erőss & Fehér, 2015	*Stagnicola montenegrinus* Glöer & Pešić, 2009
Lanzaia rudnicae Bole, 1992	Family: Physidae
Physella acuta (Draparnaud, 1805)	*Gyraulus piscinarum* (Bourguignat, 1852)
Physa fontinalis (Linnaeus, 1758)	*Gyraulus shasi* Glöer & Pešić, 2009
Family:Planorbidae	*Planorbis vitojensis* Glöer & Pešić, 2010
Ancylus fluviatilis Müller, 1774	Family: Hydrocenidae
Ancylus recurvus Martens, 1873	*Hydrocena cattaroensis* (Pfeiffer, 1841)
Gyraulus albus (O. F. Müller, 1774)	Family: Neritidae
Gyraulus albidus Radoman, 1953	*Theodoxus fluviatilis* (Linnæus, 1758)
Gyraulus crenophilus Hubendick & Radoman, 1959	*Theodoxus subterrelictus* Schütt, 1963
Gyraulus fontinalis Hubendick & Radoman, 1959	Family: Amnicolidae

(continued)

Table 11.1 (continued)

Gyraulus ioanis Glöer & Pešić, 2008	*Marstoniopsis croatica* Schütt, 1974
Gyraulus laevis (Alder, 1838)	*Marstoniopsis vrbasi* Bole & Velkovrh, 1987
Gyraulus meierbrooki Glöer & Pešić, 2007	

summarized in a monograph [31], which is still the most exhaustive source of knowledge of the anatomy and distribution of the Balkan hydrobiids. The next important scientist is the slovenian malacologist Jože Bole, who described a number of species of subterranean aquatic snails of the Western Balkan [33–38].

Recent studies on the diversity of freshwater snails found 65 new species to the Western Balkans—Albania and the former Yugoslavia, except Slovenia (Table 11.2).

Bythinella species occur predominantly in springs and spring-fed brooks, where they can form large populations and the most of the species are locally endemic, especially in mountainous regions [39]. Bank [40] noticed that ten species of the genus *Bythinella* occur in Western Balkan region. Glöer [41] and Glöer and Pešić [42] added several new *Bythinella* species - *B. nonveilleri*, *B. pesterica*, *B. marici* and *B. istoka* in the list. In Montenegro the species of the latter genus are absent from the Adriatic drainage area, with the exception of one locality in the upper part of the River Morača, near the watershed of the two drainage sea areas [43].

Among the new hydrobiids to the northwestern Bosnia and Herzegovina are endemic species—*Bosnidilhia vreloana*, *Islamia dmitroviciana* and *Belgrandiella bozidarcurcici.* First two species were found in springs near the Banja Luka city, while the third species described in the rheocrene and rheopsammocrene springs situated of the canyon of the Cvrcka River in the Bosnia and Hercegovina [42, 44]. Grego [45] and Glöer and Grego [42] added a significant number of newly described species (genus *Plagigeyeria*, *Paladilhiopsis* and *Bythiospeum*) in the list freshwater gastropod for the Bosnia and Hercegovina.

A two new species of genus *Kerkia*, *K. briani*, is described from the spring Polički Studenac Vrelo (Crkvina), adjacent to the Trebišnjica River (Bosnia and Herzegovina) and *Kerkia kareli*, from an old well near Povljana on Pag island (Croatia) [47, 48]. In the last two decades a several new hydrobiids have been found in the specificity habitats of the Republic of North Macedonia—*Sumia macedonica*, *Bythinella golemoensis*, *Bythinella melovskii* and *Pseudobithynia ambrakis.* The first two species are found in karstic waters from Jablanica mountains caves [49], third species was found in small streams near Belička River, also on Jablanica mountains [50], while the fourth hydrobiids, *Pseudobithynia ambrakis*, were found in littoral region of the Macedonian part of the Dojran Lake [51]. The most of new species described from Montenegro are restricted to subterranean habitats of the Skadar Lake catchment area, with about 30 species of hydrobiid gastropods recorded so far (see [43] for the list), of which around 82% are known only from type localities, while 33% are known only from empty shells (Fig. 11.2). Most of these subterranean species have been in strong springs such as Vitoja large karstic spring located on the northeastern

Table 11.2 List of newly described freshwater gastropod species of the Western Balkans in the last two decades (Gastropoda: Prosobranchia: Hydrobidae-Bythinellidae-Moitessieriidae and Lymnaeidae-Planorbidae-Bythiniidae-Valvatidae), SRB (Serbia), MNE (Montenegro), BIH (Bosnia and Hercegovina), HR (Croatia), NMK (Republic of North Macedonia) and ALB (Albania)

Name of species	SRB	MNE	BIH	HR	NMK	AL
Belgrandiella bozidarcurcici Glöer &Pešić, 2014						
Belgrandiella serbica Glöer, 2008	*					
Bithynia hambergerae Reischütz, Reischütz & Reischütz, 2008		*				
Bithynia radomani Glöer & Pešić, 2007		*				
Bithynia skadarskii Glöer & Pešić, 2007		*				
Bithynia zeta Glöer & Pešić, 2007		*				
Bosnidilhia vreloana Boeters, Glöer & Pešić, 2013		*	*			
Bracenica vitojaensis Glöer, Grego, Erőss & Fehér, 2015		*				
Bythinella golemoensis Glöer & Mrkvicka, 2015					*	
Bythinella istoka Glöer & Pešić, 2014	*					
Bythinella marici Glöer & Pešić, 2014			*			
Bythinella melovskii Glöer & Slavevska-Stamenković, 2015					*	
Bythinella nonveilleri Glöer, 2008	*					
Bythinella pesterica Glöer, 2008	*					
Bythinella taraensis Glöer & Pešić, 2010		*				
Bythiospeum hrustovoense Glöer and Grego, 2015			*			
Bythiospeum petroedei Glöer and Grego, 2015			*			
Bythiospeum plivense Glöer and Grego, 2015			*			
Bythiospeum szarowskae Glöer, Grego, Erőss & Fehér, 2015						*
Grossuana maceradica Boeters, Glöer & Stamenković, 2017					*	
Gyraulus ioanis Glöer & Pešić, 2007		*				
Gyraulus meierbrooki Glöer & Pešić, 2007		*				
Gyraulus shasi Glöer & Pešić, 2007		*				
Horatia ozimeci Grego & Falniowski, 2021				*		
Horatia stygorumina Grego & Rysiewska, 2021				*		
Islamia dmitroviciana Boeters, Glöer &Pešić, 2013			*			
Islamia montenegrina Glöer, Grego, Erőss & Fehér, 2015		*				
Islamia steffeki Glöer and Grego, 2015			*			
Karucia sublacustrina Glöer & Pešić, 2013		*				
Kerkia briani Rysiewska & Osikowski, 2020			*			
Kerkia kareli Beran, Bodon & Cianfanelli, 2014				*		

(continued)

Table 11.2 (continued)

Name of species	SRB	MNE	BIH	HR	NMK	AL
Lanzaia pesici Glöer, Grego, Erőss & Fehér, 2015		*				
Montenegrospeum bogici Pešić & Glöer, 2012		*				
Montenegrospeum sketi Grego & Glöer, 2018				*		
Paladilhiopsis blihensis (Glöer & Grego, 2015)			*			
Paladilhiopsis falniowskii Grego, Glöer, Erőss & Fehér, 2017						*
Paladilhiopsis insularis Cindrić & Slapnik, 2019				*		
Paladilhiopsis lozeki Grego, Glöer, Erőss & Fehér, 2017						*
Paladilhiopsis maroskoi Glöer & Grego, 2015			*			
Paladilhiopsis prekalensis Grego, Glöer, Erőss & Fehér, 2017						*
Paladilhiopsis szekeresi Grego, Glöer, Erőss & Fehér, 2017						*
Paladilhiopsis wohlberedti Grego, Glöer, Erőss & Fehér, 2017						*
Plagigeyeria angyaldorkae Grego, 2020			*			
Plagigeyeria erossi Grego, 2020			*			
Plagigeyeria jakabi Grego 2020			*			
Plagigeyeria jalzici Cindrić & Slapnik, 2019				*		
Plagigeyeria konjicensis Grego, 2020			*			
Plagigeyeria lewarnei Grego, 2020			*			
Plagigeyeria listicaensis Grego, 2020			*			
Plagigeyeria ljutaensis Grego, 2020			*			
Plagigeyeria olsavskyi Grego, 2020			*			
Plagigeyeria ozimeci Grego, 2020			*			
Plagigeyeria pseudocostellina Grego, 2020			*			
Plagigeyeria reischuetzorum Grego, 2020			*			
Plagigeyeria steffeki Grego, Glöer, Erőss & Fehér, 2017						*
Plagigeyeria vriosticaensis Grego, 2020			*			
Planorbis vitojensis Glöer & Pešić, 2010		*				
Pseudamnicola krumensis Glöer, Grego, Erőss & Fehér, 2015						*
Pseudobithynia ambrakis Glöer, Falniowski & Pešić, 2010					*	
Radix skutaris Glöer & Pešić, 2008		*				*

(continued)

Table 11.2 (continued)

Name of species	SRB	MNE	BIH	HR	NMK	AL
Sumia macedonica Glöer & Mrkvicka, 2015					*	
Travunijana djokovici Grego & Pešić, 2020		*				
Travunijana gloeri Grego, 2020			*			
Travunijana vruljakensis Grego & Glöer, 2019			*			
Valvata montenegrina Glöer & Pešić, 2008		*				

Fig. 11.2 Endemic gastropod species occurring in springs of the Lake Skadar basin. **a** *Montenegrospeum bogici* (Pešić & Glöer, 2012). **b** *Travunijana djokovici* Grego & Pešić, 2021; **c** *Docleiana tabanensis* (Boeters, Glöer & Pešić, 2014); **d** *Bracenica spiridoni* Radoman, 1973; **e** *Zeteana ljiljanae* Glöer & Pešić, 2014; **f** *Iverakia hausdorfi* Glöer & Pešić, 2013; **g** *Radomaniola curta curta* (Küster, 1853); **h** *Antibaria notata* (Frauenfeld, 1865); **i** *Vinodolia zetaevalis* (Radoman, 1973). Photos by P. Glöer (Figs. **a**, **c**–**i**) and J.Grego (Fig. **b**)

shore of Lake Skadar and spring in village Pričelje (located in the vicinity of the Zeta river; Fig. 11.3) where they were washed and brought together from geographically isolated areas [43]. Five new species are described from Montengro and Albania, namely: *Bracenica vitojaensis*, *Islamia montenegrina*, *Lanzaia pesici*, *Bythiospeum szarowskae* and *Pseudamnicola krumensis* [52]. Five of the new freshwater gastropod species from northern Albania are assigned to the genus *Paladilhiopsis* Pavlović, 1913, namely *P. prekalensis*, *P. lozeki*, *P. szekeresi*, *P. wohlberedti.*, *P. falniowskii* and one to the genus *Plagigeyeria* Tomlin, 1930, namely *P. steffeki* [53].

A several new stygobiotic species is described from spring in Croatia [54, 55]. Three species was found in Dalmatia in Split district: *Montenegrospeum sketi* was

Fig. 11.3 **a** Spring Vitoja at northeastern shore of the Lake Skadar (Montenegro). This spring is locus typicus of *Bracenica vitojaensis*, *Islamia montenegrina*, *Bosnidilhia vitojaensis* and *Lanzaia pesici*, and home to *Anagastina matjasici*, *Plagigeyeria zetaprotogona* and *Anagastina scutarica*. **b** Spring in village Pričelje (Podgorica, Montenegro). This spring is locus typicus of Iverakia hausdorfi, *Zeteana ljiljanae*, *Plagigeyeria lukai* and *Bythiospeum demattia*. Photos by V. Pešić

found in several springs feeding the Cetina River in the SW part of Sinj Basin [54], *Horatia ozimeci* was described from spring Ruda—Beguša and *H. stygorumina* was found inside the cave Mali Rumin in village Rumin [55]. Two new subterranean gastropod species, *Plagigeyeria jalzici* and *Paladilhiopsis insularis,* were found in the cave Rudnica VI located in central Croatia, near the town of Ogulin [56].

The unevenness in the number of newly described species of Gastropods of small water bodies and springs in the Western Balkans indicates a deficiency of research in certain parts of the area. Still, a significant number of newly described species in the last two decades indicates the great importance of these microhabitats (Fig. 11.4).

Anthropogenic changes in the directions of karst-water flows as a result of construction of dams and diversion of river flows, with the aim of diverting water to power plants, make the understanding on knowledge of the distribution of stygobionts even more difficult [57]. Continuous research of this specific fauna increases the possibility of implementation of protection and conservation measures these important microhabitats as well as the whole Western Balkans region.

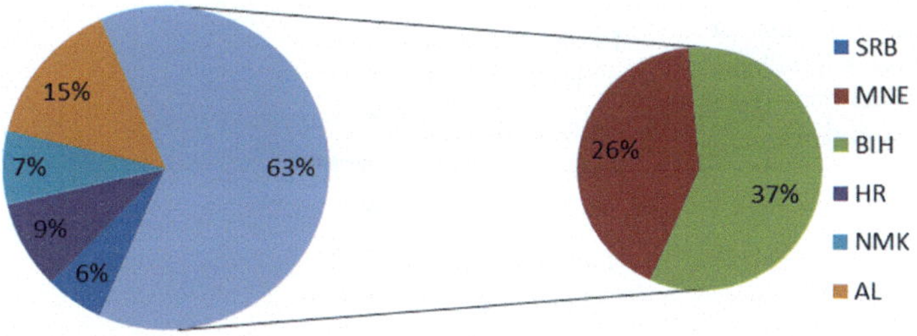

Fig. 11.4 The percentage of newly described hydrobiid gastropods by the mentioned countries of the Western Balkans in the twenty-first century

11.4 Indicator Species of Gastropods in Different Water Body Types

The distribution and structure of freshwater gastropod assemblages are influenced climate-related factors, such as temperature and precipitation as well as physiographical factors [58]. The environmental factors, such as hydrological, physicochemical, and biological factors, are important within the same climate region. In addition, within-microhabitat differences (e.g., differences between flow regimes, substrate composition, and riparian vegetation) contribute to variations among gastropod assemblages. Substrate composition is closely related to habitat complexity and resource availability for gastropods (e.g., the number of algae and aquatic macrophytes).

Although partially neglected as biological quality element for ecological status assessment due to complex and long lasting identification of many taxa, Gastropods are potentially very important component in biological water monitoring. They can be effectively used as important typological indicators of different types of waters [6, 59–61].

Indicator species are species that are used as ecological representatives of a certain biogeographical region that should be found in that specificity habitat [62].

Based on an overview of the available literature data in Table 11.1, the percentage of gastropods families in the three water body types of the Western Balkans is shown in Fig. 11.5.

Different hydrobiid gastropods are present in all types of small water bodies and thus could be considered as bioindicators, species from Bythinellidae and Moitessieriidae families can be considered as potentially confident indicators of stress, specifically general degradation in springs and groundwater habitats. Beside them, certain species from the Lymnaeidae, Planorbidae, Bythiniidae and Valvatidae families are very common in small water bodies [6]. While, the large and medium sized rivers at the mid-altitude are characterized by typical snail communities—families Viviparidae, Physidae and Lithoglyphidae as the most common species [6, 63, 64].

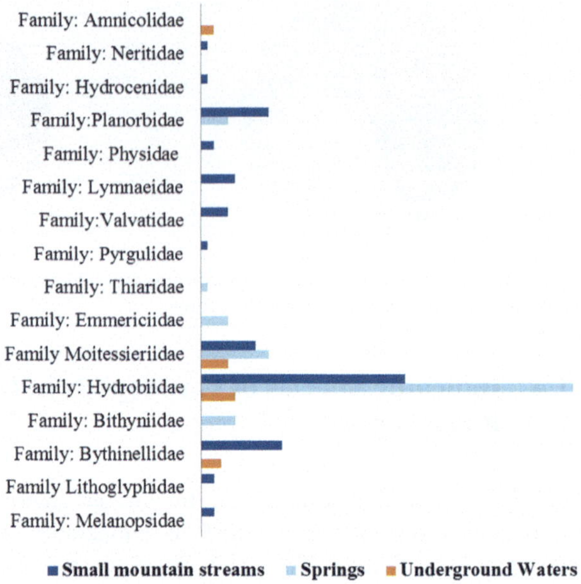

Fig. 11.5 The percentage of gastropods taxa by families in the three water body types of the Western Balkans

11.5 The Most Significant Pressures to the Diversity of Gastropods in Small Water Bodies

Aquatic ecosystems are under considerable anthropogenic impact, they suffer various pressures [65–67], have various use, origin, and water quality what reflects in the diversity and abundance of present biota.

Industrialization, agriculture as well as irresponsible waste disposal lead to groundwater pollution indirectly through land or directly, and through sources they are transmitted through watercourses. The presence of pollutants directly affects the primary production in aquatic ecosystems and reduces the possibility of survival of organisms; there is a disturbance of the balance of complex biotic relationships.

Many settlements, pastures and agricultural areas have been created right next to springs and rivers due to the available water for irrigating plants and watering livestock. In that case, the springs are blocked and ponds or ponds are formed, which further leads to a huge disturbance of aquatic ecosystems and a change in the composition of the living world [68].

According to Cuttelod et al. [18], declining of water quality in the freshwater rivers and lakes, throughout Europe is observed mainly due to intensification of agriculture (affecting 36% of the species) and urbanisation (impacting 29% of the species). Threats to freshwater biodiversity include also overexploitation which impacts 33% of freshwater species; invasion of exotic species (impacts less than 5% of the threatened species) as well as the decline in habitat quality in the freshwater rivers and lakes which is a problem throughout Europe.

Global climate change and changes in land use with more drought periods and floods will take place more frequently, and they become important threat for macroinvertebrate assemblages especially for endemic, crenobiontic species restricted to non-flooded springs [69]. Springs are places of discharge from groundwater flow, and the stability of the flow mostly depends on the volume and storage capacity of the flow systems that supply it [70].

The Balkan Penninsula shows an exceptional wealth in terms of springs, most of them were formed on a limestone base, by the action of groundwater (carst type) so there is a need to emphasize the importance of assessing the quality of groundwater, which also can be loaded with organic and inorganic substances as a result of anthropogenic action [71].

Considering that hydrobiid gastropods often restricted to a freshwater springs, lakes or single groundwater catchment system, which could be easily impacted by off-take of water for domestic and agricultural supplies, the highest number of threatened species belong to Hydrobiidae group [18, 72].

Gastropods have a very limited ability to avoid unfavorable environments, making it difficult to recover the heterogeneity of a freshwater ecosystem once it has been disrupted.

11.6 Conclusions

Hydrological changes, local pollution, agriculture, global warming and the introduction of invasive species have serious impacts on aquatic ecosystems. However, excessive water use and the construction of small hydropower plants appear to be one of the most devastating anthropogenic disturbances of the small water bodies. Consequently, these pressures can leave incorrigible consequences on the biodiversity of flora and fauna to the Western Balkan. In order to prevent ecosystem disturbances and improve species conservation strategies, the distribution and abundance of rare and endangered species in this very important area should be specifically assessed. For future conservation efforts of biodiversity, especially in areas where many springs and small mountain streams are already captured, conservation of the remaining natural habitats should be the priority.

Acknowledgements This work was supported by the Ministry of Education, Science and Technological Development of Republic of Serbia, Contract No.: 451-03-9/2021-14/200007.

References

1. Vannote RL, Minshall GW, Cummins KW, Sedell JR, Cushing CE (1980) The river continuum concept. Can J Fish Aquat Sci 37:130–137
2. Williams DD, Blair WF (1992) Aquatic insects. CAB International, p 358

3. Downes BJ, Lake PS, Schreiber ESG (1995) Habitat structure and invertebrate assemblages on stream stones: a multivariate view from the riffle. Aust J Ecol 20(4):503–514
4. Brooks N, Adger WN, Kelly PM (2005) The determinants of vulnerability and adaptive capacity at the national level and the implications for adaptation. Glob Environ Chang 15:151–163
5. Pešić V, Glöer P (2018) The diversity and conservation status of the molluscs of lake Skadar/Shkodra. In: Pešić V, Karaman G, Kostianoy A (eds) The Skadar/Shkodra lake environment. The handbook of environmental chemistry, vol 80. Springer, Cham, pp 295–310
6. Raković M, Paunović M, Tomović J, Popović N, Csányi B, Jovanović M, Glöer P, Pešić V (2020) Do molluscs assemblages reflect river typology: a case study of Montenegro. In: Pešić V, Paunović M, Kostianoy AG (eds) The Rivers of Montenegro. The Handbook of Environmental Chemistry. Springer, Cham, pp 265–285
7. Hewitt D (1999) Arbitrary and necessary: part 1, a way of viewing the mathematics curriculum. Learn Math Int J Math Educ 19(3):2–9
8. Pešić V, Grabowski M, Hadžiablahović S, Marić D, Paunović M (2020) The biodiversity and biogeographical characteristics of the River basins of Montenegro. In: Pešić V, Paunović M, Kostianoy A (eds) The Rivers of Montenegro. The handbook of environmental chemistry, vol 93. Springer, Cham, pp 157–200
9. Strong EE, Gargominy O, Ponder WF, Bouchet P (2008) Global diversity of gastropods (Gastropoda: Mollusca) in freshwater. Hydrobiologia 595:149–166
10. Van Cappellen P, Maavara T (2016) Rivers in the Anthropocene: Global scale modifications of riverine nutrient fluxes by damming. Ecohydrol Hydrobiol 16(2):106–111
11. Schwarz U (2015) Hydropower Projects on the Balkan Rivers—Update. River Watch Euro Nat 33
12. Lehner B, Czisch G, Vassolo S (2005) The impact of global change on the hydropower potential of Europe: a model-based analysis. Energy Policy 33(7):839–855
13. Stanton BMC, Dessai S, Paavola J (2016) A systematic review of the impacts of climate variability and change on electricity systems in Europe. Energy 109:1148–2115
14. Rezaei A, Zare M, Raeisi E, Ghanbari RN (2013) Interaction of afreshwater lake and a karstic spring via a syncline fold. Ground Water 51(2):305–312
15. Rezaei A, Karimi H, Zhan H (2017) The importance of understanding the hydrogeology and geochemistry of karst terrains for safely siting dams. J Cave Karst Stud 79(1):48–58
16. Common implementation strategy for the Water Framework Directive (2000/60/EC) (2003) WFD CIS guidance document No. 10. Rivers and Lakes—Typology, Reference Conditions and Classification Systems, p 87
17. Graf DL (2013) Patterns of freshwater bivalve global diver sity and the state of phylogenetic studies on the Unionoida, Sphaeriidae, and Cyrenidae. Am Malacol Bull 31:135–153
18. Cuttelod A, Seddon M, Neubert E (2011) European Red List of Non-marine Molluscs. Publications Office of the European Union, Luxembourg, p 97
19. Lydeard C, Cummings KS (2019) Freshwater mollusks of the world: a distribution atlas. JHU Press, p 255
20. Miller JP, Ramos MA, Hauffe T, Delicado D (2018) Global species richness of hydrobiid snails determined by climate and evolutionary history. Freshw Biol 63:1225–1239
21. Bouchet P, Rocroi J, Hausdorf B, Kaim A, Kano Y, Nützel A, Parkhaev P, Schrödl M, Strong E (2017) Revised classification, nomenclator and typification of gastropod and monoplacophoran families. Malacologia 61:1–526
22. Davis GM (1979) The origin and evolution of the gastropod family Pomatiopsidae, with emphasis on the Mekong River Triculinae. Acad Nat Sci Philadelphia Monogr 20:1–120
23. Seddon MB, Kebapçi Ü, Lopes-Lima M, Damme D van, Smith KG (2014) Freshwater mollusks. In KG Smith,V Barrios, WRT Darwall, Numa C (eds) The statusand distribution of freshwater biodiversity in the easternmediterranean. Cambridge, UK, Malaga, Spainand Gland. Switzerland, IUCN, pp 43–56
24. Szarowska M (2006) Molecular phylogeny, systematic and morphological character evolution in the Balkan Risooidea (Caenogastropoda). Folia Malacologica 14(3):99–168

25. Radoman P (1965) Hydrobioidea a Superfamily of Prosobranchia (Gastropoda) II. Origin, Zoogeography, Evolution in the Balkans and Asia minor, Faculty of Science, Depart. of Biology, Monographs Institute of Zoology. Belgrade 1:1–173
26. Radoman P (1973) New classification of fresh and brackish water Prosobranchia from the Balkans and Asia Minor, Posebna Izdanja. Prirodnjacki muzej u Beogradu. 32:1–30
27. Radoman P (1975) Specijacija u okviru roda Bellgrandiella i njemu srodnih rodova na Balkanskom poluostrvu. Glasnik Prirodnjakog muzeja u Beogradu B 32:29–69
28. Radoman P (1976) Speciation within the family Bythinellidae on the Balkans and Asia Minor. Zeitschrift für zoologische und systematische Evolutionsforschung 14(2):130–152
29. Radoman P (1977) Hydrobiidae auf der Balkanhalbinsel und in Kleinasien. Archiv für Molluskenkunde 107:203–223
30. Radoman P (1978) Neue Vertreter der Gruppe Hydrobioidea von der Balkanhalbinsel. Archiv für Molluskenkunde 109:27–44
31. Radoman P (1983) Hydrobioidea a superfamily of Prosobranchia (Gastropoda) I Systematics. Serbian Academy of Sciences and Arts, Monographs. Department of Sciences 547(57):1–256
32. Radoman P (1985) Hydrobioidea a superfamily of Prosobranchia (Gastropoda) II. Origin, zoogeography, evolution in the Balkans and Asia Minor, Faculty of Science, Department of Biology. Monogr Inst Zool 1:1–173
33. Bole J (1961) Nove hidrobide (Gastropoda) iz podzemaljskih voda zahodnega Balkana. Biološki vestnik 9:59–69
34. Bole J (1970) Beitrag zur kenntnis der anatomie und taxonomie der unterirdischen hydrobiiden (Gastropoda, Prosobranchia) Razprave IV. razreda. SAZU 13(2): 87–111
35. Bole J (1985) Recent subterranean snails and development of certain river basins in the Dinaric Karst Razprave IV. razreda. SAZU 26:315–328
36. Bole J (1992) Neue arten der unterirdischen schnecken Westbalkans Razprave IV razreda. SAZU 33(1):3–20
37. Bole J, Velkovrh F (1986) Mollusca from continental subterranean aquatic habitats. In: Botosaneanu L (ed) Stygofauna Mundi, pp 177–208
38. Bole J, Velkovrh F (1987) Neue arten der unterirdischen schnecken Jugoslawiens Razprave IV razreda. SAZU 28(3):69–83
39. Glöer P, Georgiev D (2011) Bulgaria, a hot spot of biodiversity (Gastropoda: Rissooidea)? J Conchol 40(5):1–16
40. Bank RA (2011) Fauna Europaea Project: Checklist of the land and freshwater Gastropoda of Albania and former Yugoslavia, p 59
41. Glöer P (2008) Three new hydrobiod species from Serbia (Mollusca, Gastropoda, Hydrobiidae). In: Pavićević D, Perreau M (eds) Advances in the studies of the fauna of the Balkan Peninsula. Glöer P, Pešić V (2014) Ecol Mont 1(4):249–255. Papers dedicated to the memory of Guido Nonveiller. Institute for Nature Conservation of Serbia, Monograph 22:349–356
42. Glöer P, Pešić V (2014) New subterranean freshwater gastropods of Montenegro (Mollusca: Gastropoda: Hydrobiidae). Ecologica Montenegrina 1:82–88
43. Pešić V, Karaman GS, Sket B (2018) The diversity and endemism of Aquatic subterranean Fauna of the Lake Skadar/Shkodra Basin. In: Pešić V., Karaman G., Kostianoy A. (Eds). The Skadar/Shkodra Lake environment. The Handbook of Environmental Chemistry, vol 80. Springer, Cham, pp 339–361
44. Boeters HD, Glöer P, Pešić V (2013) Some new Freshwater Gastropods from Southern Europe (Mollusca: Gastropoda: Truncatelloidea). Folia Malcologica 21(4):225–235
45. Grego J (2020) Revision of the stygobiont gastropod genera Plagigeyeria Tomlin, 1930 and Travunijana Grego & Glöer, 2019 (Mollusca; Gastropoda; Moitessieriidae and Hydrobiidae) in Hercegovina and adjacent regions. Eur J Taxon 691:1–56
46. Glöer P, Grego J (2015) New subterranean freshwater Molluscs from Bosnia & Hercegovina (Mollusca: Hydrobiidae). Ecol Monten 2(4):307–314
47. Beran L, Bodon M, Cianfanelli S (2014) Revision of "*Hauffenia jadertina*" Kuščer 1933, and description of a new species from Pag island, Croatia (Gastropoda: Hydrobiidae). J Conchol 41:585–601

48. Hofman S, Rysiewska A, Osikowski A, Falniowski A (2020) A new species of *Kerkia* Radoman, 1978 (Caenogastropoda, Hydrobiidae) from Bosnia and Herzegovina. ZooKeys 973(3):17–33
49. Glöer P, Mrkvicka A (2015) New taxa of freshwater snails from Macedonia (Gastropoda: Hydrobiidae, Amnicolidae). Ecologica Montenegrina. 3:13–18
50. Glöer P, Slavevska-Stamenković V (2015) Bythinella melovskii n.sp., a new species from R. Macedonia (Gastropoda: Hydrobiidae). Ecologica Montenegrina 2(2):150–154
51. Slavevska–Stamenković V, Glöer P, Hinić J, Mitić–Kopanja D, Pešić V (2021) *Pseudobithynia ambrakis* Glöer, Falniowski and Pešić 2010 (Gastropoda: Rissooidea: Bithyniidae), a new species from the Macedonian part of Dojran Lake. Ecologica Montenegrina 41:84-90
52. Glöer P, Grego J, Erőss ZP, Fehér Z (2015) New records of subterranean and spring molluscs (Gastropoda: Hydrobiidae) from Montenegro and Albania with the description of five new species. Ecologica Montenegrina 4:70–82
53. Grego J, Glöer P, Erőss ZP, Fehér Z (2017) Six new subterranean gastropod species from northern Albania and some new records from Albania and Kosovo (Mollusca, Gastropoda, Moitessieriidae and Hydrobiidae). Subterranean Biol 23:85–107
54. Grego J, Glöer P, Rysiewska A, Hofman S, Falniowski A (2018) A new *Montenegrospeum* species from south Croatia (Mollusca: Gastropoda: Hydrobiidae). Folia Malacologica 26(1):25–34
55. Falniowski A, Grego J, Rysiewska A, Osikowski A, Hofman S (2021) Two new stygobiotic species of *Horatia* Bourguignat, 1887 (Hydrobiidae) from Croatia. Subterranean Biol 37:89–104
56. Cindrić K, Slapnik R (2019) Two new subterranean freshwater gastropod species (Gastropoda: Truncatelloidea) from the Rudnica VI cave in Croatia. Natura Croatica 28:45–50
57. Delicado D, Pešić V, Ramos MA (2021) *Arganiella* Giusti & Pezzoli, 1980 (Caenogastropoda: Truncatelloidea: Hydrobiidae): a widespread genus or several narrow-range endemic genera? European J Taxonomy 750:140–155
58. Heino J, Muotka T, Paavola R, Hämäläinen H, Koskenniemi E (2002) Correspondence between regional delineations and spatial patterns in macroinvertebrate assemblages of boreal headwater streams. J N Am Benthol Soc 21:397–413
59. Paunović M (2007) Composition of macro-invertebrate communities as indicator of running waters types in Serbia, PhD thesis. Faculty of Biology, University of Belgrade, p 218
60. Tubić B, Simić V, Zorić K, Gačić Z, Atanacković A, Csányi B, Paunović M (2013) Stream section types of the Danube River in Serbia according to the distribution of macroinvertebrates. Biologia 68(2):294–302
61. Tomović J, Paunović M, Atanacković A, Marković V, Gačić Z, Csányi B, Simić V (2014) Biotic typology of the Danube River based on distribution of Mollusc Fauna as revealed by the second joint Danube Survey. Acta zoologica bulgarica 66(4):527–537
62. Dufrene M, Legendre P (1997) Species assemblages and indicator species: the need for a flexible asymmetrical approach. Ecol Monogr 67:345–366
63. Martinovic-Vitanovic V, Raković M, Popovic N, Kalafatić V (2013) Qualitative study of Mollusca communities in the Serbian Danube stretch (river km 1260–863.4). Biologia 68(1):112–130
64. Raković M (2015) Diversity of Molluscs in the Danube River (1260–863.5 rkm) and taxonomic analysis of the genera *Planorbarius*, *Radix*, *Physella* and *Ferrissia* (Pulmonata: Basommatophora). PhD thesis. Faculty of Biology, University of Belgrade, p 147
65. Marinković N, Karadžić B, Pešić V, Gligorović B, Grosser C, Paunović M, Nikolić V, Raković M (2019) Faunistic patterns and diversity components of leech assemblages in karst springs of Montenegro. Knowl Manag Aquatic Ecosyst 420:26
66. Popović N, Raković M, Đuknić J, Csányi B, Szekeres J, Borza P, Slobodnik J, Liška I, Dj M, Kolarević S, Simić V, Tubić B, Paunović M (2019) The relationship between river basin specific (RBS) pollutants and macroinvertebrate communities. J Limnol 79(1):59–69
67. Karaman BJ (2009) Endangerment of the fauna of Gastropoda (Mollusca) in Crna Gora-Montenegro and the problem of its protection. Zaštita prirode. Institute for Nature Conservation of Serbia. Belgrade 60(1–2):223–228

68. Marković Z (1998) Izvori brdsko-planinskih područja Srbije, ekološka studija makrozoobentosa. University of Belgrade , Faculty of Biology, p 318
69. Von Fumetti S, Nagel P (2012) Discharge variability and its effect on faunistic assemblages in springs. Freshw Sci 31(2):647–656
70. Van der Kamp RO (1995) The hydrogeology of springs in relation to the biodiversity of spring fauna: a review. J Kans Entomol Soc. 68(2):4–17
71. Brkić Ž, Kuhta M, Larva O, Gottstein S (2019) Groundwater and connected ecosystems: an overview of groundwater body status assessment in Croatia. Environ Sci Eur 31(1):75
72. Lajtner J, Štamol V, Slapnik R (2013) Red list of freshwater and terrestrial snails in Croatia. State Institute for Nature Protection, p 31

Chapter 12
Importance of Small Water Bodies for Diversity of Leeches (Hirudinea) of Western Balkan

Nikola Marinković, Momir Paunović, Maja Raković, Milica Jovanović, and Vladimir Pešić

Abstract The Balkan Peninsula is known as a biodiversity hotspot. This area is characterized by diverse climate, orographic and hydrological conditions which represent a good basis for the diversity of flora and fauna. The fauna of leeches (Hirudinea) is no exception. The majority of European species are recorded in this area, many of which are endemic to the Balkans. Contemporary studies have described several species new to the science, and a few subspecies have been risen to the species level after detailed investigation (*Croatobranchus maestrovi, Dina krasensis, D. dinarica, D. minuoculata, D. sketi, D. prokletijaca, Trocheta dalmatina, Glossiphonia balcanica*). The area of Dinaric Alps (Karst) with its springs, streams and glacial lakes are home to almost all of these newly described species but also other common European species. These habitats are under various anthropogenic pressure, springs are driven in to pipes for irrigation or for human use, and on many small rivers mini hydropower plants are being built which further endanger these fragile habitats and animal communities that inhabit them. The Mediterranean medicinal leech (*H. verbana*) was exported in large numbers through history form the area of Balkan Peninsula to Western Europe for medicinal use. Today, species of medicinal leeches are under numerous international and national protection acts and its trade and commercial use is very limited or prohibited. The main threat is the destruction and deterioration of suitable habitats, mainly wetland areas, marshes and ponds in the flood zones of rivers.

N. Marinković (✉) · M. Paunović · M. Raković
Institute for Biological Research "Siniša Stanković"–National Institute of the Republic of Serbia, University of Belgrade, Despota Stefana Bvld. 142, 11000 Belgrade, Serbia
e-mail: nikola.marinkovic@ibiss.bg.ac.rs

M. Paunović
e-mail: mpaunovi@ibiss.bg.ac.rs

M. Raković
e-mail: rakovic.maja@ibiss.bg.ac.rs

M. Jovanović · V. Pešić
Department of Biology, University of Montenegro, Cetinjski put b.b, 81000 Podgorica, Montenegro
e-mail: milicaj@ucg.ac.me

V. Pešić et al. (eds.), *Small Water Bodies of the Western Balkans*, Springer Water,
https://doi.org/10.1007/978-3-030-86478-1_12

Keywords Leeches · Hirudinea · Diversity · Endemism · Springs

12.1 Introduction

The sentence spoken by the common man whenever the topic of Hirudinea is brought up is *Leeches, they drink blood!?* It is fairly unknown that large number of leeches don't drink blood of vertebrates (sanguivorous), but because of their use in traditional and modern medicine, association with this trait is what first comes to mind. Hirudinea are a class of ringed worms which alongside Polychaeta and Oligochaeta make phylum Annelida. This group is comprised of approximately 700 species [1]. Theoretical ancestor of leeches was probably freshwater organism, such is the case with most of the leeches today, 15% of species are marine, and even less are adopted to terrestrial habitats [1]. Although often associated with stagnant murky water, leeches can be found in all kinds of freshwater ecosystems and moist terrestrial habitats surrounding them. Leeches are highly specialized ringed worms, whose body is built of constant number of segments (34) that are superficially divided into smaller rings called annuli, and on each end of the body a sucker is present.

Systematics of leeches is mainly based on morphological traits (build of pharynx, crop and annulation), feeding type and geographical distribution [2, 3]. Apart from already mentioned species that suck blood of mammals and birds, many representatives are predators of invertebrates which they swallow whole, or feed on their bodily fluids by sucking [1, 2]. Feature that leeches share with Oligochaeta is clitellum, a glandular belt like structure with function in creation of cocoons, in which the fertilized eggs are placed. Many freshwater species deposit cocoons in the moist soil, out of the water. Although all leeches are hermaphrodites, cross-fertilization is needed for reproduction. What sets them apart from the Oligochaeta is the absence of parapodia and chetae.

Since leeches lack discernible morphological traits, as in many invertebrate groups, structure, shape and position of genital systems are main taxonomic characters for delimiting species and higher taxa levels. Leeches are dependent on their food source, and being predators or ecto-parasites of other animals can be good indicator of the state of ecosystem they are found in. The main factors endangering the freshwater leeches are habitat destruction, pollution and in case of medicinal leeches excessive collection from nature [4, 5].

12.2 Taxonomy Discordance and Problems that Arise

Regarding the taxonomy and classification of leeches (Euhirudinea sensu Sawyer 1986) scientists are yet to come to consensus. The only thing firmly settled is that leeches are clitellate ringed worms, and that they belong to the group Clitellata alongside Oligochaeta, Acanthobdellidae and Branchibodellidae. Exact relations of these groups are still debated [2, 3, 6–8]. In this text the term leeches refers to class Hirudinea according to Nesemann and Neubert [3].

Build of pharynx, presence or absence of proboscis or pharyngeal teeth have been the criteria for classification within the leeches. The main division is based on the presence of proboscis (order Rhynchobdellida), or its absence (order Arhynchobdellida) [1–3]. Further classification is rather fluid, as some new studies of phylogenetic relations recommend reclassification, migration of species to other genera or their fusion [9–15]. One of the most influenced groups of leeches are erpobdellid leeches. Their taxonomical composition is the topic of discussion in the scientific community [1–3, 9, 14, 16] These predatory leeches that have no proboscis or pharyngeal teeth are native to Holarctic region. The main character used for distinction of genera through history was annulation patterns [9].

Molecular genetic studies have brought more complexity in this already problematic group. Common European genera *Dina* and *Trocheta* have shown to be para, or polyphyletic groups [16]. Based on this, and some additional analyses Siddall [9] recommended moving all European and North American species of Erpobdellidae to the genus *Erpobdella* until final resolution of this problem. Since this would cause confusion with too many taxa in one genus, this proposition has seen limited acceptance [16–18]. Taxonomy and classification of leeches remains to be major problem of the group, and still much work is to be done to clear all conflicting stances and conclusions. Clear, widely accepted classification and revision and validation of the recently described species is a basis for further studies of other aspects of biology of these invertebrates. It is of utmost importance for one to know on which taxon a study is being conducted, in that way the results obtained can be compared to results of other studies.

12.3 History of Leech Investigation in Western Balkans

From the publications of Blanchard, Augener and Remy, through Sket and Šapkarev, to the works presented in recent publications by several authors one can conclude that the history of leech investigation in the Balkans is long lasting [17, 19–52]. Although spanning over one century investigations of leeches were accomplishments of individuals and are rather scattered. Aforementioned Blanchard wrote about leeches of Montenegro in 1905 [24], while Augener [21–23] studied leeches of broader area (Balkan Peninsula) in third decade of XX century. Their work was followed by publications of Remy in 1934, 1937 and 1953 [35–37] which dealt with leeches of the same geographical area.

After a long break, in 1968, Sket published a comprehensive publication "Towards the knowledge of the leeches (Hirudinea) of Yugoslavia - Zur Kenntnis der Egelfauna (Hirudinea) Jugoslawiens", in which he provides detailed information on presence, distribution, morphological characteristics and ecological preferences of leeches in former Yugoslavia. Several new subspecies were described *(e.g., D. lineata dinarica, T. subviridis dalmatina, T. bykowskii krasense*), some of which were later elevated to the species level [47]. In the same period, famous Macedonian zoologist Jonce Šapkarev published a series of papers on the leeches of Macedonia, Montenegro, Bosnia and Herzegovina and Serbia and together with Sket investigated a very interesting, highly endemic fauna of the relict Ohrid Lake [38–46, 48, 50]. These two investigators have made a stepping stone for further research on the fauna of leeches of not only the Balkan Peninsula but that of Europe and wider.

Again, there was a slowdown that lasted for almost two decades. Research of leeches has undergone a revival with description of several species new to the science accompanied by several first records of leeches in the area. These new data provided a base for building and updating of checklists of leeches in countries of Western Balkans. Grosser and Pešić with their associates have compiled checklists for Montenegro (2014) and Serbia (2015), and Dmitrović and Pešić made an updated checklist of leeches of Bosnia and Herzegovina [25, 28, 29] (Table 12.1). New insight on the ecology of leeches their preferences, communities and impact of ecological factors on them, with special attention on spring habitats, was provided by Marinković et al. [32, 33].

New age brings with itself new technologies and new opportunities. Molecular genetic technics, although not so new, have shown to be a useful tool in clarifying relationship and taxonomical position of leech taxa. Trontelj and Sket [16] proved, using these technics, that morphological characters are not reliable determinants when dealing with some leeches. On the other hand, Živić et al. [4] showed, using the same technics that the only species of medicinal leeches present in Serbia is *Hirudo verbana*.

12.4 Contribution of Small Water Bodies to the Diversity of Leeches of Balkans

Leech fauna of The Balkan Peninsula is a part of European fauna. The Balkan Peninsula with its complex geological history and various climatological conditions is considered as one of the diversity hotspots of Europe [1, 53]. The leech fauna contributes to this with significant number of endemic taxa, but also with presence of majority of common European species [1, 25, 28, 29]. Families Erpobdellidae, Glossiphoniidae and Piscicolidae are the most diverse groups of leeches in Europe [1]. Similar is the case with the Hirudinean fauna of the Balkan. All the common European species are readily encountered in waters of the region. Generalist species such as *Erpobdella octoculata, Glossiphoniia complanata* and *Helobdella stagnalis* inhabit various types of water ecosystems, and their presence is noted regularly in studies of aquatic macroinvertebrate fauna.

Table 12.1 List of freshwater leech taxa recorded in the area of Western Balkan, based on Sket [47], Grosser et al. [28, 29] and Marinković [34]

	Serbia	Montenegro	Bosnia and Herzegovina	Croatia
Glossiphoniidae Vaillant, 1890				
Alboglossiphonia hyalina Lukin, 1976	+	+		
Alboglossiphonia striata (Apáthy, 1888)	+	+		
Alboglossiphonia heteroclita (Linnaeus, 1761)	+	+	+	+
Glossiphonia complanata complanata (Linnaeus, 1758)	+	+	+	+
Glossiphonia concolor (Apáthy, 1888)	+	+		
Glossiphonia nebulosa Kalbe, 1964	+	+		
Glossiphonia paludosa (Carena, 1824)	+	+		
Glossiphonia cf. pulchella Sket, 1968		+		
Helobdella stagnalis (Linnaeus, 1758)	+	+	+	+
Hemiclepsis marginata (O. F. Müller, 1774)	+	+	+	+
Placobdella costata (Fr. Müller, 1846)		+	+	+
Batracobdelloides moogi Nesemann & Csanyi, 1995		+		
Theromyzon tessulatum (Müller, 1774)	+	+		+
Piscicolidae Johnston, 1865				
Caspiobdella fadejewi (Epshtein, 1961)	+			
Cystobranhus fasciatus (Kollar, 1842)	+			+
Piscicola pawlowskii (Sket, 1968)		+		
Piscicola geometra (Linnaeus, 1761)	+	+	+	
Piscicola respirans Troschel, 1850		+	+	
Piscicola hadzii Sket, 1985			+	

(continued)

Table 12.1 (continued)

	Serbia	Montenegro	Bosnia and Herzegovina	Croatia
Haemopidae Richardson, 1969				
Haemopis elegans Moquin–Tandon, 1846	+			
Haemopis sanguisuga (Linnaeus, 1758)	+	+	+	+
Hirudinidae Whitman, 1886				
Hirudo verbana Carena, 1820	+	+	+	+
Limnatis nilotica (Savigny, 1820)	+			
Erpobdellidae R. Blanchard, 1894				
Dina absoloni Johansson, 1913	+	+	+	
Dina apathy Gedroyc, 1916	+			+
Dina lineata lineata (O. F. Müller, 1774)	+			
Dina dinarica (Sket, 1968)	+	+	+	+
Dina cf. lineata montana Sket, 1968	+	+		
Dina krasensis (Sket, 1968)				
Dina minuoculata Grosser, Moritz & Pešić, 2007	+	+	+	
Dina prokletijaca Grosser & Pešić, 2016	+			
Dina sketi Grosser & Pešić, 2014			+	
Erpobdella octoculata (Linnaeus, 1758)	+	+	+	+
Erpobdella nigricollis (Brandes, 1900)	+	+		
Erpobdella testacea (Savigny, 1822)	+	+	+	+
Erpobdella vilnensis (Liskiewicz, 1925)	+	+	+	
Trocheta cylindrica Örley, 1886	+		+	
Trocheta dalmatina (Sket, 1968)	+	+		+
Trocheta danastrica Stschegolew, 1938	+			

(continued)

Table 12.1 (continued)

	Serbia	Montenegro	Bosnia and Herzegovina	Croatia
Trocheta subviridis Dutrochet, 1817				
Trochaeta haskonis Grosser, 2000	+		+	
Salifidae Johansson, 1910				
Barbronia weberi (R. Blanchard, 1897)	+			

Fig. 12.1 Collection of the leeches in the field. **a, b** Komovi Mt., Montenegro; **c** Bukovica River, Montenegro; **d** Vradar River, North Macedonia; **e** Morača River, Montenegro; **f** Komarnica Canyon, Montenegro (Photo: N. Marinković and M. Raković)

12.4.1 Erpobdellidae

Representatives of the family Erpobdellidae are the most diverse with high degree of endemic taxa. This family is present with members of four genera, three of them being the common *Dina*, *Erpobdella* and *Trocheta*, and the endemic troglobiont *Croatobranchus* with single species *C. maestrovi*.

Sket through his extensive research of leeches described several new taxa endemic to the area of former Yugoslavia [47–50]. Mainly, these were representatives of Erpobdellidae family, but also Piscicolidae and Glossiphoniidae.

In the course of his work, he discovered hidden diversity of leeches of ancient Lake Ohrid. The Lake Ohrid, being the oldest freshwater lake in Europe harbors a very diverse flora and fauna with high degree of endemic taxa [52, 54]. The composition of leech fauna is a bit unusual, the ratio of Glossiphoniidae and Erpobdellidae representatives is, unlike other antient lakes (i.e., Baikal), shifted heavily towards family Erpobdellidae [55]. Al the endemic species of Erpobdellidae of Lake Ohrid belong to the genus *Dina*. A study of the phylogenetic relations of erpobdellid leeches of Lake Ohrid has shown that this species flock has derived from one common ancestor, and that lake springs had a large role in speciation of these leeches to over 10 separate taxa [52].

Apart from study of Lake Ohrid leeches, Sket [47] did an extensive research of erpobdellid leeches in other parts of the region, mainly the western parts, while territories of Montenegro, Serbia and Northern Macedonia were to a degree neglected. During this work he managed to describe three subspecies of the widespread species *D. lineata*, and two subspecies of two *Trocheta* species. *Dina lineata*, a widespread and common species of leeches in western Palearctic, shows variability in morphology and anatomy, and based on these traits several subspecies were described. The three subspecies endemic to Balkan that Sket [47] described were *D. lineata dinarica*, *D. l. lacustris* and *D. l. montana*.

The Dinaric leech, the common name given to *D. l. dinarica,* is the most widespread of them. This subspecies differs morphologically and genetically from typical *D. lineata* and should be treated as a separate species *D. dinarica* [29, 34]. This species was described as typical crenophilous, inhabiting springs, and small watercourses in hilly and mountainous areas of Dinaric Alps [47] (Fig. 12.2). Contemporary studies have shown that it inhabits a larger area, apart from territories of Bosnia and Herzegovina and Montenegro, the Dinaric leech was found in most of the watercourses of Serbia and Northern Macedonia. Sket [47], based on limited data, suspected that in these territories the typical *D. l. lineata* would be common, but it was shown that it has very limited presence. New studies of erpobdellid leeches [33, 34] have shown that *D. dinarica* is very common inhabitant of Balkan watercourses, and is one of the most common leech species of the region. Although it is found in all types of water ecosystems it is usually found in fast flowing mountainous springs and brooks, where it is seldomly accompanied by other leech species [32, 34].

The other two subspecies of *D. lineata* are rather less common. *Dina l. lacustris*, is restricted to glacial lakes in the western parts of Northern Macedonia, while *D. l.*

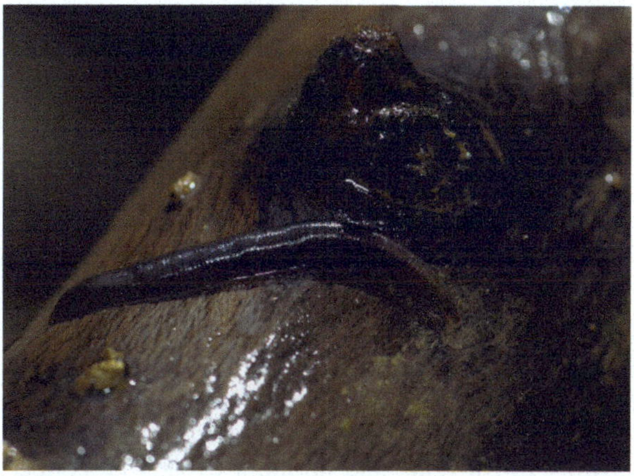

Fig. 12.2 *D. dinarica*-Rogozna Mt. Serbia (Photo: N. Marinković)

montana is known from the high mountain springs in northern parts of Montenegro [17, 34, 47] with one uncertain record on Mt. Rtanj Eastern Serbia [29]. Taxonomical position of these two subspecies is unclear, and although they differ morphologically form the nominotypic subspecies, the level of their genetical differences is still undetermined.

Beside the taxa described by Sket, twenty-first century gave descriptions of three new species of Erpobdellidae. Firstly, in a spring, a tributary of the Tara River in northern Montenegro, *Dina minuoculata* was found. These leeches are characterized by yellow spots on dorsal side of the body, which are never found in *D. dinarica* [27, 56]. Later, these leeches were found in few more springs in Montenegro and south western parts of Serbia [28, 29]. Marinković [34] also noted records of this species for Montenegro and south western Serbia, with addition of two records in streams on the territory of National Park Sutjeska in Bosnia and Herzegovina.

Another species new for science *Dina sketi*, named in appreciation of Boris Sket, was found in 2014 in north western part of Bosnia and Herzegovina in the springs surrounding the Cvrcka River [17].

In 2016 a new species of erpobdellid leech, *Dina prokletijaca* was described from the mountain range of Prokletije where it was found in several springs [30]. This leech species is most similar to *D. dinarica*, and *D. l. montana*, from which it differs the combination of the small and stocky body, dorsal surface with two wide and dark paramedian longitudinal stripes and ovisacs reaching the fourth somite after the female genital pore, and curled in their entire course [30].

Not many leeches are obligate cave dwellers, but in the area of western Balkans two species are recorded. One of them is *D. absoloni* Johansson, 1913, a leech that lacks pigmentation and eyes, and with body shape that is same as other representatives of the genus *Dina*. This species is found in caves of the Dinaric Karst, mainly in Bosnia and Herzegovina and Montenegro [47]. Leeches that were identified as *D. absoloni* were also found in Bulgaria and Georgia (described as *D. absoloni ratschaensis*) but it is not clear if this identification is valid, or are these some new species. Another obligate troglobiont present in the area of Balkan Peninsula is *Croatobranchus mestrovi*. This peculiar leech, an inhabitant of the Mountain Velebit cave system, is characterized by absence of pigmentation and presence of tentacles surrounding the cranial sucker and multiple pairs of fingerlike processes on the posterior part of the body [31]. These features are rather atypical for members of the family Erpobdellidae, but internal morphology and genetic similarities clearly classify this leech in the aforementioned group [51].

Leeches of the genus *Trocheta* are seldom recorded in the studies of macroinvertebrate fauna of the region (Fig. 12.4b). Sket [47] noted the presence of two species in former Yugoslavia, *T. cylindrica* (formaly known as *T. bykowskii*) and *T. subviridis*. Beside typical specimens, he also noted geographical variability of morphological traits of these leeches [16, 47]. Based on this, he described two subspecies. In the territory of Slovenia and northern Croatia he collected specimens of *T. bykowskii* that differed in morphology based their position. Ones collected in the northwestern parts were typical, but the ones collected in the southeast, showed some traits of the genus *Dina*. He described this variant as *T. bykowskii krasense* [47]. Since these leeches showed unusual annulation patterns, it sparked an interest in researchers to do a genetic comparison of these taxa. The results showed that this subspecies is more closely related to genus *Dina*, which led to its renaming to *Dina krasensis*. *Dina krasensis* mainly inhabits caves and springs in southern parts of Slovenia and northern parts of Croatia [16]. Another group of leeches that were different from European populations Sket [47] named *T. subviridis dalmatina*. These leeches occupy a narrow area on the Adriatic coast, from Dubrovnik to Skadar Lake [28, 47]. Since these leeches differ both morphologically and genetically from *T. subviridis*, it is considered today as standalone species *T. dalmatina*. Its main habitat are springs and streams on the Adriatic coast near Dubrovnik and Kotor (Fig. 12.3). *Trocheta haskonis* relatively new species to the science and one of the largest leeches of Europe was recorded in region. In Serbia the leech was found in a cave, and in Bosnia and Herzegovina it was found in a utility vault [26, 57]. Further investigation of similar habitats such as caves could yield more records of this leech.

Regarding the genus *Erpobdella*, almost all European species are recorded in the region of Balkan Peninsula. The only species absent, is *E. monostriata* which inhabits northern Europe and Russia and is not expected to be found [3]. *Erpobdella octoculata* a very common European leech, is also one of the most frequently recorded species in the region (Fig. 12.4a). It can be found in all kinds of water ecosystems, from large lowland rivers, lakes and streams to dystrophic ponds and heavily polluted irrigation canals [58]. Second species of *Erpobdella* genus that is often mentioned in studies is *E. testacea*. This is a common species in Europe, although not as numerous as previously mentioned one [3]. Based on the studies of macroinvertebrate communities, and water quality assessments done in the region, it could be concluded that this is a very common species in the area. Marinković [34] concluded that this species is not that frequent, and that this false picture of its presence should be attributed to misidentification during routine monitoring, since publications that deal with leeches

Fig. 12.3 *T. dalmatina* (Photo: N. Marinković) and its habitat, Široka Rijeka stream, Kotor, Montenegro (M. Raković)

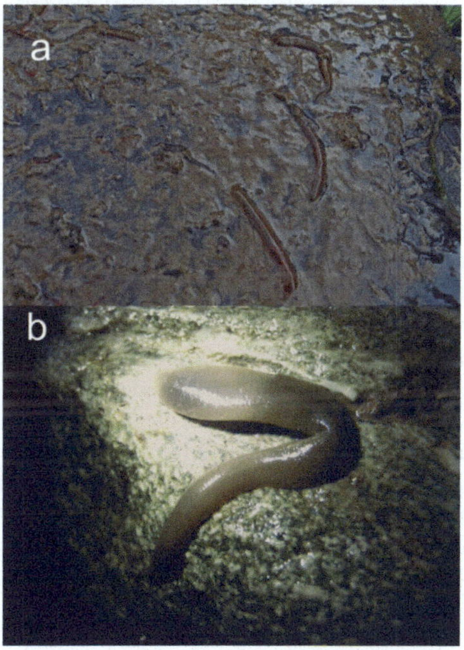

Fig. 12.4 **a** Individuals of *E. octoculata* on a stone from Topčiderska reka, a heavily polluted river, Belgrade Serbia; **b** sp. on a wet stone near a hilly stream on Jelica Mt., central Serbia (Photo: N. Marinković)

in particular rarely note its presence. These leeches are usually found in lowland rivers and standing waterbodies [59, 60].

Erpobdella vilnensis is a leech rarely mentioned in the publications from the region, but is proven to be more common (33, 34). Due to its coloration, it has probably been mistaken for *D. lineata* which is readily mentioned in papers. These leeches are often found in spring and small lakes located in high elevations and in some smaller hilly rivers and streams. The fourth species recorded mainly in the basins of the Sava and the Danube Rivers is *E. nigricollis* with addition of sublacustrine springs in Skadar Lake basin [17, 32, 47]. This species is central European species and the Balkan Peninsula represents border of its area of distribution [3].

12.4.2 Glossiphoniidae

The family Glossiphoniidae is the second most diverse in Europe [1]. This family includes several genera, some of which are ectoparasites of water fowl, pond turtles and amphibians, but also representatives that suck bodily fluids of mollusks using their proboscis [3]. These leeches are closely related to their hosts and prey organisms and are found in ecosystems which provide adequate food sources. All European genera are present in the Balkan Peninsula and they are represented with common species, such as *G. complanata, Helobdella stagnalis, Alboglossiphonia heteroclita, Theromyzon tessulatum and Placobdella costata.* There are only few endemic taxa known to the science. Two of them, native to Lake Ohrid, described by Sket [47] as a subspecies of *G. complanata*, the *G.complanata maculosa*, and a form *G.complanata f. pulchella*, and the latter one raised to the species level by Sket [48]. The only known species of Glossiphoniidae endemic to Balkan Peninsula, *G. balcanica* was described by Grosser and Pešić in 2016 [30]. These leeches were found in small springs in southern Serbia (Kosovo and Metohija province) alongside other leeches that were at that time identified as *G. nebulosa* (Fig. 12.5a, b). This leech species is not only restricted to southern Serbia but was also found in numerous karstic springs in Montenegro [30, 32] (Fig. 12.5c, d). Phylogenetical analyses of various leeches from the genus *Glossiphonia*, collected in the area of Western Balkans have shown that morphologically identified taxa as *G. nebulosa* and *G. concolor* each consists of more than one clearly different phylogenetic clade. Grosser et al. [30] noticed morphological differences between the populations of *G. nebulosa* from its *locus typicus* and those from Western Balkans, which has been confirmed by recent molecular-genetic analyses by Jovanović et al. [61] in press. These genetic differences, and morphological features indicate that populations of *G. nebulosa* from the West Balkans may represent a new species for science.

Fig. 12.5 **a** *G. balcanica*; **b** *G. cf. nebulosa* (Photo: V. Pešić); **c** sublacustrine spring Karuč; **d** karstic spring Crno oko (Photo: M. Jovanović)

12.4.3 Piscicolidae

Out of most diverse European leech families, Piscicolidae are the least studied in the Balkan. Representatives of the Piscicolidae family are ectoparasites on fish [3]. This lifestyle means that they are rarely collected during investigations of aquatic macroinvertebrates. Implementation of morphometry analyses of the body shape have allowed the description of numerous species in the genus *Piscicola* in central Europe [62, 63]. Such studies haven't been conducted on the leeches of the region and only few common European species have been recorded, mainly *Piscicola geometra* (Linnaeus, 1761) and *P. respirans* Troschel, 1850. Being dependent on host fish species they are rarely found in small water bodies (e.g., in a large karstic springs) that are seldomly inhabited by fishes. *Piscicola hadzii* was described from the large karstic spring of the Buna spring in Herzegovina [49]. Regarding potential diversity and endemic taxa within this family the conclusion cannot be made due to severe lack of data on these leeches in the region.

12.4.4 Hirudinidae and Haemopidae

The only medicinal leech found in the area is *Hirudo verbana*. Although there are numerous records of medicinal leeches under the name *Hirudo medicinalis*, it has been proven that only *H. verbana* is present [4]. These are not misidentifications but rather inadequate nomenclature used. Several studies have proven that morphological variants of the European medicinal leech are valid species, and that area of southern and southeastern Europe is inhabited by *H. verbana* [64, 65]. Throughout history these leeches were intensively collected and exported to Western Europe which led to the severe reduction in numbers. Today these species are protected by national and international legislations and are no more endangered by collection in nature [4, 66, 67]. Main factors that impact its survival are degradation of habitat and pollution. These leeches prefer wetland areas and backwaters in the flood zones of large rivers, but are also found in other places [3]. These wetland areas are under constant pressure due to irrigation measures, deforestation and pollution through agriculture and other human activities.

Another leech that feeds on blood of mammals, *Limnatis nilotica* has been reported to be present in the Balkans [39, 68, 69]. Grosser et al. [28] suggest that all of these findings should be attributed to *Haemopis elegans*. Unlike these leeches, which were recorded in only a few cases, the horse leech *Haemopis sanguisuga* is frequently encountered in fresh water ecosystems but also in the surrounding terrestrial habitats where it preys on earthworms.

12.5 Ecology of Leeches in Small Water Bodies of Western Balkan

Relations between leeches and their environment have been studied intensively in central and western Europe [70–76]. The ecology of leeches in the Western Balkans, unlike other invertebrate groups (e.g. Ephemeroptera, Trichoptera, Chironomidae, Mollusca) has sparked very little interest. Studies of communities that leeches make, their habitat preferences and influence of ecological factors on them, were absent until recent.

Two studies should be mentioned in this regard. The first one conducted by Marinković et al. [33] showed how the common erpobdellid species differ in respect to their preference to various water body types and altitude [33]. It showed that three most common species are rarely found together, and have a clear preference to particular habitats. Regarding the small water bodies, the main topic of this book, they are mostly inhabited by *D. dinarica*, especially if located at higher altitudes. These kinds of ecosystems are also inhabited by *E. vilnensis*, but less frequently in comparison to aforementioned species. Third species included in this study, *E. octoculata*, one of the commonest species of Europe, and also the Balkans, is found in all types of water bodies, but rarely occupies small oligotrophic springs and streams. It prefers lowland areas, larger rivers, canals, and lakes that are characterised by higher trophic state (Fig. 12.6).

The second study on the ecology of leeches of the region published by Marinković et al. [32] was focused on the communities of leeches in karst springs of Montenegro and factors that influence them [32]. This study, although somewhat limited in the area it covers, gives a good insight on the variables that dictate the distribution and composition. During the study more than 60% of the species present in Montenegro were collected. Since it encompasses various types of karstic springs, from sublacustrine in the basin of Lake Skadar, to the reocrene in high mountains, the conclusions drawn form it could be extrapolated to similar habitats and leech communities of the region [32]. Sublacustrine springs that function as parts of large standing water bodies such as Skadar Lake have the greatest diversity, limnocrenes with a large diversity

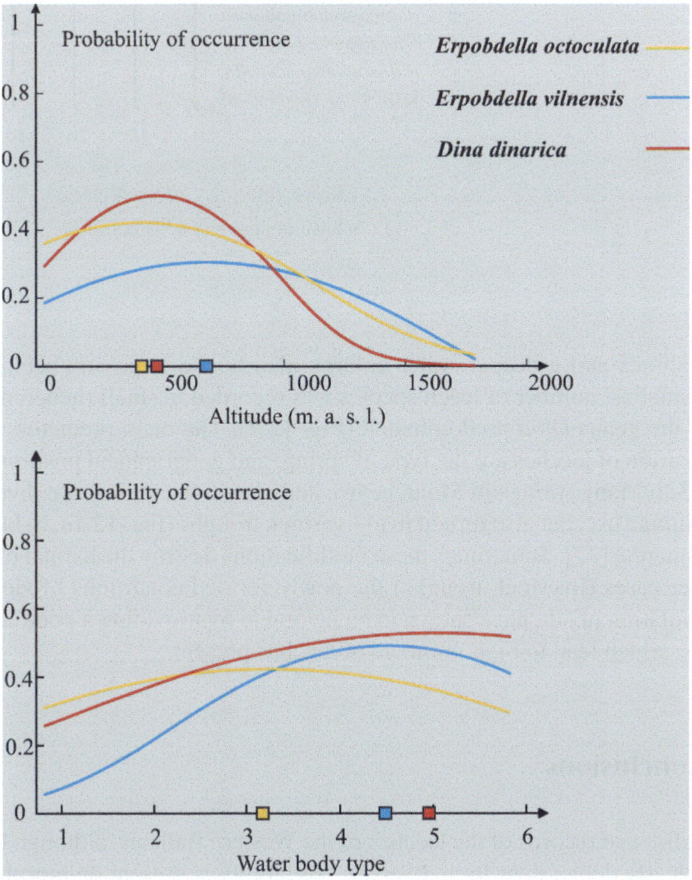

Fig. 12.6 Ecological differentiation of three most common erpobdellid species with respect to altitude and water body types; squares denote optimal values; For water body types description see Marinković et al. [33]

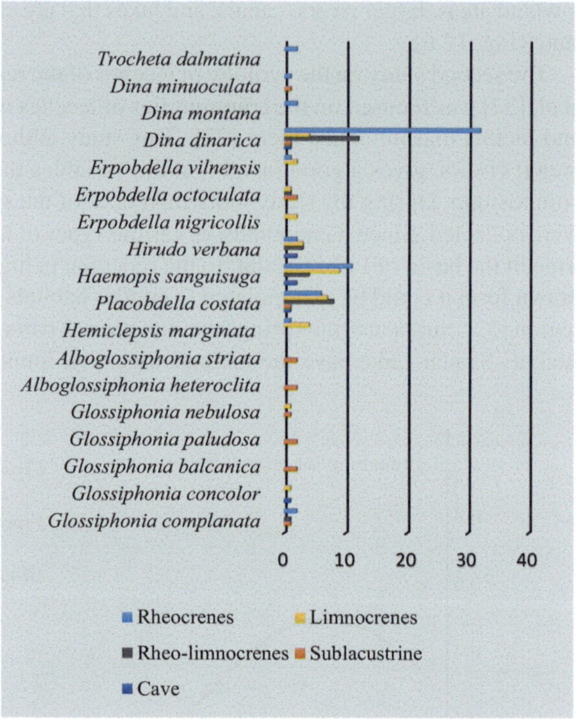

Fig. 12.7 Distribution of leeches in karst springs of Montenegro; bars represent the number of springs particular species was found in

of microhabitats and a deeper water column also host a large number of species while the smallest number of leech species was recorded in small rheocrenes where species of the genus *Dina* predominated (Fig. 12.7). The main predictors dictating the composition of leeches are the type of spring, and geographical position (mainly altitude) [32]. Many springs in Montenegro, and also in the region, are diverted into pipes for human use, but also turned in to livestock troughs (Fig. 12.1a, b) built out of wood or concrete [77]. Sometimes these modifications destroy the habitat of leeches, but in other cases (livestock troughs) the newly formed conditions of slower flow and accumulation of silt, have shown to be favorable by providing a good habitat for food items, which lead to high numbers of leeches present.

12.6 Conclusions

All the studies and records of the leeches of the Western Balkans, although limited in numbers clearly depict their high diversity. Variability is present on morphological, genetical and ecological level. Small water bodies play a significant role in providing habitat and ecological conditions for all the numerous species found in them. The Dinaric Alps with its karstic springs are certainly the center of diversity of leeches in the Balkan Peninsula. Species that inhabit these springs like *D. dinarica* show a degree of genetic variability and these populations that live in different springs could potentially harbor new species given enough time. Much work is upon scientists to be done regarding the leeches of the region. It is apparent that each effort made by investigators has yielded quite interesting results, brought new species, but also raised new questions.

Acknowledgements This work was supported by the Ministry of Education, Science and Technological Development of Republic of Serbia, Contract No.: 451-03—9/2021-14/200007.

References

1. Sket B, Trontelj P (2007) Global diversity of leeches (Hirudinea) in freshwater. Hydrobiologia 595:129–137. https://doi.org/10.1007/s10750-007-9010-8
2. Sawyer RT (1986) Leech biology and behaviour. Clarendon Press
3. Nesemann H, Neubert E (1999) Süßwasserfauna von Mitteleuropa, Bd. 6, Annelida, 2, Clitellata: Branchiobdellida, Acanthobdellea, Hirudinea; Spektrum Akademischer. Verlag, Berlin/Heidelberg, Germany, p 178
4. Živić I, Radosavljević T, Stojanović K, Petrović A (2015) The first molecular characterization of the genus Hirudo on the territory of Serbia: Estimation of endangerment. Aquat Ecol 49:81–90
5. Saglam N (2018) The effects of environmental factors on leeches. Adv Agr Environ Sci 1(1):00001
6. Siddall ME, Burreson EM (1998) Phylogeny of leeches (Hirudinea) Based on mitochondrial cytochromec oxidase Subunit I. Mol Phylogenet Evol 9(1):156–162
7. Apakupakul K, Siddall ME, Burreson EM (1999) Higher level relationships of leeches (Annelida: Clitellata: Euhirudinea) based on morphology and gene sequences. Mol Phylogenet Evol 12(3):350–359
8. Siddall ME, Apakupakul K, Burreson EM, Coates KA, Erséus C, Gelder SR, Källersjö M, Trapido-Rosenthal H (2001) Validating Livanow: molecular data agree that leeches, branchiobdellidans, and Acanthobdella peledina form a monophyletic group of oligochaetes. Mol Phylogenet Evol 21(3):346–351
9. Siddall ME (2002) Phylogeny of the leech family Erpobdellidae (Hirudinidae: Oligochaeta). Invertebr Syst 16(1):1–6
10. Borda E, Siddall ME (2004) Arhynchobdellida (Annelida: Oligochaeta: Hirudinida): phylogenetic relationships and evolution. Mol Phylogenet Evol 30(1):213–225
11. Utevsky S, Trontelj P (2004) Phylogenetic relationships of fish leeches (Hirudinea, Piscicolidae) based on mitochondrial DNA sequences and morphological data. Zoolog Scr 33(4):375–385
12. Siddall ME, Budinoff RB, Borda E (2005) Phylogenetic evaluation of systematics and biogeography of the leech family Glossiphoniidae. Invertebr Syst 19(2):105–112
13. Williams JI, Burreson EM (2006) Phylogeny of the fish leeches (Oligochaeta, Hirudinida, Piscicolidae) based on nuclear and mitochondrial genes and morphology. Zoolog Scr 35(6):627–639
14. Oceguera-Figueroa A, Phillips AJ, Pacheco-Chaves B, Reeves WK, Siddall ME (2011) Phylogeny of macrophagous leeches (Hirudinea, Clitellata) based on molecular data and evaluation of the barcoding locus. Zoolog Scr 40(2):194–203
15. Nakano T, Ramlah Z, Hikida T (2012) Phylogenetic position of gastrostomobdellid leeches (Hirudinida, Arhynchobdellida, Erpobdelliformes) and a new family for the genus Orobdella. Zoolog Scr 41(2):177–185

16. Trontelj P, Sket B (2000) Molecular re-assessment of some phylogenetic, taxonomic and biogeographic relationships between the leech genera Dina and Trocheta (Hirudinea: Erpobdellidae). Hydrobiologia 438(1):227–235
17. Grosser C, Pešić V, Dmitrović D (2014) Dina sketi n. sp., a new erpobdellid leech (Hirudinida: Erpobdellidae) from Bosnia and Herzegovina. Zootaxa 3793(3):393–397
18. Ahmed RB, Bielecki A, Cichocka JM, Tekaya S, Harrath MGAH (2013) Erpobdellid leeches (Annelida, Clitellata, Hirudinida) from Tunisia: new records with the description of a new Trocheta species. Zootaxa 3681(4):440–454
19. Augener H (1925) Blutegel von der Balkanhalbinsel. Zoologischer Anzeiger Leipzig 62:161–173
20. Augener H (1926) Nachtrag zur Blutegelfauna der Balkanhalbinsel nebst Bemerkungen über Egel aus anderen Gebieten. Zoologischer Anzeiger Leipzig 68(9/10):239–247
21. Augener H (1925) Blutegel von der Balkanhalbinsel. Zool Anz 62:161–211
22. Augener H (1926) Nachtrag zur Blutegelfauna der Balkanhalbinsel nebst Bemerkungen über Egel aus anderen Gebieten. Zool Anz 68:229–247
23. Augener H (1937) Hirudineen aus jugoslavischen Seen. Festschrift zum 60:403–413
24. Blanchard R (1905) Hirudineen aus Montenegro. Sitzungsber Königl Böhm Ges Wiss Prag, pp 1–3
25. Dmitrović D, Pešić V (2020) An updated checklist of leeches (Annelida: Hirudinea) from Bosnia and Herzegovina. Ecologica Montenegrina 29:10–19
26. Grosser C (2013) First record of Trocheta haskonis Grosser, 2000 (Hirudinea: Erpobdellidae) in Serbia. Lauterbornia 76:111–113
27. Grosser C, Moritz G, Pešić V (2007) Dina minuoculata sp. nov. (Hirudinea: Erpobdellidae)–eine neue Egelart aus Montenegro. Lauterbornia 59:7–18
28. Grosser C, Pešić V, Gligorović B (2014) A checklist of the leeches (Annelida: Hirudinea) of Montenegro. Ecologica Montenegrina 2(1):20–28
29. Grosser C, Pešić V, Lazarević P (2015) A checklist of the leeches (Annelida: Hirudinida) of Serbia, with new records. Fauna Balkana:2
30. Grosser C, Pešić V, Berlajolli V, Gligorović B (2016) Glossiphonia balcanica n. sp. and Dina prokletijaca n. sp. (Hirudinida: Glossiphoniidae, Erpobdellidae)-two new leeches from Montenegro and Kosovo. Ecologica Montenegrina 8:17–26
31. Kerovec M, Kučinić M, Jalžić B (1997) Croatodobranchus mestrovi sp. n. predstavnik nove endemske podzemne vrste pijavica (Hirudinea, Erpobdellidae). Speleolog 44(1):35–36
32. Marinković N, Karadžić B, Pešić V, Gligorović B, Grosser C, Paunović M, Nikolić V, Raković M (2019) Faunistic patterns and diversity components of leech assemblages in karst springs of Montenegro. Knowl Manag Aquat Ecosyst 420:26
33. Marinković N, Karadžić B, Slavevska Stamenković V, Pešić V, Nikolić V, Paunović M, Raković M (2020) Chorological and ecological differentiation of the commonest leech species from the suborder Erpobdelliformes (Arhynchobdellida, Hirudinea) on the Balkan Peninsula. Water 12(2):356
34. Marinković NS (2020) Taksonomska diferencijacija, diverzitet i distribucija vrsta podreda Erpobdelliformes (Annelida; Hirudinea) područja zapadnog Balkana-Taxonomic differentiation, diversity and distribution of species from the suborder Erpobdelliformes (Annelida; Hirudinea) in the western Balkans. Doctoral dissertation, University of Belgrade, Faculty of Biology
35. Remy PA (1953) Description des grottes yougoslaves (Herzegovine, Dalmatie, Crna Gora et ancien Sandjak de Novi Pazar). Glasnik Prirodnjačkog Muzeja Srpske Zemlje, B 5–6:175–233
36. Remy P (1934) Sur quelques Hirudinees des Balkans. Publications de la Société Linnéenne de Lyon 77(1):17–24
37. Rémy P (1937) Sangsues de Yougoslavie. Bull Soc Zool Fr 62:140–148
38. Šapkarev JA (1978) New contribution to the knowledge of leech distribution (Hirudinea) in Bosnia and Herzegowina. Glasnik Zemaljskog muzeja Bosne i Hercegovine u Sarajevu 17:197–205

39. Šapkarev J (1975) Contribution to the knowledge the earthworms (Lumbricidae) and leeches (Hirudinea) of Kosovo, Yugoslavia. Annuaire de la Faculté des Sciences de l'Université de Skopje 27:39–54
40. Šapkarev J (1984) Katalog faune pijavica Crne Gore. Glasnik odjeljenja prirodnih nauka Crnogorske Akademije nauka i umjetnosti 4:183–190
41. Šapkarev J (1990) New leeches (Hirudinea) of the ancient lake Ohrid. Fragmenta Balcanica 14:155–162
42. Šapkarev JA (1963) Die Fauna Hirudinea Mazedoniens. I. Systematik und Ökologie der Hirudinea des Prespa-Sees. Bull Sci Conseil Acad RSFY 8(1–2):7–8
43. Šapkarev JA (1964) Faunata na Hirudinea vo Makedonija. Folia Balcanica 2(3):1–8
44. Šapkarev JA (1970) The fauna of Hirudinea of Macedonia: the taxonomy and distribution of leeches of Aegean lakes. Internationale Revue der gesamten Hydrobiologie und Hydrographie 55(3):317–324
45. Sket B, Sapkarev J (1992) Distribution of Hirudinea (Annelida) in the ancient Ohrid Lake region. Arch Hydrobiol 124(2):225–237
46. Šapkarev JA (1975) Sistematika i rasprostranjenje pijavica (Hirudinea) Makedonije (Taxonomy and distribution of leeches (Hirudinea) from Macedonia). Biosistematika 1:87–99
47. Sket B (1968) K Poznavanju Favne Pijavk (Hirudinea) v Jugoslaviji, Zur Kenntnis der Egel-Fauna (Hirudinea) Jugoslawiens. Academia Scientiarum et Artium Slovenica Classis IV: Historia Naturalis et Medicina, Diss 9(4):127–197
48. Sket B (1981) Rhynchobdellid leeches (Hirudinea, Rhynchobdellae) in the relic Ohrid lake region: Rilčaste pijavke (Hirudinea, Rhynchobdellae) v območju reliktnega Ohridskega jezera. Biološki vestnik 29(2):67–90
49. Sket B (1985) Piscicola hadzii sp. n. (Piscicolidae, Hirudinea), a probably endemic species of leeches from Hercegovina, Yugoslavia. Biološki Vestnik 33(2):89–93
50. Sket B (1989) Intralacustrine speciation in the genus Dina (Hirudinea, Erpobdellidae) in Lake Ohrid (Yugoslavia). Hydrobiologia 182(1):49–59
51. Sket B, Dovč P, Jalžić B, Kerovec M, Kučinić M, Trontelj P (2001) A cave leech (Hirudinea, Erpobdellidae) from Croatia with unique morphological features. Zoolog Scr 30(3):223–229
52. Trajanovski S, Albrecht C, Schreiber K, Schultheiß R, Stadler T, Benke M, Wilke T (2010) Testing the spatial and temporal framework of speciation in an ancient lake species flock: the leech genus Dina (Hirudinea: Erpobdellidae) in Lake Ohrid. Biogeosciences 7(11):3387–3402
53. Griffiths HI, Krystufek B, Reed JM (2004) Balkan biodiversity. Kluwer Academic Publishers, Dordrecht, p 357
54. Albrecht C, Wolff C, Glöer P, Wilke T (2008) Concurrent evolution of ancient sister lakes and sister species: the freshwater gastropod genus Radix in lakes Ohrid and Prespa. In: Patterns and processes of speciation in ancient lakes (pp 157–167). Springer, Dordrecht
55. Epshtein VM (2004) On the origin of the Hirudinea fauna, especially Piscicolidae, in ancient lakes. Lauterbornia 52:181–193
56. Grosser C (2015) Differentiation of some similar species of the subfamily Trochetinae (Hirudinida: Erpobdellidae). Ecologica Montenegrina 2(1):29–41
57. Grosser C, Šukalo G, Pešić V (2018) Monster from the Vault: a new finding of one of the largest European leech Trocheta haskonis Grosser, 2000 from Bosnia and Herzegovina. Ecologica Montenegrina 19:69–72
58. Kutschera U (2003) The feeding strategies of the leech Erpobdella octoculata (L.): a laboratory study. Int Rev Hydrobiol J Cover Aspects Limnol Marine Biol 88(1):94–101
59. Van Haaren T, Hop P, Soes M, Tempelman D (2004) The freshwater leeches (Hirudinea) of the Netherlands. Lauterbornia 52:113–131
60. Westendorff M, Kalettka T, Jueg U (2008) Occurrence of leeches (Hirudinea) in different types of water bodies in northeast Germany (Brandenburg). Lauterbornia 65:153–162
61. Jovanović M, Haring E, Sattmann H, Grosser C, Pešić V (2021/in press) DNA barcoding for species delimitation of the freshwater leech genus Glossiphonia from the Western Balkan (Hirudinea, Glossiphoniidae). Biodivers Data J

62. Bielecki A, Palińska K, Cichocka J (2009) Body form of leeches (Hirudinida: Piscicolidae) parasitizing fishes1. Wiad Parazytol 55(4):359–365
63. Cichocka JM, Bielecki A (2015) Phylogenetic utility of the geometric model of the body form in leeches (Clitellata: Hirudinida). Biologia 70(8):1078–1092
64. Elliott JM, Kutschera U (2011) Medicinal leeches: historical use, ecology, genetics and conservation. Freshwater Rev 4:21–41
65. Kutschera U (2012) The Hirudo medicinalis species complex. Naturwissenschaften 99(5):433–434
66. Official Gazette of the Republic of Serbia, No. 23/14 (2014) Order on prohibiting the collection of certain protected species of wild flora and fauna in 2014 (in Serbian)
67. Utevsky S, Zagmajster M, Atemasov A, Zinenko O, Utevska O, Utevsky A, Trontelj P (2010) Distribution and status of medicinal leeches (genus Hirudo) in the Western Palaearctic: anthropogenic, ecological, or historical effects? Aquat Conserv Mar Freshwat Ecosyst 20(2):198–210
68. Phillips AJ, Siddall ME (2009) Poly-paraphyly of Hirudinidae: many lineages of medicinal leeches. BMC Evol Biol 9(1):1–11
69. Nakano T, Dujsebayeva T, Nishikawa K (2015) First record of Limnatis paluda (Hirudinida, Arhynchobdellida, Praobdellidae) from Kazakhstan, with comments on genetic diversity of Limnatis leeches. Biodivers Data J 3:1–16
70. Adamiak-Brud Ż, Jabłońska-Barna I, Bielecki A, Kobak J (2018) Factors shaping leech (Clitellata, Hirudinida) assemblages on artificial and natural substrata in urban water bodies. Limnologica 69:125–134
71. Cichocka J, Jabłońska-Barna I, Bielecki A, Buczyńska E, Buczyński P, Stryjecki R, Pikuła D (2015) Leeches (Clitellata: Hirudinida) of an upland stream: Taxonomic composition in relation to habitat conditions. Oceanol Hydrobiol Stud 44(2):245–253
72. Kazancı N, Ekingen P, Dugel M, Turkmen G (2015) Hirudinea (Annelida) species and their ecological preferences in some running waters and lakes. Int J Environ Sci Technol 12(3):1087–1096
73. Koperski P (2006) Relative importance of factors determining diversity and composition of freshwater leech assemblages (Hirudinea; Clitellata): a metaanalysis. Arch Hydrobiol 166(3):325–341
74. Koperski P (2010) Urban environments as habitats for rare aquatic species: The case of leeches (Euhirudinea, Clitellata) in Warsaw freshwaters. Limnol Ecol Manag Inland Waters 40(3):233–240
75. Kubová N, Schenková J, Horsák M (2013) Environmental determinants of leech assemblage patterns in lotic and lenitic habitats. Limnologica 43(6):516–524.'Kubová N, Schenková J (2014) Tolerance, optimum ranges and ecological requirements of free-living leech species (Clitellata: Hirudinida). Fundam Appl Limnol/Archiv für Hydrobiologie 185(2):167–180
76. Mann KH (1959) On Trocheta bykowskii Gedroyé, 1913, a leech new to the British fauna, with notes on the taxonomy and ecology of other Erpobdellidae. Proc Zool Soc Lond 132(3):369–379
77. Pešić V, Karaman GS, Kostianoy AG, Vukašinović-Pešić V (2018) Conclusions: recent advances and the future prospects of the Lake Skadar/Shkodra environment. In: Pešić V, Karaman G, Kostianoy A (eds) The Skadar/Shkodra lake environment. The Handbook of Environmental Chemistry, vol 80. Springer, Cham, pp 481–500

Chapter 13
Karst Springs: Isolated Ecosystem Ecology from the Water Mite Perspective

Ivana Pozojević and Vladimir Pešić

Abstract Karst springs are known to host a great diversity of water mites that include euryvalent generalists, but also a highly specialized crenobiont species (exclusively found in springs). A total of 44 crenobiontic water mite species has been reported in the springs of the Western Balkans, which is about 32% of the crenobiontic species known in Europe. Most of the crenobiontic species of the Dinaric karst are inhabitants of helocrene and rheohelocrene springs, and to a lesser extent inhabit rheocrenic springs, while only two species can be considered characteristic representatives of limnocrenic springs. Recent studies on the karstic springs of the Western Balkan have revealed that water mite richness as well as the number of crenobionts both decrease as we move from alpine to Mediterranean springs. The response of crenobiontic water mite assemblages in karstic springs is primarily driven by the crenic types and environmental variability, in particular in the geological and geographical context. Further research on the crenobiontic water mite fauna of karstic springs may help to assess the ecological challenges facing these fragile ecosystems, given increased water demand in the future and ongoing climate change.

Keywords Springs · Karst · Balkan · Water mites · Crenobionts · Diversity

13.1 Introduction

Springs are unique ecosystems that are home to different groups of water mite species: from euryvalent generalist species, to highly specialized stenovalent crenobiont species (that are exclusively found in springs), as well as species that inhabit the border between groundwater/surface water and eucrenal/hippocrenal (springhead/spring brook) habitats [1–3]. Karst springs are different from other spring types

I. Pozojević (✉)
Faculty of Science, University of Zagreb, Rooseveltov Trg 6, 10000 Zagreb, Croatia
e-mail: ivana.pozojevic@biol.pmf.hr

V. Pešić
Faculty of Sciences, University of Montenegro, Džordža Vašingtona bb., 81000 Podgorica, Montenegro

V. Pešić et al. (eds.), *Small Water Bodies of the Western Balkans*, Springer Water,
https://doi.org/10.1007/978-3-030-86478-1_13

as they usually have a higher degree of discharge and water level fluctuations that are a result of the permeability of karstic geological layers [4, 5]. The annual temperature oscillations usually do not exceed 1 or 2 °C, but are dependent on both season and discharge. Water properties such as dissolved calcium carbonate ($CaCO_3$) concentration govern the other physio-chemical interactions in karst springs [5]. Higher concentrations of dissolved $CaCO_3$ consequently lead to increased values of alkalinity and conductivity, stable pH values and the specific assemblages that inhabit these ecosystems.

Springs are probably the most common type of aquatic habitats in the Dinaric Karst. As a result of the highly karstified substrate, a large number of springs appear, from those whose activity lasts only a few days after heavy rains (periodic springs) to those that form streams and rivers (Fig. 13.1). Temporary springs are usually active during the rainy season, while they dry up during the summer. This type of spring appears at the borders between limestone and flysch, or limestone and dolomite, and is especially common in the canyons of rivers. Studies of the water mites of these springs, although rare, have not shown that there is a species characteristic of temporary springs [6]. Most water mites found in these habitats were present as single specimens, including species associated with subterranean waters (stygophilous) (Fig. 13.2).

A large number of studies have shown that the water mite fauna of the permanent springs of the Dinaric Karst is extremely diverse [6–12]. Based on the classical limnological spring classification, springs are usually categorized by morphotype [13]. This is considered to be the key defining factor in determining the crenal communities, especially the water mite assemblages [12]. The three main morphotypes include limnocrenes (lake-forming springs, Fig. 13.1), rheocrenes (river-forming springs) and helocrenes (swamp-forming spring).

Water mites (Hydrachnidia) numbering more than 7,500 species [14] inhabit a wide range of aquatic habitats, including running and standing waters, and temporary and permanent habitats, that are mainly associated with bottom sediment or aquatic macrophytes [15]. Moreover, water mites significantly contribute to the biodiversity of hyporheic interstitial waters as well as spring habitats [15]. Recent studies

Fig. 13.1 **a** Karst limnocrene called Glavaš, a spring of the River Cetina in Croatia. **b** Spring Vukovo Vrelo in Nikšić municipality, Montenegro. Photo by: I. Pozojević (**a**) and V. Pešić (**b**)

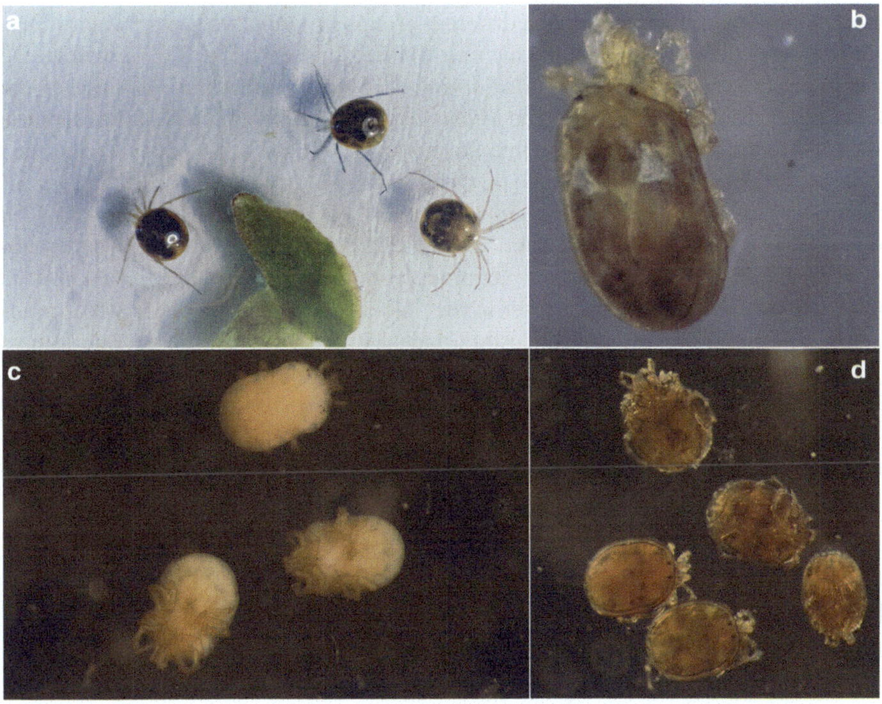

Fig. 13.2 The diverse water mites from the springs of the Western Balkans. **a** Generalist species (*Hygrobates fluviatilis*); **b** the stygophilous / crenophilous species *Torrenticola elliptica;* **c** the crenobiont species *Partnunia steinmanni* and **d** the stygophilous *Ljania macilenta*. Photos by: I. Pozojević

have shown that they can be a good indicator of environmental changes to globally endangered spring habitats [16]. The susceptibility of water mites as an indicator is based on the fact that it is the group of crenic invertebrates with the greatest share of crenobiontic species [17]. Gerecke [17] listed 137 crenobiontic water mite species in European fauna found in springs and adapted to the environmental conditions in these habitats, which represents 14% of the European Hydrachnidia fauna.

Springs play an important refuge role for diverse mite species and groups, which has probably influenced the vast water mite speciation in these habitats [15, 17]. The island-like nature of karstic springs promote the isolation of their inhabitants, which therefore leads to an increased speciation rate within the crenic communities. The dispersal abilities of water mites are highly dependent on the dispersal abilities of their insect hosts, whose diversity and abundance in karstic springs may be exceptionally high and may include not only crenobiontic but also a large number of generalists host taxa [18, 19].

The impact on macroinvertebrate assemblages of karstic springs have been well-studied and many researchers [20–22] have emphasized that different abiotic factors such as temperature, pH, discharge, and substrate composition affect the species composition and the abundance of different groups of fauna, including water mites. Furthermore, disturbance factors such as flooding and drought [23] as well as the hydromorphological modification of springs [16] may affect the response of the water mite crenic assemblages. The impact of these factors structuring the response of crenobiontic water mites, however, has not been studied sufficiently. At first, this response can be expected to depend on the crenic types as well as on certain environmental variables, particularly the geographic and geological context. Gerecke [17] identified two groups of crenobiontic taxa, i.e., "paleocrenobiont" and "neocrenobiont". Paleocrenobionts include water mites from the superfamily Hydryphantoidea and represent in an evolutionary sense the "older" species [17]. These water mites have developed from terrestrial ancestors, and they predominate in helocrenic springs [17]. Neocrenobiontic species are thought to derive from rhithrobiont species that have secondarily inhabited spring ecosystems by upstream migration from lower river reaches and usually predominated in rheocrenes [17]. However, some neocrenobiont such as *Hygrobates norvegicus* may show a preference for weakly seeping (rheo) helocrenes [17].

13.2 The Diversity of Crenobiontic Water Mites in the Western Balkans

Based on the checklist of water mites on the Balkan Peninsula provided by Pešić et al. in 2010 [6] and later supplemented in 2018 [7], the number of water mites recorded for the Balkan countries totals 390 species [7]. This number is just over 50% of the approximately 745 species inhabiting Central Europe [24]. As pointed out by Pešić et al. [7] there is no reason not to believe that the expected number of water mites in the Balkans is at least at the level of Central Europe. The main problem in terms of achieving better knowledge of the diversity of water mites in the Balkans is primarily the lack of comprehensive faunistic studies that hamper our knowledge of this limnofaunistic group in the region. In particular, habitats such as springs and hyporcheic interstitial are poorly explored [7].

Based on the available literature on the water mite fauna of the Western Balkans, in total 44 crenobiontic species have been reported from the Western Balkans (Table 13.1). This represents only about 32% of the species that Gerecke [17] listed for European fauna. As noted by this author, the habitat preferences of some species changes with geographical latitude [17], so their preference at the regional level should be documented by additional sampling. Two species, *Lebertia variolata* Gerecke, 2009 and *Woolastookia minuta* Pešić, Gerecke & Smit, 2010 should be excluded from the list provided by Gerecke [17] because they are rhitrobionts [6, 7].

Most crenobiontic species in Table 13.1 have been collected from helocrenic and rheohelocrenic (helocrenes with areas of stronger water movement) springs, and to a somewhat lesser extent in rheocrenes. These two groups of springs represent two ends of the spectrum that different assemblages of crenobiontic water mites inhabit.

Table 13.1 The Western Balkans crenobiontic water mites. Legend: AL = Albania, CR = Croatia, BiH = Bosnia and Herzegovina, MC = Macedonia, MN = Montenegro, SR = Serbia

	AL	BiH	CR	MC	MN	SR
Family Hydrovolziidae						
Hydrovolzia placophora (Monti, 1905)				+	+	
Family Hydryphantidae						
Hydryphantes (s. str.) armentarius Gerecke, 1996		+		+		
Panisopsis curvifrons (Walter, 1907)				+		
Panisopsis setipes (K.Viets, 1911)				+		
Panisus michaeli Koenike, 1896		+	+	+		
Trichothyas jadrankae Pešić, 2018					+	
Thyopsis cancellata (Protz, 1896)						+
Partnunia naprintua Gerecke, 1996					+	+
Partnunia steinmanni Walter, 1906		+	+			
Partnunia puritana Gerecke, 1996						+
Protzia squamosa squamosa Walter, 1908					+	
Protzia squamosa paucipora (K. Viets, 1955)				+	+	+
Tartarothyas romanica Husiatinschi, 1937					+	
Family Sperchontidae						
Sperchon (s. str.) longissimus K.Viets, 1920		+				
Sperchon (s. str.) mutilus Koenike, 1895				+	+	
Sperchon (s. str.) squamosus Kramer, 1879		+			+	
Sperchon (s. str.) thienemanni Koenike, 1907				+	+	
Sperchon (s. str.) vesnae Pesic, 2003					+	
Family Anisitsiellidae						
Bandakia concreta Thor, 1913				+		
Nilotonia longipora (Walter, 1925)			+	+		
Family Rutripalpidae						
Rutripalpis limicola Sokolow, 1934					+	
Family Lebertiidae						
Lebertia (Mixolebertia) bracteata K.Viets, 1925					+	
Lebertia (Mixolebertia) crenophila K.Viets, 1920					+	
Lebertia (Mixolebertia) cuneifera Walter, 1922					+	
Lebertia (Mixolebertia) holsatica K.Viets, 1920					+	
Lebertia (Mixolebertia) longipalpis K.Viets, 1936				+		
Lebertia (Mixolebertia) mediterranea Gerecke, 2009					+	
Lebertia (Mixolebertia) sefvei Walter, 1911					+	
Lebertia (Mixolebertia) separata Lundblad, 1930				+	+	+

(continued)

Table 13.1 (continued)

	AL	BiH	CR	MC	MN	SR
Lebertia (Mixolebertia) stigmatifera Thor, 1900						
Lebertia (*Lebertia*) *glabra* Thor, 1897			+		+	+
Lebertia (*Lebertia*) *guttata* K, Viets, 1952					+	
Lebertia (Lebertia) jadrensis K.Viets, 1936			+			
Lebertia (Lebertia) schechteli Thor, 1913				+	+	
Family Hygrobatidae						
Atractides (s. str.) anae Pešić, 2020					+	
Atractides (s. str.) fonticolus (K. Viets, 1920)			+	+	+	
Atractides (*s. str.*) *graecus* K.Viets, 1950					+	
Atractides (s. str.) pennatus (K. Viets, 1920)		+	+	+	+	+
Atractides (s. str.) protendens K.O.Viets, 1955				+	+	+
Hygrobates (*s. str.*) *marezaensis* Pešić & Dabert, 2017	+			+	+	
Hygrobates (*s. str.*) *limnocrenicus* Pešić, 2020				+	+	
Hygrobates (Rivobates) norvegicus (Thor, 1897)				+	+	
Family Feltriidae						
Feltria (s. str.) drilonensis K.Viets, 1936				+		
Feltria (s. str.) zschokkei Koenike, 1896				+		

Other types, such as rheohelocrenic and rheopsammocrenic springs, are transitional forms, which sometimes offer a wide spectrum of different microhabitats that have led to diverse water mite assemblages.

As far as limnocrenes are concerned, their water mite assemblages are mainly dominated by crenoxenes and it has long been believed that no true limnocrenobiont water mite species are present in Europe [17]. However, recent research (see [25, 26]) suggests that limnocrenes, at least in the Dinaric Karst, may be inhabited by true limnocrenobionts, such as *Hygrobates marezaensis*. This is a common and often abundant species found in many of the large karstic limnocrenes across the Western Balkans [25, 27].

Recent studies on the springs of Dinaric karst [9–12] have shown that limnocrenes are characterized by diverse water mite assemblages. Pešić et al. [10] investigated the springs in Central Montenegro and found that water depth is the main predictor of taxon richness and the abundance of generalist species in limnocrenic springs. Springs with a deeper water column, such a large karstic limnocrenes, are mostly inhabited by species that are characteristic of standing waters, but which show a preference for low temperatures [8, 28].

Among the crenobiontic species inhabiting the Western Balkan springs, the families Lebertidae (with the genus *Lebertia*), Hydryphantidae, Hygrobatidae (genera *Atractides* and *Hygrobates*) and Sperchontidae (with the genus *Sperchon*) are the most diverse, but species of Anisitsiellidae, Hydrovolzidae, Rutripalpidae and Feltridae have also adapted to such habitats.

Family Hydrovolzidae

Hydrovolzia placophora: This species was found both in the moss carpet of springs on hard bedrock and in helocrenic habitats rich in organic debris [6].

Family Hydryphantidae

The family in the West Balkans includes 12 water mite species considered to be crenobiontic. Most species such as *Panisus michaeli,* and *Parathyas palustris* show a preference for helocrenes rich in organic residues, preferably in mountainous areas [6]. *Partnunua naprintua*, and *P. aprutina*, as well as both subspecies of *Protzia squamosa* (*P. squamosa squamosa* and *P. squmosa paucipora*) show a preference for shaded rheohelocrenes rich in gravel and mosses [6]. *Hydryphantes armentarius* prefers weakly flowing rheohelocrenes exposed to sunlight, while *Panisops curvifrons* show a preference for springs with enlarged hygropetric areas [6].

Family Anisitsiellidae

Bandakia concreta shows a preference for rheohelocrenic and helocrenic springs [6]. The members of the genus *Nilotonia* prefer weakly seeping helocrenes, with full exposure to sunlight [6].

Family Rutripalpidae

Rutripalpis limicola, reported in Montenegro, was found in a weakly seeping helocrene at a higher altitude [29].

Family Lebertiidae

The genus *Lebertia* includes 11 crenobiont species that inhabit the springs of the Western Balkans. Most of these species (*L. crenophila, L. holsatica, L. cuneifera,* and *L. mediterranea)* prefer helocrenes and rheohelocrenes rich in macrophytes and fine detritus [6, 7, 30]. Some species such as *L. schechteli*, and *L. stigmatifera* show no clear preference for a selected spring type [6, 30]. The habitat preference of some species such as *L. jadrensis* (known only from the River Jadro near Split in Croatia [6, 30]), *L. guttata* (known only from the type locality in the French Alps and a rheocrenic spring in Montenegro [10]) and *L. longipalpis* (known only from its type locality, a rheocrenic spring, in North Macedonia [6, 30]) needs to be verified by additional findings.

Lebertia glabra: Gerecke et al. [17] noticed differences in the regional preference of this species, with populations from northern Germany and the British Isles most commonly inhabiting low order streams [31] and its Mediterranean populations found almost exclusively in springs. The status of these populations should be tested by using molecular methods.

Lebertia sefvei and *L. crenophila*: As noticed by Gerecke et al. [17], these two species are subject to contrary discussion with regard to their habitat preference. Most records of this species in the Western Balkans, where it is apparently to be rare and scattered, come from springs [30], but there is also a recent record of *L. sefvei* from the River Tara in Montenegro [7]. The record of *L. crenophila* in Montenegro comes from the River Biogradska [32].

Family Sperchontidae

Sperchon thienemanni is widespread and often the most common crenobiont in springs in the mountainous regions of Montenegro [32], preferring (as do *S. mutillus* and *S. vesnae*) rheohelocrenes with fine detritus.

Family Hygrobatidae

Atractides pennatus: This species prefers rheocrenes at low and middle elevations. In the Lake Skadar catchment area, this species is the most common crenobiont in rheocrenes [8]. Recently, combining traditional morphological techniques with the analysis of partial mtDNA COI sequences, Pešić et al. [33] described a new species (*Atractides anae*) from certain rhecrenic springs in Montenegro and Bulgaria. This species morphologically resembles *A. pennatus* with which it lives syntopically. This recent study has highlighted the use of molecular methods in detecting possible cryptic water mite species in spring habitats [33].

Hygrobates norvegicus has mainly been found in rheohelocrenes rich in macrophytes and fine detritus [6].

Hygrobates marezaensis: This species is a limnocrenobiont, and prefers large karstic limnocrenes dominated by *Berula erecta* [25]. *Hygrobates limnocrenicus* prefers cold karstic limnocrenes, but it is also found in deeper fast flowing water, such as lake outlets [26].

Family Feltriidae

A large number of species of the genus *Feltria*, such as *F. armata*, *F. cornuta* , *F. longispina* and *F. brevipes*, which have been registered in the Western Balkans (see [6]), are often reported as crenophiles. Most of these species have been found in rheocrenic springs, but also in a first-order stream rich in moss [6, 7].

13.3 The Environmental Drivers of Crenobiontic Water Mite Diversity in Karstic Springs

A number of recently published studies (e.g., [6–12]) have identified water mites as an important limnofaunistic component of springs in the Dinaric Karst. These studies have emphasized that the diversity and abundance of water mites in karst springs depend on the spring size as well as the crenic types [9, 10, 12]. Pozojević et al. [9] found 11 water mite species in six large undisturbed karst springs in the Dinaric area of Croatia. Pešić et al. [10] who studied the water mite distribution along the eu-/hypocrenon gradient in 14 karstic springs in central Montenegro, found 17 species. The latter study has shown that water mite assemblages in karstic springs do not show a clear difference between springhead and springbrook sections [10].

A study conducted by Pozojević et al. [9] in the karstic springs of Croatia showed that crenic water mite richness decreases as we move from alpine to Mediterranean springs. The two investigated regions of Croatia in the latter study [9], the inland continental area and the coast, while both parts of the West Dinaric ecoregion, differ climatologically: the first region is characterized by a temperate- humid to humid-boreal climate, with warm summers, whereas the coastal region is for the most part characterized by a Mediterranean climate [34]. This study showed that region with temperate-humid to humid-boreal climate (and cooler spring water) have a greater water mite richness and β (regional scale) diversity compared to the springs of the coastal region, characterized by a Mediterranean climate (and thus warmer spring water) [9].

A similar trend, a decrease moving from alpine to Mediterranean springs, is also noticeable in the number of crenobiontic water mites. In the study conducted by Pešić et al. [10], less than half of the studied springs (43%) of Central Montenegro were inhabited by crenobiontic water mites. These authors detected only four crenobiontic species (*Lebertia guttata, Hygrobates marezaensis, Atractides pennatus* and *A. fonticulus*) [10], with the average number of crenobiont species in these springs being 1.8. The springs examined in this study were located at low and middle elevations and were mostly rheocrenic, with a few limnocrenes. These two types of springs predominated at both lower and middle elevations in the Dinaric Karst, in particular in its Mediterranean and sub-Mediterranean regions [8].

On the other hand, in alpine springs the helocrenic and rheohelocrenic type predominated. Pešić [32] investigated seven springs located in the territory of the Biogradska Gora National Park at altitudes between 1100 and 1800 m above sea level, and found that they are home to eight crenobiontic species (*Panisus michaeli, Protzia squamosal, Partnunia naprintua, Sperchon thienemanni, S. mutillus, Lebertia schechteli, Hygrobates norvegicus* and *Feltria zschokkei*) [32]. All the investigated springs were inhabited by crenobiontic species, with the average number of crenobionts per spring being 2.1.

The factors affecting the diversity and abundance of crenbiontic water mites are still not sufficiently well known. In the study on water mite assemblages of 14 karstic springs located in the Mediterranean part of Montenegro, Pešić et al. [10] did not find significant predictors for the number of crenobiontic species. The number of non-crenobiontic taxa was predicted mainly by water depth [10]. This study showed that the abundance of crenobionts was most strongly associated with temperature: colder springs support larger populations of crenobionts, but without impact on the species richness [10].

On the other hand, a study by Pozojević et al. [12], revealed significant correlations between the presence of *Atractides pennatus* and *Partnunia angusta* (Fig. 13.3) and higher water temperature, questioning the idea that crenobiont species are dependent on low temperatures. Gerecke et al. [17] first suggested that the crenobiont water mite species are, in fact, "warm stenothermic" meaning that relatively warm winter spring temperatures probably play a more important role in restricting the presence

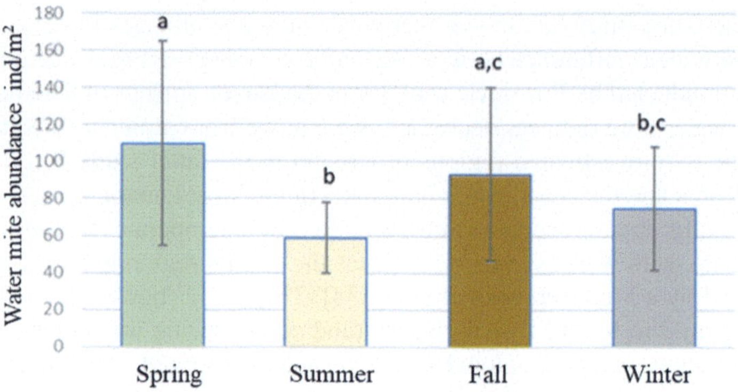

Fig. 13.3 Seasonal dynamics of water mite abundance in the karst springs of Croatia (authors unpublished data) showing two significant peaks—in the spring and the autumn

of crenobiontic mites. As pointed out by Pozojević et al. [12] water temperature alone is certainly not the only predictor of the water mite diversity in springs: a number of crenobiont taxa have been documented in tropical regions, where the temperature remains unchanged between the spring area and the lower river reaches [35]. In both of the studies of karstic springs in Croatia, the contribution of the physico-chemical water properties in explaining the total water mite variability was up to 48%, indicating that, although undoubtedly important, abiotic water properties and conditions are not the only predictors of water mite diversity in karstic springs.

The temporal (seasonal) dynamics of water mite abundances also play a significant role in their assemblage dynamics. In the study by Pozojević et al. [9], two peaks of water mite abundance appeared: in the spring and again in the autumn. This is in accordance with the author's personal findings (Fig. 13.3) and is most likely linked to insect emergence patterns. The emergence apex of most freshwater insects in springs in the Dinaric Karst has been recorded in the summer [36, 37]. This means that in the spring, prior to their emergence, mites thrive and proliferate in their surroundings where a vast array of both prey (for deutonymph and adult individuals) and hosts (for larvae) are available.

13.4 Conclusion

In the recent limnological literature, water mites are often considered to be the group of fauna with the highest share of crenobiontic species, which makes them a potentially good indicator of environmental changes to their habitats. This study has revealed that the karstic springs of the Western Balkan are home to 44 crenobiontic species which represents about 32% of the total European crenobiontic Hydrachnidia. As can be seen from Table 13.1, our knowledge of the distribution of crenobiontic species in the Balkans is uneven: the largest number of species was reported from Montenegro, while only one species is known from Albania.

Given the increased demand for water in the future and global climate change, it can be expected that springs in karstic areas such as the Western Balkans will be exposed to even greater pressure in the future. Future research on the crenobiontic water mite assemblages of these karstic springs may help assess the ecological changes to these fragile ecosystems and help in proposing adequate conservation measures. The main goal of future research should be to develop a comprehensive list of crenobiont taxa for each landscape unit, and their compilation on a wider scale. In this way, we will obtain a valuable tool for characterizing spring ecosystems and assessing their ecological status.

References

1. Gerecke R, Meisch C, Stoch F et al (1998) Eucrenon-hypocrenon ecotone and spring typology in the Alps of Berchtesgaden (Upper Bavaria, Germany). A study of microcrustacea (Crustacea: Copepoda, Ostracoda) and water mites (Acari: Halacaridae, Hydrachnellae). In: Botosaneanu L (ed) Studies in crenobiology. The biology of springs and springbrooks. Backhuys Publishers, Leiden, pp 167–182
2. Knight RL, Notestein SK (2008) Effects of nutrients on spring ecosystems. In: Summary and synthesis of the available literature on the effects of nutrients on spring organisms and systems. Florida Department of Environmental Protection, Tallahassee FL, USA, pp 271–304
3. Stoch F, Gerecke R, Pieri V et al (2011) Exploring species distribution of spring meiofauna (Annelida, Acari, Crustacea) in the south-eastern. Alps J Limnol 70(1):65–76
4. Bonacci O (1993) Karst springs hydrographs as indicators of karst aquifers. Hydrol Sci J 38:51–62
5. Prelovšek M (2010) Hydrology. In: Mihevc A, Prelovšek M, Zupan HN (eds) Introduction to the Dinaric Karst. Karst Research Institute, Postojna, Slovenija, pp 14–19
6. Pešić V, Smit H, Gerecke R, Di Sabatino A (2010) The water mites (Acari: Hydrachnidia) of the Balkan peninsula, a revised survey with new records and descriptions of five new taxa. Zootaxa 2586:1–100
7. Pešić V, Bańkowska A, Goldschmidt T et al (2018) Supplement to the Checklist of water mites (Acari: Hydrachnidia) from the Balkan peninsula. Zootaxa 4394:151–184
8. Zawal A, Pešić V (2018) The diversity of water mite assemblages (Acari: Parasitengona: Hydrachnidia) of Lake Skadar/Shkodra and its catchment area. In: Pešić V, Karaman GS, Kostianoy AG (eds) The Skadar/Shkodra lake environment. Springer, Berlin, Heidelberg, pp 311–323
9. Pozojević I, Brigić A, Gottstein S (2018) Water mite (Acari: Hydrachnidia) diversity and distribution in undisturbed Dinaric karst springs. Exp Appl Acarol 76:123–138
10. Pešić V, Savić A, Jabłońska A et al (2019) Environmental factors affecting water mite assemblages along eucrenon-hypocrenon gradients in Mediterranean karstic springs. Exp Appl Acarol 77:471–486
11. Pozojević I, Pešić V, Gottstein S (2019) Two water mite species (Acari: Hydrachnidia) from karst springs new for the fauna of Croatia with notes on distribution and environmental preferences. Natura Croatica 28:417–424
12. Pozojević I, Pešić V, Goldschmidt T, Gottstein S (2020) Crenal habitats: sources of water mite (Acari: Hydrachnidia). Diversity 12:316. https://doi.org/10.3390/d12090316
13. Glazier DS (2009) Springs. In; Likens GE (ed) Encyclopedia of inland waters, vol 1. Oxford, UK, Elsevier, pp 734–755

14. Smit H (2020) Water mites of the world, with keys to the families, subfamilies genera and subgenera (Acari: Hydrachnidia). Monogr Ned Entomol Ver 12:1–774
15. Davids C, Di Sabatino A, Gerecke R et al (2007) Acari: Hydrachnidia I. In: Gerecke R (ed) Süßwasserfauna von Mitteleuropa, vol 7/2–1. Spektrum Akademischer Verlag, Munich, Elsevier GmbN, pp 241–376
16. Pešić V, Dmitrović D, Savić A et al (2019) Application of macroinvertebrate multimetrics as a measure of the impact of anthropogenic modification of spring habitats. Aquat Conserv 29:341–352
17. Gerecke R, Martin P, Gledhill T (2018) Water mites (Acari: Parasitengona: Hydrachnidia) as inhabitants of groundwater-influenced habitats–considerations following an update of Limnofauna Europaea. Limnologica 69:81–93
18. Martin P, Stur E (2006) Parasite-host association and live cycles of spring-living water mites (Hydrachnidia, Acari) from Luxembourg. Hydrobiologia 573:17
19. Martin P, Stur E, Wiedenburg S (2010) Larval parasitism of spring-dwelling alpine water mites (Hydrachnidia, Acari): a study with particular reference to chironomid hosts. Aquat Ecol 44:431–448
20. Hahn H-J (2000) Studies on classifying of undisturbed springs in Southwestern Germany by macrobenthic communities. Limnologica 30:247–259
21. Von Fumetti S, Nagel P, Scheifhacken N, Baltes B (2006) Factors governing macrozoobenthic assemblages in perennial springs in north-western Switzerland. Hydrobiologia 568:467–475
22. Dumnicka E, Galas J, Koperski P (2007) Benthic invertebrates in karst springs? Does substratum or location define communities? Int Rev Hydrobiol 92:452–464
23. Zawal A, Stryjecki R, Buczyńska E et al (2018) Water mites (Acari, Hydrachnidia) of riparian springs in a small lowland river valley: What are the key factors for species distribution? PeerJ 6:e4797
24. Gerecke R, Gledhill T, Pešić V, Smit H (2016) Chelicerata: Acari III. In: Gerecke R (ed) Süßwasserfauna von Mitteleuropa, vol Bd. 7/2–3. Springer, Heidelberg, pp 1–429
25. Pešić V, Asadi M, Cimpean M et al (2017) Six species in one: evidence of cryptic speciation in the *Hygrobates fluviatilis* complex (Acariformes, Hydrachnidia, Hygrobatidae). Syst Appl Acarol 22(9):1327–1377
26. Pešic V, Jovanović M, Manović A et al (2020) Two new species from the *Hygrobates nigromaculatus*-complex (Acariformes, Hydrachnidia, Hygrobatidae), based on morphological and molecular evidence. Acarologia 60(4):753–768
27. Pešić V, Jovanović M., Manović A et al (2020) Molecular evidence for two new species of the *Hygrobates fluviatilis* complex from the Balkan Peninsula (Acariformes, Hydrachnidia, Hygrobatidae) Syst Appl Acarol 25(9):1702–1719
28. Gerecke R, Di Sabatino A (2007) Water mites (Hydrachnidia and Halacaridae) in spring habitats: a taxonomical and ecological perspective. In: Cantonati M, Bertuzzi E, Spitale D (eds) The spring habitat: biota and sampling methods. Museo Tridentino di Scienze Naturali, Trento (Monografi e del Museo Tridentino di Scienze Naturali 4). Marco Cantonati at Museo delle Scienze, Trento, pp 193–216
29. Gerecke R, Tuzovskij P (2001) The water mite *Rutripalpus limicola* Sokolow, 1934: new data on morphology and biology, and considerations on the systematic position of the monotypic family Rutripalpidae (Acari, Hydrachnidia). J Nat Hist 35:931–944
30. Gerecke R (2009) Revisional studies on the European species of the water mite genus *Lebertia* Neuman, 1880 (Acari: Hydrachnidia: Lebertiidae). Abh Senckenb Natforsch Ges 566:1–144
31. Martin P, Rückert, M (2011) Die Quellfauna Schleswig-Holsteins und ihre regionale Stenotopie. Faun.-Ökol Mitt 9:171–224
32. Pešić V (2004) Water mites (Acari: Hydrachnidia) of the Biogradska Gora National Park (Serbia and Crna Gora). In: Pešić V (ed) The biodiversity of the Biogradska Gora National Park. University of Montenegro & Centre for Biodiversity of Montenegro Monographies I, Department of Biology, pp 65–86
33. Pešić V, Zawal A, Bankowska A et al (2020) A new crenobiontic water mite species of the genus *Atractides* Koch, 1837 from Montenegro and Bulgaria, based on morphological and molecular data (Acariformes, Hydrachnidia, Hygrobatidae). Syst Appl Acarol 25(10):889–1900

34. Šegota T, Filipčić A (2003) Köppenova podjela klima i hrvatsko nazivlje. Geoadria 8(1):17–37 (in Croatian)
35. Goldschmidt T (2009) Water mites (Acari, Hydrachnidia) in tropical springs—Diversity, specificity, monitoring possibilities. Verh Des Int Ver Limnol 30:669–672
36. Previšić A, Kerovec M, Kučinić M (2007) Emergence and composition of trichoptera from karst habitats, Plitvice Lakes region Croatia. Int Rev Hydrobiol 92(1):61–83
37. Pozojević I, Ivković M, Cetinić KA, Previšić A (2021) Peeling the layers of caddisfly diversity on a longitudinal gradient in karst freshwater habitats reveals community dynamics and stability. Insects 12:234

Chapter 14
Large Branchiopods in Small Water Bodies: A Case Study of the Ramsar Site "Bardača Wetland" (NW Republic of Srpska, Bosnia and Herzegovina)

Dragana Miličić, Goran Šukalo, and Dejan Dmitrović

Abstract Ramsar site "Bardača Wetland" has been on the list of wetlands of international importance since 2007. It is situated in the northern part of Bosnia and Herzegovina, at the territory of Republic of Srpska. This site has a wide range of water bodies with various hydrological regimes that support a diverse locally adapted flora and fauna. The investigations of the large branchiopods have been carried out since 2016. The existence of this evolutionary very old group of crustaceans has previously not been known for this territory. Overall, seven taxa of large branchiopods have been recorded in temporary aquatic habitats. The notostracan *Lepidurus apus* and spinicaudatan *Leptestheria dachalacensis* were rediscovered, while the *Cyzicus* sp., *Eoleptestheria ticinensis* and *Limnadia lenticularis* were recorded for the first time in the territory of Bosnia and Herzegovina. *Linderiella* sp. and *Eubranchipus* (*Siphonophanes*) *grubii* have been recorded for the first time both in the Bosnia and Herzegovina and in the territory of Western Balkans. In recent years, significant part of "Bardača Wetland" suitable for development of large branchiopods was under drainage regime and has been used as arable land. Populations of the large branchiopods are potentially extremely vulnerable due to the landscape fragmentation, pollution and the global climate change. Consequently, conservation measures are needed to ensure their continuous and sustainable existence.

Keywords "Bardača Wetland" · Branchiopoda · Diversity · Disturbance · Conservation

---

D. Miličić (✉)
Faculty of Biology, University of Belgrade, Studentski trg 16, 11000 Belgrade, Serbia
e-mail: draganam@bio.bg.ac.rs

G. Šukalo · D. Dmitrović
Faculty of Natural Sciences and Mathematics, Department of Biology and Department of Ecology and Environment Protection, University of Banja Luka, Mladena Stojanovića 2, 78000 Banja Luka, Bosnia and Herzegovina
e-mail: goran.sukalo@pmf.unibl.org

D. Dmitrović
e-mail: dejan.dmitrovic@pmf.unibl.org

V. Pešić et al. (eds.), *Small Water Bodies of the Western Balkans*, Springer Water,
https://doi.org/10.1007/978-3-030-86478-1_14

14.1 Introduction

Wetland ecosystems share a number of features, including geographical, climate, hydrology, water and soil physical–chemical and biological characteristics. They are classified as the non-tidal freshwater ecosystems that vary from shallow, seasonally ephemeral pools to the medium-sized basins of more permanent nature [1]. Some of them hold water only for a few days after rain, but most water bodies remain permanently wet during the season. Small, shallow water bodies of rather temporary nature are most characterized by alternating phases of drought and flood. Based on the definition of the Ramsar Convention, those small lentic ecosystems have a self-contained hydrology and generally do not exceed an area of 10 ha [2]. Wetland ponds include a wide variety of temporary water bodies, and are also found in cultural landscapes and arable fields. Since the wetlands are generally more ephemeral than the other lentic ecosystems, their ecological characteristics are highly dependent on the hydrological regime, geology, soil type, and physicochemical characteristics of the water. They are mostly formed by the flooding, after heavy rains, or after snow melting. Whether and how long the water will stay on the surface depends on several factors: the presence of vegetation, the type of the substrate, the level of groundwater, etc.

Given the specific nature of these aquatic ecosystems, temporary pools are also well known for their highly specialized biota. First colonizers in these small aquatic ecosystems are usually detritivores feeding on the recently dead and decaying plants and animals. The filter-feeders further provide support to the arrival of predaceous species [3]. Both predatory pressure from the carnivorous invertebrates and vertebrates significantly affect the ephemeral water communities. Many species in ephemeral environments are ecological generalists; however, there are also species with a unique life cycle and very specific requirements and ways to respond to certain conditions. Such are the large branchiopod crustaceans belonging to the Class Branchiopoda. At the microhabitat level, the hydrological characteristics of their habitats are largely controlled by rainfall and evaporation, so the water level and physical–chemical properties fluctuate significantly during the season. The maximum pond volume occurs after snow melting during the spring, while during summer ponds are exposed to strong solar radiation and heat, drying off quickly and completely. Since the ponds are mostly of shallower depths, they may freeze up to the bottom during the winter. Also, the presence of some animals, such notonecta, turbellaria, odonata larvae, mosquitoes and tadpoles in their habitats may largely or even completely eliminate them from a certain area [4]. This is also true when it comes to larger predators, i.e. fish. Populations are strongly dependent on flooding and can be locally abundant in suitable habitats on a seasonal basis, but at the same time have an extremely restricted distribution on a larger geographical scale. Individuals feed on particles of different diameters and origins: phytoplankton, bacteria, protists, rotifers, crustacean larvae and small cladocerans. They can also consume organic detritus and inorganic particles, lifting them thanks to the generation of turbulent movements of water by tiny body limbs. During the swimming, the breathing, filtration and nutrition processes are performed practically continuously. Some groups (such as the tadpole shrimps) may also actively hunt the larvae of tadpoles and aquatic insects.

Large branchiopods are famous for various adaptations for living in such quite unpredictable environments. Since large branchiopods live in relatively small habitats with usually short water-filled phases and pronounced astatism, they have many attributes which are particularly relevant for survival. They exhibit a considerable physiological and morphological plasticity and spend the drought season as encysted embryos in the soil [5]. In this way they make the most optimal use of existing resources and reach the reproductive maturity and high fertility before environmental conditions become unfavorable. A stochasticity of ephemeral pond ecosystems can also play an important role for species distribution, via the transport of dehydrated diapausing cysts to other suitable habitats.

A particularly characteristic of this crustacean group is that large branchiopods can be used as indicators for the certain wetland habitat quality and as the reliable indicators of the general environmental conditions, as well. As the group with the succession of different life stages, they can respond more sensitively to different, even subtle changes in the environment and to ecosystem threats, than other species. Even though the large branchiopod crustaceans are generally an insufficiently known group of invertebrates, they are increasingly recognized as one of the flagship groups in the inland water and temporary pool ecosystems. Since they are able to reveal minor changes in the ephemeral habitat ecology, their monitoring and protection can safeguard the future of other species (such as amphibians and other vulnerable groups) in ephemeral pools [6]. Large branchiopods are already used for assessment of ephemeral wetland habitat functions and values in some countries [7], but in general, their sentinel and bioindicator role in wetland ecosystems remains poorly employed.

The Ramsar site "Bardača Wetland" has a wide range of small water bodies with various hydrological regimes and substrates that support a diverse locally adapted flora and fauna. This Chapter presents a brief overview of the investigation of this group in Bosnia and Herzegovina, and the large branchiopod taxa recorded so far. The Chapter also brings information about the new findings of large branchiopods in the Republic of Srpska. The last part is concerned with disturbing factors, the loss of the original habitats and the government policies related to protection of freshwater biodiversity and associated endangered animal species in the territory of Bosnia and Herzegovina.

14.2 Characterization of the Ramsar Site "Bardača Wetland"

Ramsar site "Bardača Wetland" is situated in the northern part of Bosnia and Herzegovina at the territory of Republic of Srpska (Fig. 14.1). According to data available online at the official Ramsar site [2], "Bardača Wetland" has surface area of

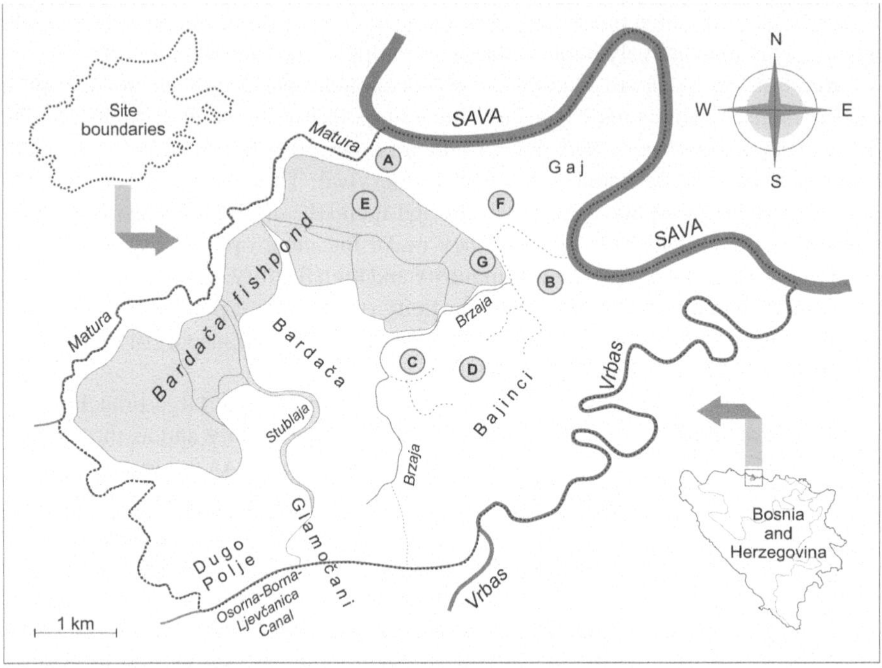

Fig. 14.1 Map of the Ramsar site "Bardača Wetland"; A–G, localities where the large branchiopods were recorded

3,500 ha (coordinates: 45°4′–45°8′ N, 17°24′–17°30′ E). It is an integral part of Lijevče field (originally: Lijevče Polje) and represents its lowest point (elevation: 85–95 m). Ramsar site "Bardača Wetland" is surrounded by three rivers: Sava at the north, while northeastern and eastern boundary is formed by river Vrbas and the western boundary by the river Matura. The southwestern boundary stretches along agricultural areas, and the southern and southeastern boundaries are formed by the canal Osorna-Borna-Ljevčanica. The Ramsar site "Bardača Wetland" encompasses several settlements: Gaj to the north, Bardača and Bajinci to the east, Dugo Polje and part of the settlement Glamočani to the south (Fig. 14.1).

Concise overview of the most important geological and geohydrological characteristics of the Ramsar site "Bardača Wetland" is based on data provided by Marković and Begović [8]. According to the mentioned authors, alluvial sediments, swamp sediments and facies sediments participate in the geological structure of this area. Alluvial sediments are dominant. The shallowest layer of these sediments composed of clays and subclays permeated with muds. Stretched clay sands are in the central layer, while deposits of sandy gravels are below. Swamp sediments are sludgy sediments with admixtures of organogenic fractions. They are present in all depressed parts of the terrain. As water runoff is prevented and infiltration of water into the soil is weak, they are covered with water for most of the year. The facies sediments are composed of mud and clay material, with an admixture of organic matter. They are present in former or periodically flooded riverbeds or meanders. The wider area, i.e. Lijevče field, is predominantly composed of Quaternary age sediments. Geological structure in this area provides hydrogeological environment characterized by a confined aquifer.

The hydrological network of the Ramsar site "Bardača Wetland", as well as the entire Lijevče field, is intricately developed. Ramsar site "Bardača Wetland" is delimited but also intersected by streams, excavated canals and artificial lakes – cyprinid ponds (Fig. 14.1). The Vrbas River, which borders this area to the northeast and east, is a right tributary of the Sava River. Within the Ramsar site "Bardača Wetland", the Sava River (a tributary of the Danube River) has the characteristics of a lowland river, which often floods the surrounding belt together with river Vrbas. In order to protect the area against floods, embankments and canals were built along the above watercourses. The largest canal is Osorna-Borna-Ljevčanica. The lowland river Matura is a right tributary of river Sava. It belongs to Lijevče field for its entire course. East of the Matura River there is the Brzaja watercourse, and between them a number of artificial lakes – cyprinid ponds have been built (Bardača fishpond), which have recently been supplied with water from the Matura River. The former meander of the river Stublaja was dammed, forming a lake that is supplied with water from the nearby wells.

According to the data provided by Marković and Begović [8], the dominant soils in this area are hydromorphic valley soils characterized by excessive humidity (eugley, fluvisols and humofluvisols). Eugley or swamp-gley soils in this area are characterized by high groundwater levels with minor fluctuations. These soils are clay-based and situated in relief depressions, so their surface is often covered in stagnant water. Fluvisols or alluvial soils are subjected to periodic flooding, which sometimes lasts for a month. The layers in the profiles of these soils are mainly clay and argilio mild soil. Alluvial soils are distributed in the valleys of rivers Sava and Vrbas as well as their tributaries. The dynamics of changes in groundwater levels in these soils coincide with the fluctuations of water levels in rivers. Humofluvisols or semigley soils occupy southern and southeastern positions, mainly around the Vrbas River. According to the mechanical composition of the horizons these soils are mostly clay soils exposed to flooding which lasts only for a short time, when the groundwater oscillates at a depth of more than one meter.

The climatic characteristics of the wider area, i.e. Lijevče field, were systematically studied by Trbić [9]. According to his data the area of Lijevče field has moderate climate characteristics, where summers are sunny and warm with low precipitation, sometimes with showers and possible appearance of hail, while winters are mostly cold and dry, even drier than summers. Heavy rainfall is characteristic of spring season (May and June), as well as autumn (September). Trbić [9] emphasized the impact of global climate change on many monitored parameters, of which temperature, humidity and precipitation will be discussed in detail in further text. Thus, there was an observed increase in mean annual air temperatures from 10.6 °C (period 1955–1965) to 11.7 °C (period 1992–2002). The trend of increase in average air temperature was also recorded at the seasonal level and particularly pronounced during spring and summer. The average annual relative humidity was slightly reduced, from 78%

(1955–1965) to 77% (1992–2002), and is the lowest in the spring. The average annual amount of precipitation has increased from 920 to 1078 mm, with the amount of precipitation in more recent times (1992–2002) being unevenly distributed, with a maximum in September.

An important source of data on the physicochemical properties of water in aquatic ecosystems situated within the Ramsar site "Bardača Wetland" was provided by the research conducted by Lolić et al. [10], Maksimović et al. [11], Lolić et al. [12] and Maksimović et al. [13]. Through seasonal monitoring of selected physical and chemical properties of water in river Matura within the period March-September 2010, Lolić et al. [10] determined the variation range for values of several parameters, including: water temperature (10.1–26.9 °C), pH (7.57–7.80), electrical conductivity (575–657 μS/cm), concentration of dissolved oxygen (4.75–12.23 mg/l), concentration of nitrate-bound nitrogen (0.3–7.7 mg/l) and orthophosphate concentration (0.07–0.63 mg/l). The same physicochemical parameters of water were simultaneously monitored at a fishpond that gets its water supply from the river Matura. The results have shown higher concentration of organic matter and a greater range of variation for most of the above-mentioned water parameters, except for nitrate-bound nitrogen [10]. The application of agricultural measures affects the values of water parameters in fishponds in Bardača area, with the intention of matching the water properties to the production needs of these fishponds [10, 14].

In addition to the permanent water bodies, the "Bardača Wetland" is also characterized by the presence of seasonally, astatic habitats, formed by rain and snow melting (commonly called vernal pools). Small ephemeral ponds can also be situated in depressions on the pastures where the groundwater emerges at the surface. As a rule, they are small in size and shallow in depth, seasonally filled after the rain periods and periodically dry up completely or partially (Fig. 14.2). Because of the high degree of sediment–water interface, the shallow waters favor critical biogeochemical processes at the landscape level [15]. The repeated episodes of soil wetting and drying can also affect the soil microbial fauna [16]. Once filled, pools hold water for varying durations, with alternating episodes of wetting and drying. During the warmer season, water temperature and salinity increase rapidly as the pools desiccate. The percent of oxygen, the variation range of pH, turbidity and nutrient concentrations also change significantly (Table 14.1). All of this indicates that in terms of predictability and stability, ephemeral ponds are very stressful habitats. On the other hand, small ephemeral water bodies are surprisingly biodiverse, and can support varying locally adapted flora and fauna [17]. They are populated by species with various adaptations for living in such an unpredictable environment. In terms of aquatic animals, this represents a strategy to avoid predation, particularly by fish. Also, most of these highly adapted species spend the drought season as seeds (plants) or resting eggs/cysts (animals) in the soil. Connectivity among ephemeral pools may also be important for survival of some species.

Fig. 14.2 Typical localities of large branchiopods in "Bardača Wetland": A – Shallow pond with macrophytes; B – pasture; C_{P2} – Puddle on a pasture filled with rainwater; C_{P4} – Puddle on a pasture filled with rainwater; D – Ephemeral pond on pasture rich in vegetation; E – Canal at the bottom of a drained fishpond; F – Shallow depressions (tracks) made by the passage of vehicles; G – Standing water at the bottom of a drained fishpond. Letters on images correspond to the tags in Fig. 14.1 and Table 14.1

Table 14.1 Data collected in the field during the study of large branchiopods at the Ramsar Site "Bardača Wetland" (localities A, B, C, D, E, F and G correspond to the map on Fig. 14.1)

Sample No	Locality	Date	Characteristics of habitat	Physico-chemical parameters of water	Elevation	Coordinates
1	A	26.04.2016	Ephemeral pond, with macrophytes, at the edge of the forest in a flooded area	Water temp. 17.7 °C, pH 7.7, Cond. 7 cF	90 m	45.12204° 17.45221°
2	B	14.04.2016	Ephemeral ponds and puddles on flooded pasture (the number of temporary water bodies varies, from a unique "lake-like" during the floodings, through numerous remaining after water withdrawal, to the drying up of the deepest ponds which are located a few tens of meters away from the point given in coordinates)	Water temp. 24.6 °C, pH 8.2, Cond. 5 cF, O_2 15.0 mg/l	88 m	45.10883° 17.47670°
3		26.04.2016		Water temp. 13.9 °C, pH 8.0, Cond. 3 cF		
4				Water temp. 24.7 °C, pH 8.3, Cond. 4 cF, O_2 14.6 mg/l		
5		24.04.2018		Water temp. 28.8 °C, pH 8.1, Cond. 5 cF, O_2 19.5 mg/l		
6		28.04.2018		Water temp. 24.6 °C, pH 8.2, Cond. 5 cF, O_2 15.0 mg/l		
7				Water temp. 25.5 °C, pH 8.2, Cond. 5 cF, O_2 17.8 mg/l		
8		14.11.2020		Water temp. 9.3 °C, pH 8.0, Cond. 8 cF, O_2 8.0 mg/l		
9		02.04.2021		Water temp. 15.3 °C, pH 8.5, Cond. 5 cF, O_2 8.7 mg/l		
10		03.05.2021		Not measured		

(continued)

Table 14.1 (continued)

Sample No	Locality	Date	Characteristics of habitat	Physico-chemical parameters of water	Elevation	Coordinates
11	C	06.04.2018	Ephemeral pond between pasture and forest filled with rainwater (Pond 1 – C_{P1})	Water temp. 12.3 °C, pH 8.2, Cond. 2 cF, O_2 4.3 mg/l	90 m	45.10092° 17.45519°
12		13.04.2018		Water temp. 17.5 °C, pH 8.2, Cond. 3 cF, O_2 4.3 mg/l		
13		06.04.2018	Puddle on a pasture filled with rainwater (Pond 2 – C_{P2})	Water temp. 16.7 °C, pH 8.2, Cond. 3 cF, O_2 10.8 mg/l	89 m	45.10046° 17.45437°
14		06.04.2018	Puddle on a pasture filled with rainwater (Pond 3 – C_{P3})	Water temp. 18.6 °C, pH 8.2, Cond. 1 cF, O_2 11.6 mg/l	88 m	45.09957° 17.45398°
15		03.06.2019	Puddle on a pasture filled with rainwater (Pond 4 – C_{P4})	Water temp. 26.4 °C, pH 8.3, Cond. 3 cF, O_2 9.6 mg/l	87 m	45.10126° 17.45387°
16	D	02.04.2021	Ephemeral pond on pasture, with vegetation	Water temp. 13.9 °C, pH 7.7, Cond. 6 cF, O_2 4.1 mg/l	86 m	45.09872° 17.46430°
17		11.05.2021		Water temp. 22.4 °C, pH 7.5, Cond. 5 cF, O_2 6.9 mg/l		
18	E	12.04.2021	Standing water in the canal at the bottom of a drained fishpond which is used as arable land	Not measured	87 m	45.11783° 17.44850°
19	F	18.05.2021	Shallow depressions (tracks) made by the passage of vehicles, between forest and arable land	Water temp. 29.7 °C, Cond. 6 cF, O_2 7.0 mg/l	88 m	45.1175° 17.4684°
20	G	25.05.2021	Standing water at the bottom of a drained fishpond which is used as arable land	Water temp. 31.3 °C, pH 8.9, Cond. 2 cF, O_2 10.3 mg/l	87 m	45.11303° 17.46622°

14.3 Main Characteristics of Plant and Animal Communities of "Bardača Wetland"

Data on qualitative composition of phytoplankton in different aquatic habitats, which are now located within the Ramsar site "Bardača Wetland", were first provided by Matavulj et al. [18]. The authors reported the presence of 94 taxa, subdivided within the divisions of Chrysophyta, Bacillariophyta, Pyrrophyta, Euglenophyta, Chlorophyta, and Cyanobacteria. Research on phytoplankton diversity at several fishponds of Bardača also indicated a presence of representatives of the Xanthophyta division, as well as the fact that the largest number of taxa belongs to divisions Chlorophyta and Bacillariophyta [12, 19, 20]. In some fishponds in Bardača, the first record of the invasive cyanobacterium *Cylindrospermopsis raciborskii* (Woloszynska, 1912) Seenayya and Subba-Raju, 1972 for the area of Bosnia and Herzegovina was reported in 2011 [19, 20].

Research on the diversity of lignicolous macrofungi conducted by Matavulj et al. [21] indicates that 21 species of these group are present in the wider area of Bardača (today within the Ramsar site "Bardača Wetland"), while lichenological research indicated presence of 15 species of lichens in the same area [22]. Authors believe that such a rather low diversity of these two groups of organisms is associated with air pollution in some parts of the analyzed area and degradation of autochthonous plant communities by anthropogenic impact.

The vascular plants of the Ramsar site "Bardača Wetland" has been researched in great detail. According to the results provided by Kovačević [23], 316 species of vascular plants were recorded in this area. Davidović et al. [24] published the first record of the species *Potamogeton rutilus* Wolfg. in Bosnia and Herzegovina in one of the fishponds of Bardača. Vascular plant species are classified into 83 families, where the largest number (131 species) belong to the families Asteraceae, Poaceae, Fabaceae, Lamiaceae, Ranunculaceae and Cyperaceae, while genera with the large number of species are *Carex*, *Polygonum* and *Potamogeton* [23]. Documented flora of this area includes 11 relict representatives of the Tertiary flora as well as eight allochthonous plant species [25]. Regarding the vegetation, the most characteristic types of vegetation for this area are: aquatic vegetation, marsh vegetation, floodplain meadow vegetation and floodplain forest vegetation [26]. According to Nedović et al. [25] vegetation of this area was differentiated into more than 40 plant communities.

Invertebrate fauna within the Ramsar site "Bardača Wetland" has been insufficiently explored. Data on qualitative composition of zooplankton at different aquatic habitats in this area were presented by Matavulj et al. [18], citing a large number of invertebrate taxa (Protozoa, Rotatoria, Cladocera and Copepoda). Analyzing the diet of ichthyophagous birds in this area, Obratil [26] listed the following taxa of invertebrates: Platyhelminthes (Cestodes), Nematoda, Annelida (Oligochaeta), Mollusca, and Arthropoda (Arachnida, Crustacea and Insecta). Within Crustacea, the Isopoda and Amphipoda (Gammaridae) have been registered, while within the class of insects the following taxa have been recorded: Collembola, Ephemeroptera

(Potamanthidae and Ephemeridae), Odonata (Agrionidae, Aeschnidae and Libellulidae), Orthoptera (Gryllotalpidae and Tetrigidae), Hemiptera (Corixidae, Notonectidae and Gerridae), Megaloptera, Coleoptera (Carabidae, Dytiscidae, Staphylinidae, Hydraenidae and Hydrophilidae), Hymenoptera, Trichoptera, Diptera (Tipulidae, Chironomidae, Ceratopogonidae, Tabanidae and Syrphidae) and Lepidoptera. During the more recent studies, 16 species of dragonflies have been recorded as well [27]. The first record of lepidopteran species *Apatura metis* Freyer, 1829 in Bosnia and Herzegovina was recorded in this area [28], as well as several species of hard ticks [29–31], including the first finding of the species *Haemaphysalis concinna* Koch, 1844 [31].

According to the literature sources [32, 33] there are 26 species of fish recorded in the Bardača fishponds and natural watercourses belonging to the area (rivers Matura, Brzaja and Stublaja), within seven families (Esocidae, Cyprinidae, Cobitidae, Siluridae, Ictaluridae, Percidae and Centrarchidae). They include six allochthonous species, of which Prussian carp (*Carassius gibelio* (Bloch, 1782)) and brown bullhead (*Ameiurus nebulosus* (Lesueur, 1819)) are dominant fish species in the natural watercourses of this area [33]. Recently, European mudminnow (*Umbra krameri* Walbaum, 1792), one of the rarest fish species in Bosnia and Herzegovina, was also recorded in the Matura River and its tributaries [34]. It is endemic to the Danube and Dniester basins, but also the only indigenous member of the Umbridae family in Europe [35]. An invasive species of fish, *Neogobius melanostomus* (Pallas, 1814), recently recorded in rivers Una and Sava [36], was also found in temporary water bodies in the flood zone of the Sava River in Bajinci (Šukalo and Dmitrović pers. obs.). According to Radević and Vuković [37] characteristic bred species in fishpond Bardača are common carp (*Cyprinus carpio* Linnaeus, 1758), silver carp (*Hypophthalmichthys molitrix* (Valenciennes, 1844)), northen pike (*Esox lucius* Linnaeus, 1758), grass carp (*Ctenopharyngodon idella* (Valenciennes, 1844)), Wels catfish (*Silurus glanis* Linnaeus, 1758) and zander (*Stizostedion lucioperca* Linnaeus, 1758).

The herpetofauna of this area is represented by 11 species of amphibians and six species of reptiles [26, 38], Šukalo pers. obs.), while an individual of allochthonous turtle species *Pelodiscus sinensis* (Wiegmann, 1835) was also caught in this area in 2010 [39]. According to original research (Šukalo pers. obs.), the most abundant reptiles within the Ramsar site "Bardača Wetland" are grass snakes (*Natrix natrix* (Linnaeus, 1758)) and European pond turtle (*Emys orbicularis* (Linnaeus, 1758)), while presence of a lowland population of European adders (*Vipera berus* (Linnaeus, 1758)) was also recorded. Population studies of grass snake were systematically performed [40] and some unusualness related to the diet of this species were identified [41]. Among the amphibians, particularly important are three species recorded in the far north along the Sava River (Šukalo pers. com.): Danube crested newt (*Triturus dobrogicus* (Kiritzescu, 1903)), European fire-bellied toad (*Bombina bombina* (Linnaeus, 1758)) and common spadefoot toad (*Pelobates fuscus* (Laurenti, 1768)).

Ramsar site "Bardača Wetland" is a habitat of special importance for nesting, feeding and resting of migratory birds. Obratil [26] listed 185 bird species from 44 families, while Gašić and Dujaković [38] reported presence of 204 bird species from 17 orders and 50 families. In relation to the seasonal aspect, 144 species were recorded during autumn migration, 116 species during the nesting period, 113 species during spring migration and 81 species during the winter period [26]. A total of 84 species of nesting birds was registered [42], and according to data provided by Kotrošan et al. [43] the most important species is the globally endangered ferruginous duck (*Aythya nyroca* (Güldenstädt, 1770)). Thanks to the great diversity of wetland birds, Bardača was declared a protected Important Bird Area in 1969 [38]. However, according to the data provided by Panić and Nagradić [44], the number of bird species in this area is decreasing, which is related to the disturbance of the environment.

According to data provided by Radević [45], mammal fauna includes foxes, rabbits, moles, hedgehogs and weasels.

14.4 Large Branchiopod Crustacean Communities in Temporary Ponds (Fairy Shrimps, Tadpole Shrimps, Clam Shrimps)

The large branchiopods are crustaceans with flattened, leaf-like (phyllopodous) trunk limbs. The phyllopods are biramous and used for locomotion, feeding, and gas-exchange (the name 'branchiopod' means 'gill-foot') and in some groups, legs are also used in the process of reproduction. The majority of branchiopods are filter feeders, who filter nutrient particles from the water column, where the water currents are produced by the metachronal beating of the thoracic appendages. Except the omnivorous feeding, some species are carnivorous, feeding on zooplankton. Only a few species can be typical predators feed on fish fry or other small vertebrates. During the hunting for prey, the species showed unusual body flexibility for Crustacea [46].

Branchiopoda is normally reproduced bisexually. However, there are several species that can reproduce parthenogenetically. After egg hatching, the full development usually precedes through several nauplii and metanauplii larval stages. Development is anamorphic, with gradual adding the segments and limbs, as they moult and develop into adults [47]. The growth rate is very rapid, and sexual maturity in the local climate can be reached within ten days to two weeks after hatching [48]. Large branchiopod species quite rely on habitats dependent on rainfall, such as the vernal pools formed by rain and snow-melting, fishless lakes, alkali and salt lakes. The most important feature of these habitats is that they are strictly seasonal, and drained for a significant portion of the year. The period for which they hold water ranges from a few days to few months. It's kind of a 'race against time' to produce the number of resting eggs (cysts), before the pools dry up, or frost in winter. Such behavior of large branchiopods is considered as adaptation to unpredictable and unstable habitats. The eggs lay dormant in the soil until the drying phase, forming the 'egg bank' that can remain viable for decades, possibly centuries [49], thus providing a potentially long-term genetic diversity. Since the cysts are resistant to desiccation, it is likely that they are easily dispersed by wind. The other kind of efficient egg dispersal is via zoochory, by the freshwater animals and by terrestrial vertebrates visiting water sources [50–53], or by humans.

Fossil records of large branchiopod crustaceans extending back to the about 500 million years ago. Many of Carbonaceous ancestors known as Fossil Stem-Branchiopods (such as the Upper Cambrian *Rehbachiella kinnekullensis* Müller, 1983) were found in rocky deposits in Sweden, Canada and other places around the world [54, 55]. Recent species have a worldwide distribution, including the Polar Regions.

The crustacean class Branchiopoda represents a morphologically and ecologically fairly diverse group. The group includes orders Anostraca (fairy and brine shrimps), Notostraca (tadpole shrimps) and Diplostraca (clam shrimps and cladocerans). Clam shrimps, tadpole shrimps and cladocerans are closely related to each other, which is strongly supported by both morphological and molecular data (for review see in Martin and Davis [56]). Together they belong to the Subclass Phyllopoda. On the other side, fairy and the brine shrimps belong to the Subclass Sarsostraca (named after Norvegian marine and freshwater biologist and branchiologist Georg Ossian Sars). All above-mentioned groups, except cladocerans, belong to the Eubranchiopoda or large branchiopods, which will be the further subject of this Chapter.

Anostraca lacks a carapace. This feature distinguishes them from the other members of the class Branchiopoda. The body is elongated, with eleven phyllopodous thoracic limbs. This makes them elegant swimmers who are constantly swimming on their backs in the water column. Males use their well-developed antennae to hold the female during mating. The shape of male antennae is compatible with an egg-bearing ovisac on the ventral side of the female body [57]. Males possess the paired penes with a basal rigid, and apical retractable part. Populations are mostly bisexual, with an equal share of males and females. Anostracans have almost worldwide distribution, missing only from the Polar Regions.

The Order Notostraca (tadpole shrimps) is composed of the single family Triopsidae that includes two genera: *Triops* Schrank, 1803 and *Lepidurus* Leach, 1819. Notostraca is characterized by the presence of a dorsal shield (carapace). The abdomen is elongated, with chitinous ring-like segments which can be incomplete or fusioned in spiral form [58]. The thorax has 11 segments, each bearing a pair of leaf-like swimming appendages. However, the number of abdominal pairs of limbs does not match the number of abdominal segments. The exopodite of the eleventh pair of thoracopods is modified into brooding chamber for holding eggs in females. Males could be rare or completely absent from populations. The tadpole shrimps are predominantly benthic filter-feeders, feeding on detritus. Adults can be raptorial, feeding on small-sized living or dead animals. Notostracans inhabit a variety of small inland waters, shallow lakes, and fishponds and irrigation canals, worldwide. They can reach a larger body size when environmental conditions are favorable.

The Order Diplostraca recognizes three suborders within large branchiopods, commonly called clam shrimps: Spinicaudata (the large-tailed clam shrimps), Laevicaudata (small-tailed clam shrimps) and Cyclestherida. The group is formerly defined as Conchostraca [56]. Clam shrimps are small shell-shaped crustaceans with a laterally compressed segmented body. Each body segment bearing a pair of foliaceous limbs, used for locomotion, respiration and feeding. The body is completely enclosed by a bivalved carapace. Female glues the fertilized eggs between the body and the carapace. In males, the first trunk limbs bear hook-like claspers with the function in holding females during the mating process. Reproduction is generally bisexual, but some species are parthenogenetic or hermaphrodite. Clam shrimps are non-selective feeders on algae and detritus. They live in rain-water pools where they burrow into the substratum or lie within a bunch of macrophytes. Species occur on all continents except Antarctica.

14.5 Diversity of the Large Branchiopods in Bosnia and Herzegovina

First record of large branchiopod species on the territory of Bosnia and Herzegovina was published by Dr. K. Marcus Jena in 1912 [59]. In the "Contributions to the knowledge of the freshwater fauna of the northwestern Balkan Peninsula" Marcus [59] reported about his field trips in Bosnia, Herzegovina and Dalmatia during August and September 1912. These countries were generally well known in the past in terms of fauna, but with exception of the freshwater animals. He captured several specimens of both sexes in an alpine puddle about 10 m in diameter in the Treskavica Mt. (south of Sarajevo) above the Crno Jezero, at an altitude of about 1650 m. According to the Monograph by Daday [60], Marcus ascribed those specimens to the anostracan Family Chirocephalidae Daday, 1910 and to the species *Chirocephalus stagnalis* Daday, 1910 (= *Chirocephalus diaphanus* Prévost, 1803, sensu Rogers [61]). The second finding of this species was in the Zelengora Mt. at an altitude of about 1700 m. He caught juveniles and a sexually mature female which could also be identified with certainty as *Chirocephalus stagnalis* (after Daday [60]). This was not surprising, since this species was also found in the neighboring countries. The nearest localities were: the Mountain range Komovi in eastern Montenegro [62], and Gospić and Vrhovina in Croatia [60].

In the second trip to the Treskavica Mt., Marcus found 4 males and 12 females belonging also to the genus *Chirocephalus* Prevost, 1803. Since the individuals morphologically resembled to *C. stagnalis*, Marcus wanted to place those specimens as a variety of *C. stagnalis*. However, after closer examination it turned out to be so different that the establishment of a new species seemed necessary. Dr. Marcus Jena called them *Chirocephalus reiseri* Marcus 1913, "...according to the well-known ornithologist and curator at the Bosnia-Herzegovinian State Museum, Othmar Reiser, as a small token of my gratitude that he was always at my side with advice and action, and thus contributed to the success of the excursion in an outstanding way" (cited from Marcus, K. J. [59], page 410).

After an extensive hydrobiological and plankton studies during year 1926, the Bosnian hydrobiologist and botanist Đorđe Protić reported species *Branchipus stagnalis* (L. 1758) in Borilovačko Lake (the local toponymy: Jezero pod Ljeljenom) on Zelengora Mt. in south-eastern Bosnia and Herzegovina, at an altitude of about 1500 m [63]. Dedijer and Grđić [64] characterized Borilovačko lake as a glacier type. The lake was 3–4 m in depth, overgrown with *Potamogeton*, and inhabited by insect larvae and amphibians, but lacking fish [63]. At the beginning of September 1926, Protić [63] visited the Uloško Lake (on Mt. Crvanj), where he found individuals from the group Laevicaudata Linder, 1945, Family Lynceidae Stebbing, 1902, *Lynceus* sp. *Lynceus* was caught together with zooplankton in the littoral zone (these crustaceans are usually 2–4 mm in size). The Uloško Lake lies in the southeastern part of the country (on Mt. Crvanj) at an altitude about 1100 m. The shore of the lake was overgrown with a wide belt of macrophytes. According to Protić [63] there was no fish in the lake but only the larvae of several insect groups. Water was drained from the lake into the River Neretva. To better understanding of the characteristics of mountain standing waters in which large branchiopods were found, Protić underlined that "…all these mountain lakes are more or less small and shallow; some of them do not deserve the name 'lake', but a pond or swamp" … and that… "only the locals call them a lake" (Protić [63], pages 4 and 7).

First data of Notostraca in Bosnia and Herzegovina was published by the Bosnian entomologist Mirjana Tanasijević [65]. In May 1979 she found a population of *Lepidurus apus* (Linnaeus, 1756) formed by the spring flooding of River Jaruga, in the southeastern part of Livanjsko Polje (Livno karst field). Near the village of Golubović, the abundant female-biased population occurred in large stagnant water areas covered with macrophytes.

In the last decade of 20th Century, the Former Yugoslavia investigators Petrov and Marinček [66] reported the occurrence of populations of *Branchipus stagnalis* (L. 1758) (synonym of *Branchipus schaefferi* Fischer, 1834, according to Rogers [61]) near the villages Batković and Crneljevo Donje, and *Branchipus* sp. near Brodac. All of the investigated localities are situated in the vicinity of the city of Bijeljina in Eastern Bosnia and Herzegovina. The same investigators surveyed also the populations of Conchostraca (Spinicaudata according to the current nomenclature) and reported the occurrence of the species *Leptestheria saetosa* Marinček & Petrov, 1992 (a synonym of *Leptestheria dahalacensis* (Rüppell, 1837) according to Rogers [67]) around the city of Brčko in north-east of Bosnia and Herzegovina.

Research of the large branchiopod fauna in Bosnia and Herzegovina have been also carried out by the famous branchiologists Graziella Mura, Paola Zarattini and Svetozar Petkovski [68]. They analyzed morphological variation in resting eggs of different populations of *Chirocephalus diaphanus carinatus* Daday, 1910 from the Balkan area, including the populations sampled from the temporary ponds near Belo Lake and Crno Lake in Treskavica Mt. (at altitude over 1700 m).

Research in the Ramsar site "Bardača Wetland" has been carried on since year 2016. The data so far is revealing a presence of several taxa of large branchiopods (Class Branchiopoda Latrielle, 1817) previously unknown to exist in this territory (Table 14.2, Fig. 14.3). In the following, a list and the general characteristics of the large branchiopod taxa discovered in "Bardača Wetland" will be presented.

Eubranchipus (Siphonophanes) grubii belongs to the Holarctic subfamily Eubranchipodinae. This spring species occurred in dense populations with almost

Table 14.2 List of taxa recorded from habitats in "Bardača Wetland". The sample numbers (No) are referred to Table 14.1

Group/Taxon	Sample No
CLASS BRANCHIOPODA Latrielle, 1817	
SUBCLASS SARSOSTRACA Tasch, 1969	
Order Anostraca Sars, 1867	
Family Chirocephalidae Daday, 1910	
Subfamily Eubranchipodinae Daday, 1910	
Eubranchipus Verrill, 1870	
Eubranchipus (*Siphonophanes*) Simon, 1886	
Eubranchipus (*Siphonophanes*) *grubii* (Dybowski, 1860)	11, 12, 13, 14
Family Linderiellidae Brtek, 1964	
Linderiella Brtek, 1964	15
SUBCLASS PHYLLOPODA Preuss, 1951	
Order Notostraca Sars 1867	
Family Triopsidae Keilhack, 1909	
Lepidurus Leach, 1819	
Lepidurus apus (Linnaeus, 1758)	2, 4, 5, 7
Order Diplostraca Gerstaecker, 1866	
Suborder Spinicaudata Linder, 1945	
Family Cyzicidae Stebbing, 1910	
Cyzicus Audouin, 1837	1, 3, 6, 7, 8, 9, 10, 16, 17, 18, 19
Family Leptestheriidae Daday, 1913	
Leptestheria Sars, 1898	
Leptestheria dahalacensis (Rüppell, 1837)	20
Eoleptestheria Daday, 1913	
Eoleptestheria ticinensis (Balsamo-Crivelli, 1859)	15
Family Limnadiidae Burmeister, 1843	
Limnadia Brongniart, 1820	
Limnadia lenticularis (Linnaeus, 1761)	15

equal numbers of both sexes. Species was recorded in ephemeral ponds and puddles filled with rainwater during April (water temperature 12.3 –18.6 °C, pH 8.2). This species has never been found in the co-occurrence with other large branchiopods in Bardača.

Linderiella sp. was found at the beginning of June in the puddle with slightly alkaline water (temperature 26.4 °C, pH 8.3). Only one specimen (male) was caught in 2019 together with spinicaudatans *Limnadia lenticularis* and *Eoleptestheria ticinensis.* Due to insufficient sample and the very restricted and disjunct range of known species [69], this finding requires additional research and clarification.

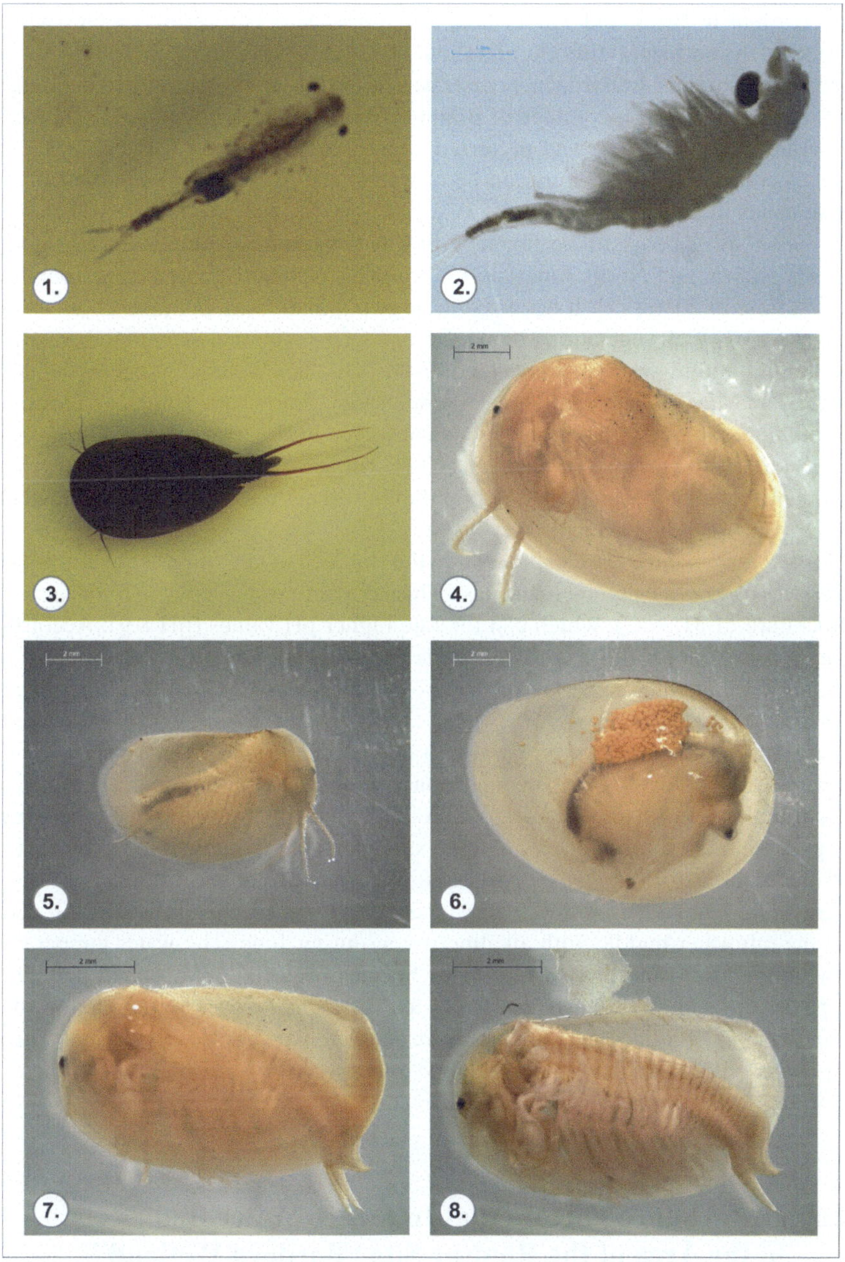

Fig. 14.3 Large branchiopods (Branchiopoda) found in the Ramsar site "Bardača Wetland": 1 – *Eubranchipus* (*Siphonophanes*) *grubii* (Dybowski, 1860) (not scaled); 2 – *Linderiella* Brtek, 1964; 3 – *Lepidurus apus* (Linnaeus, 1758) (not scaled); 4 – *Cyzicus* Audouin, 1837; 5 – *Eoleptestheria ticinensis* (Balsamo-Crivelli, 1859); 6 – *Limnadia lenticularis* (Linnaeus, 1761); 7 – *Leptestheria dahalacensis* (Rüppell, 1837); 8 – *L. dahalacensis* (internal morphology)

According to Brtek and Thiéry [70] *Lepidurus apus* is a spring species widely distributed in continental Europe, and should be sought particularly in the flood zones of rivers and streams. In Bardača, *L. apus* was found during the April 2016 in flooding vegetated pasture near the village of Bajinci [71]. This species occurred in the water temperature up to 28.8 °C and preferred slightly alkaline water (pH 8.1–8.3). The Bardača community of *L. apus* can be characterized as highly female-biased with sparse males in the population.

Cyzicus sp. appears to be taxon with the geographical ranges encompassing Eurasia, Africa, and North America. In "Bardača Wetland" *Cyzicus* was found for the first time in 2016 [72]. It occurred in large populations in ephemeral ponds and puddles mostly with vegetation (28. 04. 2018. *Cyzicus* recorded there together with *L. apus*) or in standing water in arable land. In the "Bardača Wetland", species occurred in April and May (in the vicinity of settlements Gaj, Bajinci and Bardača). However, the autumn/wintering population was also observed in November 2020. The individuals can tolerate temperature between 9.3 °C (in autumn) and 29.7 °C (in spring), and the pH between 7.7 and 8.5. The population still has undetermined species status and further research is needed.

Leptestheria dahalacensis appeared in the Bardača in late May, at the arable land which was used as a fishpond in the past. This is a warm water species (found at water temperature of 31.3 °C), which can tolerate higher pH values (pH 8.9). According to Brtek and Thiéry [70] *L. dahalacensis* is eurytopic species widespread in Europe, which also inhabits large parts of Asia, from the Caucasus region to Mongolia and China.

Eoleptestheria ticinensis is a warm water species found in late spring (water temperature 26.4 °C, at pH 8.3) in puddle on a pasture filled with rainwater, near River Brzaja. In the biotope, *E. ticinensis* was found in cohabitation with two other Branchiopods: anostracan *Linderiella* and the spinicaudatan *L. lenticularis*. The species distribution area of *E. ticinensis* covers Europe, Asia Minor and China [73].

Limnadia lenticularis is a widespread and eurytopic species. In Bardača it occupied the rainwater puddle with alkaline water, during June. The water temperature in the moment of sampling was 26.4 °C. Species occurred in vegetated puddle on the pasture in syntopy with *E. ticinensis* and *Linderiella* sp. The population of *L. lenticularis* in "Bardača Wetland" was strongly female-biased. Brtek and Thiéry [70] consider the wood-steppe and steppe zones of Europe as important zones for the origin of this species.

14.6 Disturbance Factors that Influence Water Bodies in the Area of the Ramsar Site "Bardača Wetland"

Begović [74] provided a wealth of data on the original conditions and the reclamation activities in this area, stating that before the flood-protection activities most of the wider area of Bardača was commonly flooded, either with water from the Sava

River which flooded a 3–6 km wide belt, or water from Vrbas River that flooded a belt 2–8 km wide. Smaller watercourses in this area, such as Matura, Stublaja and Brzaja rivers, also took part in the flooding. The same author stated that since the end of the nineteenth century numerous land reclamation works have been carried out in order to protect land against floods. Embankments and artificial fishponds were built, while numerous canals were excavated. According to the above facts, it may be assumed that the potential habitats suitable for the *Lepidurus* (the species which mostly inhabiting the flooded areas in river valleys) used to cover more than two thirds of today's Ramsar site "Bardača Wetland" territory. However, nowadays this habitat is reduced to a narrow belt in the northern part of the area, between Sava River and the man-made embankments constructed at the recent time. An additional prospective threat for the survival of large branchiopods is the embankment maintenance works and ongoing reconstruction. Thus, during the recent maintenance works on the embankment between the Fishpond Bardača and the Sava River, one of the *Cyzicus* habitats (Locality A) was destroyed by backfilling. One of the potential major problems in the future may also include the uncontrolled accumulation of improper waste disposal sites. Ephemeral pond between pasture and forest (Locality C, Pond 1) has recently been disrupted in that sense. Waste is improperly disposed of several tens of meters away from Locality B (Fig. 14.41, but there is also possibility that waste can be brought to this locality by Sava River during the flooding periods.

As previously mentioned, the Ramsar site "Bardača Wetland" includes several settlements: Gaj, Bardača, Bajinci, Dugo Polje and one part of the settlement Glamočani (Fig. 14.1). The structure of agricultural land in this area is dominated by pastures (Gaj and Bajinci), fishponds (Bardača), meadows (Dugo Polje) and small areas under forests. The exception is the settlement Glamočani where the largest share in the total agricultural land area includes plowed fields and gardens [75]. A large part of Bardača, over 700 ha, is under fishponds [76]. Although the fishponds are recognized as possible habitats of some representatives of large branchiopods [77] this was not confirmed by the field studies in the Ramsar site "Bardača Wetland" so far. In addition, in the most recent period only a few of the former Bardača fishponds still contained water, while the rest of them were recalled from the irrigation

Fig. 14.4 Sections with waste material (1) and Canadian poplar (2) in the Ramsar site "Bardača Wetland"

regime and drained. Nowadays, their bed is mainly used as arable land that most likely involving use of pesticides that are known to have a negative effect on large branchiopods [78]. Although the *Cyzicus* and *L. dahalacensis* were found in standing water at the bottom of a drained fishponds (Localities E and G), a long-term drying of the fishponds, with the simultaneous application of agricultural measures, could have negative effects on large branchiopods in the future. Recently, certain parts of Ramsar "Bardača Wetland" have been afforested with plantations of Canadian poplar (Fig. 14.42), an allochthonous species of woody plants which further affects the draining of the area. The negative impact of these plantations on the state of wetland ecosystems in the region of the Western Balkans is well-documented in some previous studies [79, 80].

In the future, climate change may have an adverse impact on large branchiopods and their temporary habitats [81]. Growth of the average annual air temperature and the average annual precipitation in the area of Lijevče field, observed during the survey from 1955–1965 and 1992–2002, was linked to the global climate change [9]. The previously mentioned increase in the annual precipitation is characterized by an uneven distribution of precipitation during the year, with the highest precipitation shifted toward autumn and reduced precipitation in the spring, when is the main season for the occurrence of large branchiopod populations. According to the predictions of climate change for Bosnia and Herzegovina provided by Bajić and Trbić [82], the average annual air temperature at the Ramsar site "Bardača Wetland" may increase to 12 °C by 2030 and to 15 °C in the period between 2071 and 2100. According to those predictions, the average annual precipitation in the same area may reach the 1000 mm by 2030 and around 900 mm in the period between 2071 and 2100, with the highest average precipitation in the summer period. The level of impact of these climate changes on populations of large branchiopods at the Ramsar site "Bardača Wetland" is yet unknown, but it may be supposed that an increase in temperature accompanied with same or reduced levels of rainfall will likely to shorten the duration of temporary, generally shallow aquatic habitats. This may jeopardize the possibility of completing the life cycle of representatives of this specific group of crustaceans, and thus would endanger their survival in this protected area.

14.7 The Freshwater Biodiversity Protection in Bosnia and Herzegovina (Specifically Refers to the Fauna of Large Branchiopods)

As the Ramsar convention stated, wetlands are characterized as habitats by a self-contained hydrology and are vital to the health and maintaining good condition of the wildlife and humans [2]. Based on the Ramsar definition, temporary water bodies are small, shallow habitats characterized by alternating phases of drought and flood. They have an indispensable role in primary productivity, chemical and biological cycling of nutrients, and water purification.

Brtek and Thiéry [70] emphasized the region of Balkans as a center of diversification of Branchiopod families Branchipodidae, Chirocephalidae and Limnadiidae in Europe. As a significant part of the Balkan Peninsula, Bosnia and Herzegovina is characterized by high geomorphological and hydrological diversity, climate diversity and ecological heterogeneity. A very rich large branchiopod community was found in the Ramsar Wetland site "Bardača Wetland", formerly unknown for large branchiopods. Overall, seven species of large branchiopods (two Anostraca, one Notostraca and four Spinicaudata) have been recorded in temporary aquatic habitats. All of them except *L. apus* and *L. dahalacensis*, are new records for Bosnia and Herzegovina [83]. *Linderiella* sp. and *Cyzicus* sp. found in "Bardača Wetland" still represent undefined taxons and further research of their taxonomy is needed.

Maintaining open land areas, such as pastures and sustainably handled arable land (i.e. without filling the depressions and/or use of pesticides), has a positive effect on large branchiopods [78]. This is encouraging piece of news, given that pastures make up a large share of the structure of agricultural land within the Ramsar site "Bardača Wetland". Habitats of *Eubranchipus*, *Linderiella*, *Eoleptestheria* and *Limnadia* are actually depressions at rain-filled pastures used to sheep grazing (Locality C), while the habitats of *Lepidurus* and *Cyzicus* are maintained in the open state by grazing cattle (Localities B and D). Flooding of the terrain with water from rivers Sava and Vrbas probably plays an important role in maintaining open areas in Localities A, B and F (Figs. 14.1 and 14.2). One of the important factors in maintaining the proper habitat structure for the growth of large branchiopods may include the flocks of domestic geese kept by local people, which are common in Localities B, C and D (Fig. 14.1). Visits by geese to localities inhabited by large branchiopods have a positive effect on the growth and fecundity of these crustaceans. Namely, enrichment of water with bird feces, resulting in development of dense populations of bacteria and algae, upon which the large branchiopod crustaceans feed [84]. On the other side, domestic geese and other birds that use crustaceans for feeding may appear as potential vectors for dispersal of large branchiopod eggs over a wider area [85, 86]. These facts reflect the necessity for maintaining a varied mosaic of habitats even when it comes to small water bodies such as the examined ponds. On the other side, all of the above points to a realistic possibility of unhindered coexistence of people and large branchiopods within the boundaries of the Ramsar Site "Bardača Wetland". However, at the level of local states wetland areas are often overlooked as important natural resources and receive only sparse and occasional attention when it comes to protection. Some of the large branchiopod crustaceans do not hatch every year and can therefore easily be overlooked during examination. This is one reason why species living in temporary water bodies are occasionally neglected even by ecologists and conservationists, and decision makers [87].

According to the modern concept of wetland management, wetland protection focuses on a single or a few species representing the needs of the majority of other species ('umbrella species'). Species that typically inhabit such habitats (such as the large branchiopods) are also termed a 'flagship' species, chosen to raise support for biodiversity conservation [88]. The monitoring and measures taken for their survival may also benefit the ecosystem as a whole and the protection of those freshwater crustaceans can safeguard the other species in ephemeral ponds [6, 89].

In 2007, "Bardača Wetland" was included in the list of wetlands of international importance according to the methodology used by Ramsar Convention. In addition to the Ramsar site "Bardača Wetland", there are two other Ramsar sites in Bosnia and Herzegovina: Livanjsko Polje and Hutovo Blato [90]. The "Bardača Wetland" is the only Ramsar site located in the administrative territory of Republic of Srpska, while the remaining two sites are situated in the administrative territory of the Federation of Bosnia and Herzegovina. The first record of *Lepidurus apus* from the floodplain of the intermittent river Jaruga in Bosnia and Herzegovina, as published by Tanasijević [65], de facto originated from a site which is presently within the boundaries of the Ramsar site Livanjsko Polje in the Federation of Bosnia and Herzegovina. In addition to the listed areas protected according to the Ramsar Convention, there are numerous additional areas in Bosnia and Herzegovina protected according to the legal regulations of the Republic of Srpska and the Federation of Bosnia and Herzegovina (according to the data provided by Crnić-Babić and Pušić-Babić [90]). In addition to the listed sites protected by the Ramsar Convention, there are numerous additional areas in Bosnia and Herzegovina protected according to the legal regulations of the Republic of Srpska and the Federation of Bosnia and Herzegovina [90]. Panić and Nagradić [44] reported 25 areas protected under local regulations on the territory of Republic of Srpska, including nature reserves, national parks, natural monuments, protected habitats, parks of nature, and areas with sustainable use of natural resources. However, it is important to note that the "Bardača Wetland" has never been granted the status of a protected area in the Republic of Srpska, although the natural values of this area are recognized worldwide.

There are still no state-level Red Lists in Bosnia and Herzegovina. The Red Lists were adopted at the entity level, separately for the territory of the Republic of Srpska and separately for the Federation of Bosnia and Herzegovina. In the Republic of Srpska, the "Decree on the Red List of Protected Species of Flora and Fauna of the Republic of Srpska" [91] was adopted in 2012, but it did not specify the categories of threat status for each taxon. Two years later a legal act entitled "Red List of Wild Species and Subspecies of Plants, Animals and Fungi" [92] became official in the Federation of Bosnia and Herzegovina. Categories of threat status were provided for each species and subspecies and listed in this document. Recently, the "Decree on Strictly Protected and Protected Wild Species" came into force in the Republic of Srpska [93]. All of the above documents include representatives of invertebrates of inland waters. However, none of the previously mentioned documents included representatives of crustaceans from the group of large branchiopods or proposed them for entity or state endangered status. Only one species (*Chirocephalus reiseri*) is endemic to Bosnia and Herzegovina [94]. However it has not been discovered in recent times and there are no more detailed studies on the current status of population and distribution of this endemic species. In the IUCN Red Data List, *C. reiseri* was categorized as a 'Vulnerable' (VU D2, ver. 2.3) [95]. The state or entity lists in these documents should definitely be supplemented in the future with inclusion of large branchipod taxa, and particularly the endemic and endangered species.

14.8 Conclusions

In this Chapter, we presented the species richness, local distribution, endangerment and conservation status of the crustaceans of Class Branchiopoda (commonly called 'large branchiopods') with explaining the more recent data and new records from Bosnia and Herzegovina. Existence of very rich large branchiopod fauna in the Ramsar site "Bardača Wetland" confirms that flooding areas and other temporary water bodies represent the prospective hot spots of the rare crustacean populations. The large branchiopod species *Lepidurus apus* and *Leptestheria dahalacensis* were rediscovered in Bosnia and Herzegovina after almost four decades. On the other hand, some of the taxa were recorded for the first time in the territory of Bosnia and Herzegovina (*Cyzicus* sp., *Eoleptestheria ticinensis* and *Limnadia lenticularis*) or even for the first time in the territory of the Western Balkans (such as *Linderiella* sp. and *Eubranchipus* (*Siphonophanes*) *grubii*). Large branciopods at the Ramsar site "Bardača Wetland" can be used as bioindicators of environmental conditions as it preferred small and eutrophic ponds, particularly those with complex macrophyte cover. Also, this group is also nominated as the flagship group for these ecosystems and for temporary pool animals, in general.

Large branchiopods are used for assessment of ephemeral wetland habitat functions and values. The greatest threat nowadays is habitat destruction through agriculture, urbanization, and pollution.The main potential threats to large branchiopods are alteration of the water level of ponds, in particular by the pumping of the groundwater and the drying up. A significant part of this area was, or is currently under drainage regime, and has been largely used as arable land. The use of pesticides has also a negative effect on large branchiopods. Another threat is the introduction of fish which feed on crustaceans, as well as disturbance of the water regime in floodplains by planting allochthonous forests.

Populations are potentially extremely vulnerable due to the landscape fragmentation and the global climate change, and need the measures which allow their continuous and sustainably existence. On the positive side, the 'egg bank' of dormant eggs in the soil can provide a potentially long-term genetic diversity, while domestic and migratory birds that visit the Ramsar site "Bardača Wetland" may appear as potential vectors for their dispersal. The phenologically occurrence of autumnal populations (*Cyzicus* sp.) could be also one of the good indicators of wetland conditions. All of the above points to a realistic possibility of sustainable coexistence of people and large branchiopods within the boundaries of the wonderful "Bardača Wetland" Ramsar Site.

Distribution of large branchiopod species in Bosnia and Herzegovina is currently exhaustively being investigated. Many astatic water bodies were screened in the other parts of the country. In that sense, the anostracan *Branchipus schaefferi* Fischer, 1834 and spinicaudatans *Cyzicus* sp. and *E. ticinensis* were noticed to be present in nearby ponds situated outside the administrative borders of Ramsar site "Bardača Wetland". A similar is the case with species *B. schaefferi* found on the right bank of Una River in western Bosnia and Herzegovina (unpubl. data).

The conservation status of large branchiopod species in Bosnia and Herzegovina remains unresolved, mainly due to the lack of studies concerning their distribution. A good start for protection might be to put so far known large branchiopods to the state or entity Lists of endangered and protected species and assign them the highest category of protection.

Acknowledgements Advice and the appropriate literature provided by colleagues Dr. Federico Marrone (Italy), Dr. Dani Boix Masafret and Dr. Jordi Sala Genoher (Spain) was highly appreciated and useful for the interpretation of some sampling data, and we very grateful for that. This publication was partly supported by the Ministry of Education, Science and Technological Development of Serbia (Grant No 451-03-9/2021-14/ 200178.).

References

1. Thorp JH, Covich AP (eds) (2009) Ecology and classification of North American freshwater invertebrates, 3rd edn. Academic Press, Amsterdam, Boston, Heidelberg, London, New York, Oxford Paris, San Diego, San Francisco, Singapore, Sydney, Tokyo
2. Ramsar (2021) Ramsar Sites Information Service. https://www.ramsar.org. Accessed 9 Feb 2021
3. Wiggins GB, Mackay RJ, Smith IM (1980) Evolutionary and ecological strategies of animals in annual temporary pools. Arch Hydrobiol Suppl 58(1):97–206
4. Brendonck L, Michels E, De Meester L, Riddoch B (2002) Temporary pools are not "enemy-free." Hydrobiologia 486(1):147–159
5. Brendonck L (1996) Diapause, quiescence, hatching requirements: what can we learn from large branchiopods (Crustacea: Branchiopoda: Anostraca, Notostraca, 'Conchostraca'). Hydrobiologia 320:85–97
6. Pullin AS (2002) Conservation biology. Cambridge University Press, Cambridge
7. Rogers DC (1998) Aquatic macroinvertebrate occurrences and population trends in constructed and natural vernal pools in Folsom, California. In: Witham CW, Bauder ET, Belk D, Ferren WR Jr, Orduff R (eds) Ecology, conservation, and management of vernal pool ecosystems. Proceedings from a 1996 conference, Sacramento, June 1996. California Native Plant Society, pp 224–235
8. Marković M, Begović P (2007) Fizičke karakteristike područja Bardače. In: Marković M (ed) Restauracija i rehabilitacija močvarnog područja Bardača, Bosna i Hercegovina. Finalni izvještaj. Poljoprivredni fakultet, Banja Luka, Institut za agroekologiju i zemljište, Banja Luka, pp 9–25
9. Trbić G (2004) Lijevče polje: klimatske odlike. Geografsko društvo Republike Srpske, Banja Luka
10. Lolić S, Matavulj M, Maksimović T (2012) Komparativna analiza kvaliteta vode rijeke Mature i bazena Prevlaka na Bardači. Skup 4(2):121–128
11. Maksimović T, Lolić S, Lubarda B, Baltić M (2012) Koncentracija K i Na u tkivu *Phragmites communis* Trin., *Typha latifolia* L. i *Typha angustifolia* L. na području Bardača. Skup 4(1):48–56
12. Lolić S, Matavulj M, Dekić R, Maksimović T (2014) Kvalitativni i kvantitativni sastav fitoplanktona u vodi bazena Sinjak (ribnjak Bardača). Skup 6(1):39–47
13. Maksimović T, Lolić S, Kukavica B (2020) Seasonal changes in the content of photosynthetic pigments of dominant macrophytes in the Bardača fishpond area. Ekológia (Bratislava) 39(3):201–213

14. Lolic S, Matavulj M, Dekic R, Ivanc A, Manojlovic M (2014) The impact of agricultural measurements on water quality parameters at Fishpond Bardača. Paper presented at the 5th international scientific agricultural symposium "Agrosym 2014", Faculty of Agriculture, University of East Sarajevo, Jahorina, 23–26 Oct 2014
15. Cronk JK, Fennessy MS (2001) Wetland plants: biology and ecology. CRC Press, Boca Raton
16. Killham K (1994) Soil ecology. Cambridge University Press, Cambridge
17. Simovich MA (1998) Crustacean biodiversity and endemism in California's ephemeral wetlands. In: Witham CW, Bauder ET, Belk D, Ferren WR Jr, Orduff R (eds) Ecology, conservation, and management of vernal pool ecosystems. Proceedings from a 1996 Conference, Sacramento, June 1996. California Native Plant Society, pp 107–118
18. Matavulj M, Nemeš K, Đurđević S, Radević M (2008) Biološka raznovrsnost planktona u vodama regiona Bardača (BiH). Skup 2:221–229
19. Lolić S, Matavulj M, Dekić R, Maksimović T (2013) Biološka raznovrsnost i distribucija fitoplanktona u vodama bazena Necik (Bardača). Skup 5(1):3–13
20. Lolić S, Matavulj M, Dekić R (2015) Distribution of phytoplankton at fishopond Prevlaka (Bardača). Paper presented at 7th International conference, Water & Fish, University of Belgrade, Belgrade, 10–12 June 2015
21. Matavulj M, Karaman M, Gojković I, Đurđević S (2005) Lignicolus macrofungi of the Bardača floodplain region. Proc Nat Sci Matica Srpska Novi Sad 109:161–167
22. Matavulj M, Đurđević S (2008) Lišajevi močvarnog regiona Bardača (BiH). Skup 2:215–219
23. Kovačević Z (2015) Vascular Flora of Bardača Complex. Agroznanje 16(2):193–214
24. Davidović B, Blaženčić J, Stevanović V (2006) Report 29. In: Vladimirov V, Dane F, Nikolić T, Stevanović V, Tan K (eds) New floristic record in the Balkans: 2. Phytol Balc 12(2):283
25. Nedović B, Lakušić R, Kovačević Z, Marković B (2004) Raznoliki živi svijet. In: Šarić Ž, Maksimović Č, Stanković M, Butler D (eds) Život u močvari. Urbanistički zavod Republike Srpske, Banja Luka, pp 88–97
26. Obratil S (1980/1981) Ekološki pristup utvrđivanju štetnosti ihtiofagnih ptica u ribnjacima Bardača. Glasnik Zemaljskog muzeja Bosne i Hercegovine u Sarajevu 19–20:139–256
27. Kulijer D, Miljević I, Jakovljević J (2016) Contribution of the participants of 4th Balkan odonatological Meeting to the knowledge of Odonata distribution in Bosnia and Herzegovina. J Int Dragonfly Fund 95:1–25
28. Lelo S (2009) Freyer's purple emperor, *Apatura metis* Freyer, 1829 (Papilionoidea: Nymphalidae, Apaturinae), a new species in the butterfly fauna of Bosnia and Herzegovina. Prilozi fauni Bosne i Hercegovine 5:7–13
29. Bilbija B, Dmitrović D, Šnjegota D, Lukač M (2014) Polni dimorfizam adultnih krpelja vrste *Dermacentor reticulatus* sa područja Bardače. Paper presented at Drugi naučno-stručni skup sa međunarodnim učešćem ''5. juni—Svjetski dan zaštite okoliša", Univerzitet u Bihaću, Bihać, 4–5 June 2014
30. Dmitrović D, Bilbija B, Lukač M, Šnjegota D (2013) Polni dimorfizam adultnih krpelja vrste *Dermacentor marginatus* sa područja Bardače. Paper presented at Naučno-stručna konferencija sa međunarodnim učešćem "Zaštita životne sredine između nauke i prakse—stanje i perspektive", Institut za zaštitu i ekologiju Republike Srpske, Banja Luka, 13 Dec 2013
31. Dmitrović D, Bilbija B, Lukač M, Šnjegota D (2014) First record of *Haemaphysalis concinna* Koch, 1844 (Acari: Ixodidae) in Bosnia and Herzegovina. Ecol Mont 1(2):89–91
32. Radević M (2000) Ekološki i cenotički odnosi faune riba u srednjem i donjem toku Vrbasa i ribnjaku Bardači. Prirodno-matematički fakultet, Banja Luka
33. Vuković D, Tursi A, Carlucci R, Dekić R (2008) Ichthyofauna of the wetland ecosystem in the Bardača area (Bosnia and Herzegovina). Ribarstvo 66:89–103
34. Marić S, Stanković D, Šanda R, Ćaleta M, Čolić S, Šukalo G, Snoj A (2019) Genetic characterisation of European mudminnow (*Umbra krameri*) populations from the Sava River system. Knowl Manag Aquat Ecosyst 420(46):1–9
35. Delić A, Grlica ID, Razlog-Grlica J (1997) Nova nalazišta crnke (*Umbra krameri* Walbaum, 1972) u Hrvatskoj. Ribarstvo 55:93–98

36. Čolić S, Šukalo G, Čolić V, Kerkez V (2018) First record of round goby *Neogobius melanostomus* Pallas, 1814 (Pisces: Gobiidae) in Bosnia and Herzegovina. Ecol Mont 16:108–110
37. Radević M, Vuković D (2004) Ihtiofauna Bardače—ribe i ribnjaci. In: Šarić Ž, Maksimović Č, Stanković M, Butler D (eds) Život u močvari. Urbanistički zavod Republike Srpske, Banja Luka, pp 98–103
38. Gašić B, Dujaković G (2009) Ptice Bardače. Republički zavod za zaštitu kulturno-istorijskog i prirodnog nasljeđa, Banja Luka
39. Jelić L, Jelić D (2015) Allochthonous species of Turtles in Croatia and Bosnia and Herzegovina. Hyla 1:53–64
40. Šukalo G (2017) Populaciono-ekološke odlike i morfološka varijabilnost bjelouške (*Natrix natrix*) i ribarice (*Natrix tessellata*) na području sjeverozapadnog dijela Republike Srpske. Dissertation, University of Banja Luka
41. Šukalo G, Đorđević S, Dmitrović D, Lj T (2012) Introduced fish *Ameiurus nebulosus* (Le Sueur, 1819): hazard to the Grass snake, *Natrix natrix* (Laurenti, 1768). Hyla 2:41–42
42. Obratil S (1974) Ornitofauna ribnjaka Bardača kod Srpca. Glasnik Zemaljskog muzeja Bosne i Hercegovine u Sarajevu 11–12:153–193
43. Kotrošan D, Drocić N, Trbojević S, Šimić E, Dervović I (2012) Program IBA (Međunarodno značajna područja za ptice) u Bosni i Hercegovini. Ornitološko društvo „Naše ptice", Sarajevo
44. Panić G, Nagradić S (2019) Zaštita prirodnog nasljeđa u Republici Srpskoj. Republički zavod za zaštitu kulturno-istorijskog i prirodnog nasljeđa Republike Srpske, Banja Luka
45. Radević M (2007) Ostali predstavnici faune na Bardači. In: Marković M (ed) Restauracija i rehabilitacija močvarnog područja Bardača, Bosna i Hercegovina. Finalni izvještaj. Poljoprivredni fakultet, Banja Luka, Institut za agroekologiju i zemljište, Banja Luka, pp 63–66
46. Boudrias MA (2013) Swimming Fast and Furious: Body and Limb Propulsion at Higher Reynolds Numbers. In: Watling L, Thiel M (eds) The natural history of Crustacea. Functional Morphology and Diversity, vol 1. Oxford University Press, Oxford, pp 319–337
47. Martin JW, Olesen J, Høeg JT (eds) (2014) Atlas of Crustacean Larvae. Johns Hopkins University Press, Baltimore
48. Petrov B, Cvetković MD (1997) Community structure of Branchiopods (Anostraca, Notostraca and Conchostraca) in the Banat province in Yugoslavia. Hydrobiologia 359:23–28
49. Brendonck L, De Meester L (2003) Egg banks in freshwater zooplankton: evolutionary and ecological archives in the sediment. Hydrobiologia 491:65–84
50. Bilton DT, Freeland JR, Okamura B (2001) Dispersal in freshwater invertebrates. Annu Rev Ecol Evol Syst 32:159–181
51. Beladjal L, Dierckens K, Mertens J (2007) Dispersal of Fairy Shrimp *Chirocephalus diaphanus* (Branchiopoda: Anostraca) by the Trout (*Salmo trutta*). J Crustacean Biol 27(1):71–73
52. Beladjal L, Mertens J (2009) Diaspore dispersal of anostraca by flying insects. J Crustacean Biol 29(2):266–268
53. Rogers DC (2014) Larger hatching fraction in avian dispersed anostracan eggs (Branchiopoda). J Crustacean Biol 34(2):135–143
54. Walossek D (1993) The Upper Cambrian *Rehbachiella* and the phylogeny of Branchiopoda and Crustacea. Fossils Strata 32:1–202
55. Gueriau P, Rabet N, Gaël C, Lagebro L, Vannier J, Briggs DEG, Charbonnier S, Olive S, Béthoux O (2016) A 365-million-year-old fresh water community reveals morphological and ecological stasis in branchiopod crustaceans. Curr Biol 26:383–390
56. Martin JW, Davis GE (2001) An updated classification of the recent crustacea. Natural History Museum of Los Angeles County, California
57. Brendonck L, Belk D (1997) On potentials and relevance of the use of copulatory structures in anostracan taxonomy. Hydrobiologia 359:83–92
58. Šaganović I, Makarov S, Lučić L, Pavković-Lučić S, Tomić V, Miličić D (2019) Developmental and other body abnormalities in the genus *Lepidurus* Leach, 1819 (Crustacea: Notostraca) from Serbia. Arthropoda Sel 28(1):65–72

59. Marcus KJ (1912) Beiträge zur Kenntnis der Süßwasserfauna der nordwestlichen Balkanhalbinsel I. Zool Anz 41:409–412
60. Daday J (1910) Monographie systématique des Phyllopodes anostracés. Ann Sci Nat 11(9):91–489
61. Rogers DC (2013) Anostraca Catalogus (Crustacea: Branchiopoda). Raffles B Zool 61(2):525–546
62. Zograf N (1907) Phyllopodenstudien. Z wiss Zool 86:446–522
63. Protić Đ (1927) Hidrobiološke i plankton-studije na jezerima Bosne i Hercegovine. Četvrti deo—Planinska jezera. Glasnik Zemaljskog muzeja u Bosni i Hercegovini 39(1):3–42
64. St DJ, Grđić V (1904) Glacijalni tragovi na Zelen-Gori, Tovarnici i Magliću. Glas Srp Kralj Akademije 27(69):169–189
65. Tanasijević M (1981) The first find of *Lepidurus apus* (Linné, 1758) (Notostraca) in Yugoslavia. Godišnjak Biološkog Instituta Univerziteta u Sarajevu 34:162–163
66. Petrov B, Marinček M (1991) On the Anostraca (Crustacea) of Yugoslavia. Hydrobiologia 212:267–272
67. Rogers DC (2020) Spinicaudata catalogus (Crustacea: Branchiopoda). Zool Stud 59:45. https://doi.org/10.6620/ZS.2020.59-45
68. Mura G, Zarattini P, Petkowski S (2002) Morphological variation among *Chirocephalus diaphanus carinatus* populations (Anostraca) from the Balkan area. J Crustacean Biol 22(1):162–172
69. Brtek J, Mura G (2000) Revised key to families and genera of the Anostraca with notes on their geographical distribution. Crustaceana 73(9):1037–1088
70. Brtek J, Thiéry A (1995) The geographic distribution of the European Branchiopods (Anostraca, Notostraca, Spinicaudata, Laevicaudata). Hydrobiologia 298:263–280
71. Miličić D, Šukalo G, Dmitrović D, Šaganović I, Čolić S (2018) Rediscovery of genus *Lepidurus* (Crustacea, Notostraca)-New data of occurrence in Bosnia and Herzegovina. Paper presented at the 6th International scientific conference ''June 5th–World environment day" University of Bihać, Biotechnical faculty, Bihać, 18–19 June 2018
72. Dmitrović D, Šukalo G, Miličić D (2020) First records of *Cyzicus* sp. (Crustacea, Spinicaudata) in Bosnia and Herzegovina. Paper presented at the 4th symposium of biologists and ecologists of Republic of Srpska with international participation "SBERS2020", University of Banja Luka, Faculty of Natural Sciences and Mathematics, Banja Luka, 12–14 Nov 2020
73. Straškraba M (1966) Taxonomical studies on Czechoslovak Conchostraca. III, Family Leptestheriidae with some remarks on the variability and distribution of Conchostraca and a key to the Middle-European species. Hydrobiologia 21:571–589
74. Begović P (2007) Fizičke karakteristike slivnog područja. In: Marković M (ed) Restauracija i rehabilitacija močvarnog područja Bardača, Bosna i Hercegovina. Finalni izvještaj. Poljoprivredni fakultet, Banja Luka, Institut za agroekologiju i zemljište, Banja Luka, pp 26–34
75. Marković M (2004) Zemljišta Bardače. In: Šarić Ž, Maksimović Č, Stanković M, Butler D (eds) Život u močvari. Urbanistički zavod Republike Srpske, Banja Luka, pp 59–66
76. Vuković D (2004) Ribnjačarstvo. In: Šarić Ž, Maksimović Č, Stanković M, Butler D (eds) Život u močvari. Urbanistički zavod Republike Srpske, Banja Luka, pp 134–135
77. Sukop I (2007) Zoobentos Lednických rybníků. Acta univ agric et silvic Mendel Brun 55(2):85–94
78. Schernhammer T, Wessely J, Eder E, Straka U, Essl F (2020) Modelling the distribution of large branchiopods (Crustacea: Notostraca, Anostraca & Spinicaudata) for predicting occurrences in poorly sampled regions. Glob Ecol Conserv 23:e01083
79. Petrov B, Petrov I (1997) The status of Anostraca, Notostraca and Conchostraca in Yugoslavia. Hydrobiologia 359:29–35
80. Miličić D, Lukić D, Nahirnić A, Šćiban M, Marković A (2014) Current status of ephemeral habitats of ancient Crustaceans (Class Branchiopoda) in Serbia with varying degrees of human impact. J BioSci Biotech (Special edition):165–170
81. Pyke CR (2005) Assessing climate change impacts on vernal pool ecosystems and endemic branchiopods. Ecosystems 8:95–105

82. Bajić D, Trbić G (2016) Klimatski atlas Bosne i Hercegovine: Temperature i padavine (1961–1990, A1B 2001–2030, A1B 2071–2100, A2 2071–2100). Univerzitet u Banjoj Luci, Prirodno-matematički fakultet, Banja Luka
83. Cvetković-Miličić D, Petrov B (2001) A Review of the diversity of large branchiopods (Anostraca, Notostraca and Conchostraca) in countries of former Yugoslavia and some other parts of Europe—status endangerment and conservation. Ekologija 36(2):101–123
84. Vekhoff NV (1997) Large hranchiopod Crustacea (Anostraca, Notostraca, Spinicaudata) of the Barents Region of Russia. Hydrobiologia 359:69–74
85. Incagnone G, Marrone F, Barone R, Robba L, Naselli-Flores L (2015) How do freshwater organisms cross the "'dry ocean"'? A review on passive dispersal and colonization processes with a special focus on temporary ponds. Hydrobiologia 750:103–123
86. Lukić D, Nahirnić A, Marković A, Karan Žnidaršič T, Šćiban M, Miličić D (2012) An Updating of Large Branchiopods (Crustacea: Branchiopoda) Distribution in Serbia. Acta Zool Bulg 4:19–23
87. Boix D, Biggs J, Céréghino R, Hull AP, Kalettka T, Oertli B (2012) Pond research and management in Europe: "Small is Beautiful." Hydrobiologia 689:1–9
88. Marrone F, Korn M, Stoch F, Naselli-Flores L, Turki S (2016) Updated checklist and distribution of large branchiopods (Branchiopoda: Anostraca, Notostraca, Spinicaudata) in Tunisia. Biogeographia 31:27–53
89. Colburn EA (2004) Vernal pools: natural history and conservation. McDonald and Woodward Publishing Company, Granville
90. Crnić-Babić D, Pušić-Babić O (2013) Međunarodno pravna zaštita močvarnih staništa u Bosni i Hercegovini. Svarog 7:328–333
91. Službeni glasnik Republike Srpske (2012) Uredba o Crvenoj listi zaštićenih vrsta flore i faune Republike Srpske. Vlada Republike Srpske, no 124/12
92. Službene novine Federacije Bosne i Hercegovine (2014) Crvena lista divljih vrsta i podvrsta biljaka, životinja i gljiva. Federalno ministarstvo okoliša i turizma, no 7/14
93. Službeni glasnik Republike Srpske (2020) Uredba o strogo zaštićenim i zaštićenim divljim vrstama. Vlada Republike Srpske, no 65/20
94. Belk D, Brtek J (1995) Checklist of the Anostraca. Hydrobiologia 298:315–352
95. Sket B (1996) *Chirocephalus reiseri*. In: The IUCN red list of threatened species. https://www.iucnredlist.org/species/4662/11070997. Accessed 30 Apr 2021

Chapter 15
How Important are Small Lotic Habitats of the Western Balkans for Local Mayflies?

Marina Vilenica, Ana Petrović, Biljana Rimcheska, Katarina Stojanović, Bojana Tubić, and Yanka Vidinova

Abstract Mayflies are amphibious insects which represent an important link in food and energy transfer from aquatic to terrestrial habitats. They constitute a large proportion of the aquatic ecosystems' biomass. Although our knowledge about mayfly (bio)diversity in the Balkan Peninsula is still far from complete, more extensive systematic studies have been conducted within the last decade. In this chapter we explore development of mayfly research in the area of the Western Balkans, mayfly species richness in various small lotic habitats, including springs, streams and rivers, and determine the importance of such habitats for conservation of local mayflies. We discuss the value of mayflies as bioindicators of freshwater ecosystems' health, as well as the influences of various anthropogenic activities on mayfly assemblages in small lotic habitats of the Western Balkans. Moreover, we present current gaps in research and we give recommendations for future directions of mayfly research in the area of the Western Balkans.

M. Vilenica (✉)
Faculty of Teacher Education, University of Zagreb, Trg Matice Hrvatske 12, 44250 Petrinja, Croatia
e-mail: marina.vilenica@ufzg.hr

A. Petrović
Faculty of Science, Department of Biology and Ecology, University of Kragujevac, Radoja Domanovića 12, 34000 Kragujevac, Serbia
e-mail: ana.petrovic@pmf.kg.ac.rs

B. Rimcheska · Y. Vidinova
Department of Aquatic Ecosystems, Bulgarian Academy of Sciences, Institute of Biodiversity and Ecosystem Research, 1 Tsar Osvoboditel Blvd, 1000 Sofia, Bulgaria
e-mail: rimceska@gmail.com

K. Stojanović
Faculty of Biology, University of Belgrade, Studentski trg 16, 11000 Belgrade, Serbia
e-mail: k.bjelanovic@bio.bg.ac.rs

B. Tubić
Institute for Biological Research Siniša Stanković–National Institute of the Republic of Serbia, University of Belgrade, Bulevar despota Stefana 142, 11060 Belgrade, Serbia
e-mail: bojana@ibiss.bg.ac.rs

V. Pešić et al. (eds.), *Small Water Bodies of the Western Balkans*, Springer Water,
https://doi.org/10.1007/978-3-030-86478-1_15

Keywords Research development · Species richness · Biodiversity · Endemic species · Springs · Streams · Bioindicators · Anthropogenic threats

15.1 Introduction

15.1.1 Mayfly Biology and Ecology

Mayflies (Ephemeroptera) are relict insects, originating back to the late Carboniferous or early Permian periods (290 Mya) [1]. They differ from all other insect orders by having two winged-adult stages (subimago and imago) [2]. Mayflies are amphibious insects (i.e., with aquatic larva and short-living terrestrial adults) with larvae being able to colonize all kinds of freshwater habitats [3]. Higher species diversity is a characteristic of lotic habitats, especially the upper reaches of fast flowing streams and rivers, highland rivers, and lower reaches of slow-flowing natural lowland rivers, while springs and lentic habitats represent suitable habitats for much lower number of mayfly species [4–13].

A high number of species is adapted to a particular microhabitat, whereas aquatic vegetation and lithal substrates are recorded to be the most attractive to the most of grazer and collector feeding mayflies [3, 14]. While choosing their habitats, mayflies respond to multiple environmental factors, such as water temperature, water velocity, oxygen content and pH [e.g. 3, 15–20]. Each mayfly species has its own phenology, influenced by altitude and latitude [21]. Moreover, each species emerges to terrestrial habitats at its own characteristic time of year [2]. This is primarily driven by photoperiod and water temperature [20, 22].

Mayflies are often one of the most abundant groups of aquatic macroinvertebrates in all kinds of freshwater habitats, contributing to approximately 25% of the total benthic macroinvertebrate production [23]. Also, they have an important role in secondary production providing a substantial food source for various freshwater and terrestrial predators [2, 24].

15.1.2 Development of Mayfly Research in the Western Balkans and Current Knowledge

Balkan Peninsula is one of the biodiversity hotspots, both in freshwater and terrestrial ecosystems [25–27] hence it is not surprising that various European researchers were incited to study mayflies of the Balkan freshwaters [e.g. 28–32]. Many have even described new species based on materials collected from rivers and streams of the Balkan Peninsula [e.g. 33–43]. So far, 369 mayfly species are known for Europe [3] of which about 150 taxa have been recorded from the Balkans [32 and references herein], where currently Bulgaria represents one of the thoroughly researched countries [eg. 16, 44–50].

The extensive research of mayfly taxonomy and ecology in the area of the Western Balkan Peninsula (Fig. 15.1) began in the second half of the twentieth century by the Balkan researchers, such as Ikonomov [e.g. 51–53], Tanasijević [e.g. 54–56] and Filipović [57]. In some countries, such as Croatia, mayflies were until the beginning of the twenty-first century, mostly studied only as part of limnological studies [e.g. 58]. At the moment, due to the efforts of local researchers, mayfly fauna is relatively well known in Croatia [e.g. 18, 59–61], Macedonia [62] and Serbia with Kosovo* [17, 63, 64], while the data are still rather scarce from Albania [32] Bosnia and Herzegovina [3, 65, 66] and Montenegro [28, 40, 41, 67, 68] (Table 15.1).

Fig. 15.1 Map of Europe with position of the Western Balkan region

Table 15.1 The Western Balkans mayflies. Legend: A = Albania, B&H = Bosnia and Herzegovina, C = Croatia, M = Macedonia, MN = Montenegro, S/K = Serbia with Kosovo. Taxonomy follows Bauernfeind and Soldán [3]

Taxa/Country	A	B&H	C	M	MN	S/K	
Ameletidae							
Ameletus inopinatus Eaton, 1887			x	x		x	
Metreletus balcanicus (Ulmer, 1920)		x	x	x			x
Siphlonuridae							
Siphlonurus (Siphlonurus) abraxas Jacob, 1986					x		
Siphlonurus (Siphlonurus) aestivalis (Eaton, 1903)		x	x	x		x	
Siphlonurus (Siphlonurus) armatus (Eaton, 1870)			x	x			
Siphlonurus (Siphlonurus) croaticus Ulmer, 1920		x	x	x			
Siphlonurus (Siphlonurus) lacustris (Eaton, 1870)		x	x		x		
Ametropodidae							
Ametropus fragilis Albarda, 1878			x				
Baetidae							
Baetis (Acentrella) hyalopterum (Bogoescu, 1951)				x			
Baetis (Acentrella) sinaicus Bogoescu, 1931						x	
Baetis (Baetis) sp. nov. (near *B. nexus* Navás, 1918)	x						
Baetis (Baetis) alpinus (Pictet, 1843)		x	x	x	x	x	x
Baetis (Baetis) melanonyx (Pictet, 1843)			x	x		x	x
Baetis (Baetis) buceratus Eaton, 1870		x	x			x	x
Baetis (Baetis) nexus Navás, 1918*			x				x
Baetis (Baetis) beskidensis Sowa, 1972	x						
Baetis (Baetis) fuscatus (Linnaeus, 1761)		x	x	x		x	
Baetis (Baetis) scambus Eaton, 1870			x			x	
Baetis (Baetis) lutheri Müller-Liebenau, 1967	x		x	x		x	
Baetis (Baetis) meridionalis Ikonomov, 1954	x			x		x	
Baetis (Baetis) vardarensis Ikonomov, 1962		x	x	x		x	
Baetis (Baetis) pavidus Grandi, 1949						x	
Baetis (Baetis) kozufensis Ikonomov, 1962				x		x	
Baetis (Baetis) liebenauae Keffermüller, 1974			x			x	
Baetis (Baetis) tracheatus Keffermuller & Machel, 1967						x	
Baetis (Baetis) vernus Curtis, 1834		x	x	x		x	

(continued)

Table 15.1 (continued)

Taxa/Country	A	B&H	C	M	MN	S/K	
Baetis (*Labiobaetis*) *balcanicus* Müller-Liebenau & Soldan, 1981				x			
Baetis (Labiobaetis) tricolor Tshernova, 1928			x	x		x	
Baetis (Nigrobaetis) digitatus Bengtsson, 1912				x			
Baetis (Nigrobaetis) muticus (Linnaeus, 1758)	x	x	x	x	x	x	x
Baetis (Nigrobaetis) niger (Linnaeus, 1761)			x	x		x	
Baetis (Rhodobaetis) gemellus Eaton, 1885					x		
Baetis (Rhodobaetis) rhodani (Pictet, 1843)	x	x	x	x	x	x	x
Baetopus (Raptobaetopus) tenellus (Albarda, 1878)		x	x				
Centroptilum luteolum (Müller, 1776)	x	x	x	x	x	x	
Centroptilum pirinense Ikonomov 1962				x			
Cloeon (Cloeon) dipterum (Linnaeus, 1761)	x	x	x	x	x	x	x
Cloeon (*Similicloeon*) *praetextum* Bengtsson, 1914				x			
Cloeon (Similicloeon) simile Eaton, 1870			x	x		x	
Procloeon (Procloeon) bifidum (Bengtsson, 1912)	x		x	x		x	
Procloeon (Pseudocentroptilum) macronyx (Kluge & Novikova, 1992)						x	
Procloeon (Pseudocentroptilum) nana (Bogoescu, 1951)			x				
Procloeon (Pseudocentroptilum) pennulatum (Eaton, 1870)	x		x	x		x	
Procloeon (*Pseudocentroptilum*) cf. *pulchrum* (Eaton, 1885)							x
Procloeon (*Pseudocentroptilum*) ? *romanicum* (Bogoescu, 1951)	x						
Isonchiidae							
Isonychia (*Isonychia)ignota* (Walker, 1853)				x			
Oligoneuriidae							
Oligoneuriella pallida Hagen, 1855				x		x	
Oligoneuriella rhenana (Imhoff, 1852)	x	x	x	x		x	x
Heptageniidae							
Ecdyonurus (Ecdyonurus) aurantiacus (Burmeister, 1839)	x	x	x	x		x	
Ecdyonurus (Ecdyonurus) dispar (Curtis, 1834)		x	x	x		x	

(continued)

Table 15.1 (continued)

Taxa/Country	A	B&H	C	M	MN	S/K	
Ecdyonurus (Ecdyonurus) graecus Braasch, 1984							x
Ecdyonurus (Ecdyonurus) insignis (Eaton, 1870)			x	x		x	x
Ecdyonurus (Ecdyonurus) macani Thomas & Sowa, 1970			x				x
Ecdyonurus (Ecdyonurus) puma Jacob & Braasch, 1986	x				x		x
Ecdyonurus (Ecdyonurus) starmachi Sowa, 1971			x				x
Ecdyonurus (Ecdyonurus) submontanus Landa, 1969			x	x		x	x
Ecdyonurus (Ecdyonurus) torrentis Kimmins, 1942			x			x	
Ecdyonurus (Ecdyonurus) venosus (Fabricius, 1775)			x	x		x	x
Ecdyonurus (Helvetoraeticus) austriacus Kimmins, 1958						x	
Ecdyonurus (Helvetoraeticus) epeorides Demoulin, 1955		x		x		x	x
Ecdyonurus (Helvetoraeticus) helveticus Eaton, 1883				x		x	
Ecdyonurus (Helvetoraeticus) cf. *krueperi* (Stein, 1863)							x
Ecdyonurus (Helvetoraeticus) picteti (Meyer-Dür, 1864)				x			
Ecdyonurus (Helvetoraeticus) siveci Hefti, Tomka & Zurwerra, 1986			x				x
Ecdyonurus (Helvetoraeticus) subalpinus Klapalek, 1907				x		x	x
Ecdyonurus (Helvetoraeticus) vitoshensis Jacob & Braasch, 1984			x				x
Ecdyonurus (Helvetoraeticus) zelleri (Eaton, 1885)		x	x	x		x	
Electrogena affinis (Eaton, 1883)			x	x		x	
Electrogena lateralis (Curtis, 1834)		x	x			x	
Electrogena mazedonica (Ikonomov, 1954)			x	x		x	x
Electrogena quadrilineata (Landa, 1970)						x	
Electrogena ujhelyii (Sowa, 1981)			x				
Heptagenia (Dacnogenia) coerulans Rostock, 1878		x	x	x		x	

(continued)

Table 15.1 (continued)

Taxa/Country	A	B&H	C	M	MN	S/K	
Heptagenia (Heptagenia) flava Rostock, 1878		x	x			x	
Heptagenia (Heptagenia) longicauda (Stephens, 1835)	x		x	x		x	
Heptagenia (Heptagenia) sulphurea (Müller, 1776)	x	x	x			x	
Heptagenia (Kageronia) fuscogrisea (Retzius, 1783)			x	x		x	
Epeorus (Epeorus) assimilis Eaton, 1885			x	x	x	x	x
Epeorus (Ironopsis) yougoslavicus (Šamal, 1935)		x		x	x	x	x
Rhithrogena neretvana Tanasijević, 1985	x	x					
Rhithrogena gr. *diaphana* Navàs, 1917			x	x		x	x
Rhithrogena beskidensis Alba-Tercedor & Sowa, 1987						x	
Rhithrogena bulgarica Braasch, Soldán & Sowa, 1985	x						x
Rhithrogena savoiensis Alba-Tercedor & Sowa, 1987				x			
Rhithrogena zernyi Bauernfeind, 1991	x						
Rhithrogena germanica Eaton, 1885			x			x	x
Rhithrogena gr. *hybrida* Eaton, 1885			x	x			
Rhithrogena fiorii Grandi, 1953						x	
Rhithrogena hercynia Landa, 1969							x
Rhithrogena gr. *semicolorata* (Curtis, 1834)			x	x	x		x
Rhithrogena braaschi Jacob, 1974		x	x	x		x	x
Rhithrogena iridina (Kolenati, 1839)		x	x				
Rhithrogena semicolorata (Curtis, 1834)		x	x			x	
Rhithrogena gr. *sowai* Puthz, 1972							x
Rhithrogena marinkovici Tanasijević, 1985		x					
Leptophlebiidae							
Choroterpes (*Choroterpes*) *picteti* (Eaton, 1871)	x	x	x			x	
Choroterpes (*Euthraulus*) *balcanica* (Ikonomov, 1961)				x			
Habroleptoides confusa Sartori & Jacob, 1986		x	x	x		x	x
Habrophlebia eldae Jacob & Sartori, 1984	x						x
Habrophlebia fusca (Curtis, 1834)		x	x	x		x	

(continued)

Table 15.1 (continued)

Taxa/Country	A	B&H	C	M	MN	S/K	
Habrophlebia lauta Eaton, 1884		x	x	x		x	
Leptophlebia marginata (Linnaeus, 1767)		x	x				
Leptophlebia vespertina (Linnaeus, 1758)			x				
Paraleptophlebia cincta (Retzius, 1783)				x		x	
Paraleptophlebia lacustris Ikonomov, 1962**				x			
Paraleptophlebia ruffoi Biancheri, 1956		x		x			
Paraleptophlebia submarginata (Stephens, 1835)	x	x	x	x		x	x
Paraleptophlebia werneri Ulmer, 1920			x	x		x	
Ephemeridae							
Ephemera (Ephemera) danica Müller, 1764		x	x	x	x	x	
Ephemera (Ephemera) hellenica Demoulin, 1955		x		x		x	
Ephemera (Ephemera) lineata Eaton, 1870			x	x		x	
Ephemera (Ephemera) cf. *parnassiana* Demoulin, 1958							x
Ephemera (Ephemera) vulgata Linnaeus, 1758		x	x	x		x	x
Ephemera (Ephemera) zettana Kimmins, 1937		x	x				
Ephemera (Sinephemera) glaucops Pictet, 1843		x	x			x	
Polymitarcyidae							
Ephoron virgo (Olivier, 1791)			x	x		x	
Potamanthidae							
Potamanthus luteus (Linnaeus, 1767)		x	x	x		x	x
Ephemerellidae							
Ephemerella ignita (Poda, 1761)	x	x	x	x		x	x
Ephemerella maculocaudata (Ikonomov, 1961)	x			x			
Ephemerella mucronata (Bengtsson, 1909)		x	x	x	x	x	x
Ephemerella notata Eaton, 1887				x		x	
Serratella ikonomovi (Puthz, 1971)	x	x		x		x	x
Torleya major (Klapalek, 1905)	x	x	x	x		x	x
Eurylophella karelica Tiensuu, 1935			x				
Caenidae							
Brachycercus harrisellus Curtis, 1834	x		x			x	
Caenis beskidensis Sowa, 1973			x				
Caenis horaria (Linnaeus, 1758)		x	x	x		x	x

(continued)

Table 15.1 (continued)

Taxa/Country	A	B&H	C	M	MN	S/K	
Caenis lactea (Burmeister, 1839)			x			x	
Caenis luctuosa (Burmeister, 1839)		x	x			x	
Caenis macrura Stephens, 1835	x	x	x	x	x	x	x
Caenis pseudorivulorum Keffermuller, 1960	x			x		x	
Caenis pusilla Navàs, 1913	x		x			x	
Caenis rivulorum Eaton, 1884	x		x			x	
Caenis robusta Eaton, 1884		x	x			x	
Caenis strugaensis Ikonomov, 1961				x			x
Neoephemeridae							
Neoephemera maxima (Joly, 1870)	x			x			
Palingeniidae							
Palingenia longicauda (Olivier, 1791)			x	x		x	
Prosopistomatidae							
Prosopistoma pennigerum (Muller, 1785)	x			x			
Total number of taxa = 138	34	50	86	80	15	103	

*Sartori and Soldán [69] propose precedence of the name *Baetis pentaphlebodes* Ujhelyi, 1966 over *B. nexus*, yet the taxonomy used in compiling this species list follows Bauernfeind and Soldán [3], who used the *B. nexus* species name

**Details regarding *Paraleptophlebia lacustris* are discussed in Rimcheska [62]

15.1.3 Problems in Species Identification and Taxonomically Interesting Taxa

The most frequently used mayfly identification keys are appropriate for the identification of West European [e.g. 70] and Central European mayflies [71, 72], while they do not include mayfly taxa that are endemic to the Balkan Peninsula (e.g. Fig. 15.2) [32]. Therefore, the use of publications with species description is essential, and very often the only possible option when identifying mayfly samples from freshwater habitats of the Balkans.

Moreover, mayfly identifications in the region showed to be problematic for several more reasons [32, 59, 73]. For instance, many taxa are known only from original descriptions that do not have sufficient taxonomic finesse essential for adequate and correct identification. In some taxa only one of the life stages is known—larva or imago (e.g. *Rhithrogena neretvana* Tanasijević, 1985, *Rhithrogena jacobi* Braasch and Soldán, 1988, *Rhithrogena zernyi* Bauernfeind, 1991, where only adult stage is described, or *Baetis kozufensis* Ikonomov, 1962 and *Caenis strugaensis* Ikonomov, 1961 where only larval stage is described). Furthermore, for some species, such as *Baetis gemellus* Eaton, 1885, *B. kozufensis* Ikonomov, 1962, or *Centroptilum pirinense* Ikonomov 1962, unclear identities were recorded, therefore they still have the status of the *species inquirenda* requiring further revision [3].

Fig. 15.2 Examples of mayfly species endemic for the Balkan Peninsula (all in dorsal view) with their typical habitats: (**a**) *Rhithrogena braaschi* Jacob, 1974, and Bijela river in Croatia; (**b**) *Rhithrogena bulgarica* Braasch Soldán and Sowa 1985, and Kulidzhinska river in Bulgaria; (**c**) *Rhithrogena thracica* Sowa, Soldán and Braasch, 1988, and Elesnhitsa river in Bulgaria

Moreover, Vilenica et al. [74] showed that within the same species, even in the same population, a high morphological variability could be present, such as in *Rhithrogena braaschi* Jacob, 1974, which makes the identification even more challenging. On the other hand, some species could be morphologically remarkably similar, but genetically different, and represent so-called cryptic species [75]. Based on previous population-genetic research using the mitochondrial cytochrome C oxidase (COI) as a molecular marker, cryptic species have been registered within the *Baetis* genus. Different authors analysed the barcoding region of mitochondrial DNA in *Baetis alpinus* (Pictet, 1843) from different populations and came to the same conclusions of the existence of multiple cryptic taxa within this species [e.g. 76, 77]. Similar was observed in *B. alpinus* specimens from two different streams in Serbia [73]. Molecular analyses revealed that the sequences of the barcode region were identical between those specimens, but when compared with reference specimens [77] deposited in the gene bank (NCBI–National Center for Biotechnology Information) they differed over 20% [73].

15.2 Mayflies of Small Lotic Habitats in the Western Balkans

15.2.1 Mayflies of Springs

Springs are natural ecotones between groundwater and surface water [78]. They are characterized by stable environmental conditions and high heterogeneity of the aquatic, semi-aquatic and terrestrial microhabitats [79, 80]. Due to their isolation and generally small dimensions, springs are particularly sensitive to human pressures. In the Balkans, they are highly endangered by numerous anthropogenic activities, such as water abstraction, sedimentation, removal of the surrounding vegetation, nutrient deposition and the effects of climate change [81, 82]. Nevertheless, many studies emphasized their importance in maintaining the regional biodiversity of freshwater ecosystems [19, 66, 82, 83].

Some studies showed that mayfly species richness in springs is strongly linked to the persistence of aquatic habitats, rather than structure, i.e. mayfly assemblages are much diverse and abundant in perennial springs compared to the intermittent ones [84]. Nevertheless, it is possible that a higher abundance of some mayflies is present in intermittent springs, due to their ability to rapidly colonise water bodies after the flow continuation [85]. Such patterns are still to be investigated in the Western Balkan springs, as in this area only perennial ones were systematically studied [e.g. 19, 20, 65, 66, 74, 81].

Although mayflies are still relatively poorly investigated in springs of the Western Balkans (Fig. 15.3), several studies conducted in Croatia [19, 20, 59, 74, 81], Bosnia and Herzegovina [65, 66] and Serbia [8, 86] confirmed their low species richness, characteristic for the spring habitats [3, 10]. Karst springs in the region were so far recorded to harbour between one and six mayfly taxa, including the endemic *Rh. braaschi* [3]. Such low diversity could be a result of the specific environmental factors determined for karst springs, for instance, low nutrient resources, high alkalinity, and low and constant water temperature year-round [19, 20, 66, 81, 83, 87]. Therefore, such habitats are characterized by the presence of cold-water preferring species, such as *Baetis alpinus* (Pictet, 1843) and *B. melanonyx* (Pictet, 1843), or eurytherm

Fig. 15.3 Examples of the Western Balkans springs: (**a**) Cetina river spring in Croatia, (**b**) Mlava river spring, (**c**) Radovanska river spring in Serbia, (**d**) Matica river spring in Macedonia (photo by Nikola Paskali)

species, such as *B. rhodani* (Pictet, 1843) or *Habroleptoides confusa* Sartori and Jacob, 1986 [19, 20, 60, 65, 81, 88].

Savić et al. [65] showed that Western Balkan karst springs with higher discharge, higher microhabitat heterogeneity (especially the ones with present macrophytes), and without anthropogenic impact, tend to have more speciose mayfly assemblages. Nevertheless, Vilenica et al. [81, 66] recorded higher abundance of mayfly larvae in karst springs with certain level of anthropogenic impact, i.e., at springs with higher inflow of organic matter which could have resulted in higher periphyton and algal development thus providing more food resources for mayflies. Yet, mayfly diversity there was rather low (i.e. three or four species per spring).

Marković [8] investigated 234 springs in Serbia and recorded a total of 28 mayfly taxa. The most common were species within the genera *Baetis* (*B. vernus* Curtis, 1834, *B. fuscatus* (Linnaeus, 1761), *B. alpinus* and *B. rhodani*) and *Ecdyonurus* (*E. venosus* (Fabricius, 1775), *E. helveticus* (Eaton, 1885), *E. dispar* (Curtis, 1834), *E. torrentis* Kimmins, 1942)) together with *H. confusa* and *Rhithrogena semicolorata* (Curtis, 1834). These findings corroborate the results of Ikomonov [51] who studied springs of North Macedonia and found *H. confusa* and *B. rhodani* to be the most common species.

15.2.2 Mayflies of Streams and Rivers

Many studies showed that pristine lotic habitats of Balkan Peninsula (Fig. 15.4), both planar and submontane, represent highly valuable habitats for local mayfly species [e.g. 17–20, 49, 59, 65, 74, 81]. Especially important are highly heterogeneous habitats with a wide range of available microhabitats, that can provide food resources and shelter for various mayfly species [14]. Together with microhabitat characteristics (i.e., current velocity and substrate type) and the availability of different food resources, physicochemical properties of the water, such as water temperature and oxygen content, also play an important role in mayfly occurrence a particular stream or river [18, 19, 60, 81]. Vilenica et al. [81] recorded a peculiarity of karst rivers in the Western Balkans: even if the river were at some segment altered and anthropogenically impacted, it could still have had relatively healthy segments, where mayfly assemblages were well established and diverse. Studying mayflies of the Cetina river in Croatia, Vilenica et al. [74, 81] found no obvious influence of the river regulation on the mayfly assemblages at the sites that were not directly influenced by upstream or downstream alterations (e.g. damming, pollution from urban areas), i.e. only anthropogenically impacted segments of the river had impoverished mayfly assemblages. The reasons for that most probably lie in the interplay of general karst hydromorphology, with high influence of the numerous lateral springs along the course of the river, and specific environmental conditions, such as suitable substrate composition, that enable recovery of mayflies at sites away from the alterations. Small karst streams are often characterized by low water temperature, especially if located in mountainous regions. Moreover, the water can be oligotrophic, providing relatively poor food resources for mayflies. For such reasons, they can be characterized by relatively low diversity as well as low mayfly abundances [19, 20]. Nevertheless, such habitats are highly important for mayfly conservation, as they can also contain some rare and endemic species (e.g. *Rh. braaschi,* Fig. 15.2a) [19, 20, 59].

Study of Živić [89] included a high number of rhithral stream sections of lotic habitats in Serbia (i.e., 76), and showed the dominance of the representatives of the Baetidae and Heptagenidae. The *Baetis* genus was the most frequently recorded, with *Baetis alpinus* and *B. rhodani* as the most frequent species. Another common species was *Ephemera danica* Müller, 1764, generally collected from microhabitats with psammal and akal, rarely pelal substrates. The most frequently recorded heptageniid was *Epeorus assimilis* Eaton, 1885, sometimes present in high abundances. Small lotic habitats in Serbia also provide home for some rare mayflies such as *Epeorus yougoslavicus* (Šamal, 1935) (Fig. 15.5). Although *E. yougoslavicus* shares the same (micro) habitat preferences as *E. assimilis*, i.e., lithal in fast current areas [88], Živić [89] found it in only one stream (Vrla stream, southeastern Serbia, the South Morava basin). Petrović et al. [67] recorded the species at two more rivers in Serbia (Gobeljska

(a) (b) (c) (d) (e) (f)

Fig. 15.4 Examples of small pristine lotic habitats of the Western Balkans: (**a**) Jankovac stream, (**b**) Krčić river in Croatia, (**c**) Beleshnicka river, (**d**) Obednichica river in Macedonia, (**e**) Visočica stream, (**f**) Vrla stream in Serbia

(a) (b)

Fig. 15.5 An example of threatened and rare species recorded in the streams of the Balkan Peninsula: (**a**) *Epeorus yougoslavicus* (Šamal, 1935) and (**b**) the species' typical habitat (stream in Bulgaria)

river and Masurička Reka river, the basin of South and West Morava rivers), also with exceptionally low population densities. Based on the available data, *E. yougoslavicus* inhabits rhithral section (upper reaches) of xenosaprobic mountainous and foothill streams and rivers, where it can be found in microhabitats characterized by boulders and cobbles and fast current velocity [16, 67]. This stenovalent species (especially regarding temperature, oxygen and water velocity parameters) can be considered as endangered in Serbian lotic habitats due to the small number of populations and relatively mutually distant and isolated biotopes.

15.3 Mayflies as Bioindicators—Influences of Various Anthropogenic Activities on Mayfly Communities

Increased urbanization and industrialization, together with significant growth of human population have led to various problems in natural habitats. The increased demand for drinking water, and for land use for purposes of agriculture, forestry, irrigation activities and wetland drainage have resulted in dramatic habitats loss, habitat morphology and hydrological regime alterations, degraded water quality, pollution and increased sediment erosion into lotic systems [90, 91]. Such anthropogenic pressures represent severe threats to aquatic systems in the Balkan Peninsula (Fig. 15.6), their biodiversity and make lotic habitats among the most endangered ones [92, 93].

Mayflies are valuable descriptors of their environment, due to their wide distribution, importance in food webs, specific requirements for habitats and microhabitats of natural state and for high water quality. They are one of the aquatic insects that are most sensitive to habitat alterations [e.g. 61, 81, 94, 95], and most often, their absence indicates certain disturbance in a particular habitat [96]. Due to having such particular demands for environmental conditions, many species rapidly disappear

Fig. 15.6 Examples of anthropogenic threats to small lotic habitats of the Western Balkans: small hydropower plans at (**a**) Gradska stream in Serbia, and (**b**) stream in Papuk Nature Park in Croatia; trout farm outlet at (**c**) Studenica stream in Serbia; destruction of the stream banks and waste disposal at (**d**) Vetovka stream in Croatia

when challenged even with small-scale disturbance in their habitat [61, 97]. Therefore, they are widely used as an important indicator of freshwater habitats' health in biomonitoring programs worldwide [2, 98]. The larvae are much more informative for biomonitoring purposes since they have a longer lifespan and are good indicators of water quality, especially given the extent of tolerance to organic pollution among species within a particular genus [99]. Moreover, larvae are more reliably identifiable, as they have more distinguishing morphological characteristics compared to the adults [72].

In the Balkans, lotic habitats are seriously threatened by array of human pressures. Amongst the outstanding ones is the wave of several hundred planned hydropower stations that are known to have a significant negative impact on the river ecosystem and the longitudinal continuum for living organisms and sediments [100]. Vilenica et al. [61, 74, 81] showed that river regulations, such as damming and canalization, lead to substantial habitat loss for mayflies at altered river segments. Altering the hydro-morphological and physico-chemical conditions dramatically reduce mayfly diversity and change composition and structure of their assemblages, where only a few species can cope with such altered environment. Mayfly assemblages of such habitats often show signs of "potamization" and are inhabited with high abundances of relatively low number of widespread generalists and species characteristic for lower reaches and lentic habitats [61, 81, 95]. High abundances of the two rare mayfly species in the Balkan region, *E. yougoslavicus* (Fig. 15.5) and *Baetis pavidus* Grandi, 1949, were recorded in mountainous rivers in the Sutjeska National Park (Southeastern Bosnia and Herzegovina). The construction of several hydropower plants on these rivers (Hrčavka, Jabušnica and Sutjeska rivers), might have endangered the existence of those fragile species [101], which remains to be re-evaluated by the future research.

Vilenica et al. [95] confirmed high mayfly sensitivity to water pollution in lotic habitats, where at severely polluted sites, only the euryvalent *Cloeon dipterum* (Linnaeus, 1761) could thrive, or such rivers were completely unsuitable for any mayfly species. Marković et al. [102] recorded *Caenis horaria* (Linnaeus, 1758) in heavy polluted Turija stream, near Belgrade in Serbia, where extremely high concentrations of As, Cu, Ni and Zn in sediment samples were measured, which could indicate the species' resistance to such pollution. This species was also common in polluted man-made lakes in Croatia [61]. Studies of Vilenica et al. [74, 81] showed that when oligotrophic karst habitats are exposed to increased inflow of organic matter into the habitat, algal growth is enabled, which provides more food for inhabiting mayfly species, resulting in more abundant populations. Nevertheless, this is not sustainable in long-term as increased trophy leads to decrease in oxygen content [103], which will surely have a negative impact on mayflies in those habitats.

Several authors reported negative impacts of trout farming on freshwater ecosystems [73, 104]. The stream exposure to trout farms pollution, depending on trout farms capacities, leads to their eutrophication. The effects of such pollution are most obvious during the dry months (i.e., in summer in a temperate climate) when the water level and current velocity are generally lower [73, 104]. Sensitive aquatic organisms react to such perturbation by the reduction in their abundance or by their replacement with less sensitive taxa [105]. In the study where the effects of nine trout farms were monitored, a decrease in mayfly relative abundance was recorded at localities below the trout farm outlets, compared to control localities [73]. However, on several occasions within the same study, the relative abundance of Ephemeroptera was higher at those localities, due to the dominance of euryvalent generalist *B. rhodani*, adapted to various environmental conditions [106]. So, quantification of the data could be expressed through index based on the ratio of the Baetidae abundance to total Ephemeroptera abundance [Bae/Eph, 104]. Bae/Eph index showed to be a good metric to obtain information about the fine impact of organic pollution (delivering from trout farms) on the recipient streams [73, 104]. Also, an index that provides information on the ratio of Baetidae abundance to the sum of Heptagenidae and Baetidae (Bae/Bae + Hep, designed in the study by Stojanović [73]) also showed the similar trend in the two studied streams (Crnica and Mlava streams, Eastern Serbia). Namely, the values of the index were lower in the control localities and localities at a certain distance from the trout farms, indicating the dominance of the more sensitive Heptagenidae family, while higher values at the impacted localities showed a dominance Baetidae (especially *B. rhodani* larvae). Both indices are estimates of evenness, where effects of organic pollution increase the relative abundance of more tolerant taxa [107].

The mentioned human activities and impacts on watercourses are among the main causes for disturbed balance in aquatic ecosystems in a long-term scale, even causing local or regional extinction of sensitive mayfly species [108].

15.4 Current Gaps and Recommendations for Future Research

As previously mentioned, despite the more research focus has been dedicated to mayflies in many countries of the Western Balkans during the last several decades, our knowledge is still far from complete. Many countries are still reporting new species findings for their fauna [e.g. 18, 61, 109], while there is also a possibility of discovering new species for science, especially in families Baetidae and Heptageniidae [59, 64]. Nevertheless, there is still vast number of regions that are poorly investigated, and future studies should consider filling those gaps.

Studying the Balkans as a center of taxonomic diversity of the aquatic fauna in Europe by different approaches, i.e., by combining the methods of integrative taxonomy should improve the understanding of intrageneric and intraspecific genetic variability within the remarkably diverse genera *Ecdyonurus*, *Electrogena, Rhithrogena* and *Baetis*. Among the particularly important topics is cryptic mayfly species identification, which includes the method of DNA barcoding [110], already proven as effective in mayfly identification [111]. Besides the previously mentioned cryptic taxa within the *B. alpinus*, many taxonomists considered *Baetis rhodani* to be a complex of several forms of unclear taxonomic status. Williams et al. [112] analysed the barcoding region (mtDNA-COI) and identified two monophyletic groups (consisting of seven haplogroups), with significant evolutionary distances indicating the existence of cryptic species within the *B. rhodani* complex. Lucentini et al. [113] identified 25 different haplotypes in *B. rhodani* populations in Italy alone, including as many as 11 potential cryptic species. Molecular tools showed to be a particularly useful method in species delimitation also within the *Baetis vernus* species group [114] since morphological characteristics of larvae are variable and as such are assumed to be constantly exposed to selective environmental pressures, which overly complicates their identification [72]. Also, cryptic diversity was recorded among *Rhithrogena* species [115]. Having in mind that these aquatic insects are frequent inhabitants of the Western Balkan streams, the more attention should be dedicated to the effort of revealing cryptic taxa to increase overall mayfly diversity. Mayfly barcoding studies will facilitate the compiling of a reference DNA barcode library for the Balkans, as well as to compose an identification key which would successfully integrate both native species and the other European taxa.

We strongly recommend a more research effort to be given to the spring habitats, as they showed to be potentially important for some rare mayflies [19, 20, 66]. Another interesting point of mayfly ecology, that is highly understudied in this region, are the mayfly emergence periods. Vilenica et al. [20] and Vilenica and Ivković [22] gave a good basis for such studies. Vilenica and Ivković [22] even conducted a first long-term study on mayfly emergence from small lotic habitats of the Balkan Peninsula, that gave an important contribution to our knowledge of biology and ecology of several European (including Balkan) mayflies, for which our knowledge is still rather scarce. Such knowledge might contribute to the understanding of the durable and long-term changes in the aquatic environment.

15.5 Conclusions

Mayfly research in the Western Balkan region intensified during the past two decades, providing new and interesting data on mayfly distribution and ecology. Nevertheless, there are still relatively large gaps in our knowledge on Balkan mayflies. As one of the biodiversity hotspots, freshwater habitats of the Balkan peninsula harbour many interesting and rare mayfly taxa. Many of them are still poorly studied, in taxonomic and ecological sense, which represents an interesting subject of future scientific work. Even though the small lotic habitats of the Western Balkans provide habitat for numerous endemic species, those habitats are severely endangered by current and planned anthropogenic threats, which jeopardizes the survival of inhabiting mayflies. Future management and conservation practices should focus on restoration of altered habitats and preservation of pristine ones, in order to protect many of the species important for sustainable functioning of lotic systems in the Western Balkan region.

References

1. Barber-James H, Gattolliat JL, Sartori M, Hubbard MD (2008) Global diversity of mayflies (Ephemeroptera, Insecta) in freshwater. Hydrobiologia 595:339–350
2. Brittain JE, Sartori M (2003) Ephemeroptera (Mayflies). In: Resh VH, Cardé RT (eds) Encylopedia of insects. Academic Press, Amsterdam, pp 373–380
3. Bauernfeind E, Soldán T (2012) The mayflies of Europe (Ephemeroptera). Apollo Books, Ollerup, Denmark
4. Simić V (1993) Saprobiological valorisation of the Svljiški and Trgoviški Timok based on the structure of macrozoobenthos. MSc thesis. Faculty of Biology, University of Belgrade [in Serbian]
5. Simić V (1995) Possibilities of ecological monitoring of river systems in Serbia based on macrozoobenthos coomunities. Ph.D. thesis. Faculty of Biology, University of Belgrade [in Serbian]
6. Simić V, Simić S (1999) Use of the river macrozoobenthos of Serbia to formulate a biotic index. Hydrobiologia 416:51–64
7. Marković Z (1995) Reka Đetinja. Makrozoobentos u oceni kvaliteta vode. Ministarstvo za zaštitu životne sredine Republike Srbije, Beograd, 117 [in Serbian]
8. Marković Z (1998) Springs in mountainous regions of Serbia: ecological study of the macrozoobenthos. University of Belgrade, Faculty of Biology, Belgrade [in Serbian]
9. Marković Z, Mitrović-Tutundžić V (1997) Fauna Ephemeroptera izvorišta Dubašnice. Naša Ekološka istina, V Naučno stručni skup o prirodnim vrednostima i zaštiti životne sredine. Zbornik radova, pp 294–298 [in Serbian]
10. Bauernfeind E, Moog O (2000) Mayflies (Insecta: Ephemeroptera) and the assessment of ecological integrity: a methodological approach. Hydrobiologia 422:71–83
11. Živić I (2005) Faunistical and ecological study of macrozoobenthos in the rivers of the Southern Morava River basin with an emphasis on taxonomy of Trichoptera larvae (Insecta). Ph.D. thesis. Faculty of Biology, University of Belgrade [in Serbian]
12. Paunović M, Jakovčev-Todorović D, Simić V, Stojanović B, Veljković A (2006) Species composition, spatial distribution and temporal occurrence of mayflies (Ephemeroptera) in the Vlasina river (southeast Serbia). Arch Biol Sci 58:37–43
13. Petrović A (2014) Possibilities of using a database in the strategy of conserving the biodiversity of inland water macroinvertebrates at a national level. Ph.D. thesis. Faculty of Science, University of Kragujevac [in Serbian]

14. Vilenica M, Brigić A, Sartori M, Mihaljević Z (2018) Microhabitat selection and distribution of functional feeding groups of mayfly larvae (Ephemeroptera) in lotic karst habitats. Knowl Manag Aquat Ec 419(17):1–12
15. Russev B, Vidinova Y (1994) Verbreitung und Ökologie der Vertreter einiger Familien der Ordnung Ephemeroptera (Insecta) in Bulgarien. Lauterbornia 19:107–113
16. Vidinova Y, Russev B (1997) Distribution and ecology of the representatives of some Ephemeropteran families in Bulgaria. In: Landolt P, Sartori M (eds) Ephemeroptera & Plecoptera: Biology-Ecology-Systematics.MTL, Fribourg, Switzerland, pp 139–146
17. Petrović A, Milošević Dj, Paunović M, Simić S, Djordević N, Stojković M, Simić VM (2015) New data on the distribution and ecology of the mayfly larvae (Insecta: Ephemeroptera) of Serbia (central part of the Balkan Peninsula). Turk J Zool 38:195–209
18. Vilenica M, Brigić A, Kerovec M, Gottstein S, Ternjej I (2016) Spatial distribution and seasonal changes of mayflies (Insecta, Ephemeroptera) in a Western Balkan peat bog. ZooKeys 637:135–149
19. Vilenica M, Mičetić Stanković V, Sartori M, Kučinić M, Mihaljević Z (2017) Environmental factors affecting mayfly assemblages in tufa-depositing habitats of the Dinaric Karst. Knowl Manag Aquat Ec 418(14):1–12
20. Vilenica M, Ivković M, Sartori M, Mihaljević Z (2017) Mayfly emergence along an oligotrophic Dinaric karst hydrosystem: spatial and temporal patterns, and species–environment relationship. Aquat Ecol 51:417–433
21. Brittain JE (1982) Biology of mayflies. Annu Rev Entomol 27:119–147
22. Vilenica M, Ivković M (2020) A decade-long study on mayfly emergence patterns. Mar Freshwater Res 72(4):507–519
23. Williams DD (1980) Applied aspects of mayfly biology. In: Flannagan JF, Marshall KE (eds) Advances in Ephemeroptera Biology. In: Proceedings of 3rd International Conference on Ephemeroptera, Winnipeg, Manitoba, 1979. Plenum Press, New York and London, pp 1–17
24. Merten EC, Snobl ZR, Wellnitz TA (2014) Microhabitat influences on stream insect emergence. Aquat Sci 76:165–172. https://doi.org/10.1007/s00027-013-0326-3
25. Kenyeres Z, Rácz IA, Varga Z (2009) Endemism hot spots, core areas and disjunctions in European Orthoptera. Acta Biol Cracov Zool 52:189–211
26. Previšić A, Walton C, Kučinić M, Mitrikeski PT, Kerovec M (2009) Pleistocene divergence of Dinaric Drusus endemics (Trichoptera, Limnephilidae) in multiple microrefugia within the Balkan Peninsula. Mol Ecol 18:634–647
27. Ivković M, Plant A (2015) Aquatic insects in the Dinarides: identifying hotspots of endemism and species richness shaped by geological and hydrological history using Empididae (Diptera). Insect Conserv Divers 8:302–312
28. Puthz V (1974) Einige Ephemeropteren aus Jugoslawien, vorwiegend aus Montenegro und Serbien. Acta Rer Natur Mus Nat Slov 19(2):147–156
29. Puthz V (1980) Ergebnisse der Albanien-Expedition 1961 des Deutschen Entomologischen Institutes. 94. Beitrag: Ephemeroptera. Beitr Entomol Berlin 30(2):343–355
30. Studemann D, Landolt P, Tomka I (1989) Contribution to the study of European Ephemerellidae (Ephemeroptera), II. Description of the winged stages of *Ephemerella ikonomovi* Puthz, 1971, and *Serratella albai* Gonzalez del Tanago & Garcia de Jalon, 1983. Mitt Schweiz Entomol Ges 62:119–127
31. Kovács T, Murányi D (2013) New country data of some mayflies (Ephemeroptera) from Europe. Fol Hist Nat Mus Matr 37:15–19
32. Bauernfeind E (2018) Mayflies (Ephemeroptera) of the River Vjosa. Albania. Acta Zoo Bot Austria 155(1):155–162
33. Jacob U (1974) *Rhithrogena braaschi* n. sp., eine neue Heptageniidae aus Bulgarien (Insecta, Ephemeroptera). Ent Nachr 18(11–12):167–173
34. Müller-Liebenau I, Soldán T (1981) *Baetis balcanicus* sp. n., a new species of the genus *Baetis* from Bulgaria and Greece (Insecta, Ephemeroptera, Baetidae: atrebatinus group). Spixiana, 4(3):291–295

35. Jacob U, Braasch D (1984) Neue und statusrevidierte Taxa der *Ecdyonurus helveticus*-Großgruppe (Ephemeroptera, Heptageniidae). Ent Abh Staatl Mus Tierk Dresden 48(6):53–61
36. Braasch D, Soldán T (1985) Ein neuer *Ecdyonurus* der *ruffii*-Gruppe aus Bulgarien (Heptageniidae, Ephemeroptera). Ent Nachr Ber 29(2):67–68
37. Braasch D, Soldán T, Sowa R (1985) *Rhithrogena bulgarica* n. sp. und zwei für die Fauna Bulgariens neue Eintagsfliegenarten (Ephemeroptera, Heptageniidae). Faun Abh 12(12):125–127
38. Braasch D, Soldán T (1988) *Rhithrogena jacobi* n. sp. aus Bulgarien (Insecta, Ephemeroptera, Heptageniidae). Reichenbachia 25(24):117–118
39. Jacob U (1988) *Thraulus thraker* sp. n. aus Bulgarien (Insecta, Ephemeroptera: Leptophlebiidae). Ent Abh Staatl Mus Tierk Dresden 26(1):1–3
40. Jacob U (1986) *Siphlonurus abraxas* n. sp.–ein neues Taxon der *lacustris*- Gruppe aus Montenegro (Jugoslawien). Reichenbachia 23(25):139–143
41. Jacob U, Braasch D (1986) *Ecdyonurus puma* sp. n. aus Montenegro (Jugoslawien) (Ephemeroptera, Heptageniidae). Reichenbachia 23(31):177–180
42. Sowa R, Soldán T, Braasch D (1988) *Rhithrogena thracica* sp. n.–a new species of mayfly from Bulgaria with a description of subimago and larva of *Rh. bulgarica* Braasch et al. (1985) (Ephemeroptera, Heptageniidae). Acta Zool Bulg (36):31–38
43. Bauernfeind E (2001) *Rhithrogena lisettae* sp. n.–A new representative of the *R. diaphana* species–group from Greece (Insecta: Ephemeroptera. In: Gaino E (ed) Research update on Ephemeroptera & Plecoptera 2003. University of Perugia, Perugia, Italy, pp 109–112
44. Russev B (1971) New representatives of Ephemeroptera and Plecoptera (Insecta) for the fauna of Bulgaria. Bull Zool Inst Mus 33:111–114 [in Bulgarian]
45. Russev B (1992) Threatened species of Ephemeroptera (Insecta) from Bulgaria. Gefährdete Ephemeroptera-Arten (Insecta) aus Bulgarien. Lauterbornia 9:13–17
46. Russev B (1993) Review of literature and established mayfly species (Ephemeroptera, Insecta) from Bulgaria. Lauterbornia 14:71–77
47. Russev B, Vidinova Y (1994) New representatives of the order Ephemeroptera (Insecta) for the fauna of Bulgaria. Lauterbornia 15:85–87
48. Vidinova Y, Janeva I, Tyufekchieva V (2000) Ephemeroptera and Plecoptera from glacial waters in Rila Mountain. In: Golemanski V, Naidenow W (eds) Biodiversity and evolution of glacial water ecosystems in the Rila Mountains. Drinov M Academic Publishing House, Prof, pp 51–55
49. Vidinova Y (2003) Contribution to the study of mayfly fauna (Ephemeroptera) in Bulgaria. In: Gaino E (ed) Research update on Ephemeroptera and Plecoptera. University of Perugia, Perugia, Italy, pp 159–163
50. Vidinova Y, Yaneva I, Tyufekchieva V (2006) Ephemeroptera and Plecoptera (Insecta) from the Bulgarian part of the Struma River. Acta zool bulg 58(1):125–130
51. Ikomonov P (1959) Ephemeroptera na Makrdonija. Sistematika i faunistika. Doktorska disertacija. Univerzitet Skopje (in Macedonian)
52. Ikonomov P (1960) Die Verbreitung der Ephemeroptera in Mazedonien. Acta Mus Macedonici Sci Nat 7(3):41–74
53. Ikonomov P (1970) Repartition des larves Éphéméroptères et Plécoptères dans les coursd'eau de Macédoine. Annuairede la Faculté des Sciences d'Université de Skopje 22:5–11
54. Tanasijević M (1970) Fauna Ephemeroptera na području planina Maglić, Volujak i Zelengora. Glasnik zemaljskog muzeja N. S. 14 Prirodne nauke, Sarajevo, 180–184
55. Tanasijević M (1974) Nalaz vrsta roda *Siphlonurus* Eaton 1868 (Ephemeroptera) u nekim područjima Jugoslavije. Glasnik zemaljskog muzeja N. S. 13 Prirodne nauke, Sarajevo, 287–292
56. Tanasijević M (1975) *Heptagenia ozrensis* sp. n. und der Fund von fünf Ephemeroptera-Arten in Bosnien und der Hercegowina. Wiss Mitt Bosn-herzeg Landesmus 4–5:243–246
57. Filipović D (1975) Fauna Ephemeroptera SR Srbije. Zbornik radova o entomofauni SR Srbije 1:211–219 [in Serbian]

58. Matoničkin I (1959) Faunistička istraživanja reikotopnih biotopa na Plitvičkim jezerima. Ljetopis 63:355–360 [in Croatian]
59. Vilenica M, Gattolliat J-L, Mihaljević Z, Sartori M (2015) Croatian mayflies (Insecta, Ephemeroptera): species diversity and distribution patterns. ZooKeys 523:99–127
60. Vilenica M, Ergović V, Mihaljević Z (2018) Mayfly (Ephemeroptera) assemblages of a Pannonian lowland mountain, with first records of the parasite *Symbiocladius rhithrogenae* (Zavrel, 1924) (Diptera: Chironomidae). Ann Limnol - Int J Lim 54(31):1–10
61. Vilenica M, Vučković N, Mihaljević Z (2019) Littoral mayfly assemblages in South-East European man-made lakes. J limnol 78(1):47–59. https://doi.org/10.4081/jlimnol.2019.1853
62. Rimcheska B (2020) A check list of the mayfly fauna (Insecta; Ephemeroptera) of the Republic of North Macedonia with a short literature review. Acta Mus Macedonici Sci Nat 23:51–60
63. Marković Z, Živić I (2002) Fauna of Ephemeroptera in the running waters of West Serbia. Arch Biol Sci 54(3–4):117–124
64. Xërxa BL, Sartori M, Gashi A, Gattolliat JL (2019) First checklist of mayflies (Insecta, Ephemeroptera) from Kosovo. Zookeys 874:69–82
65. Savić A, Dmitrović D, Pešić V (2017) Ephemeroptera, Plecoptera and Trichoptera assemblages of karst springs in relation to environmental factors: a case study in central Bosnia and Hercegovina. Turk Zool Derg 41:119–129. https://doi.org/10.3906/zoo-1512-31
66. Vilenica M, Bilić M, Mičetić Stanković V, Kučinić M (2018) Mayfly ecological traits in a European karst spring: species, microhabitats and life histories. Community Ecol 19(3):248–258
67. Petrović A, Simić V, Paunović M, Stojanović B (2006) A new records of *Epeorus yougoslavicus* (Šamal, 1935) [Ephemeroptera] in Serbia and Montenegro. Biotechnol Biotec Eq 20:67–71
68. Pešić V, Gadawski P, Gligorović B, Glöer P, Grabowski M, Kovács T, Murányi D, Płóciennik M, Šundić D (2018) The Diversity of the Zoobenthos Communities of the Lake Skadar/Shkodra Basin. In: Pešić V, Karaman G, Kostianoy A (eds) The Skadar/Shkodra Lake environment. The Handbook of environmental chemistry, vol 80. Springer, Cham, pp 255–293
69. Sartori M, Soldán T (2006) *Baetis pentaphlebodes* Ujhelyi, 1966 (Insecta: Ephemeroptera: Baetidae): proposed precedence of the specific name over *Baetis nexus* Navás, 1918. Bull zool nomencl 63:23–27
70. Elliott JM, Humpesch UH (2010) Mayfly Larvae (Ephemeroptera) of Britain and Ireland: keys and a review of their ecology. Scientific Publications, Ambleside, Cumbria, UK, Freshwater Biological Association
71. Studemann D, Landolt P, Sartori M, Hefti D, Tomka I (1992) Ephemeroptera. Insecta Helvetica, vol 9, Mauron, Tinguely & Lachat SA, Moncor, Fribourg, Switzerland
72. Bauernfeind E, Humpesch UH (2001) Die Eintagsfliegen Zentraleuropas (Insecta: Ephemeroptera): Bestimmung und Ökologie. Naturhistorisches Museum Wien, Wien, Austria
73. Stojanović K (2017) Influence of trout farms on macrozoobenthos communities with special emphasis on larvae of the genus *Baetis* (Ephemeroptera, Insecta) [dissertation]. University of Belgrade, Faculty of Biology, Belgrade
74. Vilenica M, Previšić A, Kučinić M, Gattolliat JL, Sartori M, Mihaljević Z (2016) Distribution and autecology of mayflies (Insecta, Ephemeroptera) in a Mediterranean river in the Western Balkans. Entomol News 126(1):19–35
75. Bickford D, Lohman DJ, Sodhi NS, Ng PKL, Meier R, Winker K, Ingram KK, Das I (2007) Cryptic species as a window on diversity and conservation. Trends Ecol Evol 22(3):148–155
76. Finn DS, Zamora-Muñoz C, Múrria C, Sáinz-Bariáin M, Alba-Tercedor J (2014) Evidence from recently deglaciated mountain ranges that Baetis alpinus (Ephemeroptera) could lose significant genetic diversity as alpine glaciers disappear. Freshw Sci 33(1):207–216
77. Leys M, Keller I, Räsänen K, Gattolliat JL, Robinson CT (2016) Distribution and population genetic variation of cryptic species of the Alpine mayfly *Baetis alpinus* (Ephemeroptera: Baetidae) in the Central Alps. BMC Evol Biol 16:77

78. Krešić N (2010) Types and classifications of springs. In: Krešić N, Stevanović Z (eds) Groundwater hydrology of springs. Engineering, theory, management and sustainability. Elsevier Inc. BH, Amsterdam, pp 31–85
79. Barquín J, Scarsbrook M (2008) Management and conservation strategies for cold-water springs. Aquat Conserv 18:580–591
80. Kløve B, Ala-aho P, Bertrand G, Boukalova Z, Ertürk A, Goldscheider N, Ilmonen J, Karakaya N, Kupfersberger H, Kvœrner J, Lundberg A, Mileusnić M, Moszczynska A, Muotka T, Preda E, Rossi P, Siergieiev D, Šimek J, Wachniew P, Angheluta V, Widerlund A (2011) Groundwater dependent ecosystems. Part I: Hydroecological status and trends. Environ Sci Policy 14:770–781
81. Vilenica M, Previšić A, Ivković M, Popijač A, Vučković I, Kučinić M, Kerovec M, Gattolliat J-L, Sartori M, Mihaljević Z (2016) Mayfly (Insecta: Ephemeroptera) assemblages of a regulated perennial Mediterranean river system in the Western Balkans. Biologia 71(9):1038–1048
82. Pešić V, Gligorović B, Savić A, Buczyński P (2017) Ecological patterns of Odonata assemblages in karst springs in central Montenegro. Knowl Manag Aquat Ecosyst 418(3):1–20
83. Pozojević I, Pešić V, Goldschmidt T, Gottstein S (2020) Crenal habitats: sources of water mite (Acari: Hydrachnidia) diversity. Diversity 12(9):316
84. Smith H, Wood PJ, Gunn J (2003) The influence of habitat structure and flow permanence on invertebrate communities in karst spring systems. Hydrobiologia 510:53–66
85. Glazier DS, Gooch JL (1987) Macroinvertebrate assemblages in Pennsylvania (U.S.A.) springs. Hydrobiologia 150:33–43
86. Filipović D (1968) Limnološka karakteristika izvorskog regiona Lisinskog potoka na Kopaoniku. Zbornik radova 11(24):325–337 [in Serbian]
87. Prelovšek M (2010) Hydrology. In: Mihevc A, Prelovšek M, Zupan Hajna N (eds) Introduction to the Dinaric karst. Karst Research Institute, ZRC SAZU, Postojna, pp 14–19
88. Buffagni A, Armanini DG, Cazzola M, Alba-Tercedor J, López-Rodríguez MJ, Murphy J, Sandin L, Schmidt-Kloiber A (2020) Dataset "Ephemeroptera". www.freshwaterecology.info - the taxa and autecology database for freshwater organisms, version 7.0 (accessed on 25.01.2021)
89. Živić I (2020) Istraživački monitoring bioloških, fizičko-hemijskih i hidromorfoloških parametara u cilju definisanja referentnih lokaliteta na području Republike Srbije. Ministarstvo poljoprivrede, šumarstva i vodoprivrede republike Srbije - Republička direkcija za vode, Beograd [in Serbian]
90. Waters TF (1995) Sediment in streams: sources, biological effects and control. American Fisheries Society symposium, Monograph, p 7
91. Woodward G, Gessner MO, Giller PS, Gulis V, Hladyz S, Lecerf A, Malmqvist B, McKie BG, Tiegs SD, Cariss H, Dobson M, Elosegi A, Ferreira V, Graca MAS, Fleituch T, Lacoursiere JO, Nistorescu M, Pozo J, Risnoveanu G, Schindler M, Vadineanu A, Vought LBM, Chauvet E (2012) Continental-scale effects of nutrient pollution on stream ecosystem functioning. Science 336:1438–1440
92. Hering D, Johnson RK, Kramm S, Schmutz S, Szoszkiewicz K, Verhonschot PFM (2006) Assessment of European streams with diatoms, macrophytes, macroinvertebrates and fish: a comparative metric-based analysis of organism response to stress. Freshw Biol 51:1757–1785
93. Carpenter SJ, Stanley EH, Vander Zanden MJ (2011) State of the world's freshwater ecosystems: physical, chemical, and biological changes. Annu Rev Env Resour 36:75–99
94. Zedková B, Rádková V, Bojková J, Soldán T, Zahrádková S (2014) Mayflies (Ephemeroptera) as indicators of environmental changes in the past five decades: a case study from the Morava and Odra River Basins (Czech Republic). Aquat Conserv 25(5):622–638
95. Vilenica M, Kerovec M, Pozojević I, Mihaljević Z (2020) Mayfly response to different stress types in small and mid-sized lowland rivers. ZooKeys 980:57–77
96. Jacobus LM, Macadam CR, Sartori M (2019) Mayflies (Ephemeroptera) and their contributions to ecosystem services. Insects 10:1–26
97. Firmiano KR, Ligeiro R, Macedo DR, Juen L, Hughes RM, Callisto M (2017) Mayfly bioindicator thresholds for several anthropogenic disturbances in neotropical savanna streams. Ecol Indic 74:276–284

98. Beketov MA (2004) Different sensitivity of mayflies (Insecta, Ephemeroptera) to ammonia, nitrite and nitrate: linkage between experimental and observational data. Hydrobiologia 528:209–216
99. Moog O (2002) Fauna Aquatica Austriaca. 2nd edn. Wassewirtschaftskataster, Bundesministerium für Land-und Forstwirtschaft, Umwelt und Wasserwirtschaft. Wien, Austria: Austrian Federal Ministry of Agriculture Forestry Environment & Water Management, Wasserwirtschaftskataster Vienna
100. Huđek H, Žganec K, Pusch MT (2020) A review of hydropower dams in Southeast Europe–distribution, trends and availability of monitoring data using the example of a multinational Danube catchment subarea. Renew Sustain Energ Rev 117:109434
101. Đuknić J, Tubić B, Marković V, Čanak-Atlagić J, Anđus S, Kračun Kolarević M, Paunović M (2016) Aquatic macroinvertebrate assemblages of mountainous rivers in the Sutjeska National Park (Southeastern Bosnia and Herzegovina). In: Proceedings of Paper presented at the 2nd Central European Symposium for Aquatic Macroinvertebrate Research. University of Pécs, Hungary, July, pp 3–8
102. Marković V, Petković A, Tubić B, Ilić M, Kračun-Kolarević M, Marinković N, Novaković B, Gačić Z, Paunović M (2014) Impact of heavy metals in sediments on the macroinvertebrates communities of some smaller streams in Belgrade Region (Serbia). In: Proceedings of Paper presented at the 1st Central European Symposium for Aquatic Macroinvertebrate Research Szarvas. Hungary, April, pp 10–13
103. Tammeorg O, Möls T, Niemistö J, Holmroos H, Horppila J (2017) The actual role of oxygen deficit in the linkage of the water quality and benthic phosphorus release: potential implications for lake restoration. Sci Total Environ 599–600:732–738
104. Živić I, Marković Z, Filipović-Rojka Z, Živić M (2009) Influence of a trout farm on water quality and macrozoobenthos communities of the receiving stream (Trešnjica River, Serbia). Int Rev Hydrobiol 94:673–687
105. Stojanović K, Živić M, Dulić Z, Marković Z, Krizmanić J, Milošević Dj, Miljanović B, Jovanović J, Vidaković D, Živić I (2017) Comparative study of the effects of a small-scale trout farm on the macrozoobenthos, potamoplankton, and epilithic diatom communities. Environ Monit Assess 189:403
106. Sartori M, Landolt P (1999) Fauna Helvetica 3: Atlas de distribution des Ephémeres de Suisse (Insecta, Ephemeroptera). Centre Suisse de Cartographie de la Faune and Schweizerische Entomologische Gesellschaft, Neuchâtel, Switzerland
107. Barbour MT, Gerritsen J, Snyder B, Stribling J (1999) Rapid bioassessment protocols for use in streams and wadeable rivers: periphyton, benthic macroinvertebrates and fish. Second edition. United States Environmental Protection Agency, Office of Water, Washington D.C
108. Vidinova Y (2015) Ephemeroptera. In: Golemanski et al (eds) Red data book of Bulgaria. vol 2. Animals, BAS & MEW, p 52–54, pp 113–118
109. Slavevska-Stamenković V, Rimcheska B, Vidinova Y, Tyufekchieva V, Paunović M, Ristovska M, Prelić D (2016) Contribution to the knowledge of Ephemeroptera, Plecoptera and Trichoptera from Republic of Macedonia. Acta zool bulg 68(2):199–206
110. Hebert DNP, Cywinska A, Ball SL, de Waard JR (2003) Biological identifications through DNA barcodes. Proc R Soc Lond [Biol] 270:313–321
111. Ball SL, Hebert DN (2005) Biological identifications of mayflies (Ephemeroptera) using DNA barcodes. J North Am Benthol Soc 24:508–524. https://doi.org/10.1899/04-142.1
112. Williams HC, Ormerod SJ, Bruford MW (2006) Molecular systematics and phylogeography of the cryptic species complex Baetis rhodani (Ephemeroptera, Baetidae). Mol Phylogenet Evol 40(2):370–382. https://doi.org/10.1016/j.ympev.2006.03.004
113. Lucentini L, Rebora M, Puletti ME, Gigliarelli L, Fontaneto D, Gaino E, Panara F (2011) Geographical and seasonal evidence of cryptic diversity in the *Baetis rhodani* complex (Ephemeroptera, Baetidae) revealed by means of DNA taxonomy. Hydrobiologia 673:215–228
114. Ståhls G, Savolainen E (2008) MtDNA COI barcodes reveal cryptic diversity in the Baetis vernus group (Ephemeroptera, Baetidae). Mol Phylogenet Evol 46(1):82–87
115. Vuataz L, Rutschmann S, Monaghan MT, Sartori M (2016) Molecular phylogeny and timing of diversification in Alpine *Rhithrogena* (Ephemeroptera: Heptageniidae). BMC Evol Biol 16:194

Chapter 16
Fish Communities Over the Danube Wetlands in Serbia and Croatia

Milica Stojković Piperac, Djuradj Milošević, Dubravka Čerba, and Vladica Simić

Abstract Wetland ecosystems are recognized to play a crucial role in the conservation of fish biodiversity. River-associated wetlands are characterized by a high diversity of fishes as they use that type of aquatic ecosystem as a refuge for breeding, feeding, and nesting purposes at one or the other stages of their life cycle. This chapter aims to present fish community structure and composition in several wetland ecosystems as a type of small water bodies (SWB) over the Danube floodplain in Serbia and Croatia. Also, the effect of different ecosystem types on fish community structure and diversity will be discussed. This chapter contains the first published data regarding fish fauna in the Danube floodplain on the territory of Serbia and Croatia, providing a detailed insight into the community structure and composition of fish fauna.

Keywords Fish community · Diversity · Wetlands

M. Stojković Piperac (✉) · D. Milošević
Faculty of Sciences and Mathematics, Department of Biology and Ecology, University of Niš, Višegradska 33, 18000 Niš, Serbia
e-mail: milicas@pmf.ni.ac.rs

D. Milošević
e-mail: djuradj@pmf.ni.ac.rs

D. Čerba
Department of Biology, Josip Juraj Strossmayer University of Osijek, Cara Hadrijana 8/A, 31000 Osijek, Croatia
e-mail: dcerba@gmail.com

V. Simić
Faculty of Science, Institute of Biology and Ecology, University of Kragujevac, Kragujevac, Serbia
e-mail: vladica.simic@pmf.kg.ac.rs

V. Pešić et al. (eds.), *Small Water Bodies of the Western Balkans*, Springer Water,
https://doi.org/10.1007/978-3-030-86478-1_16

16.1 Wetlands—Functions, Values, and Conservation

Wetland ecosystems are characterized as a natural resource of global significance. They provide numbers of goods and services for the human population and sustainable natural management worldwide [1, 2]. More precisely, they present a transitional zone between land and water that have a critical role in protection against extreme floods. Also, wetlands present freshwater supplies for both, drinking and irrigation. Beside many local and regional services, wetlands provide benefits on the global level, being a global resource of carbon dioxide since peatlands are recognized as major carbon storage in the world.

Biodiversity is recognized as one of the important wetland values [1, 3, 4]. Wetlands have great importance in maintaining biodiversity since they provide habitat for many rare, threatened, and endangered species. They may serve as refuges of rare and relict species and their relative isolation in the landscape may promote microevolution of specialized phenotypes. It is widely known that wetland ecosystems have a crucial role in the conservation of fish biodiversity [5]. River-associated wetlands are characterized by a rich diversity of fishes which they use as a refuge for breeding, feeding, and nesting purposes at one or the other stages of their life cycle [6].

However, during the twentieth century, many of these unique freshwater habitats in Europe were subjected to intensive human pressure [7]. More precisely, European wetlands have been exploited and managed for various purposes. Large wetland areas have been drained and reclaimed mainly for agriculture and the establishment of human settlements. Bearing all this in mind, wetlands are the most valuable but also the most vulnerable ecosystems in the world. According to the IPBES (Intergovernmental Science-Policy Platform on Biodiversity and Ecosystem Services) report, over 85% of wetlands (area) has been lost with the great majority of indicators of ecosystems and biodiversity showing rapid decline [8, 9].

Due to the high level of diversity, floodplains and wetlands have become a high priority for conservation. The legal protection of wetlands has been supported by international agreements such as the Ramsar Convention on the Conservation of Wetlands (www.ramsar.org) and the International Conservation of Biological Diversity (www.biodiv.org/convention/default.shtml). The Ramsar Convention is considered as the primary basis for the conservation of the most valuable wetlands in Europe.

Lentic ecosystems within the wetlands can be considered as small water bodies (SWB). By definition, SWB encompass a range of small standing and running freshwater ecosystems [10]. Standing waters, such as ponds and small lakes placed along the river floodplains, mainly are categorized as SWB [11].

16.2 Danube Floodplain in Western Balkans—State and Perspective

The floodplains and wetlands of the Danube Basin are unique and valuable freshwater ecosystems not only in Europe but also on global terms [12]. Environmental degradation of the Danube River Basin and its flooded area lasts for centuries and mainly causes irreversible habitat alteration and biodiversity loss. The fluctuating river discharge is the main factor that determines life along the riverbanks and adjacent floodplains. The various habitats created by the Danube and its tributaries host a unique diversity of animals and plants. Moreover, the remaining large floodplain forests and the Danube Delta are the last refuges in Europe for many species. Having in mind that many species require very special living conditions in terms of different water and habitat quality features, the available connections between side-arms and the main bed are the main prerequisite for functioning of aquatic ecosystems by providing important habitats for fish as well as other fauna. During the last two centuries, most large floodplain areas within the Danube basin have disappeared [13]. Nevertheless, few areas within Danuban floodplain are still in their natural or even near-natural state. According to the Danube River Basin District Management Plan [12], a substantial percentage of wetland areas of high ecological importance, having a potential for reconnection and/or improvement of water regime is located in Serbia and Croatia [14]. Moreover, the so-called "Amazon of Europe", the Transboundary UNESCO Biosphere Reserve that combines the group of thirteen protected areas along the Mura–Drava–Danube region, hosts an amazing biological diversity and is a hotspot of rare natural habitats such as large floodplain forests, river islands, gravel and sand banks, side branches and oxbows. In this research, we considered three of them: Kopački Rit (Croatia), Gornje podunavlje (Serbia), and Karađorđevo (Serbia).

Wetlands with a natural hydrological regime along the Danube River in Serbia and Croatia were recognized as sites of high conservation importance at the national and European levels. However, they are usually neglected from the routine monitoring programs and a convenient method for their ecological status assessment has still not been proposed.

Thus, the aim of the chapter was to present the structure and composition of fish communities with the special attention of commercial, endangered, and alien species along the Danube river floodplain in Serbia (1400–1250 r. km) as well as Croatia (1410–1383 r. km). Moreover, we investigated differences in fish community structure and diversity among different types of ecosystems within the Danube river floodplain.

16.3 Danube Floodplain in Serbia and Croatia

The Danube river flows a total length of 2,700 km, with 588 km in Serbia flowing through Serbian mountains, plains, cities, villages as well as protected nature reserves and national parks. The Croatian part of the Danube is 188 km long and forms part of Croatia's eastern border with Serbia. Fieldwork was carried out at 66 sampling sites, distributed over the Danube floodplain, on the territory of Serbia and Croatia (Fig. 16.1). The investigated wetland areas we characterized as ponds, lakes, channels, oxbow lakes, and side arms of the Danube River.

The investigated area within the territory of Serbia encompasses five Danube wetland areas (53 sampling sites) situated in Vojvodina province: (1) Special Nature Reserve 'Koviljsko-Petrovaradinski Rit' (Ramsar site no. 2028, IPA, IBA; WGS84 coordinates: 45.17778, 20.06944; stakeholder: JP 'Vojvodina šume'); (2) Special Nature Reserve 'Gornje Podunavlje' (Ramsar site no. 1737, IPA, IBA, UNESCO Biosphere Reserve; WGS84 coordinates: 45.564687, 18.944598; stakeholder: 'JP Vojvodina šume'); (3) the wetland area located near Bačko Novo Selo (the National Ecological Network code BAČ04; WGS84 coordinates: 45.308285, 19.113918, a candidate for Special Nature Reserve); (4) Nature Park 'Begečka jama' (WGS84 coordinates: 45.220991, 19.604875 WGS84; stakeholder: 'DTD Ribarstvo doo'); and (5) Special Nature Reserve 'Karađorđevo' reserve (IBA; WGS84 coordinates:45.278056, 19.237778 WGS4; stakeholder: 'Vojna ustanova Morović').

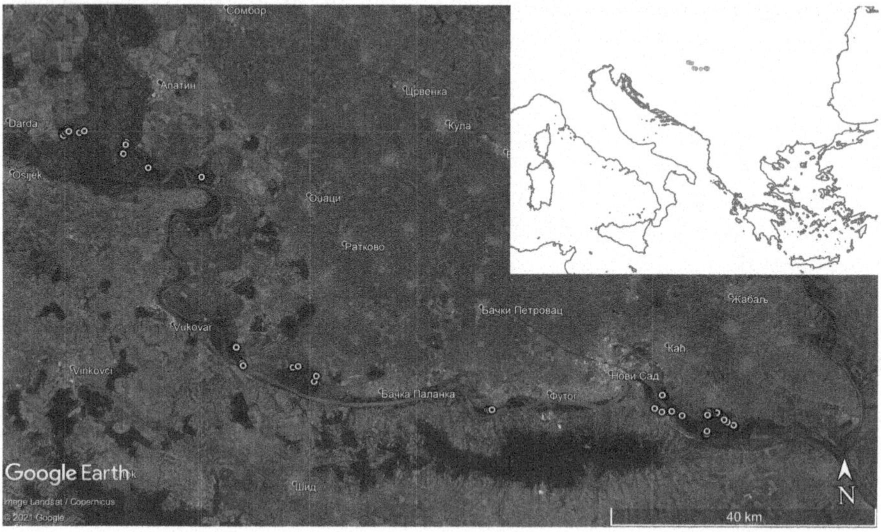

Fig. 16.1 The map showing the distribution of sampling sites over the Danube floodplain in Serbia and Croatia

In Croatia, the study was carried out in Slavonia and Baranja region, (13 sampling sites) within Nature Park 'Kopački Rit' (Ramsar site no. 583, IBA; WGS84 coordinates: 45.6407983, 18.869069; stakeholder: JP 'Javna ustanova Park Prirode Kopački Rit'), one of the Danube's largest conserved floodplains. The floodplain ecosystem, Kopački Rit, is situated in the northeast of Croatia between the River Drava and the Danube and is highly influenced by the Danube hydrologic regime.

In the Croatian part of the Danube floodplain sampling campaign was conducted in the autumn (September) of 2017, while in the Serbian part of the Danube floodplain, the fieldwork was performed during the summer (July and August) 2019. The investigated area encompasses four lentic water bodies (lake, channels, and pond) and 20 lentic water bodies (side arms of the Danube river, oxbow lakes, and ponds) in Croatia and Serbia, respectively (Fig. 16.2). The number and location of sample points per ecosystem were chosen to cover all freshwater habitat types (sensu EUNIS classification [15]. Fish communities were surveyed using the EN 14011:2003 methodology [16]. Fish were sampled from a boat using a DC Aquatech IG 1300 electrofisher (2.6 kW, 80–470 V). The constant catch-per-unit-effort (CPUE) of time (10 min) was provided. The fish density was expressed as the number of individuals caught per 1 min. Each individual fish was identified to the species level according to relevant morphological and anatomic features and the actual names of species are used according to [17].

Fig. 16.2 Different types of aquatic ecostems with in the Danube floodplain: **a** Mali Sakadaš pond (Kopački Rit wetland, Croatia); **b** Sakadaš lake (Kopački Rit wetland, Croatia); **c** Channel Čonakut (Kopački Rit wetland, Croatia); **d** Arkanj oxbow lake (Koviljsko-Petrovaradinski Rit wetland, Serbia); **e** pond in Gornje Podunavlje wetland, Serbia; **f** Koviljski Dunavac (Koviljsko-Petrovaradinski Rit wetland, Serbia). Photos by D. Čerba and D. Cvijanović

16.4 Fish Community Structure and Composition

The Danube is considered as a river with the highest species richness in Europe, with as many as 102 species of fishes recorded [18]. A high fish species richness occurs as a consequence of high habitat diversity and environmental conditions suitable for a wide variety of different fish species [19, 20] as well as the fact that the Danube river is a major migration route for a diverse Central Asian and Ponto–Caspian fauna [21]. In addition, a high level of spatio-temporal heterogeneity makes riverine floodplains among the most species-rich environments known. Indeed, the Danube floodplain area, is characterized by high productivity and high biodiversity [22].

Out of the total number of 31 fish species found in this study, 26 were caught in Serbian and 16 in the Croatian part of the Danube floodplain. The basic information about each fish species, including origin, commercial value, and conservation status on both European and national levels are presented in Tables 16.1 and 16.2.

A permutational multivariate analysis of variance (PERMANOVA) showed significant differences in fish community structure between different ecosystem types ($F = 4.6208$; $p = 0.001$). In addition, the distributional pattern of sampling sites on non-metric multidimensional scaling (NMDS) plot reveals which ecosystem types differ in terms of fish community structure (see Fig. 16.3). According to the NMDS plot, fish communities clearly differ in terms of community structure between side arms of the river and oxbow lakes as well as with other types of ecosystems except ponds. Indeed, side arms of the river are expected to be characterized by the highest species richness and species diversity since it is directly connected to the belonging river, and thus, their community structure differs from the other ecosystem that is partially and periodically separated from the main channel. Similarly, fish community data from lakes and channels are mutually similar but different from the other ecosystem types. All samples from the lake and channels, in this study, belong to the "Kopački Rit" wetland (Croatia). In contrast to the other wetlands, it is showed that the "Kopački Rit" wetland is characterized by the presence of species *Alburnus chalcoides, Leucaspius delineatus* and *Rutilus pigus* which are not found on the other wetlands. For instance, *R. pigus* is an endemic species of the River Danube catchment [26] and in Serbia mainly distributed to the Drina catchment [27]. In contrast, *R. pigus* is common and reproduces naturally in the basins of the rivers Drava and Sava and their floodplains in Croatia [28], which is the possible reason for its presence only in the Croatian part of the Danube floodplain. In addition, *A. chalcoides* and *L. delineatus* rare and strictly protected species in Serbia, and thus unlikely to be found in the Serbian part of the Danube floodplain.

The Kruskall–Wallis ANOVA test and Mann–Whitney as a post hoc test revealed statistically significant differences ($p < 0.05$) in species richness (S), species abundance (N) and biomass (B), Shannon's diversity index (H), and Simpson's index of diversity (1-D) between different ecosystem types (Table 16.3). The highest species richness was observed in side arms of the Danube River due to the direct connection with the main river flow, while the lowest number of species found in ponds and oxbow lakes (Table 16.3). This result is in accordance with founds of Holèik and

Table 16.1 List of fish species found over the Serbian part of Danube floodplain

Fish species	Origin	Commercial value	Conservation status—IUCN Red list category EU [23]	Conservation status—National legislative [24]
Cyprinidae				
Abramis brama (Linnaeus, 1758)	Native	High	LC	Protected
Abramis sapa (Pallas, 1814)	Native	High	LC	Protected
Alburnus alburnus (Linnaeus, 1758)	Native	Low	LC	/
Aspius aspius (Linnaeus, 1758)	Native	High	LC	Protected
Blicca bjoerkna (Linnaeus, 1758)	Native	Low	LC	/
Carassius auratus gibelio Bloch, 1783	Introduced	Low	/	/
Cyprinus carpio Linneaus, 1758	Native	High	VU	Protected
Hypophthalmichthys molitrix (Valenciennes, 1844)	Introduced	High	/	/
Leuciscus idus (Linneaus, 1758)	Native	High	LC	Protected
Pseudorasbora parva (Temminck and Schlegel, 1846)	Introduced	Low	/	/
Rhodeus sericeus (Pallas, 1776)	Native	Low	LC	/
Rutilus rutilus (Linneaus, 1758)	Native	Low	LC	/
Scardinius erythrophthalmus (Linneaus, 1758)	Native	Low	LC	/
Cobitidae				
Cobitis elongatoides Băcescu and Mayer, 1969	Native	Low	LC	/
Cobitis taenia Linnaeus, 1758	Native	Low	LC	Protected
Misgurnus fossilis (Linneaus, 1758)	Native	Low	LC	Strictly protected
Centrarchidae				

(continued)

Table 16.1 (continued)

Fish species	Origin	Commercial value	Conservation status—IUCN Red list category EU [23]	Conservation status—National legislative [24]
Lepomis gibossus (Linneaus, 1758)	Introduced	Low	/	/
Micropterus salmonides (Lacepède, 1802)	Introduced	Low	/	/
Esocidae				
Esox lucius Linnaeus, 1758	Native	High	LC	Protected
Percidae				
Gymnocephalus cernuus (Linnaeus, 1758)	Native	Low	LC	/
Sander lucioperca (Linnaeus, 1758)	Native	High	LC	Protected
Perca fluviatilis Linnaeus, 1758	Native	High	LC	Protected
Siluridae				
Silurus glanis Linnaeus, 1758	Native	High	LC	Protected
Ictaluridae				
Ameiurus nebulosus (Lesueur, 1819)	Introduced	Low	/	/
Gobiidae				
Neogobius gymnotrachelus (Kessler, 1857)	Introduced	Low	/	/
Neogobius fluviatilis (Pallas, 1814)	Introduced	Low	/	/

Bastl [29] who claimed that there were positive correlations between the species number and habitat size during the investigations on the backwaters in the Danube floodplain. Nevertheless, during flood events from the Danube river, the area under this study is flooded so that fish can freely move throughout the ponds and oxbow lakes and main river on the other side [30]. The rate of the movement and species exchange depend on river proximity. The closest the distance between the Danube and flooded water bodies, the highest possibility to be connected even during even a partial flooding event of the area.

The results of Indicator species analysis (IndVal) showed that the number of species with significant indicator values differs among the groups (Table 16.4). The highest number of indicator species was found in the side arms of the river, a consequence of the highest species richness due to the direct connection with the main river flow. Many species selected as the representative for this group are species characteristic for the Middle Danube river [18]. Moreover, the species composition of

Table 16.2 List of fish species found over the Croatian part of Danube floodplain

Fish species	Origin	Commercial value	Conservation status—IUCN Red list category EU [23]	Conservation status—National legislative [25]
Cyprinidae				
Alburnus chalcoides (Güldenstädt 1772)	Native	Low	LC	VU
Aspius aspius (Linnaeus, 1758)	Native	High	LC	VU
Carassius auratus gibelio Bloch, 1783	Introduced	Low	/	/
Cyprinus carpio Linneaus, 1758	Native	High	VU	EN
Leucaspius delineatus (Heckel 1843)	Native	Low	LC	VU
Pseudorasbora parva (Temminck and Schlegel, 1846)	Introduced	Low	/	/
Rutilus pigus Lacepède, 1803	Native	Low	LC	NT
Rutilus rutilus (Linneaus, 1758)	Native	Low	LC	/
Scardinius erythrophthalmus (Linneaus, 1758)	Native	Low	LC	/
Squalius cephalus (Linneaus, 1758)	Native		LC	/
Tinca tinca (Linneaus, 1758)	Native		LC	/
Cobitidae				
Cobitis taenia Linnaeus, 1758	Native	Low	LC	/
Centrarchidae				
Lepomis gibbosus (Linneaus, 1758)	Introduced	Low	/	/
Esocidae				
Esox lucius Linnaeus, 1758	Native	High	LC	/
Percidae				
Perca fluviatilis Linnaeus, 1758	Native	High	LC	/
Ictaluridae				

(continued)

Table 16.2 (continued)

Fish species	Origin	Commercial value	Conservation status—IUCN Red list category EU [23]	Conservation status—National legislative [25]
Ameiurus nebulosus (Lesueur, 1819)	Introduced	Low	/	/

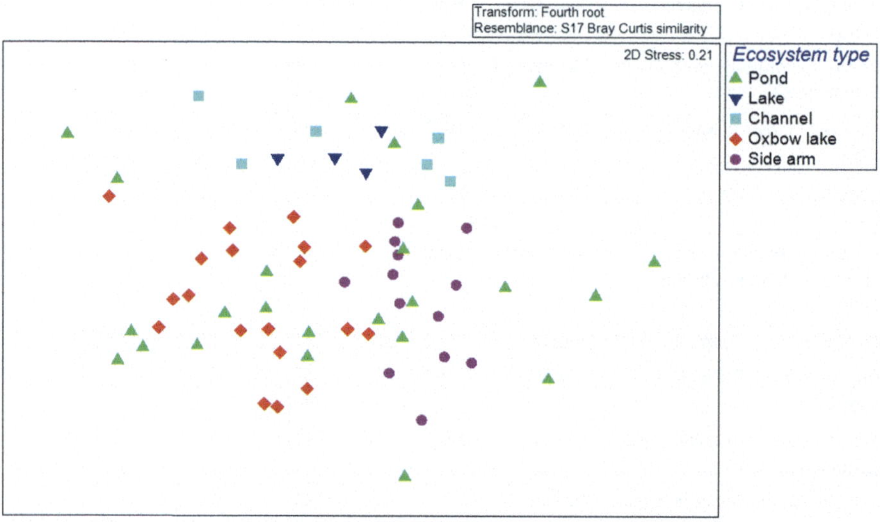

Fig. 16.3 Non-metric multidimensional scaling (nMDS) of the 66 sampling sites from six Danube wetlands in Serbia and Croatia. Ecosystem types are marked by different shapes and colors

Table 16.3 Differences in fish species richness and diversity (mean ± SD) in different types of wetland ecosystems. Values not sharing a common letter are significantly different (Mann–Whitney test): $^{a,b}p < 0.05$

	Pond	Lake	Channel	Oxbow lake	Side arm
S	5.27 ± 2.45^{a}	$6 \pm 1.633^{a,b}$	$6.67 \pm 2.658^{a,b}$	$5.42 \pm 2.388^{a,b}$	7.35 ± 3.404^{b}
H	0.991 ± 0.527^{a}	$0.851 \pm 0.429^{a,b}$	1.29 ± 0.697^{a}	0.693 ± 0.485^{b}	1.21 ± 0.610^{a}
1-D	0.512 ± 0.262^{a}	$0.391 \pm 0.191^{a,b}$	0.586 ± 295^{a}	0.326 ± 0.223^{b}	0.561 ± 0.263^{a}

this group differs from the others since it comprises mainly rheophilic and eurytopic species [18], which is obviously a result of its connection with the Danube. On the other side, ponds and lakes are characterized mainly by limnophilic and eurytopic species. Next, species *A. chalcoides, L. delineatus,* and *R. pigus* have been already found to be prepresentative for the Croatian Danube floodplain, and consequently selected as indicators of channels and lake, respectively. IndVal analysis showed that

Table 16.4 Species distinguished by the Indicator species analysis (IndVal), for previously defined ecosystem types. The bold letters indicate the species' representative for the particular ecosystem type, having IndVal values of more than 25%. The species without significant IndVal values are included at the end of columns. * Significance level: <0.05, ** Significance level: <0.01, *** Significance level: <0.001

Pond	Lake		Channel		Oxbow lake		Side arm	
H. molitrix	***R. pigus***	**39.6***	***A. chalcoides***	**78*****	***C. gibelio***	**39.4***	***L. gibbosus***	**42.2***
A. sapa	*S. cephalus*		*L. delineatus*	15.4*	*A. brama*		***L. idus***	**35.7***
S. glanis			*C. taenia*		*M. salmonides*		***P. fluviatilis***	**39.4**
T. tinca			*E. lucius*				***R. rulitus***	**42.3***
			P. parva				***S. lucioperca***	**39.9***
							A. alnburnus	
							A. aspius	
							B. bjoerkna	
							C. elongatoides	
							C. carpio	
							G. cernuus	
							A. nebulosus	
							M. fossilis	
							N. gymnotrachelus	
							N. fluviatilis	
							R. sericeus	

species *Carassius auratus gibelio* is representative species for oxbow lakes, but is important to emphasize that it is the most common and widespread species within this study. This species is widely known as highly invasive and it easily reproduces and rapidly spreads in open waters [31].

16.5 Conclusion

The results obtained in this chapter enable a comprehensive insight into the community structure and composition of fish fauna within the Danube floodplain on the territory of Serbia and Croatia. Bearing in mind that by now there is no published data regarding fish fauna in the investigated floodplain this study provides basic information for the wide audience of interest. Thus, future research should be focused on how fish communities in various types of floodplain ecosystems can be related to key environmental parameters that reflect the ecological status and water quality, which will be the first step for their inclusion in routine biomonitoring.

References

1. Mitsch WJ, Gosselink JG (2000) The value of wetlands: importance of scale and landscape setting. Ecol Econ 35:25–33
2. Čížková H, Květ J, Comin FA, Laiho R, Pokorný J, Pithart D (2013) Actual state of European wetlands and their possible future in the context of global climate change. Aquat Sci 75(1):3–26
3. Gopal B, Junk WJ (2000) Biodiversity in wetlands: assessment, function and conservation, vol 2. Backhuys Publishers, Leiden
4. Gopal B (2009) Biodiversity in wetlands. In: The wetlands handbook, vol 2. pp 65–95
5. Kottelat M, Whitten T (1996) Freshwater biodiversity in Asia: with special reference to fish. The World Bank, Washington, DC
6. Wetzel RG (2001) Limnology, 3rd edn. Academic Press, New York
7. Tockner K, Malard F, Ward JV (2000) An extension of the flood pulse concept. Hydrol Process 14:2861–2883
8. Finlayson CM, D'Cruz R, Davidson N (2005) Millennium ecosystem assessment. In: Ecosystems and human well-being: wetlands and water synthesis. World Resources Institute, Washington, DC
9. Díaz S, Settele J, Brondízio E (2019) Summary for policymakers of the global assessment report on biodiversity and ecosystem services of the intergovernmental science-policy platform on biodiversity and ecosystem services. Carneiro da Cunha M, Mace G, Mooney H (ed)
10. Riley WD, Potter EC, Biggs J, Collins AL, Jarvie HP, Jones JI, Kelly-Quinn M, Ormerod SJ, Sear DA, Wilby RL, Broadmeadow S (2018) Small water bodies in Great Britain and Ireland: ecosystem function, human-generated degradation, and options for restorative action. Sci Total Environ 645:1598–1616
11. Biggs J, Von Fumetti S, Kelly-Quinn M (2017) The importance of small waterbodies for biodiversity and ecosystem services: implications for policy makers. Hydrobiologia 793(1):3–39
12. International Commission for the Protection of the Danube River (ICPDR) (2015) The Danube River Basin Management Plan 2015–2021—Factsheet. https://www.icpdr.org/main/sites/default/files/nodes/documents/managementplansbrochure2015.pdf. Accessed 26 Aug 2019
13. Tockner K, Pusch MT, Borchardt D, Lorang MS (2010) Multiple stressors in coupled river-floodplain ecosystems. Freshw Biol 55(1):135–151
14. Tadić L, Bonacci O, Dadić T (2014) Dynamics of the Kopački Rit (Croatia) wetland floodplain water regime. Environ Earth Sci 71(8):3559–3570
15. EEA (ETC/BD) (2013) EUNIS habitat classification 2012—a revision of the habitat classification descriptions. EEA (ETC/BD)
16. European Committee for Standardization (CEN) (2003) Water quality—sampling of fish with electricity. (EN 14011:2003)
17. Kottelat M, Freyhof J (2007) Handbook of European freshwater fishes. Publications Kottelat, Berlin
18. Kováč V (2015) Current status of fish communities in the Danube. In: The Danube River Basin. Springer, Berlin, Heidelberg, pp 359–388
19. Schiemer F (2002) Fische als Bioindikatoren und Naturschutzziel von großen Fließgewa¨ssern. Verhandlungen der Gesellschaft fu¨r Ichthyologie 3:19–35
20. Schiemer F, Guti G, Keckeis H et al (2004) Ecological status and problems of the Danube River and its fish Fauna: a review. In: Welcomme RL, Petr T (eds) Proceedings of the second international symposium on the management of large rivers for fisheries sustaining livelihoods and biodiversity in the new millennium, vol I. FAO Regional Office for Asia and the Pacific, Bangkok

21. Balon EK, Crawford SS, Lelek A (1986) Fish communities of the upper Danube Germany, Austria prior to the new Rhein-Main-Donau connection. Environ Biol Fishes 15:243–271
22. Preiner S, Weigelhofer G, Funk A, Hohensinner S, Reckendorfer W, Schiemer F, Hein T (2018) Danube floodplain Lobau. In: Riverine ecosystem management. Springer, Cham, pp 491–506
23. Freyhof J, Brooks E (2011) European red list of freshwater fishes. Publications Office of the European Union, Luxembourg
24. ANONYMOUS (2010) Pravilnik o proglašenju i zaštiti strogo zaštićenih i zaštićenih divljih vrsta biljaka, životinja i gljiva.—Službeni glasnik Republike Srbije, 5/10 (In Serbian)
25. Mrakovčić M, Brigić A, Buj I, Ćaleta M, Mustafić P, Zanella D (2006) Red book of freshwater fish of Croatia. Crvena Knjiga Slatkovodnih Riba Hrvatske. Ministry of Culture, Zagreb, p 253 (In Croatian)
26. Simonović P, Povž M, Piria M, Treer T, Adrović A, Škrijelj R, Nikolić V, Simić V (2015) Ichthyofauna of the River Sava system. In: The Sava River. Springer, Berlin, Heidelberg, pp 361–400
27. Simonivić P (2006) Ribe Srbije. NNK International, Beobrad (In Serbian)
28. Povž M, Šumer S, Leiner S (1998) Sport fishing catch as an indicator of population size of the Danube roach Rutilus pigusvirgo in Slovenia (Cyprinidae). Ital J Zool 65:545–548
29. Holèik J, Bastl I (1973) Hlavné faktory limitujúce ichtyomasu v periodických jezierkach (The main factors limiting level of the ichthyomass in periodic lakes). In: Sb. Symposium k 75.na/: prof B. Kostomarova. vSZ Brno: 225–236 (in Slovak. with a summary in English)
30. Halacka K, Lusk S, Luskova V (1998) Fish communities in artificial pools in the floodplain along the lower reaches of the River Dyje. Folia Zool Praha 47:125–134
31. Tomljanović T, Fabijanić N, Treer T, Matulić D, Piria M, Šprem N, Aničić I, Safner R (2012) Utjecaj invazivne vrste riba babuške (Carassius gibelio) na sastav ihtiocenoze u Republici Hrvatskoj. In: Proceedings. 47th Croatian and 7th international symposium on agriculture, Opatija, Croatia 2012, vol 634, p 636 (In Croatian)

Chapter 17
The Importance of Small Water Bodies' Conservation for Maintaining Local Amphibian Diversity in the Western Balkans

Jelka Crnobrnja-Isailović, Avdul Adrović, Ferdinand Bego, Natalija Čađenović, Elvira Hadžiahmetović Jurida, Daniel Jablonski, Bogoljub Sterijovski, and Olga Jovanović Glavaš

Abstract In these times of aggressive and enormous anthropogenic alterations of freshwater aquatic habitats throughout the Western Balkans, including drying out, pollution or the introduction of invasive species, the value of small water bodies

J. Crnobrnja-Isailović (✉)
Faculty of Sciences and Mathematics, University of Niš, Višegradska 33, 18000 Niš, Serbia
e-mail: jelka@ibiss.bg.ac.rs; jelka@pmf.ni.ac.rs

University of Belgrade, Institute for Biological Research, Siniša Stanković—National Institute of Republic of Serbia, Despota Stefana Blvd. 142, 11000 Belgrade, Serbia

A. Adrović · E. H. Jurida
Faculty of Natural Sciences and Mathematics, University of Tuzla, Univerzitetska 4, 75000 Tuzla, Bosnia and Herzegovina
e-mail: elvira.h.jurida@untz.ba

F. Bego
Faculty of Natural Sciences, University of Tirana, "Zogu I" Blvd. 25/1, 1010 Tiranë, Albania
e-mail: ferdinand.bego@fshn.edu.al

N. Čađenović
Natural History Museum of Montenegro, Trg vojvode Bećir bega Osmanagića 16, Podgorica, Montenegro
e-mail: direktor@pmcg.co.me

D. Jablonski
Department of Zoology, Comenius University in Bratislava, Ilkovičova 6, Mlynská dolina, 842 15 Bratislava, Slovakia
e-mail: daniel.jablonski@uniba.sk

B. Sterijovski
Faculty of Natural Sciences, Macedonian Ecological Society, University of Skopje, Bvld Kuzman, Josifovski-Pitu, 28/3-7, 1000 Skopje, North Macedonia
e-mail: sterijovski@mes.org.mk

O. J. Glavaš
Department of Biology, Josip Juraj Strossmayer University of Osijek, Cara Hadrijana 8/A, 31000 Osijek, Croatia
e-mail: ojovanovic@biologija.unios.hr

V. Pešić et al. (eds.), *Small Water Bodies of the Western Balkans*, Springer Water,
https://doi.org/10.1007/978-3-030-86478-1_17

(SWB) as refugia for local amphibians is clearly increasing. A network of SWB could be crucial for the conservation of local amphibian richness, if this network is properly maintained and monitored. Together, countries of the Western Balkan harbor 29 amphibian species and only one of these species does not use SWB. Almost all those countries share the same challenges—drainage, technical flood protection measures, channelization, planned small dam construction, agriculture, deforestation, pollution, road construction and decrease of aquifers. The conservation of SWB in the region is related to the socio-economic status of the people i.e. to the stimulation of local stakeholders to continue with the traditional practices of agriculture and animal husbandry, including maintenance of SWB, and to their continuous education on importance of SWB for preservation of high values of local biodiversity.

Keywords Albania · Anurans · Bosnia and Herzegovina · Croatia · Montenegro · North Macedonia · Serbia · Small freshwater habitats · Urodeles

17.1 Introduction

Amphibians are currently the most vulnerable group of vertebrates, with 31% of species assessed for IUCN Red List status being threatened to a certain extent [1]. Their decline is happening in all continents, caused by habitat destruction, agrochemicals and chemical pollution, increased UV-B irradiation, diseases, introduced species, exploitation, climate change and complex causes [2]. Warning reviews on this issue appeared as far back as the end of 1990s [3], but this decline probably started decades before [2]. The loss of natural habitats is the major cause of amphibian declines in the entire Palearctic [4] and has been connected with the development of intensive agriculture in the mid-twentieth century [5, 6].

In 2005, 26 Palearctic amphibian genera were recognized as being dependent on aquatic habitats in at least one phase of their life history [7]. Indeed, most of European amphibians require aquatic habitats, at least for reproduction [8–10]. This dependence on a freshwater environment for the completion of a life cycle increases the risk of local population declines in habitats subjected to intense anthropogenic alteration [11, 12] which include not only the drying out of wetlands of all kinds, but also, for species which require running waters (lotic aquatic habitats), the construction of different kinds of hydroenergy power plants for the production of electric energy and/or water storage (see an example in [13]). Additionally, importing fish stock into originally fishless freshwater environments decreases the growth rate of local amphibian populations or causes their extinction [14].

Amphibians utilize a number of different freshwater habitats: lentic, such as lakes, oxbows, bogs, marshes, ponds, pools or disconnected river channels, and/or lotic, such as rivers, streams, springs and ditches. SWB include both small lentic and lotic freshwater ecosystems; lotic ones include headwaters (streams) up to 2.5 km from the source, while lentic ones are small water bodies whose size ranges from 1 m^2 to 5 ha; all categories of SWB can be man-made or natural, permanent or seasonal [15].

Ponds are considered as the most important SWB for amphibians [16], although their suitability depends on many factors [17].

Amphibian assemblages detected in European SWB are highly variable; for example, in one survey conducted in Great Britain the most represented amphibian was the common frog (*Rana temporaria*), followed by the common toad (*Bufo bufo*), smooth newt (*Lissotriton vulgaris*) and palmate newt (*L. helveticus*), while the great crested newt (*Triturus cristatus*) was not so common [18]. In the network of ponds in Western France, the common toad was the most frequent amphibian species in the assemblage of SWB, followed by the common frog, alpine newt (*Ichthyosaura alpestris*) and palmate newt, while the edible frog (*Pelophylax* kl. *esculentus*), parsley frog (*Pelodytes punctatus*) and moor frog (*R. arvalis*) were the rarest [12].

Amphibian species which inhabit Balkan peninsula present 37% of overall European amphibian species richness (30% in the Western Balkan countries) [19]. In comparison with the already infrastructurally developed and mostly devastated parts of Western Europe, Western Balkan countries are facing both challenges and opportunities in balancing anthropogenic alteration and protection of the environment, including freshwater habitats, which, coupled with the climate change, could significantly diminish amphibian species richness. Heatwole and Wilkinson [19] compiled country reports on the status of conservation and the decline of amphibians in Southern Europe and Turkey, where habitat change (fragmentation, degradation, and destruction) was mentioned as an important threatening factor in all Western Balkan countries. In this chapter the focus will be shifted from amphibian species to the national status of the SWB important for the survival of their local amphibian populations in the Western Balkans, also on the occurring and ongoing threats to SWB, and on possible ways of their mitigation. Regarding the list of amphibian species occurring in the countries of Western Balkans, we accepted those taxonomic suggestions of Dufresnes [20] and Taxonomic Committee of Societas Europaea Herpetologica [21] which were in line with taxonomy approved by Frost [22]—a taxonomic authority for IUCN Amphibian Specialist Group.

17.2 The Metapopulation Approach in Conservation of Small Water Bodies for Amphibians

Spatial population organization of most amphibian species depends on the spatial distribution of suitable aquatic habitats. For better understanding of amphibian declines it is important to check how these aquatic habitats are distributed in space and how strongly they are functionally connected, particularly in terms of metapopulation structures. In the areas where a number of ponds or water bodies of similar or different size are occurring relatively close to each other, there is always the possibility that they are a metapopulation [2].

The metapopulation approach to the spatial organization of pond-dependent amphibian species sounds logical because, at landscape level, a group of ponds or other SWB do indeed represent patches of habitats suitable for breeding, larval development, and, regarding some species, all daily activities within certain parts of the year. A metapopulation is a very dynamic spatial system, strongly dependent on the asynchrony of extinction/recolonization of its elements (local populations) and thus also on the ability and intensity of migration among suitable habitat patches; co-occurrence of metapopulations of several species, connected by ecological interactions, make a metacommunity [23]. It is particularly important to check for metapopulation organization and the possible presence of metacommunities in an area assigned to anthropogenic change—if minimum viable number of suitable habitat patches for specific metapopulation no longer exists, the entire metapopulation is doomed [23]. Therefore, short-term monitoring of just one pond in a complex landscape which harbors a number of water bodies of various sizes would not provide complete information about spatial organization of occurring amphibian species. Instead, the simultaneous presence/absence recording of several species within the entire network of aquatic habitats, established as a long-term project, helps to reveal the existence of an amphibian metacommunity [24]. This further enables proper adjustment of management and/or spatial action plans towards limiting or rejecting inadequate anthropogenic changes of the landscape.

One of the three determinants of metapopulation organization—extinction—does not necessarily occur within every network of local amphibian populations [25]. Continuous monitoring can record presence after the absence (or vice versa) of certain amphibian species in the same SWB but it would not necessarily imply extinction followed by recolonization. Members of one local population can simply colonize some other SWB after the winter pause driven by biotic or/and abiotic cues, which has been described elsewhere (see in [25]). Therefore, amphibian aggregations which can be seen in the individual breeding ponds are not always individual local populations, since normally they would move among the ponds through consecutive reproductive seasons. In that way, even without the occurrence of real extinction events, and due to frequent movements between the ponds, a network of ponds is more recommended as a management unit than an individual pond.

Another important modification of the classical metapopulation approach is that, apart from ponds, the surrounding suitable terrestrial habitat is also important to the conservation of amphibian local populations: preserving only ponds or wetlands without their surrounding terrestrial habitats will probably fail to maintain viable amphibian populations [25]. Looking at metapopulation or metacommunity levels, conservation of SWB network in an area or, in other words, of a minimum viable number of aquatic habitat patches will not contribute much to the survival of local amphibian assemblage if surrounding suitable terrestrial habitats are degraded or destroyed.

17.3 Small Water Bodies and Amphibians in Croatia

17.3.1 General Hydrographic Features of Croatia

Croatia covers an area of 56 594 km^2, but due to its position at the crossroads of several biogeographical regions (Continental, Alpine, Mediterranean, and a narrow stretch of the Pannonian region; [26]), its geology and subsequently biology, is very diverse. Hydro-geologically, Croatia is divided into two very different regions [27]: the Dinaric region predominated by karst, and the Pannonian region lying on sedimentary rocks. The border between them runs along a narrow stretch dividing Croatia into two equal parts. These regions roughly correspond to the two main basins; the Danube River basin covers 62% of its territory, and the Adriatic Sea drainage comprises 38% of the territory [28]. The main alluvial aquifers in the Pannonian region are the Sava and Drava rivers, while in between there are fine Neogene sediments without productive aquifers, some of which may act as aquifers of secondary porosity with only local significance [29].

In total, almost 44% of Croatian territory is covered by karst relief [30], but it is dominant in the Dinaric region. Its hydrology has most distinctive elements, including large underground cave systems and surface karst features. Highly permeable carbonate rocks prevail over confined impermeable, mostly flysch sediments.

Islands with a total land area of 3300 km^2 mostly lack freshwater sources. Previously, collecting rainwater in man-made ponds was of great importance for drinking and agricultural use [31]. However, many of those ponds have been abandoned in the last decades, and now there is an effort to restore some of them [32].

17.3.2 The Origin and Status of Small Water Bodies

The SWB in Croatia are largely dependent on the river ecosystems, and around 80% of the total length of rivers in Croatia are made of SWB [33]. The Sava, Drava and Danube floodplains cover the largest area in the Pannonian region in Croatia, and are relatively well ecologically preserved systems in their European context [34].

The national habitat classification of Croatia recognizes *Inland surface water and wetland habitats* as one of eleven major categories (OG 88/2014), and the majority of SWB are included [35]. *Freshwater, tree-dominated wetlands, Seasonal/intermittent freshwater lakes (including flooded karst fields)* and *Seasonal/intermittent freshwater marshes—flooded meadows* include the largest area, however, many of these are not SWB, but cover greater surfaces [35]. Other freshwater habitat types include *Permanent freshwater marshes/pools—reedbeds, Shrub-dominated wetlands, Permanent freshwater marshes/pools—ponds, Seasonal/intermittent freshwater marshes—ponds, Man-made ponds* and *Freshwater springs*, but data on their surface coverage are not detailed [35].

Islands have very few freshwater resources, yet there are in total 2739 wetlands recorded on 71 islands, of which 402 are artificial [32]. Although their surface is not large, they are of great importance for maintaining biodiversity, and the first restoration project is already in progress on Dugi otok island (pond Sakarun).

17.3.3 Amphibian Species Occurring in Small Water Bodies

Despite its small surface area, Croatia has around 25% of all European amphibian species [36]. Of 20 amphibian species recorded in Croatia, only one species never uses SWB during its lifetime (alpine salamander, *Salamandra atra*; [36]). All other species (six Caudata and 13 Anura species) (Table 17.1) can be found in SWB during their breeding season (Fig. 17.1). Nonetheless, some of those species also use larger water bodies and do not depend only on SWB [36]; i.e. the Danube crested newt (*T. dobrogicus*) is commonly found in flooded areas, the olm (*Proteus anguinus*) mostly uses large underground lakes, common and green toads (*B. bufo* and *Bufotes viridis*) can be found breeding in large lakes, as well as green frogs (*Pelophylax* sp.) and brown frogs (*Rana* sp.). Other newt and salamander species (alpine newt, smooth newt, Italian crested newt (*T. carnifex*) and fire salamander *S. salamandra*), fire- and yellow-bellied toads (*Bombina bombina* and *B. variegata*), common spadefoot (*Pelobates fuscus*), and common tree frog (*Hyla arborea*) (Fig. 17.2) are generally found in SWB [9, 36]. While some of these species prefer forest habitats (e.g. alpine newt, fire salamander, agile and Italian agile frog (*R. latastei*), yellow bellied toad), others are more opportunistic and do not show specific preferences (e.g. smooth and crested newt; [9, 36]).

17.3.4 The Main Threats for Small Water Bodies

Since most amphibian species found in Croatia use SWB, preserving these habitats is of crucial importance for their survival. However, destruction of small ecosystems is still present, which affects SWB more than other larger habitats. The anthropogenic impact varies between different types of SWB, presenting a distinct level of threats. All are susceptible to the impact of climate change, mainly through changes in the amount of rainfall and evapotranspiration [37]. Streams and flooded areas face increased pressure from drainage, technical flood protection measures, canalizing, and in a few cases, planned small dam construction [38]. Ponds are especially sensitive to eutrophication, which can result from agriculture (as a diffuse source) or a decrease of aquifers, as well as deforestation. In addition, water quality analysis of surface waters in Croatia show the significance of the catchment area, since pollution sources are sometimes difficult to recognize [39]. Underground habitats are mainly threatened by direct alteration of the habitat, and changes of quality and quantity of water [40].

Table 17.1 List of amphibian species in the countries of the Western Balkans which use SWB

Species	IUCN	BERN	HABITATS	NCS CROATIA	NCS BOSNIA AND HERZEGOVINA	NCS MONTENEGRO	NCS ALBANIA	NCS SERBIA	NCS NORTH MACEDONIA
Proteus anguinus	VU	II	II, IV	SP	SP/P				
Salamandra salamandra	LC	III	–	–	–	SP	P	SP	–
Ichthyosaura alpestris	LC	III	–	–	–	SP	P	SP	–
Lissotriton vulgaris	LC	III	–	–	–	SP		SP	–
Lissotriton graecus [1]	NA	–	–	–	SP/-	SP	P	SP	–
Triturus carnifex	LC	II	II, IV	SP	SP/-				
Triturus cristatus	LC	II	II, IV					SP	
Triturus dobrogicus	NT	II	II	SP	SP/-			SP	
Triturus ivanbureschi[2]	NA	II	II, IV					SP	P
Triturus macedonicus[3]	NA	II	II, IV		SP/-	SP	SP	SP	P
Bombina bombina	LC	II	II, IV	SP	–			SP	

(continued)

Table 17.1 (continued)

Species	IUCN	BERN	HABITATS	NCS CROATIA	NCS BOSNIA AND HERZEGOVINA	NCS MONTENEGRO	NCS ALBANIA	NCS SERBIA	NCS NORTH MACEDONIA
Bombina variegata	LC	II	II, IV	SP	SP/-	–	SP	SP	P
Bufo bufo	LC	III	–	–	–	SP	P	SP	–
Bufotes viridis	LC	II	IV	SP	–	SP	SP	SP	P
Hyla arborea	LC	II	IV	SP	–	SP	SP	SP	P
Pelobates fuscus	LC	II	IV	SP	SP/-			SP	
Pelobates balcanicus[4]	NA	II	IV				SP	SP	P
Pelophylax epeiroticus	NT	III	–	*			P		
Pelophylax kl. *esculentus*	LC	III	V	–	–			P	
Pelophylax kurtmuelleri	LC	–	–	–	–	SP	P		
Pelophylax lessonae	LC	III	IV	SP	–			P	
Pelophylax shqipericus	VU	III	–			SP	P		
Pelophylax ridibundus	LC	III	V	–	–	SP		P	–
Rana arvalis	LC	II	IV	SP					

(continued)

Table 17.1 (continued)

Species	IUCN	BERN	HABITATS	NCS CROATIA	NCS BOSNIA AND HERZEGOVINA	NCS MONTENEGRO	NCS ALBANIA	NCS SERBIA	NCS NORTH MACEDONIA
Rana dalmatina	LC	II	IV	SP	–	–	SP	SP	P
Rana graeca	LC	III	IV		SP/-	SP	SP	SP	P
Rana latastei	VU	II	II, IV	SP					
Rana temporaria	LC	III	V	–	–	–	P	SP	–

IUCN—International Union for Conservation of Nature (VU—Vulnerable; NT—Near Threatened; LC—Least Concerned; NA—Not Assessed); BERN—Bern Convention on conservation of European wild flora, fauna, and natural habitats (Appendix II—strictly protected animal species, Appendix III—protected animal species); HABITATS—Habitats Directive or European Directive on Conservation of Natural Habitats and of Wild Flora and Fauna (Annex II—vulnerable/sensitive species which could become endangered in the near future if the factors of threat continue to act, Annex IV—species that require strict protection, Annex V—species which breeding in the wild and exploitation could be a matter of management); NCS—National Conservation Status (SP—Strictly Protected; P—Protected): Croatia—OG (144/13) and OG (73/16); Bosnia and Herzegovina—Decree on strictly protected and protected wild species (Official Gazzete of Republic of Srpska entity No 65, 7.7.2020)/ Rule Book on protection measures of strictly protected species and subspecies and protected species and subspecies (Official Gazzete of Bosnian Federation No. 66/13, 13.3.2020); Montenegro—Official Gazette of the Republic of Montenegro, No. 76/2006); Albania -Official Gazette of the Republic of Albania (07/1998) and (194/2014); Serbia—Official Gazette of Republic of Serbia, No. 5/2010; North Macedonia—Official gazette of the Republic of Macedonia. No. 139. 2011.

New species or revised taxonomy: [1] considered international status and NCS as of *L. vulgaris*; [2]considered international status and NCS as of *T. karelinii;* [3] considered international status and NCS as of *T. carnifex*; [4]considered international status and NCS as of *P. syriacus*. See in [21].

Grey cell—species doesn't occur in the country.

*Species is not included in the official amphibian species list of Croatia; however, recent data suggest their presence (non-native species) [40–43].

Fig. 17.1 Forest pond in Baranja, Croatia. The aquatic habitat of *L. vulgaris, B. bombina, B. bufo, H. arborea, P. fuscus, Pelophylax* sp. and *R. dalmatina*. Photo: O. Jovanović Glavaš

17.4 Small Water Bodies and Amphibians in Bosnia and Herzegovina

17.4.1 General Hydrographic Features of Bosnia and Herzegovina

Bosnia and Herzegovina cover an area of 51 129 km^2 and have significant water resources which represent one elementary economic potential [44]. Elementary basin areas in Bosnia and Herzegovina are the Black Sea basin, encompassing ca. 38 719 km^2 (75.70%) of the total territory of Bosnia and Herzegovina and the Adriatic Sea basin, encompassing ca. 12 410 km^2 (24.30%).

The territory of Bosnia and Herzegovina comprises moderate wetlands, with a mean annual precipitation value of ca. 1 250 mm, which, given the total area of the country (51,209 km^2), amounts to 64 × 109 m^3 of water, or 2 030 m^{3}s^{-1}. Of the total precipitation amount, ca 1 155 m^3/s (57%) flows into watercourses of the country's territory. Bosnia and Herzegovina are also characterized by numerous subterranean waters found in three geographically distinct areas each with specific its own features.

Fig. 17.2 Common tree frog—*Hyla arborea*. Photo: O. Jovanović Glavaš

17.4.2 The Origin and Status of Small Water Bodies

The territory of Bosnia and Herzegovina abounds in river flows, as well as in a relatively large number of lakes of various sizes and features. A number of them belong to the SWB category. According to data available [45], natural lakes cover only 0.11% of the total territory of the country which is 10 times less than the global average. Natural mountain lakes are spread in highland and mountain areas, usually found at the borders between forests and grasslands. They have emerged around pits and upstream, owing to various erosion processes.

SWB can be found close to the larger wetlands, such as the unique swamp systems of fishponds: Bardača, Hutovo Blato, and Livanjsko polje, internationally important for their specific and rich biodiversity. Besides, a certain percentage of the total precipitation goes into watercourses, ponds and brooks [46]. Ponds can be found anywhere (Fig. 17.3), but primarily in the waterless karst, thus being the only source of water in wide areas.

Fig. 17.3 Small water body near Kordići, Bosnia and Herzegovina. The aquatic habitat of *B. variegata*, *B. bufo* and *R. temporaria*. Photo: D. Jablonski

17.4.3 Amphibian Species Occurring in Small Water Bodies

Almost all of the amphibian species of Bosnia and Herzegovina [47–49] are directly linked to SWB by their life history (Table 17.1). In the peri-Pannonian lowlands of the northern part of the country, *T. dobrogicus* can be found in irrigation channels [49], *P.* kl. *esculentus* is present in ponds, springs and fishponds [50], while *B. bombina* is found mostly in ponds [51] and irrigation channels [52] in the lowlands along the Sava River. Here are also the only records of *P. fuscus* in the country, found in permanent mesotrophic or eutrophic canals, ponds, abandoned pits, and smaller lakes, rich in vegetation and with high oxygen concentrations [50, 53]. *Pelophylax lessonae* is underexplored, but it is assumed to be widely present along the Una River in the central part of northern Bosnia and Herzegovina [54].

In the mountainous region, the endemic *P. anguinus* has been found in smaller and larger springs, subterranean waters of pits and caves, wells, closed wells, estavelles, temporary springs and karst founts in the west and south-west of the country [55]. *Salamandra salamandra* lays its larvae in the moderately cold mountain lotic waters. *Lissotriton vulgaris* reproduces in standing and very shallow bodies of water. The Greek smooth newt (*L. graecus*) is linked to the ponds and creeks of the Popovo polje and Fatničko polje [49]. *Ichthyosaura alpestris* is widely found in ponds, swamps, lakes, springs and wells at altitudes between 100 m and 2 500 m [47], while *T. carnifex* is present in standing bodies of water in the highlands of western, southwestern and

Fig. 17.4 Common frog—*Rana temporaria*. Photo: D. Jablonski

central Bosnia and Herzegovina, with the westernmost locality lying in the ponds around the stadium in the city of Bihać [56].

Bombina variegata is found in still waters and ponds next to watercourses that dry up during summer, often in ephemeral ponds in forest landscapes (in the forest road tracks, owing to precipitation). *Bufo bufo* spawns in various standing bodies of water at altitudes of up to 2 000 m. *Bufotes viridis* spawns in the lowlands, but also in the highlands at altitudes of over 2 000 m. *Hyla arborea* is found in various SWB, mainly rich in vegetation, during the reproductive season, while *R. dalmatina* is linked to water exclusively during the breeding season, preferring shallow standing bodies of water in forests or in the vicinity of forests, while *R. graeca* is present in southeastern, central and northwestern parts, at altitudes of up to 2 000 m, often close to mountain creeks with rocky bottom. *Rana temporaria* (Fig. 17.4) is a terrestrial species found in the wet meadows, swamps, canyons and mountain creeks across central and northern Bosnia. *Pelophylax ridibundus* is a widely spread aquatic species, present in fishponds, ponds, lakes, channels and slow lowland rivers.

17.4.4 The Main Threats for Small Water Bodies

The wetlands in Bosnia and Herzegovina are generally perishing because of the negligence of relevant authorities and non-compliance with domestic laws, regardless of the fact that they play an important role in the mitigation of climate changes. This will eventually lead to sudden decline in the number of amphibian species.

Due to changes in the hydrological regime, swamps are quite endangered and also threatened by water pollution and erosion [57]. Ponds disappear or are overrun owing to the development of the water supply system industry [58]. In contrast, SWB as natural springs and creeks usually change very little. However, under the influence of certain causes (erosion, landslides, earthquakes), they can significantly change their form, position, shape and size. Seasonal (intermittent) springs and creeks can change their beds and even the direction of the flow, almost every year. Most of these changes are caused by human activity [58, 59]. In the karst regions, surface watercourses (rivers and creeks) are very vulnerable to human alterations. SWB exist only in places where sections of the terrain are built of impermeable layers (such as clay, marl, compact limestone and dolomite), or, if the river bed is eroded to the karst base [59].

"Run-of-river" small hydropower plants in the region are rapidly expanding and there are currently 81 active hydroelectric power plants, plus another 186 planned in Bosnia and Herzegovina [60]. They are eroding small mountain rivers—breeding sites for some amphibian species [13].

Pollution of the aquatic systems, including the fragmentation of habitats, building roads and artificial canals, and uncontrolled fishing as the key risk factor also jeopardizing SWB. The size of facilities used in fish exploitation in Bosnia and Herzegovina for trout fishponds amounts to 91 026 m^2 and another 2 278 ha for carp fishponds [61]. In addition, during the construction of roads, water supply systems, spring captivation and many other construction facilities, many SWB are being destroyed. Environmental licenses do not require concessionaires to conduct the monitoring of local amphibian populations nor to take proper care of SWB as their reproduction sites.

17.5 Small water bodies and amphibians in Montenegro

17.5.1 General Hydrographic Features of Montenegro

Montenegro covers an area of 13 812 km^2. It is a Mediterranean country with high mountains along the coastal line, which spread inland in the form of calcareous plateau and the high mountains of the central part of the country, intersected by numerous, larger or smaller river valleys [62]. Hydrographic and hydrological features have resulted in the appearance of numerous rivers, springs, and underground water accumulations [63]. About 47.5% of the Montenegrin territory belongs to the Adriatic basin, and 52.5% to the Black Sea basin. Although the Adriatic basin has a smaller surface area in comparison to the Black Sea basin, it provides about 50% more onflow.

The relatively small Montenegrin territory contains almost all of the known forms of surface and underground waters, as well as numerous karstic hydrologic phenomena, since 75–80% of its territory is karst [64]. Montenegro has 30 natural lakes which accumulate over 4 billion m^3 of water (4 118 928 874 m^3) during high water-level. Artificial lakes, with a total volume of over 1 billion of m^3 of water, have been created mostly for the hydro-energy exploitation, while a small amount of this water is used for irrigation and fish farming.

Montenegro is among the world's richest countries with regard to water and the average specific outflow is more than six times higher than the world's average (found in only 3% of the Earth's territory). All the continental waters of Montenegro, both surface and underground ones, except for part of the river Cijevna, are formed within the country, which is an extremely valuable ecological factor.

17.5.2 The Origin and Status of Small Water Bodies

In comparison with all other Mediterranean countries, Montenegro has the largest percent of its territory under wetlands. There are 7.7% of wetland ecosystems [65], which include marshes, peatlands, natural and artificial waters (permanent or running, standing or flowing), as well as the coastal zone adjacent to wetlands. In wetlands we also include those made by human activities, such as fishponds, salinas, reservoirs, and channels [66]. Classification of habitats in Montenegro, important for the European Union, according to the cataloguing in [67], recognizes several types of surface continental water habitats: natural eutrophic waters with vegetation *Magnopotamion* i *Hydrocharition* (3150); shores of oligotrophic to mesotrophic stagnant waters with amphibious vegetation (3130) occur in all parts of Montenegro; hard oligo-mesotrophic waters with bottoms covered with Charas (3140) are relatively common habitats in all parts of the country; occasional karst lakes (turlozi); Mediterranean occasional ponds (3170); watercourses with common water crowfoot vegetation (3260) are relatively common habitats in Montenegro; transitional bogs—communities of transition mires and quaking bogs (7140); occasional flows of Mediterranean rivers (3290) present in the sub-Mediterranean-Mediterranean part of Montenegro; alkaline peat-wetland habitats are moistened by groundwater or surface waters that are very rich in bases.

On the plateau of deep karst, there are many small water reservoirs (stone cisterns, wells, pools, and snow pits) as the source of drinking water for cattle. Examples include a large number of stone ponds, wells, pools, and baths in the area of Banjani, or large numbers of isolated stone pools and wells in the area of Riječka nahija; on the karstic massif of Sinjajevina, numerous surface and subterranean forms of relief support SWB: karrens, karst sinkholes, depressions, rounded peaks and fields, formed by the activity of glaciers in the geological past [68].

17.5.3 Amphibian Species Occurring in Small Water Bodies

Fourteen different species comprising five species of Caudata and nine species of Anura registered in Montenegro [69] have been recorded in SWB (Table 17.1), particularly during the reproductive season [70, 71]. Natural and anthropogenic ponds are very important biotopes for batrachofauna. They mostly represent rounded depressions in the land that may have already existed or been excavated afterwards. For example, in the area of Sinjajevina, these ponds usually have no walls (a small number of ponds have walls) and are in the form of simple puddles filled by rainwater used for cattle watering (Fig. 17.5). Within this research area, 33 water objects of natural or human origins have been registered.

Only the Alpine salamander *(S. atra)* has not been detected in SWB. Some species, like *P. shqipericus* and *P. ridibundus/kurtmuelleri* are often found in flooded areas, as well as in oxbows and overgrown parts of slow-flowing rivers and canals [10, 72]. Green frogs predominantly inhabit freshwater basins, but their presence is also confirmed in shallow brackish habitats [72, 73]. Toads and newt species: *B. bufo*, *B. viridis*, *I. alpestris* (Fig. 17.6) and *B. variegata* are also frequently found in SWB, as well as in the large lakes [74–76]. The fire salamander *(S. salamandra)* usually prefers the vicinity of small streams with rocky bottom.

Fig. 17.5 Pond on Mountain Sinjajevina, Montenegro. The aquatic habitat of *B. varigata, I. alpestris,* and *B. bufo.* Photo: N. Čađenović

Fig. 17.6 Alpine newt—*Ichthyosaura alpestris*—frequently inhabits alpine lakes and ponds of Montenegro. Photo: N. Čađenović

The importance of SWB in maintaining rich amphibian diversity can be perceived in several examples: in the area of Banjani in western Montenegro, some species have an emphasized metapopulational organization, where the populations are spatially organized in larger number of discrete *demes* (reproduction centers) connected by a limited gene flow through migrants; hence, the genetic exchange is much larger within such reproductive units than between them [77]. This plateau of deep karst is the scarcest regarding surface waters, but has diverse SWB made for cattle in a single area, which has resulted in a great diversity of amphibians (*H. arborea*, *B. bufo*, *B. viridis*, *I. alpestris*, *L. graecus*, *B. variegata*, and *P. ridibundus/kurtmuelleri*) in these objects during the reproductive period. In the area of Riječka nahija, *T. macedonicus*, *B. variegata*, *L. vulgaris*, *B. bufo*, *I. alpestris* and *P. ridibundus/kurtmuelleri* inhabit SWB [70, 78]. At the karstic massif of Sinjajevina, seven species of amphibians (five species of Anura and two species of Caudata) were detected in and around SWB [79].

17.5.4 The Main Threats to Small Water Bodies

Since most of the registered amphibian species in Montenegro dwell in SWB, conservation and protection of these waters, as well as the protection of water and water-dependent ecosystems, should be of concern and an obligation for the state, local authorities and all economical stakeholders. Alteration, fragmentation, and habitat loss are among the greatest threats for amphibian populations [80].

Within some coastal parts of Montenegro, there has been a continued longstanding alteration of the water regime and the reduction of wetlands for the purpose of agriculture, settlements, and tourism. Filling and pollution of habitats with solid waste is very frequent. Pollution of habitats with wastewater has a negative impact on the process of amphibian metamorphosis and causes malformation of the reproductive system [81]. Particularly SWB in the lowland wetlands of southeast Montenegro, inhabited by the endangered Skadar frog (*Pelophylax shqipericus*), a stenoendemic species of the Balkan peninsula and recognized as vulnerable species (VU) by IUCN, should be carefully monitored and protected. It is important to emphasize that the vulnerability of green frogs is also augmented by multiple decades of catching for commercial purposes [82].

Small rivers are recently becoming increasingly threatened by the construction of small hydropower plants and the consequent devastation of the riverbeds [13].

Drainage systems for inundated lowlands and forest exploitation negatively affects the water quality and flow uniformity of streams and consequently the aquatic wildlife [63]. Solid waste depots are permanent air polluters; they are, unfortunately, often situated near the water streams [63]. Natural lakes are particularly threatened by erosive processes and intensive eutrophication, and uncontrolled fish re-stocking has led to the questionable survival of several subspecies of alpine newt.

17.6 Small Water Bodies and Amphibians in Albania

17.6.1 General Hydrographic Features of Albania

Albania (28 748 km^2) has a specific position in the Balkan peninsula. It is situated in the western part of the peninsula and its territory touches the surface of all three big lakes of the Balkans: Lake of Shköder in the northern part of the country, sharing its waters with Montenegro, and the Ohrid and Prespa lakes in the southeastern part, sharing waters with North Macedonia and Greece. Moreover, the country lies between two main topographical units, the Dinarides in the north and the Hellenides/Albanides in the rest of the country. These two units have a clear biogeographic border formed by the Drin River flowing approximately from the Kosovo Polje to the Adriatic Sea. It is simultaneously the longest river in the country, with a length of 285 km. Albania has more than 152 rivers or streams, the most important of which are the Buna, Drin, Mat, Ishëm, Erzen, Shkumbin, Seman, and Vjosë.

Albania is a very water—rich country, with 40 billion meters per year running to the sea [83]. This is clearly due to its topography and the variety of high elevation mountains with a mean elevation of approximately 700 m. The only lowland area is the Western Lowlands, consisting of the sea, rivers, numerous lagoons, marshes, and ponds, which form diverse habitats and ecosystems supporting a rich biodiversity.

The river Vjosë has a specific position; it belongs to very rarely seen rivers in Europe today, sometimes called the last wild river [84, 85].

In addition to the three largest lakes in the Balkans (and one of the oldest in the world), Albania is known to have around 250 other lakes of karstic or glacial origin. Its small, mountain lakes are well-known e.g. the Lura lakes situated in the Lura Mountains in northern Albania, or the Valamarë lakes in the southeast. In the Lura Mountains, twelve lakes are situated at a high elevation of between 1 500 and 1 800 m and are unique for their distinctive colors. Albania is also known for its thermal or karst waters, representing an important resource [86, 87].

The hydrographic basin of Albania has a total area of 43 305 km^2, i.e. 50% more than the territory of the country [88]. Thus, it is not surprising that one of its most important natural resources is groundwaters, including SWB. Apart from wetlands of natural origin, Albania is rich in artificial water reservoirs (623 water reservoirs) built for irrigation and hydropower production, of which more than half are SWB [89].

17.6.2 The Origin and Status of Small Water Bodies

In Albania, the majority of SWB are connected with river systems or anthropogenic activities. Around 250 lakes that have glacial or karst origin and a size smaller than 1 ha can be found throughout Albania (Fig. 17.7). More than 130 glacial lakes of Albania are formed by magmatic and terrene formations;

On the other hand, coastal lagoons and marshes are formed of Quaternary deposits with an alluvial origin [90]. Other SWB are parts of river systems or they have an artificial origin, such as water reservoirs (more than 350), excavations (3 sites), or aquaculture ponds (3 sites) [89]. The first attempt to create a national habitat classification of water bodies in Albania was made in early 2000, with the first Wetland Inventory Project, in which the Ramsar and MedWet wetland classification was applied [89]. However, water bodies and wetlands smaller than 0.36 ha were excluded from this inventory. SWB located inside the network of protected areas are protected, but all other SWB falling outside this network are facing many threats and pressures of anthropogenic origin.

17.6.3 Amphibian Species Occurring in Small Water Bodies

The rich fauna of Albania, including amphibians, has still not been completely explored [91]. Despite relatively good knowledge of the diversity and distribution of amphibians in the country [92–94], data on ecology, habitat preferences, and threats. are generally scarce.

Of the 16 species of amphibians recorded in Albania, only *S. atra* and *P. balcanicus* are not known for using SWB during its life. The distribution range of these species

Fig. 17.7 The artificial SWB in the Shëndelli mountain of southern Albania, the aquatic habitat of *L. graecus*, *T. macedonicus*, and *B. variegata*. Photo: D. Jablonski

in the country includes only the extreme north (*S. atra*) and southeast (*P. balcanicus*) parts, where they are uncommon [92, 94]. Other amphibian species regularly use SWB (Table 17.1), especially during the reproductive period, depending on the geographic location and the type of terrestrial habitat. Regarding the Salamandridae family, *I. alpestris* and *L. graecus* are typical inhabitants of SWB, especially in the mountain or forest areas. On the other hand, *T. macedonicus* inhabits mainly larger water bodies from lowlands to high mountains (e.g. Korab, Jablonski 2013, pers. obs.). For most of the year, especially in the Mediterranean parts of the country, SWB are used by *B. variegata*, *Pelophylax* spp., *B. bufo*, *B. viridis*, *H. arborea*, *R. dalmatina* and *R. graeca*. Several species, *B. variegata*, *B. bufo*, *R. graeca*, and the dubious taxon *P.* (*ridibundus*) *kurtmuelleri*, were recorded as inhabiting thermal water habitats ([95, 96] and Jablonski, pers. obs.).

17.6.4 The Main Threats for Small Water Bodies

Freshwater bodies in Albania are under pressure from different sources [91]. Albania was for most of the twentieth century a pristine country, free of major landscape changes connected with anthropogenic activities. For example, due to the number of wetlands, malaria was common for a long time in the last century. However, in the early 1950's and 1960's, Albanian wetlands were profoundly altered through the investigation of a marshland reclamation scheme and the expansion of agricultural land by alteration of riparian woodlands. More than 50% of the coastal wetlands and most of the inland marshy areas were lost due to the development of drainage projects, which negatively impacted amphibian species, although newly established irrigation reservoirs in hilly and mountain areas were used by some amphibians [97]. More recently, since 1992, when the administrative structures collapsed, following the collapse of the Communist regime, a second degradation has occurred leading to uncontrolled urban and tourism development [90].

Waters in Albania are currently threatened by phosphorus pollution, pesticides, littering, or development, especially around big cities. Lake Ohrid is a good example of how formerly clear water may change [98]. This is also the major problem for the unique local biodiversity connected with SWB. In the context of amphibians and SWB, Albania represents the main distribution area for the endemic species of water frog—*Pelophylax shqipericus* (Fig. 17.8). This species is distributed in a small area with a size of ~ 5000 km^2. Most of the species range is limited to coastal areas where massive development has occurred in the past years and the species might be threatened mainly by fragmentation and degradation of habitats. Moreover, the area is also the main agricultural region of Albania, and drainage or chemical pollution may be an important factor that will decrease populations of *P. shqipericus* as well as other amphibian species [99]. A similar threat might occur for another endemic, *P. epeiroticus*, reaching the Albanian territory in the extreme south.

Fig. 17.8 Albanian Pool Frog—*Pelophylax shqipericus*—the endemic frog species for the Western Balkans, photographed at Torovica locality in Albania. Photo: D. Jablonski

Other amphibians in Albania are also using SWB during their life. Therefore, it is absolutely necessary to preserve SWB (especially in the Mediterranean regions with limited precipitation) as an important source of water for local amphibians. On the other hand, it seems that small alpine or karst lakes are not under high pressure, but might be seasonally threatened by tourism or by cattle breeding as mountain lakes represent a source of waterholes for cattle. Furthermore, some of these glacial lakes, such as those in Lura, Balgjaj and Valamarë are under serious pressure from small scale hydropower projects that are planned or under construction in those areas (Bego, pers. comments).

17.7 Small water bodies and amphibians in Serbia

17.7.1 General Hydrographic Features of Serbia

Serbia covers an area of 88 361 km^2, and, regarding macro-relief, has three distinct areas: The Pannonian lowland in the north of the country; the Perripanonian flatlands along the Sava and Danube rivers, as well as along the lower parts of the Drina, Kolubara, Morava, Mlava and Pek rivers; the mountainous area, which is the largest of the three and includes parts of Dinaride, Karpato-Balkan and Rhodopi mountain belts, intersected by a few large river valleys and numerous small rivers and streams [100]. The most of the territory of Serbia belongs to the Danube (Black Sea) drainage basin (92%), additional 5% to the Adriatic drainage basin, and 3% to the Aegean basin.

Serbia is a hydrologically not very rich country: only 1.09% of the total surface of the country belongs to inland waters, both running and standing ones [101]. The wetlands/floodplains comprise the largest part of these inland waters, associated with the lowland rivers of the Pannonian area. Although the wetlands were heavily degraded through the process of conversion to the either arable or urban land, some remain in their original state, thanks to the international recognition of value which the wetland ecosystem services provide; these are proclaimed nationally protected areas and nature reserves [100].

Rainfall on the territory of Serbia depends on relief and climate, providing mean annual precipitation of 730 mm [102]. Serbia is, also, the poorest Balkan country in autochthonous surface waters [103]. The mostly continental pattern of precipitation is reflected in the highest amount of rain falling in May and June, while the least precipitation is in February and October; the southernmost part of the country has somewhat different precipitation regime, as there the most rain falls from October to January and least in August [104].

17.7.2 The Origin and Status of Small Water Bodies

A large number of SWB in Serbia are of anthropogenic origin: in the lowlands of Voivodina in the Pannonian part of Serbia, intensive habitat alteration started at the beginning of eighteenth century. During colonization of the area, people were predominantly using soil for building houses, and by digging the soil they made numerous small and middle-sized pits, which, in time, were filled with atmospheric water and became suitable for local amphibian species [105]. Villages in the hilly-mountain area of Serbia consisted of scattered households with relatively small plots of arable land. Every household maintained a water-collecting place for the livestock, often in the form of pit dug in the ground and filled with atmospheric water (Crnobrnja-Isailović, field notes). Over the years, those pits became breeding sites for local amphibians (Fig. 17.9). Apart from watering holes for cattle, some SWB suitable for amphibians originated from abandoned quarries, or clay and gravel pits [106].

Small water bodies in Serbia which provide aquatic habitat for Strictly Protected Wild Species are formally protected from any degradation, alteration and destruction by national legislation (see in: [107]). Out of 21 autochthonous amphibian species in the country, 20 utilize SWB, and 17 of them are Strictly Protected (Table 17.1).

Fig. 17.9 Pond on Miroč mountain, Eastern Serbia, aquatic habitat of *T. cristatus*, *L. vulgaris*, *B. bufo*, *B. variegata*, *Pelophylax* sp. and *R. dalmatina*. Photo: J. Crnobrnja-Isailović

Actually, all amphibian species in Serbia, except for water frogs, belong to this category, and this suggests that the most of SWB in Serbia are indirectly protected by the national law. However, it does not mean that SWB are not subjected to degradation and destruction, because there are discrepancies between theory and practice, or between the rules defined by the law and the efficiency of their implementation in real life.

17.7.3 Amphibian Species Occurring in Small Water Bodies

Only one of 21 amphibian species recorded in Serbia (Table 17.1) is not found in SWB: the alpine salamander (*S. atra*), which is an entirely terrestrial species. The Greek frog (*R. graeca*) spawns in running waters [108], which also includes small rivers and streams categorized under SWB.

Most amphibian species in Serbia spawn in SWB (Table 17.1), often of anthropogenic origin. Relatively soon after the construction of a water hole or a pit, a pioneer amphibian species colonizes it; in the mountain areas it is often *B. variegata*, followed by *I. alpestris* (Pešter Plateau, Crnobrnja-Isailović 1996, 1997, field notes). The structure of these man-made aquatic habitats and the general occurrence of aquatic habitats in the area shape the diversity of local amphibian assemblages.

One example from Western Serbia (Peripannonian region) is the roughly circular water collecting pit, dug in the ground on a private estate, situated near the stable, in the shady part of a courtyard, 4m^2 large, 1.5 m deep, and turbid. For decades, this pond was the breeding place for local *T. macedonicus*, *R. dalmatina* and *B. variegata* (Crnobrnja-Isailović, 2001, field note). The size of the SWB would not be a factor which predominantly determines the number of species in a local amphibian assemblage; rather it could be a combination of several parameters such as overall number of amphibian species in the area, density and composition of submersed vegetation, the presence of aquatic predators, and the degree of pollution.

In mostly forested areas or complex habitat systems in hilly-mountain area south from the Sava and Danube rivers, consisting of open deciduous forests and meadows at elevations ranging from 40 to 800 m above sea level, the predominant amphibian species in SWB are those which typically inhabit deciduous forest habitats. For example, on the territory of "Djerdap" National Park and in the buffer zone, an inventory done in 2012 revealed that 79% of potentially occurring amphibian species used an SWB as a breeding site [109]. *R. dalmatina* and *Pelophylax* species were recorded in 52 and 36% of recorded SWB, respectively, followed by *B. bufo* (Fig. 17.10), *S. salamandra* and *B. variegata* (found in 32% of known SWB). The remaining five amphibian species were detected in only 4% of recorded SWB. For comparison, 18 surveyed SWB at Pešter plateau in southwestern Serbia, where elevation varies from 800 to 1300 m above sea level, harboured 33.3% of detected amphibian species [110]. *B. variegata* and *Pelophylax* sp. were recorded in 67% of surveyed SWB, followed by *I. alpestris* and *L. vulgaris* (found in 50 and 44% of surveyed SWB, respectively),

Fig. 17.10 Common toad—*Bufo bufo*—common amphibian species in SWB of Serbia. Photo: J. Crnobrnja-Isailović

while *B. bufo* and *Ranae fuscae* (brown frogs) were found in only 6% of surveyed SWB.

17.7.4 The Main Threats for Small Water Bodies

In the past, the Pannonian and Peripannonian parts of Serbia had been subjected to the most drastic habitat changes: In the Pannonian part, fragmentation and drainage of wetlands started as far back as in the XVIIIth century, with later intensification, especially after the end of World War II; this resulted in the same or even higher levels of destruction of freshwater habitats in the Peripannonian part of Serbia—along the Morava river [105].

The Balkan part of Serbia was characterized by mass-migration of people to the urban areas, starting at the end of World War II. In time, numerous households became abandoned or without cattle. Consequently, many watering holes e.g. anthropogenically originated SWB, became abandoned. Succession and/or (recently) climate change has led to their overgrowth and desiccation [100]. Certain ponds became dumping places for all kind of waste [106, 111] or were simply drained by landowners to make orchards (for example: a pond in Gledeće village in the vicinity of Ljubovija town in Western Serbia; Crnobrnja-Isailović 1998, field note), or potato fields (for example: area of Mlekominci at the outskirts of Bosilegrad town in Southeastern Serbia [106]). A survey done in year 2003 on 53 great crested newt (*T. cristatus* species complex) breeding sites, previously recorded a few decades before, revealed that 40% of them no longer exist, as, over time, they were transformed into the agricultural land or they simply dried out; an additional 9% were transformed into private fish farms [106]). In short, a number of SWB in Serbia which lost their function of water storages or fishponds were degraded or destroyed.

17.8 Small water bodies and amphibians in North Macedonia

17.8.1 General Hydrographic Features of North Macedonia

North Macedonia covers an area of 25 713 km^2. There are different types of water bodies, such as groundwater, springs, flowing waters, streams and rivers, as well as natural lakes and reservoirs. The total water wealth of the country is estimated at about 26 billion m^3. Accordingly, North Macedonia could be categorized as a medium sized, water-rich country. Water resources are unevenly distributed on its territory, as a result of different geological compositions, relief structures, and climatic characteristics. Even 84% of the water springs originate from North Macedonia, while only 16% come from beyond the borders [112].

The rivers in North Macedonia belong to three river basins: the Aegean, Adriatic, and Black Sea. The Aegean river basin is the largest in North Macedonia and covers an area of 22 319 km^2 or 87% of the total territory of the country. It consists of the Vardar river basin (with river tributaries this includes 80% of the territory of the country).

Within the borders of North Macedonia there are 160 lakes with a total area of about 500 km^2, i.e., about 2% of the total area. Of these, 50 are natural and the remaining 111 are artificial—reservoirs. There are three tectonic lakes in North Macedonia: Ohrid Lake, Prespa Lake, and Dojran Lake [112]. Besides these tectonic lakes, there is Moklishko Lake, a coastal-spring (Island near the monastery of St. Naum), six swamp lakes, several cave lakes, and 43 glacial lakes (mostly on Shar Planina—21, and the rest are on Korab, Deshat, Jablanica, Stogovo, Pelister, and Jakupica).

Regarding the artificial lakes there are 110 reservoirs in North Macedonia. Of these, 22 have a total volume of more than 1 million m^3, and the rest have a smaller volume. Larger artificial lakes comprise Tikveš Lake on Crna Reka, Kozjak on Treska River, Mavrovo Lake on Mavrovska Reka, Spilje on Crn Drim, Kalimanci on Bregalnica River, and Strezevo on Semnica River [112].

17.8.2 The Origin and Status of Small Water Bodies

From a socio-economic point of view, North Macedonia is a country with a rich tradition of agriculture, and a special emphasis on animal husbandry. The distribution of SWB mainly refers to rural areas/villages where the peoples' primary means of subsistence is agriculture/livestock. In general, many of these SWB are man-made watering holes, as well as temporary reservoirs for irrigating small garden areas, both in lowland and mountainous areas. Also, some SWBs are formed along rivers and streams under conditions of the spring rise of water level as a result of snow melting. The snow forms small alluvials which are important habitats for aquatic animals and breeding sites for amphibians (Fig. 17.11).

Agriculture and animal husbandry are more pronounced in the southern, south-eastern, southwestern, and western parts of North Macedonia [113], where.

SWB are mainly distributed. It is interesting that a high concentration of SWB can mainly be observed in the eastern and south-eastern parts of the country (Kočani and Strumica fields) [114].

All of the above also refers to the lowland and mountainous areas where agriculture/livestock is present [114] Additionally, some SWB result from the extraction of sand and sediment, and the digging of irrigation channels. These artificial SWB are generally present in the vicinity of river flows in the lowland parts of the country as well as in the parts that have intensive agriculture (Pelagonia plain, Polog plain, Kočani field). The highest frequency of SWB is observed in Kočansko Pole where

Fig. 17.11 Small alluvial water body near near Graeska reka in North Macedonia, aquatic habitat of *B. variegata, P. ridibundus, L. vulgaris/graecus* and *T. macedonicus*. Additionaly, *H. arborea* was found in close proximity. Photo: B. Sterijovski

the cultivation of rice is traditionally the main agricultural branch, as well as in Strumica valley where agriculture is the primary economic strata for the majority of local inhabitants.

In certain parts of North Macedonia, SWB are the basic reproductive centers for amphibian species, as in the central part of the country, where, according to the climatic-vegetation conditions, the area is characterized as steppe [115]. The Slan Dol (lower course of the river Bregalnica) is characterized by dry, hot summers [116] so all water bodies (except the river Bregalnica) dry up during the summer [116]. The SWB that appear during the spring months are extremely important for the survival of the local amphibian species as their key and unique reproductive centers.

17.8.3 Amphibian Species Occurring in Small Water Bodies

Eight of the 14 species of amphibians present in North Macedonia [118], have the status of nationally protected [118]: *T. ivanbureschi, T. macedonicus, B. variegata* (Fig. 17.12), *B. viridis, H. arborea, P. balcanicus, R. dalmatina* and *R. graeca* (Table 17.1). According to the database made for the compilation of the national Red List [120], 12 species (80% of all amphibian species in North Macedonia) use SWB among the other aquatic habitats (Table 17.1). As much as 27% of the records on amphibian species in North Macedonia were done in SWB. Three amphibian species

Fig. 17.12 Yellow-bellied toad—*Bombina variegata*—in SWB of "Galičica" National Park in North Macedonia. Photo: D. Jablonski

(*H. arborea, R. dalmatina, R. graeca*) are considered as Near Threatened (NT), two species (*T. macedonicus, P. balcanicus*) are categorized as Vulnerable (VU), while four species (*I. alpestris, L. vulgaris, T. ivanbureschi, R. temporaria*) are considered as Endangered (EN) by the National Red List [120].

The fact that almost one third of the nationally recorded amphibians were observed in SWB and that 33.3% of all amphibian species observed in North Macedonia have an endangered status on the National Red List highlights the exceptional importance of protecting and conserving these water bodies. Also, local amphibian diversity hotspots overlap with the geographic position of SWB nationwide. This shows that SWB play an exceptional role in maintaining the national amphibian diversity. Additionally, in the arid region of the central part of North Macedonia, the existence of SWB is a basic prerequisite for the distribution and dispersion of amphibian species. Sterijovski and Arsovski [120] found that the high diversity areas for amphibians in North Macedonia are overlapping with the areas where agricultural/animal husbandry is present, thus connecting it with SWB.

Of course, this also overlaps with the presence of the large water bodies (such as large lakes—natural and artificial, big rivers, and streams), but 27% of all records for amphibians are in the SWB. This indicates that SWB are the key factor for wider dispersion of the amphibians in North Macedonia.

17.8.4 The Main Threats for Small Water Bodies

Modification and fragmentation of amphibian habitats is recognized as the greatest threat [121] that puts direct pressure on SWB. This process has been even more intense in the past three decades and it is still going on [122]. Historically overviewed, in North Macedonia this process started after the Second World War when most of the swamps and marshes were drained due to occupy the land for agriculture and to fight against malaria [123]. The SWB were not excluded, either. This was especially prevalent in the southern and southeastern parts of the country [123]. More recently, field research has shown the drying out of SWB. Climate change is leading to changes in climatic seasons that directly affect biodiversity at all levels, since species and ecosystems have adapted to living in a particular climate and often cannot adapt to higher temperatures, more frequent droughts and more extreme weather events [112].

In addition, the deteriorating socio-economic situation is forcing people to abandon traditional farming and cattle breeding practices. This directly affects the condition of SWB as they slowly but surely disappear [112]. Abandonment of pastoral systems and lack of grazing are recognized as the highest threat to biodiversity [112]. More recently, the practice of installing small hydropower plants in mountainous regions has indirectly affected SWB. Namely, insufficient control in situ by the competent institutions means that they have overlooked cases where an ecological minimum for water flow is not provided. This creates completely dry riverbeds and the SWB that are formed by the rivers as small alluvial enclaves are lost [112].

Threats to small waterfowl have been recognized in the latest strategy, with an action plan for the biodiversity of North Macedonia. Recommendations for the conservation of SWB are primarily related to the socio-economic situation of the people.

17.9 Conclusions

This study revealed that 97% of 29 amphibian species listed for the Western Balkans use SWB as suitable habitats during at least one phase of their life cycle. It confirmed statements from the literature [12, 15, 16, 124] that SWB (ponds, small lakes, ditches, springs and small streams) are important for conservation of local and regional amphibian diversity, especially in regions under the high impact of economic transition, which the Western Balkans certainly is. As it was pointed out in [124], SWB are often the last intact freshwater habitats in the area which provide a refuge for species have already disappeared from larger, more damaged water bodies. On the territory of the Western Balkans, freshwaters, together with forests, are treated as free capital which can be easily converted into economic benefit, which implies intensive exploitation organized by specific parties with a common interest. As a result, SWB are under continuous threat of pollution (Fig. 17.13), degradation, alien species (mostly fish) introduction, and are disappearing, either due to anthropogenic impact (drying out or abandoning, which refers to a SWB of anthropogenic origin formerly actively maintained for keeping water, mostly because of cattle raising), or due to

Fig. 17.13 A wasted SWB, aquatic habitat of endemic amphibian species. Photo: D. Jablonski

natural causes (succession and/or change of climate). Intensive habitat alterations, such are the construction of new motorways and highways, the opening of new mines for exploitation of rare or essential minerals, deforestation, establishment of small hydropower plants, and many others, additionally jeopardize already fragmented amphibian populations.

Based on the fact that amphibians are globally the most threatened vertebrate group [1], the loss of their essential habitats is not just a national issue. The conservation of SWB is in many cases related to the socio-economic status of the people i.e. to the stimulation of local stakeholders to continue with traditional practices of agriculture and animal husbandry, including maintenance of SWB, and to their continuous education on the importance of SWB for the preservation of high values of local biodiversity. Additionally, SWB that occur naturally should be regularly monitored for threats and any detected disturbance, referring either to an unusual change of water level or to the decline of populations and/or species, which should be properly mitigated.

Acknowledgements Text editing was done by Mrs Esther Helaizen. JCI was supported by the Ministry of Education, Science and Technological Development of the Republic of Serbia, Contracts No. 451-03-9/2021-14/200124 and No. 451-03-9/2021-14/200007. DJ was supported by the Slovak Research and Development Agency under the contract no. APVV-19-0076.

References

1. The IUCN Red List of Threatened Species (2020) Summary statistic tables. https://www.iucnredlist.org/resources/summary-statistics#Summary%20Tables. Accessed Aug. 8th 2020
2. Beebee TJC, Griffiths RA (2005) The amphibian decline crisis: a watershed for conservation biology? Review. Biol Cons 125:271–285
3. Barinaga M (1990) Where have all the froggies gone? Science 247:1033–1034
4. Anthony B, Arntzen JW, Baha El Din S, Böhme W, Cogalniceanu D, Crnobrnja-Isailović J, Crochet P-A, Corti C, Griffiths R, Kaneko Y, Kuzmin S, Wai Neng Lau M, Li P, Lymberakis P, Marquez R, Papenfuss T, Pleguezuelos JM, Rastegar N, Schmidt B, Slimani T, Sparreboom M, Ugurtas I, Werner Y, Xie F (2008) Amphibians of the Palaearctic realm. In: Stuart SN, Hoffmann M, Chanson JS, Cox NA, Berridge RJ, Ramani P, Young BE (eds) Threatened Amphibians of the World. Lynx Edicions, with IUCN—The World Conservation Union, Conservation International and NatureServe, Barcelona, pp 106–113
5. Kolozsvary MB, Swihart RK (1999) Habitat fragmentation and the distribution of amphibians: patch and landscape correlates in farmland. Can J Zool 77:1288–1299
6. Ray N, Lehmann A, Joly R (2002) Modelling spatial distribution of amphibian populations: a GIS approach based on habitat matrix permeability. Biodivers Conserv 11:2143–2165
7. Vences M, Köhler J (2005) Global diversity of amphibians (Amphibia) in freshwater. Hydrobiologia 595:569–580
8. Gasc J-P, Cabela A, Crnobrnja-Isailović J, Dolmen D, Grossenbacher K, Haffner P, Lescure J, Martens H, Martinez-Rica JP, Maurin H, Oliveira ML, Sofianidou TS, Veith M, Zuiderwijk A (eds) (1997) Atlas of Amphibians and reptiles in Europe. Societas Europaea herpetologica & Museum Nationall d' Histoire Naturelle (IEGB/SPN), Paris
9. Arnold EN, Ovenden DW (2002) Reptiles and amphibians of Europe. Princeton University Press, Princeton and Oxford

10. Speybroeck J, Beukema W, Book B, Van der Voort J (2016) Field guide to the Amphibians and Reptiles of Britain and Europe. Bloomsbury Natural History, London
11. Semlitsch RD (2000) Principles for management of aquatic-breeding amphibians. J Wildl Manag 64:615–631
12. Arntzen JW, Abrahams C, Meilink WRM, Iosif R, Zuiderwijk A (2017) Amphibian decline, pond loss and reduced population connectivity under agricultural intensification over a 38 years period. Biod Cons 26:1411–1430
13. Crnobrnja-Isailović J, Jovanović B, Ilić M, Ćorović J, Čubrić T, Stojadinović D, Ćosić N (2021) Small Hydropower Plants' Proliferation Would Negatively Affect Local Herpetofauna. Mini Review. Frontiers in Ecology and Evolution 9: Article 610325
14. Kats LB, Ferrer RP (2003) Alien predators and amphibian declines: review of two decades of science and the transition to conservation. Divers Distrib 9:99–110
15. Riley WD, Potter ECE, Biggs J, Collins AL, Jarvie HP, Jones I, Kelly-Quinn M, Ormerod SJ, Sear DA, Wilby RL, Broadmeadow S, Brown CD, Chanin P, Copp GH, Cowx IG, Grogan A, Hornby DD, Huggett D, Kelly MG, Naura M, Newman JR, Siriwardena GM (2018) Small water bodies in great britain and ireland: Ecosystem function, human-generated degradation, and options for restorative action. Sci Total Environ 645:1598–1616
16. Curado N, Hartel T, Arntzen JW (2011) Amphibian pond loss as a function of landscape change—a case study over three decades in an agricultural area of northern France. Biol Cons Biol Conser 144:1610–1618
17. Denoël M, Lehmann A (2006) Multi-scale effect of landscape processes and habitat quality on newt abundance: implications for conservation. Biol Cons 130:495–504
18. Wilkinson JW, Arnell AP (2013) NARRS Report 2007–2012: Establishing the Baseline (HWM Edition). ARC Research Report 13/01
19. Heatwole H, Wilkinson JW (2015) Amphibian Biology. Volume 11. Status of Conservation and Decline of Amphibians: Eastern Hemisphere. Part 4. Southern Europe and Turkey. Pelagic Publishing, Exeter, UK
20. Dufresnes C (2018) Patterns of amphibian diversity in the Western Palearctic. Herpetol Bull 145:28–30
21. Speybroeck J, Beukema W, Dufresnes C, Fritz U, Jablonski D, Lymberakis P, Martinez-Solano I, Razzetti E, Vamberger M, Vences M, Voros J, Crochet P-A (2020) Species list of the European herpetofauna—2020 update by the taxonomic committee of the societas Europaea herpetologica. Amphibia-Reptilia 41:139–189
22. Frost DR (2021) Amphibian species of the world: An online reference. Version 6.1 (*04/04/2021 07:43 am GMT*). Electronic Database accessible at https://amphibiansoftheworld.amnh.org/index.php. American Museum of Natural History, New York, USA
23. Hanski I (1999) Metapopulation ecology. Oxford University Press
24. Werner EE, Yurewicz KL, Skelly DK, Relyea RA (2007) Turnover in an amphibian metacommunity: the role of local and regional factors. Oikos 116:1713–1725
25. Marsh DM, Trenham PC (2001) Metapopulation dynamics and amphibian conservation. Conserv Biol 15:40–49
26. Anonymous (2004) Emerald Network Pilot Project in Croatia—Final Report. Convention on the conservation of European wildlife and natural habitats. Croatian Ministry of Environmental Protection and Physical Planning, Strasbourg
27. Hrvatske vode (2009) Strategija upravljanja vodama (national policy document NN 21/1998, in Croatian). [online]. voda.hr. http://www.voda.hr/sites/default/files/dokumenti/strategija_upravljanja_vodama.pdf. Accessed 17 March 2021
28. Croatian River Basin Management, 2016–2021 http://www.voda.hr/sites/default/files/plan_upravljanja_vodnim_podrucjima_2016._-_2021.pdf (Accessed 17 March 2021)
29. Hrvatske vode (2016) Plan upravljanja vodnim područjima 2016–2021 (national policy document NN 66/2016, in Croatian) [online]. voda.hr, 550 pages. Available at: http://www.voda.hr/sites/default/files/plan_upravljanja_vodnim_podrucjima_2016._-_2021.pdf (Accessed 17 March 2021)

30. Bognar A, Faivre S, Buzjak N, Pahernik M, Bočić N (2012) Recent landform evolution in the dinaric and pannonian regions of Croatia. In: Lóczy D, Stankoviansky M, Kotarba A (eds) Recent Landform Evolution. Springer Geography, Springer, Dordrecht
31. Orešić D, Čanjevac I (2019) Groundwater Resources in Croatia. In: Negm AM, Romanescu G, Zelenakova M (eds) Water Resources Management in Balkan Countries. Springer Water. Springer, Cham
32. Sučić I, Talaja M, Štih A, Burić I, Zadravec M, Koren T, Lauš B, Vujanović M, Popović Dujmović I (2018) Water peARLS OF THE Adriatic Islands. 1st systematic Island Wetland Inventory in Croatia. 9th European Regional Meeting, Olomouc, Czech Republic
33. SWD (2015) Report on the implementation of the Water Framework Directive River Basin Management Plans. Member State: CROATIA. European Commission, Brussels
34. Schneider-Jakoby M (2004) The Sava and Drava floodplains: Threatened ecosystems of international importance. In: Bloesch J (ed) River basin Management: Concepts and Transboundary Implementation (with special reference to the Danube River). Selected papers of the 35th IAD conference in Novi Sad (Serbia and lMontenegro), pp 249–289
35. Radović J, Čivić K, Topić R (2006) Biodiversity of Croatia. Ministry of Culture, State Institute for Nature Protection, Zagreb, Croatia
36. Jelić D, Kuljerić M, Koren T, Treer D, Šalamon D, Lončar M, Podnar-Lešić M, Janev-Hutinec B, Bogdanović T, Mekinić S (2012) Red book of Amphibians and Reptiles of Croatia. Ministry of Environmental and Nature Protection, State Institute for Nature Protection, Zagreb, Croatia
37. UNFCCC (2001) The First National Communication of the Republic of Croatia under the United Nations Framework Convention on Climate Change. Ministry of Environmental Protection and Physical Planning, Zagreb, Croatia
38. Pithart D, Petrov Rančić I, Kutleša P, Duplić A (2014) Study of Freshwater Ecosystem Services in Croatia. Ministry of Environmental and Nature Protection, State Institute for Nature Protection, Zagreb, Croatia.
39. Tadić L, Šperac M, Karleuša B, Rubinić J (2019) Water Quality Status of Croatian Surface Water Resources. In: Negm AM, Romanescu G, Zelenakova M (eds) Water Resources Management in Balkan Countries. Springer Water. Springer, Cham
40. Polak S, Pipan T (2011) Subterranean habitats and fauna, their threats and conservation. In: Prelovšek M, Zupan Hajna N (eds) Pressures and Protection of Underground Karst—Cases from Slovenia and Croatia. Karst Research Institute ZRC SAZU, Postojna, Slovenia
41. Jelić M, Vucić M, Klobučar G, Korlević P, Đikić D, Franjević D, Jelić D (2015) Molecular study on water frogs (genus *Pelophylax*) in Croatia—preliminary results. In: Kopjar N, Gligora Udovič M, Lukša Ž, Jelić D (eds) Zbornik sažetaka (12. Hrvatski biološki kongres s međunarodnim sudjelovanjem). Hrvatsko biološko društvo, Zagreb, pp 70–71
42. Koska S (2016) Gene diversity and degree of hybridization in water frogs of the genus *Pelophylax* Fitzinger, 1843 (Anura: Ranidae) from Crna Mlaka. https://urn.nsk.hr/urn:nbn:hr:217:999970.
43. Schmidt B (2018) Molecular phylogenetic and morphological characteristics of water frogs (genus *Pelophylax*) in Croatia. Unpublished thesis, University of Zagreb. https://urn.nsk.hr/urn:nbn:hr:217:811997
44. Hadžić E, Imamović A (2020) Water resources in Bosnia and Herzegovina. In: Negm A, Romanescu G, Zelenakova M (eds) Water resources management in Balkan countries. Springer Water. Springer, Cham. https://doi.org/10.1007/978-3-030-22468-4_8
45. https://visitmycountry.net/bosnia_herzegovina/bh/
46. https://ju-priroda.hr/2015/12/lokve-kao-izvori-vode-za-pice/
47. Đurović E, Vuković T, Pocrnjić Z (1979) Vodozemci Bosne i Hercegovine. Zemaljski Muzej Bosne i Hercegovine, Sarajevo
48. Adrović A (2015) Status of amphibians in Bosnia and Herzegovina. In: Heathvole H, Wilkinson J (eds) Amphibian Biology Vol 11. Part 4. Status of Conservation and Decline of Amphibians: Eastern Hemisphere: Southern Europe and Turkey. Pelagic Publishing, Exeter, UK, pp 62–66

49. Lelo S, Zimić A, Čengić M, Jelić D (2015) Biodiverzitet vodozemaca (Chordata: Vertebrata; Amphibia) Bosne i Hercegovine. Biosistematski prijegled podataka sa preliminarnim kartamarasprostranjenja. Udruženje za inventarizaciju i zaštitu životinja, Ilijaš, Sarajevo
50. Obratil S (1980/1981) Ekološki pristup utvrđivanju štetnosti ihtiofagnih ptica u ribnjacima Bardača. Glasnik Zemaljskog muzeja Bosne i Hercegovine u Sarajevu 19–20:139–256
51. Gašić B (2012) Zbirka "Ribe, vodozemci i gmizavci", katalozi zbirke "Ribe, vodozemci I gmizavci" i izložbe "Ribe, vodozemci i gmizavci naših krajeva." JU Muzej Republike Srpske, Banja Luka
52. Lelo S (2010) New record of *Triturus dobrogicus* (Kiritzescu, 1903) and *Bombina bombina* (Linnaeus, 1761) (Vertebrata, Amphibia) in Bosnia and Herzegovina. Contributions to the fauna of Bosnia and *Herzegovina* 6:42–47. (In Bosnian, with summary in English)
53. Šukalo G, Đekić M, Đukić D, Đorđević S, Tomović LJ (2014) New records of the Common Spadefoot Toad, *Pelobates fuscus* (Anura: Pelobatidae), in Bosnia and Herzegovina. Ecol Monten 1:92–95
54. Jelić D, Karaica D (2012) First data on the fauna of amphibians and reptiles of the lower Una River and its coastal area. HYLA 2011:22–41
55. Kotrošan D (2002) Rasprostranjenje čovječje ribice (*Proteus anguinus* Laurenti, 1768) na području Bosne i Hercegovine. Naš Krš 22:57–65
56. Zimić A, Šunje E (2015) Italian crested newt—*Triturus carnifex* Laurenti, 1768 (Amphibia, Caudata, Salamandridae, Pleurodelinae) in the batrachofauna of Bosnia and Herzegovina. Hyla 2015:52–55
57. Imamović A, Trožić-Borovac S (2013) Utjecaj klimatskih promjena na vodni režim rijeke Bosne. Naše šume 30–31:34–41
58. http://prirodahrvatske.com/hidrografija/
59. http://os-pucisca.skole.hr/upload/os-
60. https://phralipen.hr/2019/11/20/male-hidroelektrane-veliki-ekoloski-problem/
61. Golub Vuković D, Tursi A, Carlucci R (2008) Ichthyofauna of the wetland ecosystems in the Bardača area (Bosnia and Herzegovina). Croat J Fish 66:89–103
62. Fuštić B, Đuretić G, Spalević V, Knežević M (2000) Univerzitet Crne Gore-Biotehnički institut, Podgorica
63. Radojičić B (2005) Vode Crne Gore. Filozofski fakultet, Institut za Crnu Goru
64. Burić M (2010) Atlas voda Crne Gore. Crnogorska akademija nauka i umjetnosti. Podgorica
65. Plan Bleu (2016) Economic assessment of ecosystem services provided by Mediterranean wetlands in terms of climate regulation. Plan Bleu, Valbonne.
66. Ramsar Convention Secretariat (2013) The Ramsar Convention Manual: a guide to the Convention on Wetlands (Ramsar, Iran, 1971), 6th ed. Ramsar Convention Secretariat, Gland, Switzerland
67. Petrović D, Hadžiablahović S, Vuksanović S, Mačić V, Milanović Đ, Lakušić D (2019) Katalog tipova staništa od interesa za EU u Crnoj Gori. Verzija 3. Podgorica-Banja Luka-Beograd
68. Dević N, Joković P (2020) Preliminarni geološki i hidrogeološki izvještaj za prostor planiranog vojnog poligona na širem području Sinjajevine, JU Zavod za geološka istraživanja—Podgorica
69. Crnobrnja-Isailović J, Polović L, Ljubisavljević K, Čađenović N, Čubrić T, Haxhiu I (2018) Diversity and conservation status of Batrachofauna and Herpetofauna in the Lake Skadar Region. In: Pešić V, Karaman G, Kostianoy A (eds) The Handbook of Environmental Chemistry, vol XX. The Skadar/ Shkodra Lake Environment. Springer-Verlag GmbH, Heidelberg, Germany, pp 383–414
70. Ćirović R, Radović D, Vukov T (2008) Breeding site traits of European newts (*Triturus macedonicus, Lissotriton vulgaris* and *Mesotriton alpestris*, Salamandridae) in the Montenegrin karst region. Archiv Biol Sci (Belgrade) 60:459–468
71. Ćulafić G, Vuksanović S, Bubanja N, Polović L, Čađenović N, Dragićević S, Malidžan S, Burzanović K, Vizi A, Biberdžić V, Ćetković I, Karaman M (2017) Chapter 4. Ecological Context The Surroundings of Crvena Stijena. In: R. Whallon (ed.) Crvena Stijena in Cultural and Ecological Context. Multidisciplinary Archaeological Research in Montenegro. Montenegrin Academy of Sciences and Arts. Special edition (Monographies and Studies) 138, Podgorica, pp 28–44

72. Ljubisavljević K, Iković V (2020) Zaštita i očuvanje ugrožene skadarske žabe i njenih staništa na prostoru delte rijeke Bojane. Crnogorsko Društvo Ekologa, Podgorica
73. Polović L, Čađenović N (2014) The Herpetofauna of the Great Ulcinj Beach area including Ada Island (Montenegro). Turkish J Zool 38:104–107
74. Radovanović M (1951) Vodozemci i gmizavci naše zemlje. Naučna knjiga, Beograd
75. Džukić G (1991) Fauna Durmitora: vodozemci i gmizavci. In: Nonveiller G (ed) Fauna Durmitora, 4, Posebna izdanja, vol 24. Odeljenje prirodnih nauka 15. Crnogorska akademija nauka i umjetnosti, Titograd, pp 9–78
76. Lj T, Ajtić R, Crnobrnja-Isailović J (2004) Contribution to distribution and conservation of batrachofauna and herpetofauna on Bjelasica Mountain in Montenegro. In: Pešić V (ed) The biodiversity of the Biogradska Gora national park. University of Montenegro & Centre for Biodiversity of Montenegro, Monographies I, Department of Biology, pp 140–148
77. Sjogren-Gulve (1994) Distribution and extinction patterns within a northern metapopulation of the pool frog, *Rana lessonae*. Ecology 75:1357–1367
78. Čađenović N (2014) Catalogue of amphibian fauna of Montenegro (Crna Gora). Catalogues 10, Book 9. The Montenegrin Academy of Sciences and Arts, Natural Sciences, Podgorica
79. Čađenović N., 2020. Vodozemci Sinjajevine. Izrada nultog stanja biodiverziteta na području planiranog vojnog poligona na Sinjajevini. Agencija za zaštitu prirode i životne sredine Crne Gore, Podgorica
80. Cushman S (2006) Effects of habitat loss and fragmentation on amphibians: a review and prospectus. Biol Cons 128:231–240
81. Sowers A (2009) The effects of wastewater effluent on a fish and amphibian species. All Dissertations. Clemson University, Tiger Prints, p 355
82. Džukić G, Kalezić ML, Ljubisavljević KM (2003) Zaštita i očuvanje zelenih žaba u Srbiji i Crnoj Gori. Beograd: Savezni sekretarijat za rad, zdravstvo i socijalno staranje
83. National Agency of Natural Resources (2009) http://www.akbn.gov.al/
84. Shumka S, Bego F, Beqiraj S, Paparisto A, Kashta L, Miho A, Shumka L, Mali S (2018) Trading-of freshwater biodiversity and hydropower in a unique Balkan hotspot (Vjosa River Watershed, Albania). Holistic Appr Environ 8:114–123
85. Schiemer F, Beqiraj S, Drescher A, Graf W, Egger G, Essl F, Frank T, Hauer C, Hohensinner S, Miho A, Meulenbroek P, Paill W, Schwarz U (2020) Vitecek S (2020) The Vjosa River corridor: a model of natural hydromorphodynamics and a hotspot of highly threatened ecosystems of European significance. Landscape Ecol 35:953–968. https://doi.org/10.1007/s10980-020-00993-y
86. Eftimi R, Frasher A (2016) Thermal waters of Albania. In: Mineral and Thermal Waters of Southeastern Europe, pp 201–215
87. Eftimi R (2020) Karst and Karst water resources of Albania and their management. Carbonates Evaporites 35:69
88. Eftimi R (2010) Hydrological characteristics of Albania. AQU Amundi 01012:79–92
89. Mima M, Fitoka EN, Bego F (eds) (2003) Inventory of Albanian Wetlands. ECAT and Greek Biotope/Wetland Centre (EKBY); 341 p and 75 p Annexes. Thermi, Greece
90. Çullaj A, Hasko A, Miho A, Schanz F, Brandl H, Bechofen R (2005) The quality of Albanian natural waters and the human impact. Environ Int 31:133–146
91. Zeneli G, Shuka L, Bego, F, Kashta L, Shumka S, Hoxha S, Diku A, Bano E, Krasniqi F (2014) Draft document of the National biodiversity strategy of Albania for the period 2012–2020. doi: https://doi.org/10.13140/RG.2.2.31927.93606
92. Szabolcs M, Mizsei E, Jablonski D, Vági B, Mester B, Végvári Z, Lengyel S (2017) Distribution and diversity of amphibians in Albania: new data and basis for a comprehensive database. Amphibia-Reptilia 38:435–448
93. Dufresnes C, Strachinis I, Suriadna N, Mykytynets G, Cogalniceanu D, Vukov T, Arntzen JW, Wielstra B, Lymberakis P, Geffen E, Gafny S, Kumlutas Y, Ilgaz Ç, Candan K, Mizsei E, Szabolcs M, Kolenda K, Smirnov N, Geniez P, Lukanov S, Crochet P-A, Dubey S, Perrin N, Litvinchuk SN, Denoël M (2019) Phylogeography of a cryptic speciation continuum in spadefoot toads (*Pelobates*). Mol Ecol 28:3257–3270

94. Haxhiu I (1994) The herpetofauna of Albania, Amphibia: species composition, distribution, habitats. Zool. Jahrb. Syst. 121:321–334
95. Oruçi S (2010) Bio-Ecological data on amphibians of thermal water of Permeti area (south Albania). Balwois 2010—Ohrid, Republic of Macedonia—25, 29 May 2010.
96. Frank T, Saçdanaku E, Duda M, Bego F (2018) Amphibian and reptile fauna of the Vjosa River, Albania. Acta ZooBot Austria 155:323–336
97. Haxhiu I (2015) Amphibians of Albania. In: Heathvole H, Wilkinson J (eds) Amphibian biology, Vol 11. Part 4. Status of conservation and decline of Amphibians: Eastern Hemisphere: Southern Europe and Turkey. Pelagic Publishing, Exeter, UK, pp 74–79
98. Wagner B, Lotter AF, Nowaczyk N, Reed JM, Schwalb A, Sulpizio R, Valsecchi V, Wessels M, Zanchetta G (2009) A 40,000-year record of environmental change from ancient Lake Ohrid (Albania and Macedonia). J Paleolimnol 41:407–430
99. Saçdanaku E, Mesiti A, Bytyçi N, Halilaj M (2020) Species action plan. Measures for protection of Albanian frog (*Pelophylax shqipericus*) in the Bay of Vlora. EcoAlbania, Tirana
100. Crnobrnja-Isailović J, Paunović M (2015) Conservation and decline of European Amphibians: the republic of Serbia. In: Heathvole H, Wilkinson J (eds) Amphibian biology Vol 11. Part 4. Status of conservation and decline of Amphibians: Eastern Hemisphere: Southern Europe and Turkey. Pelagic Publishing, Exeter, UK, pp 45–55
101. CLC (2006) CORINE land cover 2000 data base for Serbia and Montenegro. EvroGeomatika d.o.o., Belgrade, Geological Survey of Montenegro, Podgorica
102. Dimitrijević V (2020) The impact of mini hydropower plants on water system, design and construction requirements. In: Anđelković M (ed) Environmental impact of small hydropower plants. Serbian Academy of Sciences and Arts, Belgrade, Serbia, pp 21–41 (In Serbian, with summary in English)
103. Ristić R, Malušević I, Polovina S, Milčanović V, Radić B (2020) Small derivate hydropower plants: a negligible contribution to power generation and an imparable adverse environmental impact. In: Anđelković M (ed) Environmental impact of small hydropower plants. Serbian Academy of Sciences and Arts, Belgrade, Serbia, pp 107–133 (In Serbian, with summary in English)
104. Karadžić B, Mijović A (eds) (2007) Environment in Serbia: an indicator-based review, 1st edn. Ministry of Science and Environmental Protection Agency, Republic of Serbia, Belgrade
105. Džukić G, Vukov TD, Kalezić ML (2015) Fauna tailed amphibians of Serbia. Serbian academy of sciences and arts, Belgrade, Serbia (In Serbian, with summary in English)
106. Crnobrnja-Isailović J, Aleksić I, Arntzen JW (2005) The status of great crested newt breeding sites in Serbia. Froglog 67:2–3
107. Anonymous (2010) Book of rules on declaring and protection of strictly protected wild species of plants, animals and fungi. Official Gazette RS No 5/2010
108. Kalezić ML, Tomović Lj, Džukić G (2015) Red List of Fauna of Serbia. I. Amphibians. University of Belgrade, Faculty of Biology & Institute for Nature Protection of Serbia
109. Crnobrnja Isailović J, Milojković D, Macura B (2015) Vodozemci i gmizavci Đerdapa/Amphibians and Reptiles of Đerdap. JP Nacionalni Park Đerdap/PE Djerdap National Park, Donji Milanovac, Serbia (bilingual monograph—Serbian/English)
110. Crnobrnja-Isailović J, Jovanović B, Ćorović J, Ilić M, Pejović N (2019) Amphibians and reptiles of Special Nature Reserve "Uvac"—monitoring of species and habitats in 2019. Unpublished project report
111. Jovanović B, Crnobrnja-Isailović J (2019) Fluctuations in population abundance in two anurans from Central Serbia. Herpetozoa 32:65–71
112. MOEPP (2018) National biodiversity strategy and action plan: for the period 2018–2023. Support to the Republic of Macedonia for revision of national biodiversity strategy and action plan and development of the fifth national report to the convention on biodiversity. Ministry of environment and physical planning, Skopje
113. Dimovska G (2013) Osvrt za razvojot na zemjodelstvoto vo Makedonija vo poslednite deset godini. Centar za ekonomski analizi—CEA

114. State Statistic Office of North Macedonia (2015) Annual publication of the State Statistic Office of North Macedonia. 10 Agriculture. Skopje, pp 444–497
115. Matevski V, Čarni A, Kostadinovski M, Košir P, Šilc U, Zelnik I (2008) Flora and vegetation of the Macedonian steppe. Biološki inštitut Jovana Hadžija ZRC SAZU. Ljubljana
116. Lazarevski A (1993) Klimata vo Makedonija. Kultura—Skopje
117. Dimovski A (1971) Zoocenološki istražuvanja na stepskite predeli vo Makedonija. Godišen Zbornik Prirodno-Matematičkog Fakulteta, Univerziteta u Skoplju, Skoplje, Knjiga 23(Biologija 4):25–54
118. Sterijovski B, Arsovski D (2020) National Red List assessment. Amphibia. http://redlist.moepp.gov.mk
119. Official gazette of the Republic of Macedonia. No. 139. 2011
120. Sterijovski B, Arsovski D (2019) National red list assessment. Final report. Herpetofauna. Achieving Biodiversity Conservation through Creation and Effective Management of Protected Areas and Mainstreaming Biodiversity into Land Use Planning. GEF/UN Environment project
121. Sterijovski B (2015) Conservation and protection status of Amphibians in Macedonia. In: Heatwole H, Wilkinson JW (eds) The amphibian biology series volumes on conservation and decline of amphibians. South–East Europe and Turkey. Pelagic Publishing, pp 78–84
122. MOEPP (2003) Country study of biodiversity of the republic of Macedonia. First national report. Ministry of environment and physical planning, Skopje
123. Melovski LJ, Ivanov GJ, Angelova N, Velevski M, Hristovski S (2008) Monospitovsko blato. Poslednoto mocuriste vo Makedonija. Opstina Bosilovo, Bosilovo
124. Biggs J, von Fumetti S, Kelly-Quinn M (2016) The importance of small waterbodies for biodiversity and ecosystem services: implications for policy makers. Hydrobiologia 793:3–39

Chapter 18
Human Impact Induces Shifts in Trophic Composition and Diversity of Consumer Communities in Small Freshwater Ecosystems

Oksana Y. Buzhdygan, Milica Stojković Piperac, Olivera Stamenković, Dubravka Čerba, Aleksandar Ostojić, Britta Tietjen, and Djuradj Milošević

Abstract The current global rates of human impact intensification in freshwater ecosystems and their surroundings have caused an alarming loss of freshwater biodiversity and functioning across multiple trophic levels. Particularly vulnerable to anthropogenic pressures are small water bodies (SWB). However, the majority of the evidence is limited to a single community, especially macroinvertebrates, thus leaving us uncertain about cross-community multitrophic responses of freshwater

---

Electronic supplementary material The online version contains supplementary material available at (https://doi.org/10.1007/978-3-030-86478-1_18).

O. Y. Buzhdygan (✉) · B. Tietjen
Theoretical Ecology, Institute of Biology, Freie Universität Berlin, Königin-Luise-Straße 2-4, Gartenhaus, 14195 Berlin, Germany
e-mail: oksana.buzh@fu-berlin.de

B. Tietjen
e-mail: britta.tietjen@fu-berlin.de

M. Stojković Piperac · O. Stamenković · D. Milošević
Department of Biology and Ecology, Faculty of Science and Mathematics, University of Niš, Višegradska 33, 18106 Niš, Serbia
e-mail: milicas@pmf.ni.ac.rs

O. Stamenković
e-mail: olivera.stamenkovic@pmf.edu.rs

D. Milošević
e-mail: djuradj@pmf.ni.ac.rs

D. Čerba
Department of Biology, Josip Juraj Strossmayer University of Osijek, Cara Hadrijana 8/A, 31000 Osijek, Croatia
e-mail: dcerba@biologija.unios.hr

A. Ostojić
Faculty of Science, Institute of Biology and Ecology, University of Kragujevac, Radoja Domanovića 12, 34000 Kragujevac, Serbia
e-mail: aleksandar.ostojic@pmf.kg.ac.rs

V. Pešić et al. (eds.), *Small Water Bodies of the Western Balkans*, Springer Water,
https://doi.org/10.1007/978-3-030-86478-1_18

consumers to anthropogenic pressure. Furthermore, the effects of human impact on trophic composition and diversity of consumers may depend on the nature of human activities. We therefore tested the effects of different co-occurring human activities (agricultural land use, gravel exploitation, road proximity, and waste input) and of the overall human-impact intensification on trophic trait diversity and composition of different animal communities in SWB. For this, four types of consumer communities (i.e., zooplankton, benthic macroinvertebrates, epiphytic macroinvertebrates, and fish community) were sampled in a total of 58 study sites in two geographic areas in Serbia and Croatia for two years. In addition to human impact, we also tested the effects of altered water properties, macrophyte biomass, and predation. Our results clearly show that human impact induced changes in the trophic diversity and the trophic composition of different consumer communities. The direction and strength of such functional shifts depended on the nature of human impact and on the type of consumer community. Furthermore, we found that the effects of human impacts on the distribution of trophic traits across all consumer communities significantly differed between trophic groups (i.e., herbivores, decomposers, and carnivores) and trophic levels (i.e., trophic position in the food web). Our findings point toward the urgent need to consider multiple stressors and a range of animal community types in future research and management of freshwater ecosystems. These results highlight the importance of a multitrophic perspective when predicting the consequences of human impact for ecosystem biodiversity and functioning.

Keywords Small freshwater ecosystems · Consumer communities · Functional traits · Trophic diversity · Trophic composition · Human impact · Agricultural land use · Waste input · Gravel exploitation · Road effects · Zooplankton · Benthic macroinvertebrates · Epiphytic macroinvertebrates · Fish community

18.1 Introduction

Freshwater ecosystems cover only up to 2.4% of the Earth's terrestrial surface [1], but support high biodiversity across multiple trophic levels, thus providing wide variety of ecosystem services crucial for human well-being [1–4]. Human impacts (HI) via multiple stressors are profoundly altering the natural dynamics of freshwater communities [4–8], thus threatening their functions, services, and stability [9]. Particularly vulnerable to anthropogenic pressures are small water bodies (SWB) because of their low water volume and close connection to adjacent agricultural fields and urban areas [10–12]. The most evident examples of anthropogenic stressors in SWB and their surroundings include agricultural land use, urbanization, gravel and sand mining, waste inputs, and associated changes to nutrient cycling, sedimentation,

B. Tietjen
Berlin-Brandenburg Institute of Advanced Biodiversity Research (BBIB), 14195 Berlin, Germany

altered water temperature and acidity, which affect resource quantity and quality for consumers, as well as their interactions with predators [7, 8, 13–21].

In response to human-induced changes in resource availability or predation pressure, trophic composition of consumer communities is shifted as a result of altered trophic interactions [8, 9, 20, 22]. This may have profound consequences for functioning of freshwater ecosystems, particularly the disrupted transfer of energy, matter, and nutrients within the ecosystem [9, 23, 24] but also with adjacent terrestrial or aquatic ecosystems [25]. On the other hand, changes in water properties can mediate effects of human activities on functional composition of consumer communities by filtering out vulnerable species, thus causing changes in trophic interactions [26]. Such underlying mechanisms that structure community assemblages may strongly depend on the type of human activities. Thus, considering multiple stressors and understanding the trade-offs among their simultaneous effects is essential for understanding the responses of freshwater ecosystems to human activities [5, 6]. However, HI effects on the trophic composition of freshwater consumers have majorly been assessed in previous studies by considering single community types, for example zooplankton communities [27], or macroinvertebrates [9, 13, 26, 28]. While these studies found strong effects of HI stressors on trophic diversity and composition of these communities, such single community-type approach captures only parts of the trophic network, thus limiting our ability to understand the essential changes across all consumer trophic levels of the ecosystem. This may be particularly problematic when predicting HI effects on ecosystem-level functioning (e.g., decomposition, predation, flow of energy and nutrients), because anthropogenic effects on one trophic level can cascade up or down to other trophic levels [29], or attenuate across trophic levels [30].

To better understand the multitrophic responses to HI we therefore need a mechanistic perspective on how multiple stressors of HI affect the trophic composition across all consumer communities [25] and how organisms constrain one another across trophic levels [23, 24, 31]. This can be challenging, as there is no generalized approach to address such complexity in freshwater research. One approach to close this knowledge gap is the consideration of feeding traits. Accounting for feeding traits is a universal approach for assessing the changes in functional diversity and trophic composition spanning different animal communities, and ecosystem types and scales [32]. Because feeding traits are comparable across different taxonomic groups, they provide a unified measure across communities differing in taxonomic composition [17, 33]. Furthermore, functional diversity measures, such as feeding traits, were found to be a better predictor of ecosystem functioning than species or taxa diversity measures alone [32, 34, 35]. Moreover, studies show that feeding traits are sensitive to human impact including different land-use practices [16, 33].

Here we used a feeding trait approach to test the effects of overall HI intensity and of different HI types on trophic diversity and trophic composition of different consumer communities in SWB (ponds, small lake, and channel), i.e. zooplankton, benthic macroinvertebrates, epiphytic macroinvertebrates, and fish communities. Small fresh-water bodies are particularly important sources of freshwater biodiversity because they harbour rare and threatened species by serving as species refuge [3, 36–38]. Despite high contribution to aquatic biodiversity and functions, SWB are largely underrepresented in the scientific literature as well as in water protection and legislation directives [3]. Moreover, research on context-dependencies of the effects of various HI stressors across different consumer communities is scarce for freshwater ecosystems [39]. However, evidence is emerging that impact of anthropogenic pressure and associated environmental changes on diversity and functioning of fauna communities may strongly depend on the trophic group (e.g. herbivores vs. decomposers vs. predators) [30, 40–42], on the trophic level in the food web [23], and on the nature of HI disturbances [6, 8, 9, 43, 44]. Such variations in HI effects across trophic groups and trophic levels may be attributed to differences in body size, home range and tolerance to stress, as well as to differences in species richness between trophic levels [23, 31, 45]. For example, predators generally have larger requirements for home range, lower abundances, and larger body size, that can make them more vulnerable to disturbances [45]. Some work also suggests that the vulnerability to changes in local environmental conditions increases with increasing trophic position in ecological networks [46, 47]. On the other hand, predators may be less susceptible to contaminant stress (e.g., water toxins), which affects smaller species disproportionately [45]. In contrast, because of higher species richness in lower trophic levels, there is more "insurance" against the species losses, and these species may have greater capacity for adaptive change due to their shorter life spans and faster turnover rates [45]. However evidence is scarce and majorly limited to experimental simple-structured communities, thus limiting our understanding of human-impact consequences for ecosystem multitrophic structure and diversity in real-world conditions [25].

In this study we therefore considered the following various types of human activities co-occurring in different SWB study sites and their surroundings: agricultural land use, gravel exploitation, nearby roads, and waste input. Furthermore, we tested the effects of altered water properties associated with these HI activities. This approach that accounts for the impacts of various stress on different consumer communities allowed us to test the dependencies of the HI effects on trophic groups, trophic levels, community types, and HI types. We addressed the following questions:

(1) How overall human impact (HI) intensification and different HI types in SWB affect the diversity and composition of trophic traits of the following consumer communities: zooplankton, benthic macroinvertebrates, epiphytic macroinvertebrates, and fish community?
(2) Do these consumer communities differ in their trophic-trait responses to HI intensity?

(3) Do the effects of HI on trophic diversity and composition across these communities vary with trophic group, trophic level, and HI type?
(4) Does HI intensification affect single-trait measures (i.e., community weighted mean of each individual feeding strategy, CWM) and multivariate trophic diversity measures (functional dispersion, FDisp) differently?

18.2 Methods

Study area

The research was conducted in two geographic areas. First study area is in the south-eastern part of Serbia, (43°29′–43°30′N, 21°79′–21°80′E) with a temperate continental climate, mean annual temperature of 11.9 °C, and mean annual precipitation of 580 mm. Second study area is in the north-eastern part of Croatia, (45°59′–45°62′N, 18°78′–18°84′E), with a temperate continental climate, mean annual temperature of 11.1 °C, and mean annual precipitation of 650 mm. The Serbian study area is dominated by urban areas (roads, rural settlements) and agricultural lands, with scarce broadleaved forests located in the riparian area of a river and several permanent and ephemeral ponds. The Croatian study area is dominated by broadleaved forests and different types of freshwater ecosystems, including fishponds, channels, wetlands, ponds, and shallow lakes.

Overall, we investigated ten types of freshwater ecosystems (hereafter study systems), specifically seven ponds (ponds 1–7), two slow-flowing channels named Barbara and Čonakut, and one shallow lake, named Sakadaš. Six ponds (ponds 1–6) are located in the Serbian study area near the Južna Morava River and the remaining four study systems are located in the Croatian study area. The lake Sakadaš and the Čonakut channel in Croatia are situated in a protected flooded area of the Danube River called Kopačkirit Nature Park. The channel Barbara and pond 7 in the Croatian study area are located between the Nature Park and the village of Kopačevo. All study systems are permanent and shallow (mean depth range between 0.8 and 7 m), with a surface area ranging from 0.04 to 47.57 ha. All study ponds and channel Barbara were colonized by macrophytes, with emergent vegetation present on the shores. In contrast, macrophytes were absent from the lake Sakadaš and the Čonakut channel. Study ponds 4 and 5 (in the Serbian study area) are regularly flooded by the nearby river and due to their spatial proximity they become a connected system during flooding. The water-level fluctuations in both, the Čonakut channel and the lake Sakadaš, and to a lesser extent in pond 7 and Barbara Channel are a result of inundation events from the Danube River. Therefore, these systems were considered as connected.

Sampling

We sampled four consumer communities, namely zooplankton, benthic macroinvertebrates, epiphytic macroinvertebrates, and fishes. All study communities were sampled in September from Serbian study sites and in October from Croatian sites in

2016 and 2017. Three sampling transects (study sites) were developed within each pond system and channel systems, and four sampling transects (study sites) were distributed across the lake Sakadaš.

Zooplankton community

Zooplankton communities were sampled from 28 study sites distributed across our ten study systems (i.e., seven ponds, the lake Sakadaš, and the Čonakut channel) in 2016 (Supplementary Material, Table S1). For zooplankton analyses, one water sample of 10 L was taken at each study site, at 20 cm below the water surface, filtered through 23 μm mesh net and preserved in 4% formaldehyde solution. The abundance of zooplankton was measured as the number of individuals per water volume (individuals L^{-1}). Individuals were identified to species or genus level using relevant identification keys [48–54]. According to their feeding mode traits, zooplankton taxa were grouped into the following three trophic groups: herbivores, omnivores (feeding on both phytoplankton and suspended organic solids), and carnivores [55–59], for details see Table S5 in the Supplementary Material.

Benthic and epiphytic macroinvertebrates

Benthic macroinvertebrates were sampled from 49 study sites in 2016 and 2017 (for details see Table S1 in the Supplementary Material). Three subsamples were taken along each transect using a kick-net of 250 μm mesh size or the bottom sampling dredge, depending on the depth. All three subsamples were later merged into a single sample. The abundance of benthic macroinvertebrates was expressed as the number of individuals per m^2.

Epiphytic macroinvertebrates were sampled from 39 study sites in 2016 and 2017 (Supplementary Material, Table S1). All individuals were collected from submerged macrophytes and submerged parts of emergent and floating vegetation. The abundance of epiphytic macroinvertebrates was expressed as the number of individuals per 100 g of dry macrophyte biomass. Because the macrophytes were absent in the lake Sakadaš and the Čonakut channel, epiphytic macroinvertebrates were not sampled in the study sites of these channels. For details on sampling see *Stamenković* et al. [38, 60].

Both benthic and epiphytic macroinvertebrates were identified to the lowest possible taxonomic level using relevant taxonomic keys [61–75]. Trophic traits for each species/taxa were assembled from the literature [76], for details see Table S4 in the Supplementary Material (i.e., grazers, miners, xylophages, shredders, gatherers collectors, active filter feeders, passive filter feeders, predators, and parasites).

Fish community

Fish communities were sampled at 59 study sites in 2016 and 2017 (Supplementary Material, Table S1). Sampling was performed from a boat along transects using a DC Aquatech IG 1300 electrofisher (2.6 kW, 80–470 V). For each selected transect, the constant catch-per-unit-effort (CPUE) of time (10 min) was provided. Each fish individual was identified to the species level and assigned trophic traits using

Fig. 18.1 Types of human activities included in this study. Photo credit: *O. Buzhdygan*

literature [77–79], i.e. herbivores, decomposers (scavengers), invertebrate feeders, and piscivores; for details see Table S4 in the Supplementary Material.

The details on sampling of macrophytes, abiotic variables, and on the assessment of human impact types and intensity (Fig. 18.1) can be found in the Supplementary Material ("Supplemental methods"). All statistical analyses were performed using R version 3.6.2 [5]. For details on data analyses performed in this study see the Supplementary Material ("Supplemental methods").

18.3 Results

Zooplankton communities

From 28 study sites 66,086 individuals of 106 species of zooplankton were collected (Supplementary Material, Table S5). Omnivores were the most abundant (2287 $\pm$ 369.4, mean $\pm$ SE per study site) and species rich (20.2 $\pm$ 5.9) group and were represented by rotifers (88 species) and copepods (3 species). The herbivore group was comprised of cladocerans (11 species) with abundances of 70.2 $\pm$ 47.9 individuals per site and species numbers 1.5 $\pm$ 0.2 per site. Carnivores occurred least frequently (2.9 $\pm$ 1.1 individuals and 0.4 $\pm$ 0.1 species per site) and included rotifers (3 species) and copepods (1 species).

After applying the correlation approach, the following predictor variables remained (Supplementary Material, Fig. S2): system depth, human impact intensity index (HII), water temperature, content of dissolved oxygen, ammonium, and orthophosphates, macrophyte biomass, and fish abundance (Supplementary Material, Table S9). Trophic diversity (expressed as FDisp) was not correlated with species richness (Supplementary Material, Table S9). Human impact intensity (HII) reduced trophic diversity. Among types of human impact only the waste-effluent presence had a significant effect on trophic diversity. Water oxygen content and ammonium concentration positively influenced trophic diversity, while other variables were not significant (Table 18.1).

Zooplankton communities showed shifts in trophic trait composition with presence of waste effluent, and with variations in macrophyte biomass, and water content of ammonium and orthophosphates (Supplementary Material, Tables S10–S11). Omnivore species occurred most frequently across the gradients of study variables and were therefore positioned most closely to the centre of the NMDS ordination (Fig. 18.2). On the first axis carnivores were separated from herbivores and omnivores in the same direction of increased macrophyte biomass. On the second axis waste effluent, macrophyte biomass, and orthophosphate content associated positively with herbivore group and negatively with omnivore and carnivore groups. In contrast to this, oxygen content and ammonium concentration showed positive associations with omnivore and carnivore groups and negative associations with herbivores (Fig. 18.2).

Table 18.1 Results of backward selected linear mixed effect models for effects of human impact, water content of dissolved oxygen (%), and ammonium (mg l^{-1}) on trophic diversity (i.e., functional dispersion FDisp) of zooplankton community. The study system nested in the geographic area (Serbia, Croatia) was included as a random effect. Shown are separate models testing the effects of overall human impact intensity HII (model 1) and the effects of individual human-impact components (model 2). Significant effects ($P < 0.05$) are given in bold. For significant fixed effects the direction of the effect is given: ↑ positive, ↓ negative. Abbreviations: nDF—numerator degrees of freedom; dDF—denominator degrees of freedom; R^2m and R^2c are marginal (for fixed predictors) and conditional (for fixed and random predictors) coefficients of determination (pseudo-R^2), respectively; part-R^2 are semi-partial R^2 for each particular fixed predictor

Model 1. Test for the effects of total human impact HII (minimum adequate model)					
Predictors	Part-R^2	nDF	dDF	F	Pr(>F)
HII	0.05	1	18	7.13	**0.02 ↓**
O_2	0.06	1	16	14.42	**0.002 ↑**
NH4-N	0.06	1	17	14.61	**0.001 ↑**
	$R^2m = 0.19$, $R^2c = 0.98$				
Model 2. Test for the effects of individual human-impact types					
Predictors	Part-R^2	nDF	dDF	F	Pr(>F)
Distance to agricultural lands	< 0.001	1	19	0.10	0.75
Gravel exploitation	0.03	1	6	0.57	0.48
Distance to highway	0.004	1	17	0.36	0.56
Waste input	0.03	1	15	**10.19**	**0.01 ↓**
O_2	0.03	1	13	**10.26**	**0.01 ↑**
NH4-N	0.04	1	16	**11.46**	**0.004 ↑**
	$R^2m = 0.2$; $R^2c = 0.98$				

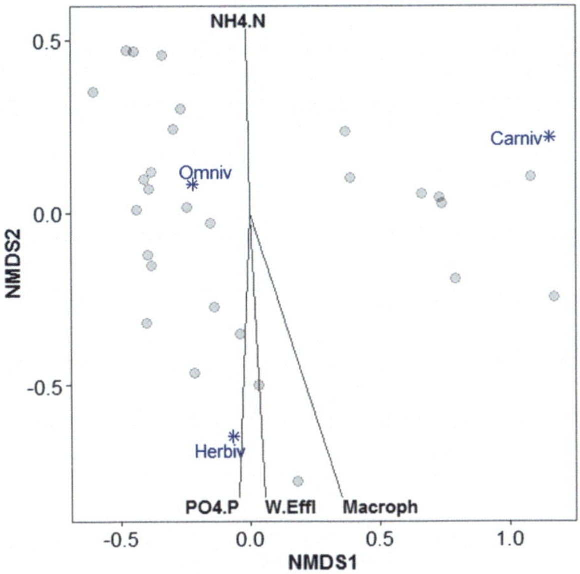

Fig. 18.2 Nonmetric multidimensional scaling (NMDS) of trophic composition of zooplankton communities: Herbiv—herbivores; Omniv—omnivores; Carniv—carnivores. Significant predictors (Supplementary Material, Tables S10–S11) are plotted post-hoc as arrows: W.Effl—waste input (effluent); NH4.N—ammonium content in water; PO4.P—orthophosphate content in water; Macroph—macrophyte biomass. Grey circles show study sites

Benthic macroinvertebrates

145 taxa with 46,752 individuals were recorded across study sites (n = 49; Table S1 in the Supplementary Material). The species number per site varied from 6 to 32 (mean: 16.24 ± 0.95), and trophic diversity FDisp ranged from 0.05 to 0.32 (mean: 0.21 ± 0.009), Table S3 in the Supplementary Material. Gatherer collectors were the most dominant trophic trait (CWM = 0.44 ± 0.02), followed by grazers and scrapers (CWM = 0.20 ± 0.02), and predators (CWM = 0.16 ± 0.02), Table S3 in the Supplementary Material. The predictor variables that remained after the correlation approach (Supplementary Material, Fig. S2) were system depth, human impact intensity index (HII), water temperature, content of dissolved oxygen, ammonium, and orthophosphates, macrophyte biomass, and fish abundance. System connectivity, distance to river, and nitrate–nitrogen content were removed because they were strongly correlated with at least one of the remaining variables (Supplementary Material, Fig. S2). Trophic diversity depended on species number in benthic macroinvertebrate community (Table 18.2). Index of human-impact intensity showed positive effect, while water temperature and macrophyte biomass showed negative effects on trophic diversity (Table 18.2).

System depth and macrophyte biomass had similar effects on community trophic composition (Fig. 18.3). Specifically increased depth and macrophyte biomass were positively associated with the CWM values for active filter feeders (Aff), grazers and

Table 18.2 Results of backward selected linear mixed effect models for effects of species richness, human impact, water temperature (T, °C), and macrophyte biomass (g m^{-2}) on trophic diversity (i.e., functional dispersion FDisp) of benthic macroinvertebrate community. The study system nested in the sampling year and geographic area (Serbia, Croatia) was included as a random effect. Shown are separate models testing the effects of overall human impact intensity HII (model 1) and the effects of individual human-impact components (model 2). Significant effects ($P < 0.05$) are given in bold. For significant fixed effects the direction of the effect is given: ↑ positive, ↓ negative. Abbreviations: nDF—numerator degrees of freedom; dDF—denominator degrees of freedom; R^2m and R^2c are marginal (for fixed predictors) and conditional (for fixed and random predictors) coefficients of determination (pseudo-R^2), respectively; part-R^2 are semi-partial R^2 for each particular fixed predictor

Model 1. Test for the effects of total human impact HII (minimum adequate model)					
Predictors	Part-R^2	nDF	dDF	F	Pr(>F)
Species richness	0.13	1	44	7.43	**0.01 ↑**
HII	0.13	1	35	5.78	**0.02 ↑**
T	0.16	1	33	7.49	**0.01 ↓**
Macrophyte biomass	0.10	1	37	4.21	**0.048 ↓**
	R^2m = 0.26; R^2c = 0.64				
Model 2. Test for the effects of individual human-impact types					
Predictors	Part-R^2	nDF	dDF	F	Pr(>F)
Species richness	0.11	1	42	7.35	**0.01 ↑**
Distance to agricultural lands	<0.001	1	18	0.04	0.83
Gravel exploitation	<0.001	1	14	0.08	0.79
Distance to highway	<0.001	1	21	0.11	0.75
Waste input	0.03	1	36	1.44	0.24
Macrophyte biomass	0.05	1	37	2.50	0.12
	R^2m = 0.17; R^2c = 0.65				

scrapers (Gra), shredders (Scr), and miners (Min) but negatively associated with the CWM for predators (Pre). The second axis in the NMDS ordination correlated positively with human-impact intensity (HII), waste deposition, ammonia–nitrogen, and negatively with distance to highway. Mean trait values CWM for gatherer collectors (Gaz) and xylophagous (Xyl) decreased with human-impact intensity (HII), waste deposition, and ammonia–nitrogen and were positively affected by larger distance to highway. The opposite effects were found for passive filter feeders (Pff) and parasites (Par) (Fig. 18.3).

Epiphytic macroinvertebrates

We collected a total of 217,018 individuals of 168 taxa of epiphytic macroinvertebrates (Supplementary Material, Table S6). Our study sites had on average 31.4 ± 1.82 species richness, ranging from 14 to 63 species (Supplementary Material, Table S3). Trophic diversity FDisp ranged from 0.13 to 0.37 (mean: 0.25 ± 0.01

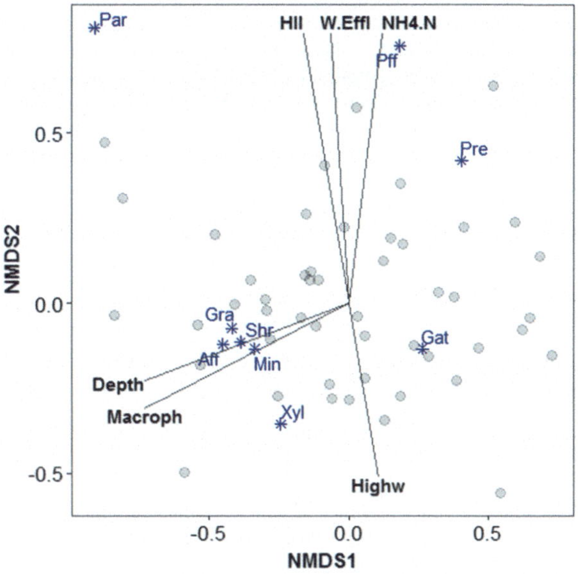

Fig. 18.3 Nonmetric multidimensional scaling (NMDS) of trophic composition of benthic macroinvertebrate communities: Gra—grazers and scrapers; Min—miners; Xyl—xylophagous taxa; Shr—shredders; Gat—gatherer collectors (detritivores); Aff—active filter feeders; Pff—passive filter feeders; Pre—predators; Par—parasites (see Supplementary Material, Table S3 for details). Significant predictors (Supplementary Material, Tables S13–S14) are plotted post-hoc as arrows: Depth—system depth; HII—human-impact intensity index; Highw—distance to highway; W.Effl—waste input (effluent); NH4.N—ammonium content in water; Macroph—macrophyte biomass. Grey circles show study sites

per study site, Table S3). Communities were dominated by gatherer collectors Gat (CWM = 0.39 ± 0.03, range 0.13 to 0.87); followed by grazers and scrapers Gra (CWM = 0.25 ± 0.02, range 0.02 to 0.43); and predators Pre (CWM = 0.17 ± 0.02, range 0.001 to 0.69), Table S3. The predictor variables that remained after the correlation approach (Supplementary Material, Fig. S3) were, system connectivity (not connected, connected), system depth, human impact intensity index (HII), water temperature, content of dissolved oxygen, ammonium, and orthophosphates, macrophyte biomass, and fish abundance. Other predictor variables were removed because they were strongly correlated with at least one of the remaining variables. Trophic diversity did not vary with changes in species richness. Only system connectivity remained after the backward selection procedure and had significant effects on trophic diversity (Table 18.3). Specifically, trophic diversity was higher in disconnected than in connected systems.

Trophic composition was affected by human impact intensity (HII), waste deposition, ammonium content, macrophyte biomass, and fish abundance (Fig. 18.5). The HII, waste deposition, and increased ammonium content were associated with higher CWM values for predators (Pre) and shredders (Shr), while negatively affecting gatherer collectors (Gat). Biomass of macrophytes and abundance of fishes had

Table 18.3 Results of linear mixed effect model for effects of system connectivity (not connected, connected) and individual human-impact components on trophic diversity (i.e., functional dispersion FDisp) of epiphytic macroinvertebrate community. The study system nested in the sampling year and geographic area (Serbia, Croatia) was included as a random effect. Connectivity was selected by backward selection procedure as the minimum adequate model and individual human-impact components were included to the minimum adequate model. Significant effects ($P < 0.05$) are given in bold. For significant fixed effects the direction of the effect is given: ↑ positive, ↓ negative. Abbreviations: nDF—numerator degrees of freedom; dDF—denominator degrees of freedom; R^2m and R^2c are marginal (for fixed predictors) and conditional (for fixed and random predictors) coefficients of determination (pseudo-R^2), respectively; part-R^2 are semi-partial R^2 for each particular fixed predictor

Predictors	Part-R^2	nDF	dDF	F	Pr(>F)
Connectivity (not connected, connected)	0.13	1	15	4.74	**0.046** ↓
Distance to agricultural lands	0.003	1	11	0.07	0.79
Gravel exploitation	0.05	1	16	1.74	0.21
Distance to highway	0.04	1	12	1.24	0.29
Waste input	0.01	1	31	0.53	0.47
	R^2m = 0.26; R^2c = 0.45				

contrasting effects on trophic composition. Specifically, grater macrophyte biomass increased the CWMs for miners (Min), grazers and scrapers (Gra), xylophagous taxa (Xyl), active filter feeders (Aff) and parasitic taxa (Par). In contrast, these groups were reduced with higher fish abundance, which, in turn, was positively associated with the CWM for grazer and scrapers (Gat) (Fig. 18.4).

Fish communities

A total of 2488 individuals of 21 fish species (Supplementary Material, Table S7) were recorded across all study site ($n = 59$; Table S1 in the Supplementary Material). The most frequently occurred species were the introduced *Carassius auratus* L. and *Lepomis gibbosus* L. Rare fish species (i.e., found only at one site) was the native *Tinca tinca* L. Feeding on invertebrates was the dominant trophic trait across the study sites (community weighted mean, CWM $= 0.59 \pm 0.01$; Table S3 in the Supplementary Material), followed by herbivory (CWM $= 0.37 \pm 0.01$; Table S3 in the Supplementary Material). Both invertebrate feeding and herbivory occurred in all 59 study sites. Feeding on other fish (piscivory trophic trait) occurred in 29 sites (49%) and was represented by three species: northern pike (*Esox lucius* L.), European perch (*Perca fluviatilis* L.), and catfish (*Silurus glanis* L.); Table S8. Pike individuals feed mainly on fish but also invertebrates, especially the yang pikes. Perch feeds on both fish and invertebrates. Catfish are both active predators (of fish, invertebrates, amphibians, birds, small mammals) and scavengers. The catfish *S. glanis* was the only species in our study assigned as scavengers (Supplementary Material, Table S8), which occurred in three study sites (5%).

The predictor variables that remained after the correlation test (Fig. S4, step 1 in Materials and methods) were system depth, human-impact intensity index, water

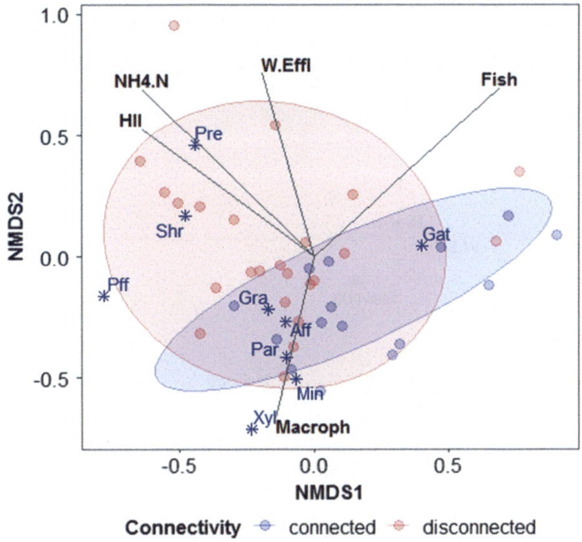

Fig. 18.4 Nonmetric multidimensional scaling (NMDS) of trophic composition of epiphytic macroinvertebrate communities: Gra—grazers and scrapers; Min—miners; Xyl—xylophagous taxa; Shr—shredders; Gat—gatherer collectors (detritivores); Aff—active filter feeders; Pff—passive filter feeders; Pre—predators; Par—parasites (see Supplementary Material, Table S3 for details). Significant predictors (Supplementary Material, Tables S16–S17) are plotted post-hoc as arrows: HII—human-impact intensity index; W.Effl—waste input (effluent); NH4.N—ammonium content in water; Macroph—macrophyte biomass; Fish—fish abundance. Circles show study sites. System connectivity is indicated by the colours of ellipses (the overlay of the system connectivity patterns in the ordination space)

content of dissolved oxygen, ammonium, and orthophosphates, and macrophyte biomass (Supplementary Material, Fig. S18). Other variables, i.e. system connectivity, river distance, water temperature, and nitrate–nitrogen content were removed due to multicollinearity (VIF > 3) because of strong correlation (Supplementary Material, Fig. S4) among connectivity and HII ($r = -0.77$); river distance and HII ($r = -0.86$); and among water contents of nitrate–nitrogen and orthophosphate ($r = 0.78$). Furthermore, water temperature was removed when we tested for the effect of individual land-use components (Supplementary Material, Table S20), because of the correlation among water temperature and the distance to highway ($r = -0.81$). Fish trophic trait diversity (FDis) was strongly associated with fish species richness. The most important predictor of trophic diversity (i.e. explaining the most variance) was system depth, showing negative effect (Table 18.4). The index of human-impact intensity (HII) and individual human-impact components had no significant effect on FDis.

Trophic trait composition of fish community was affected by system depth, water O2 and temperature (Supplementary Material, Table S19, Fig. 18.5). Among the components of human impact, waste deposition and distance to highway had the

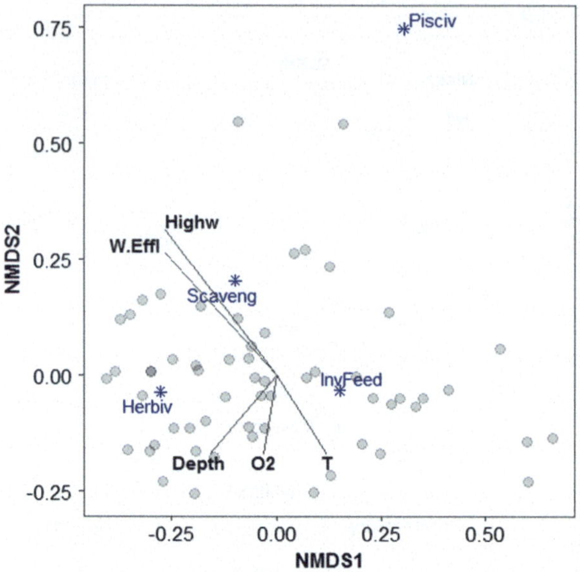

Fig. 18.5 Nonmetric multidimensional scaling (NMDS) of trophic composition of fish community: Herbiv—herbivores; InvFeed—invertebrate feeders; Pisciv—piscivores (feed on other fish); Scaveng—scavengers and others (i.e., feed on all kinds of foods indiscriminately) (see Supplementary Material, Table S3 for details). Significant predictors (Supplementary Material, Tables S19–S20) are plotted post-hoc as arrows. Abbreviations: Depth—pond depth; W.Effl—waste input (effluent); Highw—distance to the adjacent highways; O_2—dissolved oxygen content in water; T—water temperature. Grey circles show study sites

effects on trait composition (Supplementary Material, Table S20, Fig. 18.5). Scavenger trait was positively associated with waste deposition and increasing distance to highway, and negatively associated with increasing water temperature. On the first axis of the NMDS ordination plot (Fig. 18.5) herbivores were separated from piscivores in the same direction of increased system depth and water dissolved oxygen.

Dependency of human-impact effects on trophic group, trophic level, community type, and functional measure

We found significant differences between trophic groups in their responses to overall human-impact intensity (HII) (Fig. 18.6), as well as to the effects of agriculture and adjacent roads (Fig. 18.7 a, c). The effects of these HI measures were negative and relatively strong for herbivores, weaker for decomposers, and positive for carnivores (Figs. 18.6, 18.7). Effects of the HII and of each human impact type on CWM varied significantly between trophic levels (Figs. 18.6, 18.7).

The HII effects on the 2nd level (herbivore and decomposer groups, Table S4 in the Supplementary Material) ranged from negative to positive effects. The HII effects on the 4th trophic levels (i.e. macroinvertebrate filter feeders that are predominantly

Table 18.4 Results of backward selected linear mixed effect models for effects of species richness, system size (i.e., depth, m) and human impact on trophic diversity (i.e., functional dispersion FDis) of fish community. The study system nested in the sampling year and geographic area (Serbia, Croatia) was included as a random effect. Shown are separate models testing the effects of overall human impact intensity HII (model 1, HII was removed by backward selection procedure), and the effects of individual human-impact components (model 2). Significant effects ($P < 0.05$) are given in bold. For significant fixed effects the direction of the effect is given: ↑ positive, ↓ negative. Abbreviations: nDF—numerator degrees of freedom; dDF—denominator degrees of freedom; R^2m and R^2c are marginal (for fixed predictors) and conditional (for fixed and random predictors) coefficients of determination (pseudo-R^2), respectively; part-R^2 are semi-partial R^2 for each particular fixed predictor

Model 1. Test for the effects of total human impact HII (minimum adequate model)					
Predictors	Part-R^2	nDF	dDF	F	Pr(>F)
Species richness	0.14	1	52	13.35	**0.001 ↑**
Depth	0.17	1	50	8.16	**0.006 ↓**
	$R^2m = 0.28$, $R^2c = 0.79$				
Model 2. Test for the effects of individual human-impact types					
Predictors	Part-R^2	nDF	dDF	F	Pr(>F)
Species richness	0.11	1	48	12.34	**0.001 ↑**
Depth	0.11	1	49	5.84	**0.02 ↓**
Distance to agricultural lands	0.02	1	29	0.74	0.40
Gravel exploitation	0.02	1	20	0.78	0.39
Distance to highway	0.03	1	30	1.13	0.30
Waste input	< 0.001	1	51	0.03	0.86
	$R^2m = 0.32$; $R^2c = 0.79$				

decomposers but can consume prey with water filtering, Table S4) were low with the variations from negative to positive effects, Fig. 18.6. Top fish predators (7th level, piscivores, Table S4) were reduced with HII (Fig. 18.6), while other predator levels were positively affected (3rd level—carnivore zooplankton, 5th level—carnivore macroinvertebrates, and 6th level—invertebrate-feeding fish, Table S4). Responses of lower trophic levels (i.e. 2nd, 3rd, and 4th levels) showed similar patterns between different types of human impact, i.e. effects of agriculture, gravel exploitation, road proximity, and waste input (Fig. 18.7). In contrast, higher trophic groups varied strongly in their responses to different anthropogenic activities (Fig. 18.7). Carnivore macroinvertebrates (5th level) showed decreasing tendency under the effects of agriculture and gravel exploitation but increasing tendency with road proximity and waste input. Both, agriculture and road proximity had large positive effects on invertebrate-feeding fish (6th trophic level) and much weaker effects on piscivores (6th trophic level). Waste input had negative effects on invertebrate-feeding fish and weak effects on piscivores. Negative effects of gravel exploitation increased when moving up across consumer trophic levels (Fig. 18.7). Among the study types of human impact only the effects of agriculture varied significantly between different

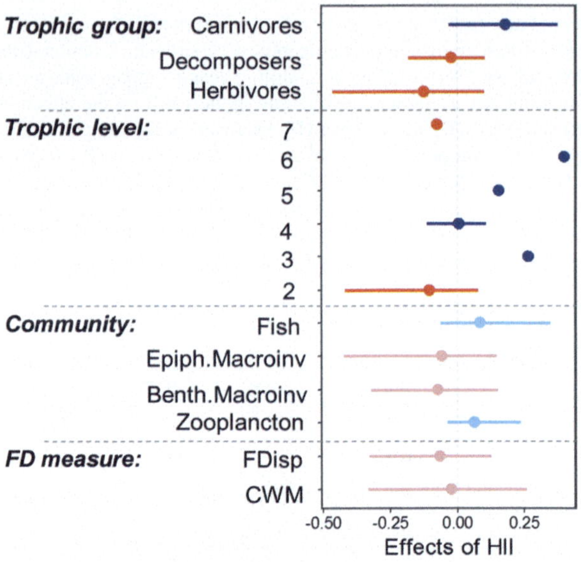

Fig. 18.6 Strength (slopes) of effects of human impact intensity HII on trophic groups, trophic levels, communities, and functional diversity (FD) measures: functional dispersion FDisp and community weighted mean CWM on trophic diversity of consumers. Shown are mean with 95% CI of the slopes obtained from the regression models. Statistically significant differences ($\alpha = 0.05$) are shown in red (negative effects) and blue (positive effects), and nonsignificant differences are shown in pink (negative effects) and light blue colours (positive effects)

communities. Specifically, the effects on zooplankton community and on epiphytic macroinvertebrates were weak and ranged from negative to positive. In contrast to this, the effects were larger and negative for benthic macroinvertebrates and positive for fish community. Effects of human activities did not vary significantly between the functional measures FDisp vs. CWM (Fig. 18.7).

18.4 Discussion

Drivers of trophic diversity and composition of zooplankton communities

Trophic trait diversity (FDisp) of zooplankton community was not correlated with species richness, suggesting functional similarity between species [80, 81]. Overall human impact intensity (HII) had negative effect on trophic diversity (Table 18.1), thus indicating higher dominance by one or few similar trophic traits in more transformed systems. When analysing the effects of the different HI types individually, we found a loss in trophic diversity with the presence of waste effluent (Table 18.1). This might be driven by the altered water properties by communal waste input, as the presence of effluent was associated with the increased content of PO_4-P and NO_3-N

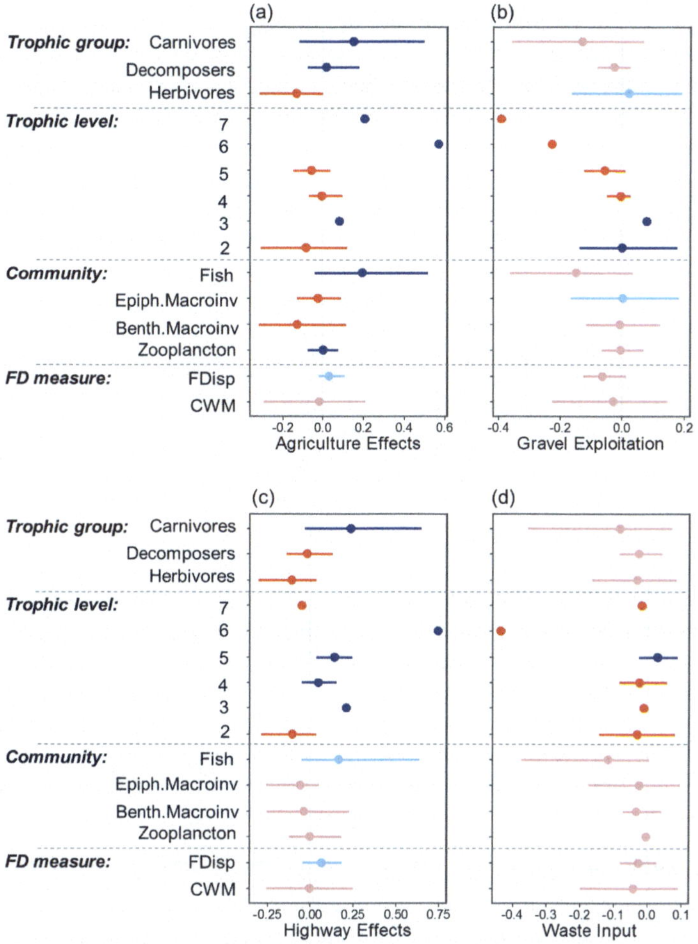

Fig. 18.7 Strength (slopes) of human impact effects on trophic groups, trophic levels, communities, and functional diversity (FD) measures: functional dispersion FDisp and community weighted mean CWM on trophic diversity of consumers. Shown are mean with 95% CI of the slopes obtained from the regression models. Statistically significant differences ($\alpha = 0.05$) are shown in red (negative effects) and blue (positive effects), and nonsignificant differences are shown in pink (negative effects) and light blue colours (positive effects). Shown are effects of different human impact components: **a** proximity to agricultural fields (reversed to distance); **b** intensity of gravel exploitation; **c** proximity to roads (reversed to distance); and **d** waste input

and reduced oxygen in water (Supplementary Material, Fig. S1). Also, there was a statistical trend for NH_4-N content to increase with waste input (Fig. S1), however, these variables showed contrasting effects on trophic diversity (negative effect of waste effluent and positive effect of ammonium, Table 18.1). These effects on trophic diversity of the zooplankton community were caused by an altered trophic composition (Fig. 18.2). Specifically, increased ammonium content was associated with the reduction of herbivore zooplankton and the increase in omnivores and carnivores, while waste effluent and orthophosphate content showed opposite effects. These results are not surprising given that the herbivore group is represented by cladocera species, while omnivore and carnivore groups are represented by rotifers and copepods. A negative effect of the N enrichment and positive effects of water P content is expected for cladocera species, which have typically higher P requirements and can be sensitive to ammonia content [82, 83]. Copepods generally have higher N requirements [82, 83] and thus the observed positive effect of increased water N on these taxa was expected. Our results agree with *Trommer* et al. [82], who found shifts in zooplankton community with ammonium load toward copepod and rotifer dominated community, containing fewer cladocerans [82]. For more mechanistic understanding of zooplankton functional composition, further studies would benefit from including into the analysis the species-specific traits of sensitivity to N and P content. The community shifts toward more herbivores, caused by waste input and increased orthophosphates, could have consequences for phytoplankton community, as well as for the higher trophic levels, such as macroinvertebrates and fishes. This can positively affect the planktivore fish, because they feed preferably on cladocera species [82], i.e. the herbivore group in our study. Macrophyte biomass associated positively with the herbivores but negatively with the omnivore and carnivore groups [84]. This could be caused by reduced predation on cladocerans in more complex habitat. Cladocerans are found to increase with greater cover of submerged macrophytes, because the plants serve as a refuge for cladocerans and favour piscivore fish at the expense of their prey planktivore fish [85]. To improve mechanical understanding of zooplankton trophic composition further studies should include functional traits of macrophytes in the analysis of fish trophic composition and their indirect effects to zooplankton.

Drivers of trophic diversity and composition of benthic and epiphytic macroinvertebrates

Some studies show no correlation between taxa richness and functional trait diversity measures of macroinvertebrates in their responses to anthropogenic activities [86]. Other studies, however, show that anthropogenic stressors may affect macroinvertebrate functional traits through altered taxonomic composition [20, 26], for instance through the dominance of taxa that are more tolerant to HI, e.g. Chironomidae [20] and Oligochaeta [14], and loss of vulnerable taxa, such as Ephemeroptera, Plecoptera, and Trichoptera [20]. In our study, the increased trophic diversity of benthic macroinvertebrates was explained by their taxa richness (Table 18.2). In contrast to benthic macroinvertebrates, the trophic diversity of epiphytic macroinvertebrates did not correlate with their taxa number (Table 18.3). Changes in species richness do not necessarily involve a change in functional composition, as some species may be functionally redundant [87].

For benthic macroinvertebrates, we found higher trophic diversity with increased HII, suggesting more evenly distributed dissimilar trophic traits in the community (Table 18.2). This may be explained by the higher proportion of predatory taxa along the HI gradient (Fig. 18.3), which, in turn, control the other dominant trophic groups via predation, such as gatherer-collectors, grazers and scrappers (Fig. 18.3), thus resulting in more evenness of trophic traits.

We found lower trophic diversity with higher macrophyte biomass (Table 18.2). Some of the dominant trophic groups (i.e., grazers and scrappers) are promoted by the greater biomass of macrophytes in expense of predators (Fig. 18.3), leading to dominance by few similar traits and, thus, reduced trophic diversity. The macrophyte-induced shifts in trophic composition toward dominance of primary consumers (herbivores and decomposers) in both benthic and epiphytic macroinvertebrate communities (Figs. 18.3–18.4) may also be associated with the higher habitat structural and functional complexity provided by macrophytes. Habitat heterogeneity promotes a greater variety of food sources for these groups and serves as a refuge from predation [18].

The decreased trophic diversity of benthic macroinvertebrates along water temperature gradient (Table 18.2) could be attributed to the dominance of few generalist taxa adapted to warm water conditions, where they outcompete taxa intolerant to water thermal changes. Furthermore, temperature may regulate trophic composition of macroinvertebrates by affecting their life history [14] and the rates of ecosystem processes, which control the trophic interactions of macroinvertebrates, such as predation rates by fish [8] and decomposition rates [14].

Trophic diversity of epiphytic macroinvertebrates was explained solely by the system connectivity, with higher FDisp in disconnected systems in contrast to connected ones (Table 18.3). This could be attributed to lower fish abundances in disconnected study systems [38], thus leading to the lower predation pressure by fishes on macroinvertebrates. The top-down control by fish predation has been found in previous studies as a strong driver of macroinvertebrate functional diversity and composition [88]. This is confirmed also by our NMDS ordination (Fig. 18.4), showing that fish abundance explained the shifts in trophic composition of epiphytic macroinvertebrates.

The human-impact intensity index HII, ammonia content, and macrophyte biomass explained changes in trophic composition in both benthic and epiphytic macroinvertebrate communities. Additionally, distance to roads and system depth explained trophic composition of benthic macroinvertebrates (Fig. 18.3) and waste input affected community composition in both benthic and epiphytic macroinvertebrates. Previous evidence shows that filterer feeders may increase in the systems undergoing human disturbances [16] followed with the increased nutrient levels [15, 89] and fine particulate organic matter [15]. We found positive association of benthic passive filterer feeders with HI intensification but no association of active filter feeders (Fig. 18.3). Epiphytic passive filter feeders were not affected by HI activities, while epiphytic active filter feeders were negatively associated with waste input (Fig. 18.4). Both benthic and epiphytic active filter feeders were strongly associated with macrophyte biomass (Figs. 18.3, 18.4). Such discrepancies in the responses of passive and active filter-feeding groups may be due to differences in their searching behaviour, i.e., while passive filter feeders rely on water velocity to bring food particles [43], mobile macroinvertebrates can actively choose habitats of higher structural complexity, protected from predators, and with greater food availability [90].

Our results of the reduced gatherer collectors along the HI gradient (Figs. 18.3, 18.4) contradict previous studies that found higher proportion of collectors in more disturbed systems [20, 22, 86]. Previous evidence suggests that due to their high feeding generalism, gatherer collectors are generally more tolerant to disturbances that may alter their food availability [20, 22]. It is likely, that the negative HI effects on gatherers in our study sites were not related to the alterations of their resources (i.e., sedimented FPOM, Table S4 in the Supplementary Material) but rather to water pollutants from waste inputs.

It is expected from previous studies that grazers and scrappers would increase with nutrient enrichment, as a result of greater availability of their food [20, 22, 89]. Some other studies, however, found no HI effects, including nutrient enrichment, on grazers and scrappers [15]. We found weak negative associations of grazers and scrappers with ammonia content, waste input, and HII (Figs. 18.3, 18.4). This may be due to weak effects of HI activities on food resources of these groups in our study sites, such as periphytic and epiphytic algae or biofilm (Supplementary Material, Table S4). Instead, grazers and scrappers were positively associated with macrophyte biomass (Figs. 18.3, 18.4), which also can serve as their food source, but also as a habitat. Furthermore, changes in water properties by waste input could explain the lower abundances of grazers and scrappers when waste effluent was present in our study sites (Figs. 18.3, 18.4). Previous research shows that scrappers are sensitive to changes in water temperature and acidity caused by waste impute [20], as well as to contamination, such as metals, which can become concentrated in their food source, such as biofilm [43].

Benthic shredders were not sensitive to HI intensity, showing weak negative associations with HII and waste input (Fig. 18.3). Instead, the increase in benthic shredders was strongly associated with the biomass of macrophytes, their primary food resource (Supplementary Material, Table S4). In contrast to benthic community, the epiphytic shredders were positively associated with HII, waste input, and ammonium content (Fig. 18.4). Our results counter to previous studies that showed strong reduction in proportion of shredders along human impact gradients [20], including nutrient enrichment [89].

Our findings of increased both benthic and epiphytic predators along the HII gradient, waste input, and ammonium content (Figs. 18.3, 18.4) are in line with the other studies that found higher proportions of macroinvertebrate predators associated with human disturbances [20], including agricultural practices [15, 91], and domestic waste water inputs [15], followed by increased salinity, acidification and heavy metals [15]. Similar to previous findings [15], we found decreasing tendency of benthic predators with system depth in our study sites.

Drivers of trophic diversity and composition of fish communities

Trophic trait diversity of fish community correlated positively with species richness, thus indicating that individual species contributed proportionally to trophic diversity at community level [80, 81]. Trophic diversity decreased with the increasing depth of the study SWB systems (Table 18.4). This could be due to the dominance of few generalist species, which can tolerate higher water depth and associated conditions, such as lower water temperature, oxygen concentration, and suspended solids (Supplementary Material, Fig. S1–S4). The variations in water depth explained also the changes in trophic composition of the fish community. Specifically, herbivores associated positively with increasing depth, which could be due to lower predation pressure on these fish group, as we found negative association of the piscivores with increasing depth (Fig. 18.5). This is in contrast to the expectation that fishes in deep waters are more protected from piscivore fish compared to in more shallow waters [92]. However, some studies suggest that top-down control might be in fact higher in shallow than in deep freshwater ecosystems due to the effects of submerged macrophytes [85]. Submerged macrophytes have been found to favour piscivore fish species, such as pike (*Esox lucius* L.) and perch (*Perca fluviatilis* L.) with a negative feedback on their prey fish, e.g. roach (*Rutilus rutilus* L.) and bream (*Abramis brama* L.) [85, 93]. In our study we did not consider the functional traits of macrophytes, however, there was a decreasing trend of macrophyte biomass with the higher depth (Supplementary Material, Fig. S4). The changes in the fish trophic structure with depths might also be a result of shifts in fish species composition based on species tolerance to temperature and oxygen levels [94–96]. In our study, water depth and oxygen content positively correlated in their effects on fish trophic composition (Fig. 18.5). Water temperature correlated negatively with depth on the first axis of the NMDS ordination (Fig. 18.5) and was associated with higher CWM values for invertebrate feeders and to a lesser degree for piscivores. Previous research shows that reduction in depth is accompanied with warmer, well oxygenated, and well mixed water conditions, which favour warm-water species, such as pike *E. lucius*, perch *P. fluviatilis*, and roach *R. rutilus* at the expense of cool-water species [94–96]. Water oxygen and temperature associated negatively with the CWM for scavenger trait (expressed only by catfish *S. glanis* in our study, Table S8 in the Supplementary Material). This is consistent with earlier findings showing that *S. glanis* can tolerate low water oxygen levels and low water temperature [97, 98]. Further studies will benefit from including the species-specific traits of temperature tolerance and oxygen demand in order to improve the mechanical understanding of the effects of depth on trophic diversity and functional composition of fish communities.

We found lower CWM values for invertebrate feeders with larger distance to highway (Fig. 18.5). This can likely be explained by water temperature, which was highly negatively correlated with distance to highway (Fig. 18.5, Fig. S4 in the Supplementary Material). This corroborates previous findings that urbanization intensity (including road proximity) can alter community composition through warming up the freshwater ecosystems [99]. Furthermore, earlier findings in our study sites show that closer distance to roads increased abundance of invasive fish species, such as *Lepomis gibbosus* L., *Carassius auratus* L., and *Pseudorasbora parva* Temminck & Schlegel [38], which are predominantly invertebrate feeders (Supplementary Material, Table S8). Waste input also explained the alterations of fish trophic composition. Specifically, the CWM score for scavengers increased with waste input while the invertebrate feeders were affected negatively (Fig. 18.5). As solid waste cover correlated highly with waste effluent effects in our study, it is difficult to disentangle the role of these two variables in the observed changes in trophic composition. Solid wastes include human food remains, that could increase the abundance of the catfish *S. glanis* (the only scavenger in our study, Table S8), which are known to feed on any alive or dead animals or plant materials of a suitable size and are extremely adaptable to new food sources [79, 100]. Furthermore, catfish is relatively tolerant to water pollution (such as communal waste), low water visibility, eutrophication, and habitat degradation and can outcompete species adapted to low nutrient and sediment levels in water [98, 101]. The increased water contents of nitrogen and phosphorus and reduced dissolved oxygen were associated with waste input in our study (Supplementary Material, Fig. S4).

General patterns of human-impact effects across multitrophic communities

Previous studies indicate that different trophic groups of consumer communities may vary in their responses to environmental changes or anthropogenic impacts [30, 40–42, 87]. Furthermore, the strengths and direction of HI effects may differ across consumer trophic levels [23, 46, 47], thus having consequences for ecosystem functioning. We tested whether the HI-induced shifts in mean trait values (abundance based CWM for each trophic trait) across the entire multitrophic community differed between trophic groups, trophic levels, consumer community types, and nature of human activities (HI type). We found significant differences in the strength of HII effects (slopes) on CWM between trophic groups: negative strong effects on herbivores, weaker effects on decomposers, and positive effects on carnivores (Fig. 18.6). The same patterns in trait responses of trophic groups were observed for the effects of agriculture and road proximity (Fig. 18.7). For the effects of gravel exploitation and waste deposition the CWM responses did not differ significantly between trophic groups (Fig. 18.7b, d). Some studies argue that predators might be more vulnerable to human-induced disturbances because, in comparison to primary consumers, they have lower abundance, larger body size, and larger requirements of habitat range [45]. Furthermore, top trophic levels are expected to be the most vulnerable to pollutants that accumulate across trophic levels through food ingestion [15]. Other studies, however, suggest that predators may be more tolerant to disturbances than lower consumer levels [20], including water contaminants [45], salinity, acidification, and heavy metals [15] resulted from agricultural land use [15, 91] and waste inputs [15]. In our study, predators were more stable to human impacts (Fig. 18.6), however, their responses varied among trophic levels and depended on HI type (Figs. 18.6, 18.7). The invertebrate-feeding and piscivore fish groups (6th and 7th levels) differed strongly in their responses to human impact. The effects of agriculture and road proximity on the invertebrate-feeding fish were strong positive, but significantly lower for the piscivore fish (Figs. 18.6, 18.7). These discrepancies may be due to the competition among these fish groups (6th and 7th trophic levels) and due to the dominance of invasive invertebrate-feeding fishes in more disturbed sites (as discussed in a section above). Negative interactions between these top predator groups (6th and 7th trophic levels) rather than cascading effects through predation could be expected in our study sites given our results of the strong contrasting patterns in their responses to HI (Figs. 18.6, 18.7). Antagonistic interactions between top predator levels in aquatic ecosystem are highly important for lower trophic levels and their functions, such as primary production and decomposition [88]. Reduction of apex predators along HII gradient (i.e., piscivore fish, 7th trophic levels) in our study (Fig. 18.6) may relax their antagonistic interactions with invertebrate-feeding fish, causing stronger exploitation of lower trophic levels by invertebrate feeders, thus simplifying trophic network and lowering rates of ecosystem functions they provide [88].

Gravel exploitation showed clear cascading effects across consumer trophic levels with strong negative effects on fish predators and attenuation of effects when moving down the trophic network (Fig. 18.7b). Gravel and send mining increase water turbidity, sediment deposition, metal contamination, and habitat fragmentation [18, 102], which may have negative consequences for predatory fish [8, 14, 21, 103]. For instance, changes in visibility due to increased sediment deposition may decrease feeding success of predatory fish [8, 21]. Also, habitat fragmentation arising from gravel exploitation practices may disturb fish spawning habitats [14, 103].

Herbivores showed generally negative responses to HII in our study, but the strength of responses varied strongly (Fig. 18.6). Agriculture and road proximity had more pronounced negative effects on mean trait values for herbivores, while gravel exploitation and waste input had less pronounced effects, shifting to positive in some cases (Fig. 18.6). We found high variation in the decomposer responses to HI, ranging from negative to positive slopes (Fig. 18.6–18.7). These variations may depend on the nature of disturbances. Increased sediment loading and dissolved organic carbon from agricultural practices or gravel mining [19] could favour decomposers [18, 19]. Water contamination with toxic pollutants arising from waste input, agriculture, gravel exploitation, may instead have negative effects on decomposers.

In line with the previous studies [6, 8, 9, 43, 44], our results provide evidence that the effects of HI on trophic diversity and composition, and thus ecosystem functioning, depend on the nature of HI disturbances, thus reinforcing the importance of including multiple stressors into study of freshwater ecosystems, especially of the SWB systems. Further studies are needed to investigate interactions among these different HI types and their indirect effects, such as through altered water properties [60]. Furthermore, we found that the effects of agriculture on mean trait values significantly differed between the consumer communities (Fig. 18.7): stronger positive effects on fish community, weak effects on zooplankton and epiphytic macroinvertebrates, and negative effects on benthic macroinvertebrates. Our results show that we need to consider different animal communities to prevent the underestimation of HI consequences for freshwater ecosystems and to better inform conservation [104].

Previous studies suggest that the functional identity measure (mean trait values CWM) and multivariate functional diversity measures (e.g., FDisp) may differently explain functioning of the ecosystem [35]. In such cases, the choice of particular measure (multivariate vs. individual feeding trait) may influence the results. In our study the slopes of the HI effects on mean trait values (CWM) and on trophic diversity (FDisp) did not differ significantly (Figs. 18.6, 18.7), thus suggesting that these functional measures complement each other in their responses to human impact. Overall, when compare results across all consumer communities we found a reduction in trophic diversity (FDisp) with increased HII (Fig. 18.6), indicating higher homogenisation in trophic structure along the human-impact intensification gradient. Previous studies suggest that such trend toward functional homogenization of consumer communities may contribute to lower ecosystem stability to variable conditions and perturbations [81].

18.5 Conclusion

Our results show that the intensification of human impacts induced shifts in the trophic composition and altered the trophic diversity of different consumer communities in freshwater ecosystems. Such functional shifts depended strongly on the nature of human impact, thus reinforcing the need to consider multiple stressors in research and management of freshwater ecosystems. Furthermore, as the consequences of human impact also varied between different consumer communities (i.e., zooplankton, benthic macroinvertebrates, epiphytic macroinvertebrates, and fish community) it is essential to simultaneously consider a range of animal community types to better understand the impact of different anthropogenic activities on functional biodiversity. We also found that the effects of human impact on the representation of trophic traits across all consumer communities significantly differed between trophic groups (herbivores, decomposers, and carnivores) and trophic levels (trophic position in the food web). These results point toward the importance of a multitrophic perspective when predicting the consequences of human impact for ecosystem biodiversity and functioning to better inform conservation and management. Trophic diversity did not always correlate with species or taxa richness in our study. This challenges the common perspective that species richness may serve as a single biodiversity proxy replacing other diversity facets, thus highlighting the importance to integrate functional traits in community and biodiversity research.

Author Contributions O.Y.B. developed the concept and analytical procedure; M.S.P, O.S., D.Č., A.O., and D.M. provided data; O.Y.B. performed data analysis, visualization, and wrote the original draft. O.Y.B., M.S.P, O.S., D.Č., A.O., B.T., and D.M. contributed substantially to review and editing.

Acknowledgements Data collected for this study were funded by the Serbian Ministry of Education, Science and Technological Development (contract number 451-03-9/2021-14/200124) and by a bilateral cooperation project between Serbia and Croatia funded by the Serbian Ministry of Education, Science and Technological Development and Croatian Ministry of Science and Education. We thank the student helpers for their work.

References

1. Lehner B, Döll P (2004) Development and validation of a global database of lakes, reservoirs and wetlands. J Hydrol 296:1–22. https://doi.org/10.1016/j.jhydrol.2004.03.028

2. Heino J, Alahuhta J, Bini LM et al (2020) Lakes in the era of global change: moving beyond single-lake thinking in maintaining biodiversity and ecosystem services. Biol Rev 92:89–106. https://doi.org/10.1111/brv.12647
3. Céréghino R, Biggs J, Oertli B, Declerck S (2008) The ecology of European ponds: defining the characteristics of a neglected freshwater habitat. Hydrobiologia 597:1–6. https://doi.org/10.1007/s10750-007-9225-8
4. van Klink R, Bowler DE, Gongalsky KB et al (2020) Meta-analysis reveals declines in terrestrial but increases in freshwater insect abundances. Science 368:417–420. https://doi.org/10.1126/science.aax9931
5. Birk S, Chapman D, Carvalho L et al (2020) Synthesizing the impacts of multiple stressors on freshwater biota across scales and ecosystems. Nat Ecol Evol 4:1060–1068. https://doi.org/10.1038/s41559-020-1216-4
6. Mondy CP, Muñoz I, Dolédec S (2016) Life-history strategies constrain invertebrate community tolerance to multiple stressors: A case study in the Ebro basin. Sci Total Environ 572:196–206. https://doi.org/10.1016/j.scitotenv.2016.07.227
7. Allan JD (2004) Landscapes and riverscapes: the influence of land use on stream ecosystems. Annu Rev Ecol Evol Syst 35:257–284. https://doi.org/10.1146/annurev.ecolsys.35.120202.110122
8. Vasconcelos FR, Diehl S, Rodríguez P et al (2019) Bottom-up and top-down effects of browning and warming on shallow lake food webs. Glob Change Biol 25:504–521. https://doi.org/10.1111/gcb.14521
9. Price EL, Sertić Perić M, Romero GQ, Kratina P (2019) Land use alters trophic redundancy and resource flow through stream food webs. J Anim Ecol 88:677–689. https://doi.org/10.1111/1365-2656.12955
10. Riley WD, Potter ECE, Biggs J et al (2018) Small water bodies in great Britain and Ireland: ecosystem function, human-generated degradation, and options for restorative action. Sci Total Environ 645:1598–1616. https://doi.org/10.1016/j.scitotenv.2018.07.243
11. Kristensen P, Globevnik L (2014) European small water bodies. Biol Environ 114B:281–287. https://doi.org/10.3318/BIOE.2014.13
12. Lorenz S, Rasmussen JJ, Süß A et al (2017) Specifics and challenges of assessing exposure and effects of pesticides in small water bodies. Hydrobiologia 793:213–224. https://doi.org/10.1007/s10750-016-2973-6
13. Burdon FJ, Bai Y, Reyes M et al (2020) Stream microbial communities and ecosystem functioning show complex responses to multiple stressors in wastewater. Glob Change Biol 26:6363–6382. https://doi.org/10.1111/gcb.15302
14. Paul MJ, Meyer JL (2001) Streams in the urban landscape. Annu Rev Ecol Syst 32:333–365
15. Statzner B, Bêche LA (2010) Can biological invertebrate traits resolve effects of multiple stressors on running water ecosystems? Freshw Biol 55:80–119. https://doi.org/10.1111/j.1365-2427.2009.02369.x
16. Dolédec S, Phillips N, Scarsbrook M et al (2006) Comparison of structural and functional approaches to determining landuse effects on grassland stream invertebrate communities. J N Am Benthol Soc 25:44–60. https://doi.org/10.1899/0887-3593(2006)25[44:COSAFA]2.0.CO;2
17. Statzner B, Bis B, Dolédec S, Usseglio-Polatera P (2001) Perspectives for biomonitoring at large spatial scales: a unified measure for the functional composition of invertebrate communities in European running waters. Basic Appl Ecol 2:73–85. https://doi.org/10.1078/1439-1791-00039
18. Yadamsuren O, Morse JC, Hayford B et al (2020) Macroinvertebrate community responses to land use: a trait-based approach for freshwater biomonitoring in Mongolia. Hydrobiologia 847:1887–1902. https://doi.org/10.1007/s10750-020-04220-2
19. Statzner B, Bady P, Dolédec S, Schöll F (2005) Invertebrate traits for the biomonitoring of large European rivers: an initial assessment of trait patterns in least impacted river reaches. Freshw Biol 50:2136–2161. https://doi.org/10.1111/j.1365-2427.2005.01447.x

20. Rawer-Jost C, Böhmer J, Blank J, Rahmann H (2000) Macroinvertebrate functional feeding group methods in ecological assessment. Hydrobiologia 422–423:225–232. https://doi.org/10.1007/978-94-011-4164-2_18
21. Kritzberg ES, Hasselquist EM, Škerlep M et al (2020) Browning of freshwaters: consequences to ecosystem services, underlying drivers, and potential mitigation measures. Ambio 49:375–390. https://doi.org/10.1007/s13280-019-01227-5
22. Barbour MT, Gerritsen J, Griffith GE, et al. (1996) A framework for biological criteria for florida streams using benthic macroinvertebrates published by : The University of Chicago Press on behalf of the Society for Freshwater Science Stable URL . http://www.jstor.org/stable/1467948. REFERENCES Linked refe. J N Am Benthol Soc 15
23. Schramski JR, Dell AI, Grady JM et al (2015) Metabolic theory predicts whole-ecosystem properties. Proc Natl Acad Sci USA 112:2617–2622. https://doi.org/10.1073/pnas.1423502112
24. Buzhdygan OY, Meyer ST, Weisser WW et al (2020) Biodiversity increases multitrophic energy use efficiency, flow and storage in grasslands. Nat Ecol Evolut 4:393–405. https://doi.org/10.1038/s41559-020-1123-8
25. Eisenhauer N, Schielzeth H, Barnes AD et al (2019) A multitrophic perspective on biodiversity–ecosystem functioning research. Adv Ecol Res 61:1–54. https://doi.org/10.1016/bs.aecr.2019.06.001
26. Li Z, Liu Z, Heino J, et al. (2020) Discriminating the effects of local stressors from climatic factors and dispersal processes on multiple biodiversity dimensions of macroinvertebrate communities across subtropical drainage basins. Sci Total Environ 711:134750. https://doi.org/10.1016/j.scitotenv.2019.134750
27. Cordero PU, Langenheder S, Striebel M, et al (2021) Functionally reversible impacts of disturbances on lake food webs linked to spatial and seasonal dependencies. Ecology 0–3. https://doi.org/10.1002/ecy.3283
28. de Castro DMP, Dolédec S, Callisto M (2017) Landscape variables influence taxonomic and trait composition of insect assemblages in Neotropical savanna streams. Freshw Biol 62:1472–1486. https://doi.org/10.1111/fwb.12961
29. Barnes AD, Allen K, Kreft H et al (2017) Direct and cascading impacts of tropical land-use change on multi-trophic biodiversity. Nat Ecol Evolut 1:1511–1519. https://doi.org/10.1038/s41559-017-0275-7
30. Filazzola A, Brown C, Dettlaff MA et al (2020) The effects of livestock grazing on biodiversity are multi-trophic: a meta-analysis. Ecol Lett 23:1298–1309. https://doi.org/10.1111/ele.13527
31. Petchey OL, Downing AL, Mittelbach GG, et al. (2004) Species loss and the structure and functioning of multitrophic aquatic systems. In: Oikos. Munksgaard, pp 467–478
32. Li Z, Heino J, Liu Z et al (2021) The drivers of multiple dimensions of stream macroinvertebrate beta diversity across a large montane landscape. Limnol Oceanogr 66:226–236. https://doi.org/10.1002/lno.11599
33. Gayraud S, Statzner B, Bady P, et al. (2003) Invertebrate traits for the biomonitoring of large European rivers: an initial assessment of alternative metrics. Freshwater Biol 48:2045–2064. APPLIED. https://doi.org/10.1111/j.1365-2427.2007.01924.x
34. van der Plas F (2019) Biodiversity and ecosystem functioning in naturally assembled communities. Biol Rev 94:1220–1245. https://doi.org/10.1111/brv.12499
35. Gagic V, Bartomeus I, Jonsson T, et al. (2015) Functional identity and diversity of animals predict ecosystem functioning better than species-based indices. Proc Royal Soc B: Biol Sci 282. https://doi.org/10.1098/rspb.2014.2620
36. Davies B, Biggs J, Williams P et al (2008) Comparative biodiversity of aquatic habitats in the European agricultural landscape. Agr Ecosyst Environ 125:1–8. https://doi.org/10.1016/j.agee.2007.10.006
37. Williams P, Whitfield M, Biggs J et al (2004) Comparative biodiversity of rivers, streams, ditches and ponds in an agricultural landscape in Southern England. Biol Cons 115:329–341. https://doi.org/10.1016/S0006-3207(03)00153-8

38. Stamenković O, Stojković Piperac M, Milošević D et al (2019) Anthropogenic pressure explains variations in the biodiversity of pond communities along environmental gradients: a case study in south-eastern Serbia. Hydrobiologia 4:65–83. https://doi.org/10.1007/s10750-019-03978-4
39. Petermann JS, Roberts AL, Hemmerling C, et al. (2020) Direct and indirect effects of forest management on tree-hole inhabiting aquatic organisms and their functional traits. Sci Total Environ 704:135418. https://doi.org/10.1016/j.scitotenv.2019.135418
40. Attwood SJ, Maron M, House APN, Zammit C (2008) Do arthropod assemblages display globally consistent responses to intensified agricultural land use and management? Glob Ecol Biogeogr 17:585–599. https://doi.org/10.1111/j.1466-8238.2008.00399.x
41. Tuck SL, Winqvist C, Mota F et al (2014) Land-use intensity and the effects of organic farming on biodiversity: a hierarchical meta-analysis. J Appl Ecol 51:746–755. https://doi.org/10.1111/1365-2664.12219
42. Lang B, Ehnes RB, Brose U, Rall BC (2017) Temperature and consumer type dependencies of energy flows in natural communities. Oikos 126:1717–1725. https://doi.org/10.1111/oik.04419
43. Pavluk TI, Bij De Vaate A, Leslie HA (2000) Development of an Index of trophic completeness for benthic macroinvertebrate communities in flowing waters. Hydrobiologia 427:135–141. https://doi.org/10.1023/A:1003911109416
44. Hooper DU, Adair EC, Cardinale BJ et al (2012) A global synthesis reveals biodiversity loss as a major driver of ecosystem change. Nature 486:105–108. https://doi.org/10.1038/nature11118
45. Raffaelli D (2004) How extinction patterns affect ecosystems. Science 306:1141–1142
46. Voigt W, Perner J, Davis AJ et al (2003) Trophic levels are differentially sensitive to climate. Ecology 84:2444–2453
47. Hines J, Eisenhauer N, Drake BG (2015) Inter-annual changes in detritus-based food chains can enhance plant growth response to elevated atmospheric CO2. Glob Change Biol 21:4642–4650. https://doi.org/10.1111/gcb.12965
48. Rylov VM (1948) Crustaceans, Freshwater Cyclopoida, Fauna of the USSR. Nauka, Moscow-Leningrad
49. Bartoš E (1959) Fauna ČSSR 15: Virnici-Rotatoria. Nakladatelství Československé Akademie věd, Praha
50. Šramek-Husek R, Straškraba M, Brtek J (1962) Fauna ČSSR 16: Lupenonožci - Branchiopoda. Praha
51. Dussart B (1969) Les Copépodes des eaux continentales d'Europe occidentale. Tome II: Cyclopoides et Biologie. Boubée et Cie, Paris
52. Flössner D (1972) Krebstiere, Crustacea: Kiemen- und Blattfüßer, Branchiopoda/Fischläuse. VEB Gustav Fischer Verlag, Jena, Branchiura
53. Koste W (1978) Rotatoria, die Rädertiere Mitteleuropas: Überordnung Monogononta. Gebrüder Borntraeger, Berlin—Stutgart
54. Petkovski S (1983) Fauna of Macedonia. Part 5: Calanoids—Calanoida (Crustacea Copepoda). Nat. Sci. Museum of Macedonia, Skopje
55. Turner JT (1984) The feeding ecology of some zooplankters that are important prey items of larval fish. NOAA Technical Report NMFS, 7.
56. Rautio M, Vincent WF (2006) Benthic and pelagic food resources for zooplankton in shallow high-latitude lakes and ponds. Freshw Biol 51:1038–1052. https://doi.org/10.1111/j.1365-2427.2006.01550.x
57. Chang KH, Doi H, Nishibe Y, Nakano SI (2010) Feeding habits of omnivorous Asplanchna: comparison of diet composition among Asplanchna herricki, A. priodonta and A. girodi in pond ecosystems. J Limnol 69:209–216. https://doi.org/10.3274/JL10-69-2-03
58. Benedetti F, Gasparini S, Ayata SD (2015) Identifying copepod functional groups from species functional traits. J Plankton Res 38:159–166. https://doi.org/10.1093/plankt/fbv096
59. Tõnno I, Agasild H, Kõiv T et al (2016) Algal diet of small-bodied crustacean zooplankton in a cyanobacteria-dominated eutrophic lake. PLoS ONE 11:1–17. https://doi.org/10.1371/journal.pone.0154526

60. Stamenković O, Simić V, Stojković Piperac M et al (2021) Direct, water-chemistry mediated, and cascading effects of human-impact intensification on multitrophic biodiversity in ponds. Aquat Ecol 55:187–214. https://doi.org/10.1007/s10452-020-09822-5
61. Moller Pillot HKM (2009) Chironomidae Larvae. Biology and Ecology of the Chironomini. KNNV, Zeist, The Netherlands
62. Moller Pillot HKM (1984) De larven der Nederlandse Chironomidae (Diptera). (Orthocladiinae sensu latu)./The larvae of the Dutch Chironomidae (Diptera). (Orthocladiinae sensu latu). Nederlandse Faunistische Mededelin 1B. Stichting European Invertebrate Survey, Leiden, Nederland
63. Glöer P (2002) Die Süßwassergastropoden Nord und Mitteleuropas: Bestimmungsschlüssel, Lebensweise. Conchbooks, Bonn, Verbreitung
64. Eiseler B (2005) Bildbestimmungsschlüssel für die Eintagsfliegenlarven der deutschen Mittelgebirge und des Tieflandes. Lauterbornia 53:1–112
65. Elliot J, Humpesch U (2010) Mayfly larvae (Ephemeroptera) of Britain and Ireland: Keys and Review of their Ecology. Freshwater Biological Association, Ambleside
66. Andersen T, Cranston PS, Epler JH (2013) The larvae of Chironomidae (Diptera) of the Holarctic region—Keys and diagnoses. Insect Systematics and Evolution, Lund, Sweden
67. Rossaro B, Lencioni V (2015) A key to larvae of Diamesa Meigen, 1835 (Diptera, Chironomidae), well known as adult males and pupae from Alps (Europe). J Entomol Acarol Res 47:123–138. https://doi.org/10.4081/jear.2015.5516
68. Moller Pillot HKM (1984) De larven der Nederlandse Chironomidae (Diptera). (Inleiding, Tanypodinae & Chironomini)/The larvae of the Dutch Chironomidae (Diptera). (Introduction, Tanypodinae & Chironomini).Nederlandse Faunistische Mededelin 1A. Stichting European Invertebrate Survey, Leiden, Nederland
69. Elliott JM, Humpesch UH, Macan TT (1988) Larvae of the British Ephemeroptera: a key with ecological notes. Freshwater Biological Association
70. Schmid PE (1993) A key to the larval Chironomidae and their instars from Austrian Danube region streams and rivers with particular reference to a numerical taxonomic approach. 1 Diamesinae, Prodiamesinae, and Orthocladiinae. Federal Institute for Water Quality, Wien-Kaisermühlen
71. Waringer J, Graf W (1997) Atlas der österreichischen Köcherfliegenlarven: Unter Einschluss der angrenzenden Gebiete. Facultas Universitätsverlag, Wien
72. Gerken B, Sternberg K (1999) Die Exuvien Europäischer Libellen (Insecta, Odonata). The exuviae of European dragonflies. Arnika & Eisvogel, Höxter, Jena
73. Timm T (1999) Eestirõngusside (Annelida) määraja. Estonian Academy Publishers, Tartu-Tallinn, A Guide to the Estonian Annelidae
74. Pfleger V (2000) A field guide in colour to Molluscs. Silverdale Books, Prague
75. Bauernfeind E, Humpesch U (2001) Die Eintagsfliegen Zentraleuropas (Insecta: Ephemeroptera): Bestimmung und Ökologie. Verlag des Naturhistorischen Museums, Wien
76. Moog O, Hartmann A (2017) Fauna Aquatica Austriaca, 3rd edn. BMLFUW, Vienna
77. Simonović P (2001) Fishes of Serbia, In Serbian. NNK International, Belgrade
78. Grenouillet G, Schmidt-Kloiber A (2006) Fish Indicator Database. EURO-LIMPACS project, Work package 7—Indicators of ecosystem health, Task 4, accessed via www.freshwaterecology.info, version 7.0 (accessed on 08.02.2021)
79. Vejřík L, Vejříková I, Blabolil P et al (2017) European catfish (Silurus glanis) as a freshwater apex predator drives ecosystem via its diet adaptability. Sci Rep 7:1–15. https://doi.org/10.1038/s41598-017-16169-9
80. Petchey OL, Gaston KJ (2002) Functional diversity (FD), species richness and community composition. Ecol Lett 5:402–411. https://doi.org/10.1046/j.1461-0248.2002.00339.x
81. Díaz S, Cabido M (2001) Vive la différence: Plant functional diversity matters to ecosystem processes. Trends Ecol Evol 16:646–655. https://doi.org/10.1016/S0169-5347(01)02283-2
82. Trommer G, Poxleitner M, Lorenz P et al (2017) Altered food-web dynamics under increased nitrogen load in phosphorus deficient lakes. Aquat Sci 79:1009–1021. https://doi.org/10.1007/s00027-017-0551-2

83. Sterner RW, Hessen DO (1994) Algal nutrient limitation and the nutrition of aquatic herbivores. Annu Rev Ecol Syst 25:1–29. https://doi.org/10.1146/annurev.es.25.110194.000245
84. Deosti S, de Fátima BF, Lansac-Tôha FM et al (2021) Zooplankton taxonomic and functional structure is determined by macrophytes and fish predation in a Neotropical river. Hydrobiologia 848:1475–1490. https://doi.org/10.1007/s10750-021-04527-8
85. Jeppesen E, Jensen JP, Søndergaard M et al (1997) Top-down control in freshwater lakes: the role of nutrient state, submerged macrophytes and water depth. Hydrobiologia 342(343):151–164. https://doi.org/10.1007/978-94-011-5648-6_17
86. Li Z, Wang J, Liu Z et al (2019) Different responses of taxonomic and functional structures of stream macroinvertebrate communities to local stressors and regional factors in a subtropical biodiversity hotspot. Sci Total Environ 655:1288–1300. https://doi.org/10.1016/j.scitotenv.2018.11.222
87. Ebeling A, Rzanny M, Lange M et al (2018) Plant diversity induces shifts in the functional structure and diversity across trophic levels. Oikos 127:208–219. https://doi.org/10.1111/oik.04210
88. O'Gorman EJ (2021) Multitrophic diversity sustains ecological complexity by dampening top-down control of a shallow marine benthic food web. Ecology 102:1–13. https://doi.org/10.1002/ecy.3274
89. Kerans BL, Karr JR (1994) A Benthic Index of Biotic Integrity (B-IBI) for Rivers of the Tennessee Valley. Ecol Appl 4:768–785
90. Hansen JP, Wikström SA, Axemar H, Kautsky L (2011) Distribution differences and active habitat choices of invertebrates between macrophytes of different morphological complexity. Aquat Ecol 45:11–22. https://doi.org/10.1007/s10452-010-9319-7
91. Reyne M, Nolan M, McGuiggan H, et al (2020) Artificial agri-environment scheme ponds do not replicate natural environments despite higher aquatic and terrestrial invertebrate richness and abundance. Journal of Applied Ecology 1–12. https://doi.org/10.1111/1365-2664.13738
92. Linehan JE, Gregory RS, Schneider DC (2001) Predation risk of age-0 cod (Gadus) relative to depth and substrate in coastal waters. J Exp Mar Biol Ecol 263:25–44. https://doi.org/10.1016/S0022-0981(01)00287-8
93. Grimm MP, Backx JJGM (1990) The restoration of shallow eutrophic lakes, and the role of northern pike, aquatic vegetation and nutrient concentration. Hydrobiologia 200(201):557–566
94. Mehner T, Holmgren K, Lauridsen TL et al (2007) Lake depth and geographical position modify lake fish assemblages of the European "Central Plains" ecoregion. Freshw Biol 52:2285–2297. https://doi.org/10.1111/j.1365-2427.2007.01836.x
95. Mehner T, Diekmann M, Brämick U, Lemcke R (2005) Composition of fish communities in German lakes as related to lake morphology, trophic state, shore structure and human-use intensity. Freshw Biol 50:70–85. https://doi.org/10.1111/j.1365-2427.2004.01294.x
96. Holmgren K, Appelberg M (2000) Size structure of benthic freshwater fish communities in relation to environmental gradients. J Fish Biol 57:1312–1330. https://doi.org/10.1006/jfbi.2000.1395
97. Hilge V (1985) The influence of temperature on the growth of the European catfish (Silurus glanis L.). J Appl Ichthyol 1:27–31. https://doi.org/10.1111/j.1439-0426.1985.tb00407.x
98. Copp GH, Robert Britton J, Cucherousset J et al (2009) Voracious invader or benign feline? A review of the environmental biology of European catfish silurus glanis in its native and introduced ranges. Fish Fish 10:252–282. https://doi.org/10.1111/j.1467-2979.2008.00321.x
99. Oertli B, Parris KM (2019) Review: toward management of urban ponds for freshwater biodiversity. Ecosphere 10. https://doi.org/10.1002/ecs2.2810
100. Syväranta J, Cucherousset J, Kopp D et al (2009) Dietary breadth and trophic position of introduced European catfish Silurus glanis in the River Tarn (Garonne River basin), Southwest France. Aquat Biol 8:137–144. https://doi.org/10.3354/ab00220
101. Castaldelli G, Pluchinotta A, Milardi M et al (2013) Introduction of exotic fish species and decline of native species in the lower Po basin, north-eastern Italy. Aquat Conserv Mar Freshwat Ecosyst 23:405–417. https://doi.org/10.1002/aqc.2345

102. Yen TP, H R, (2013) Status of water quality subject to sand mining in the Kelantan River, Kelantan. Trop Life Sci Res 24:19–34
103. Leitão RP, Zuanon J, Mouillot D et al (2018) Disentangling the pathways of land use impacts on the functional structure of fish assemblages in Amazon streams. Ecography 41:219–232. https://doi.org/10.1111/ecog.02845
104. Penone C, Allan E, Soliveres S et al (2019) Specialisation and diversity of multiple trophic groups are promoted by different forest features. Ecol Lett 22:170–180. https://doi.org/10.1111/ele.13182

Chapter 19
Pollution of Small Lakes and Ponds of the Western Balkans—Assessment of Levels of Potentially Toxic Elements

Aleksandra Milošković, Simona Đuretanović, Milena Radenković, Nataša Kojadinović, Tijana Veličković, Đurađ Milošević, and Vladica Simić

Abstract Small lakes and ponds are among the most sensitive ecosystems, vulnerable to changes that have little or no effect on larger water bodies. Therefore this topic should be given more attention around the world. Anthropogenic impacts, which have been extensively researched in larger water bodies, have been barely examined in small lakes and ponds. Considering that most of the studies are related to lakes bigger than 50 ha, a significant knowledge gap was left regarding small lakes and ponds. The same situation is in the Western Balkans, where studies related to pollutants in small lakes and ponds are limited. This chapter presents an overview of studies related to the contamination of small lakes and ponds of the Western Balkans, with special reference to pollution with potentially toxic elements PTEs. Since massive fish mortality happened several times, an assessment of PTEs in fish species *Carassius gibelio* from small Aleksandrovac Lake in Serbia is presented. The results indicate that fish muscle (meat) was exposed to the lower pressure of PTEs pollution than liver and gills, suggesting that, despite massive fish mortality, there was no risk for human health by fish consumption.

Keywords Small water bodies · Fish · Human health risk · Aleksandrovac Lake

19.1 Introduction

Small water bodies, including ponds and small lakes, are freshwater ecosystems of high ecological relevance [1]. They occur in practically all terrestrial environments,

A. Milošković (✉)
Institute for Information Technologies Kragujevac, University of Kragujevac, Kragujevac, Serbia
e-mail: aleksandra@uni.kg.ac.rs

S. Đuretanović · M. Radenković · N. Kojadinović · T. Veličković · V. Simić
Institute of Biology and Ecology, Faculty of Science, University of Kragujevac, Kragujevac, Serbia

Đ. Milošević
Department of Biology and Ecology, Faculty of Sciences and Mathematics, University of Niš, Niš, Serbia

V. Pešić et al. (eds.), *Small Water Bodies of the Western Balkans*, Springer Water,
https://doi.org/10.1007/978-3-030-86478-1_19

from polar deserts to tropical rainforests [2], and, on a global scale, cover a greater area than lakes [3]. In Europe, they constitute a significant part of the continental freshwater habitats [4].

Ponds are small standing waters that vary in size from 1 m^2 to about 5 ha in area [5]. The Ramsar Convention adopted a cut-off between ponds and lakes of 8 ha [6]. Additionally, there is no strict definition of what constitutes small or large lake. Given that in Europe the size of lakes is considered significant, according to the Water Framework Directive, 50 hectares provide a de facto cut-off point for small lakes [7]. Their typical characteristics, such as shallow waters and small size, imply a different ecological functioning [8, 9]. Ponds also vary in origin. For example, some natural ponds are the result of glacial activity, and others may be formed as oxbow ponds. There are also many man-made ponds, for example, mill ponds or ponds established for sediment retention, or water storage, or temporary ponds [10, 11].

Small lakes and ponds support higher proportions of biodiversity compared to larger freshwater systems [12, 13], and can be substantially more biologically active than large lakes [14]. A wide range of studies have confirmed that, particularly for macrophytes, aquatic micro- and macroinvertebrates, and amphibians, small water bodies are areas of high biodiversity [12]. Also, they support specific and important hydrological, chemical, and biological processes [11].

The fundamental ecological research of ponds and small lakes influenced the growing concern and awareness regarding their abundance, importance for freshwater biodiversity, their role in contributing to ecosystem services, and their sensitivity and vulnerability to anthropogenic disturbances [3, 15]. It has become clear that there is a growing interest in ponds and small lakes in the early twenty-first century, which is reflected by increasing scientific activity, especially regarding biodiversity [4]. They are ecologically very important and represent powerful model systems for studies in ecology, evolutionary biology, and conservation biology, and can be used as sentinel systems to monitor global change [2].

A number of human activities threaten small lake and pond ecosystems—draining/infilling, eutrophication, contamination, acidification, and invasion of exotic species. Still, they are also threatened by global changes, in particular by increasing temperature and UV radiation [4, 16, 17]. Furthermore, anthropogenic impacts which have been extensively researched in larger waters have been barely examined in small lakes and ponds. For instance, Phillips et al. [18] pointed out that it is still uncertain what causes one of the major impacts on small and shallow lakes—the loss of macrophytes associated with eutrophication. Small lakes and ponds are vulnerable to changes that have little or no effect on larger water bodies. Therefore, to this issue and the way these habitats are affected should be given more attention worldwide.

This chapter represents an overview of studies related to the contamination of small lakes and ponds of the Western Balkans, with special reference to pollution with potentially toxic elements (PTEs). Additionally, an assessment of PTEs in fish species from small Aleksandrovac Lake in Serbia is presented.

19.2 Small Lakes and Ponds of the Western Balkans

The precise number of ponds and small lakes in Europe cannot be estimated, nor for the area of the Western Balkans. The available literature data are contradictory. According to Kristensen and Globevnik [11], there are at least 600,000 natural lakes in Europe, with only around 100,000 having an area greater than 10 ha. Biggs et al. [7] stated that in the whole of Europe, the number of small lakes and ponds is likely between five and ten million. On the other hand, the proportion of loss of the ponds in the European landscapes is estimated higher than 50% [4].

The study of ponds was neglected in the second half of the twentieth century, only gaining momentum in the twenty-first century, particularly in Europe because of the work of the European Pond Conservation Network. The protection of small water bodies and the biodiversity they support are of concern since they are vulnerable to changes that have little effect on larger water bodies [11].

Small water bodies have not been studied systematically in the Western Balkans; nevertheless, specific research topics were related to a specific issue or group of organisms/taxa/species. A survey of ponds and their loss in the Žumberak-Samoborsko gorje Nature Park was conducted by Hutinec and Struna [19], while Sremački et al. [20] conducted environmental monitoring and assessment of protected wetland and lake water quality in Croatia and Serbia. Ranđelović et al. [21] performed phytocenological research, while Stamenković et al. [22, 23] researched the anthropogenic impact on species diversity and density across all trophic levels of Batušinačke ponds in the vicinity of Niš. Tasevska et al. [24] researched rotifers. The zooplankton communities in the small lakes of the Western Balkans were investigated by Ostojić et al. [25–27], Špoljar et al. [28, 29], Kuczyńska-Kippen et al. [30], and Mancinelli et al. [31], while Ranković et al. [32] analyzed the phytoplankton community as an indicator of water quality. The following authors studied cyanobacteria—Svirčev et al. [33], Đorđević and Simić [34], and Simić et al. [35]. Temunović et al. [36] explored the diversity of water beetles (Hydradephaga, Coleoptera) in temporary ponds of Lonjsko polje, while Vilenica et al. [37] investigated the suitability of man-made water bodies as habitats for Odonata. The diversity of aquatic insects in small water bodies has been studied by many authors [38–43], while research regarding fish was conducted by Pavlović et al. [44] and Khanom et al. [45]. Considering the above-mentioned studies, we can conclude there are still unexplored small water bodies on the Balkan Peninsula's territory.

19.3 Pollution of Small Lakes and Ponds

Certainly, shallow lakes and ponds are among the most sensitive ecosystems [46, 47]. Biggs et al. [7] indicated that pollutants might have a higher impact on ponds than larger waters because of ponds' small water volumes and less potential for pollutant dilution and retainment. Additionally, climate change can cause an increase in water temperature, decrease in water level, and these changes together can increase the effect of pollutants in small water bodies [48].

Small water bodies, small lakes and ponds, have not been omitted when it comes to threats by human activities. Those activities have increased nutrient load, acid rain volume, invasion of non-native species, and chemical contamination. Contaminants of small water bodies include agricultural and amenity pesticides, veterinary and human medicines, personal care products, biocides, polyaromatic hydrocarbons (PAHs), and metals (also termed "heavy metals" or "potentially toxic elements").

Among the most important contaminants in small water systems are PTEs, which are highly susceptible to inputs of even small amounts of these pollutants from their surroundings [7, 49, 50]. A key specificity of metals is that they are present in the earth's geological structures and enter aquatic ecosystems both by natural processes (atmospheric precipitation, geologic weathering, soil, and rock erosion), and through anthropogenic sources (industrial effluents, traffic, mining wastes, and agricultural waste products) as well as by synergistic combinations of the two.

It is a well-known fact that PTEs seriously contaminate the environment causing global ecosystem problems [7]. Within a lake or pond system, the contamination depends on both, the concentration of elements, and the processes that occur within the water, sediment, and biota. It was also observed that the intensity of water exchange, which is lower in lotic ecosystems, affects metal bioaccumulation [51]. Contamination with PTEs may have stressful effects on the ecological stability of the recipient, species richness, diversity, and may cause lethal effects to the members of biota [52]. Once released in the aquatic ecosystem, it can be distributed and accumulated in the different parts of the aquatic biota, including flora and fauna. PTEs are one of the greatest threats to aquatic biota due to possible bioaccumulation and biomagnification in food chains [53].

Fish are on the top of the food chain and accumulate the highest concentrations of PTEs in the aquatic systems [54], and therefore represent the source of PTEs in human food. The worldwide economic importance of freshwater fisheries decreases, with freshwater fisheries supply about one-fifth of the world's total fish catch [55]. Although the small size, small water bodies have the fishing potential, particularly, for the local community [2, 4]. Consequently, in humans PTEs cause toxicity at high concentrations and neurological impacts. Furthermore, some of them (e.g. arsenic) are carcinogenic [56].

19.4 Overview of the Literature Related to Pollutants in Small Lakes and Ponds of the Western Balkans

Oertli et al. [4] indicated that small lakes and ponds receive less scientific attention than other water bodies in general. Research on contamination of small lakes and ponds with different pollutants is less emphasized in recent scientific literature. Brönmark and Hansson [16] predicted that in developing countries, such as

Western Balkan countries, threats that derive from pollutants might become more of a problem in the next 25 years. Despite this, in 2021 pollutants in small lakes and ponds in Western Balkan countries are still under-examined.

There is very extensive literature on the effect of anthropogenic impacts resulting from the suite of processes leading to the phenomenon labeled "eutrophication" [18]. It may cause turbid and toxic water from cyanobacterial blooms, degradation of lake/pond ecosystems, and their biodiversity [57]. Acidification was a major focus in the scientific literature from the middle of the 1980s to the middle of the 1990s [16]. Thereafter, research papers on acidification appeared less frequently, and today acidification is considered mostly a political issue. Also, the problem of non-native species, as a result of human activities, is reflected in an increasing number of scientific papers published in this area. When it comes to PTEs, research activities are related to larger water bodies (large rivers and lakes).

In the Western Balkans, studies related to pollutants in small lakes and ponds are limited. Most studies deal with pollutants in larger lakes and reservoirs. Numerous studies are dealing with the determination of cyanotoxins and toxicity in small lakes. The largest number of studies of this type is in Serbia—Svirčev et al. [33], Đorđević and Simić [34], Simić et al. [35], Đorđević et al. [58], Svirčev et al. [59], Simić et al. [60], and Drobac et al. [61]. When it comes to persistent organic pollutants (POPs) and polycyclic aromatic hydrocarbons (PAH), studies were performed by Sakan et al. [62], Romanić et al. [63], Grba et al. [64], Drvošćak et al. [65], Kljaković-Gašpić et al. [66], Sula et al. [67], etc.

Table 19.1 summarizes available literature data on PTEs pollution studies related to lakes and ponds in the Western Balkans. It is important to emphasize that in the Western Balkans, four large lakes (Skadar, Ohrid, Prespa, and Dojran) are shared by two or more countries. Therefore, studies of pollution with PTEs in these countries (i.e., Montenegro, North Macedonia, and Albania) are mainly limited to these lakes.

As seen in Table 19.1, only the studies Djikanović et al. [52], Svirčev et al. [59], Branković et al. [91], Rašković et al. [97], Đikanović et al. [98, 99], Milošković et al. [100], and Nikolić et al. [101] included small lakes (if we consider 50 ha as the cut-off point for small lakes). Still, not a single study was related to pollution with PTEs in ponds. All these studies are related to the four small lakes in Serbia—Međuvršje, Šumarice, Aleksandrovac, and Bubanj. Study Branković et al. [91] deal with PTEs in water, sediment, and different macrophytes, while studies Svirčev et al. [59, Rašković et al. [97], Đikanović et al. [98, 99], Milošković et al. [100], and Nikolić et al. [101] are related to fish contamination with PTEs.

Table 19.1 Overview of literature data on PTEs in lakes and ponds of the Western Balkans

Study	Medium/organism	Country/location	Size (ha)
Malaj et al. [68]	Water, sediment, emergent vegetation, fish	Albania/Ohrid Lake	>50 ha
Dalo et al. [69]	Sediment, *Phragmites australis*	Albania/Ohrid Lake	>50 ha
Has-Schön et al. [51]	*Cyprinus carpio*, *Silurus glanis*	Bosnia and Herzegovina/Buško Blato	>50 ha
Has-Schön et al. [70]	*Salmo trutta*, *Leuciscus turskyi*, *Chondrostoma phoxinus*, *Aulopyge hügeli*	Bosnia and Herzegovina/Buško Blato	>50 ha
Has-Schön et al. [71]	*Cyprinus carpio*, *Tinca tinca*, *Lepomis gibbosus*, *Carassius auratus gibelio*, *Salmo dentex*, *Anguilla anguilla*	Bosnia and Herzegovina/Svtava Lake	>50 ha
Popović et al. [72]	Water, sediment, *Cyprinus carpio*, *Hypophthalmichthys molitrix*, *Silurus glanis*, *Esox lucius*	Bosnia and Herzegovina/Saničani Lake	>50 ha
Ivanković et al. [73]	Sediment	Bosnia and Herzegovina/Blidinje Lake	>50 ha
Vukosav et al. [74]	Water, sediment, *Salmo trutta*, *Oncorhynchus mykiss*, *Squalius cephalus*	Croatia/Plitvice Lake	>50 ha
Horvatinčić et al. [75]	Water, sediment	Croatia/Plitvice Lake	>50 ha
Gashi et al. [76]	Water	Kosovo/Batllava Lake	>50 ha
Malsiu et al. [77]	Water, sediment	Kosovo/Batllava Lake	>50 ha
Sahiti et al. [78]	Water, *Cyprinus carpio*	Kosovo/Batllava and Radoniqi lakes	>50 ha
Latifi et al. [79]	*Cyprinus carpio*	Kosovo/Batllava and Badovci lakes	>50 ha
Vrhovnik et al. [80]	Water, sediment, *Vimba melanops*, *Rana temporaria*	North Macedonia/Kalimanci Lake	>50 ha
Vrhovnik et al. [81]	Sediment	North Macedonia/Kalimanci Lake	>50 ha
Vrhovnik et al. [82]	Sediment	North Macedonia/Kalimanci Lake	>50 ha
Tziritis et al. [83]	Water	North Macedonia/Micro Prespa Lake	>50 ha

(continued)

Table 19.1 (continued)

Study	Medium/organism	Country/location	Size (ha)
Petrović et al. [84]	Sediment, *Trapa natans*	Montenegro/Skadar Lake	>50 ha
Kastratović et al. [85]	*Phragmites australis*	Montenegro/Skadar Lake	>50 ha
Kastratović et al. [86]	Water, sediment, *Phragmites australis*, *Ceratophyllum demersum*, *Lemna minor*	Montenegro/Skadar Lake	>50 ha
Rakočević et al. [87]	*Scardinius knezevici*, *Alburnus scoranza*, *Cyprinus carpio*, *Rutilus prespensis*, *Anguilla anguilla*, *Perca fluviatilis*	Montenegro/Skadar Lake	>50 ha
Vukašinović-Pešić et al. [88]	*Viviparus mamillatus*	Montenegro/Skadar Lake	>50 ha
Vukašinović-Pešić and Blagojević [89]	Macrophytes, molluscs, fish	Montenegro/Skadar Lake	>50 ha
Branković et al. [90]	Water, sediment, macrophytes	Serbia/Gruža Lake	>50 ha
Branković et al. [91]	Water, sediment, macrophytes	Serbia/Gruža, Bubanj*, and Šumarice* lakes	>50 ha *<50 ha
Milošković et al. [92]	Water, sediment, macrophytes, fish	Serbia/Gruža Lake	>50 ha
Milošković et al. [93]	*Sander lucioperca*, *Silurus glanis*, *Esox lucius*	Serbia/Bovan Lake	>50 ha
Milošković et al. [94]	*Sander lucioperca*, *Silurus glanis*, *Esox lucius*, *Carassius gibelio*, *Abramis brama*	Serbia/Bovan Lake	>50 ha
Sakan et al. [62]	Sediment	Serbia/Barje, Ćelije, Vrutci, Garaši, Bojnik, and Bovan lakes	>50 ha
Sunjog et al. [95]	Water, sediment, *Squalius cephalus*	Serbia/Zlatar and Garaši lakes	>50 ha
Jaćimović et al. [96]	*Perca fluviatilis*, *Ameiurus melas*	Serbia/Sava Lake	>50 ha
Rašković et al. [97]	*Squalius cephalus*	Serbia/Međuvršje* and Kruščica lakes	>50 ha *<50 ha
Đikanović et al. [98]	*Chondrostoma nasus*, *Rutilus rutilus*, *Abramis brama*, *Barbus barbus*, *Carassius gibelio*, *Squalius cephalus*, *Perca fluviatilis*, *Silurus glanis*, *Esox lucius*	Serbia/Međuvršje Lake	<50

(continued)

Table 19.1 (continued)

Study	Medium/organism	Country/location	Size (ha)
Đikanović et al. [99]	*Chondrostoma nasus*	Serbia/Međuvršje Lake	<50
Djikanović et al. [52]	Sediment, macrophytes, fish	Serbia/Međuvršje Lake	<50
Svirčev et al. [59]	Water, sediment, fish	Serbia/Aleksandrovac Lake	<50
Milošković et al. [100]	*Alburnus* spp.	Serbia/Gruža, Međuvršje*, Vlasina, Zavoj, Skadar, Ohrid, Prespa, and Dojran lakes	>50 ha *<50 ha
Nikolić et al. [101]	Sediment, *Perca fluviatilis*	Serbia/Vlasina, Perućac, Zaovine, Međuvršje*, Garaši, and Sava lakes	>50 ha *<50 ha

19.5 Creating Awareness of the Significance of Pollution of Small Lakes and Ponds Through Assessment of Levels of PTEs in Small Aleksandrovac Lake in Serbia

Looking at the literature overview, it is evident that most of the studies are related to lakes bigger than 50 ha, which leaves a significant gap in knowledge for the small lakes and ponds. Guided by this data, we decided to present the investigation of PTEs in fish species from Aleksandrovac Lake, located in Southern Serbia.

Aleksandrovac Lake (Fig. 19.1) is a man-made lake on the Aleksandrovac River, formed in 1964, at first for irrigation purposes. It is characterized as oligosaline, shallow lake (maximum depth 4 m, average 2 m) with a surface area of 12 ha. Over time, already highly eutrophic Aleksandrovac Lake suffered almost anaerobic conditions, so it was completely restored and refilled in 2010. Despite this, massive fish mortality happened several times. In December 2012, the mortality of nearly the entire fish population was recorded (more than 1.7 tons), including *Cyprinus carpio*,

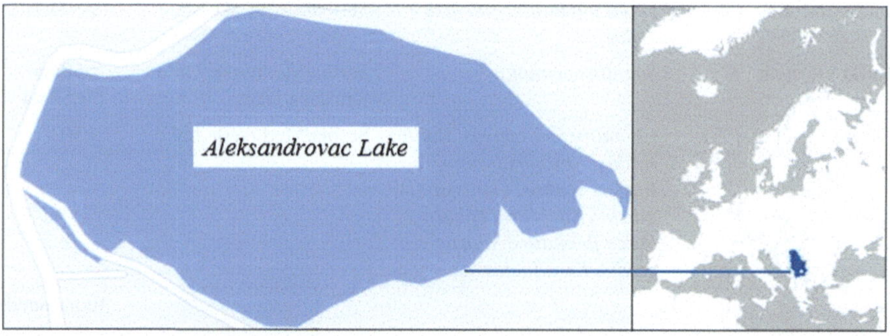

Fig. 19.1 Map of Aleksandrovac Lake

Ctenopharyngodon idella, *Hypophthalmichthys molitrix*, *Abramis brama*, *Silurus glanis*, *Carassius gibelio*, *Aspius aspius*, and *Squalius cephalus*.

19.5.1 Sampling and Methodology

Bioindicator species, distinguished by its flexibility and tolerance to environmental conditions and pollution, *Carassius gibelio* was sampled immediately after massive fish mortality at Aleksandrovac Lake on 20th December 2012. Twenty specimens were packed in special polyethylene bags and transferred on ice in a hand-held refrigerator to the laboratory. In the laboratory, fish were measured for their total body length (to the nearest cm), weighed (to the nearest g), and subsequently dissected with a decontaminated ceramic knife. The right dorsal muscle, below the dorsal fin, liver, and gills of each specimen were dissected, weighed using an electronic balance (± 0.1 g), and stored at −20 °C before analysis.

The concentrations of Al, As, Cd, Co, Cr, Cu, Fe, Hg, Mn, Ni, Pb, Se, Sn, and Zn were analyzed with inductively coupled plasma optical emission spectrometry (ICP-OES), using a Thermo Fisher Scientific iCAP 6500 Duo ICP (Cambridge, United Kingdom). The analytical procedure was described in our earlier research paper Milošković et al. [102]. The mean values and standard deviations were calculated for each group, and element concentrations were expressed as mg kg^{-1} wet weight (ww).

Although the muscle tissue (meat) of *Carassius gibelio* was used for element analysis, concentrations of the elements were compared with the maximum permitted concentrations (MPCs) in fish meat for utilization in the human diet, established by the European Union [103], FAO [104], and Serbian national legislation [105].

The metal pollution index (MPI), which refers to element concentration in fish tissues, was calculated using the following equation [106]:

$$MPI = (cf1 \times cf2 \times cf3 \times \cdots \times cfn)^{1/n} \quad (19.1)$$

where cfn = concentration of the element n in the sample.

19.5.2 Potentially Toxic Elements in Fish Tissues

Concentrations of PTEs in three tissues of *Carassius gibelio* are summarized in Table 19.2. The highest concentrations of Al, As, Cu, Cr, Mn, Ni, Pb, Se, Zn, and Cd in gills, Co, Sn, and Fe in liver, and Hg in muscle were recorded. The distribution pattern was gills–liver–muscle for Al, As, Cu, Cr, Mn, Ni, Pb, Se, and Zn, gills–liver–muscle for Cd, liver–gills–muscle for Sn and Fe, liver–muscle–gills for Co, and muscle–gills–liver for Hg.

Table 19.2 The average PTEs concentrations and standard deviation in the muscle, liver, and gills of *Carassius gibelio* in mg kg^{-1} ww and prescribed MPCs in fish meat

Elements	Muscle	Liver	Gills	MPC		
				MPC EU [103]	MPC FAO [104]	MPC RS [105]
Al	0.78 ± 0.09	ND	3.25 ± 0.8			
As	0.12 ± 0.05	ND	0.22 ± 0.09	0.1–4.0		2.0
Cd	0.005 ± 0.001	0.007 ± 0.0002	0.007 ± 0.0001	0.05	0.05	0.1
Co	0.001 ± 0.002	0.02 ± 0.005	ND			
Cr	0.33 ± 0.18	0.3 ± 0.1	0.39 ± 0.1			
Cu	0.17 ± 0.08	0.46 ± 0.1	0.56 ± 0.32	30		30*
Fe	3.03 ± 1.4	87 ± 16.25	36 ± 9.81			30*
Hg	0.10 ± 0.02	ND	0.0005 ± 0.0001	0.5	0.5	0.5
Mn	0.30 ± 0.21	0.17 ± 0.06	8.64 ± 2.44			
Ni	0.006 ± 0.006	0.002 ± 0.001	0.04 ± 0.003			
Pb	0.22 ± 0.05	0.13 ± 0.06	0.43 ± 0.31	0.3	0.5	0.4
Se	0.21 ± 0.07	0.12 ± 0.04	0.32 ± 0.13			
Sn	0.0007 ± 0.001	0.14 ± 0.05	0.12 ± 0.05			
Zn	12.7 ± 0.64	8.61 ± 1.9	75.13 ± 15.16		100	100*

ND—not detected, *MPCs in canned products

Since it was introduced in European freshwaters in the 1970s, *Carassius gibelio* has been economically important for fisheries in Serbia [55]. The negative impact of PTEs on human health was analyzed by comparing the concentrations of these elements with the appropriate MPCs [103–105]. Concentrations of As and Fe in gills were above MPC prescribed by Serbian legislation, while Pb in gills was above MPCs prescribed by Serbian and EU legislations. There were no PTEs in fish muscle (fish meat) above MPCs, suggesting no risk for humans consuming fish despite massive fish mortality.

The MPI is more reliable for expressing the pollution status of fish than accumulation patterns since it considers all metals synchronously. The concentrations of PTEs affected the MPI values (Fig. 19.2), which points out that the muscle was exposed to the lower pressure of PTEs pollution than liver and gills. The reason for higher values of MPI for gills may be the fact that gills are the primary site of element uptake from water, especially if elements are bound to particulate matters. It is recommended that gills should be removed before using fish in the human diet due to elevated concentrations of the most toxic elements—As and Pb.

Svirčev et al. [59] collected data a month before massive fish mortality happened in Aleksandrovac Lake. Their results represent concentrations of PTEs in muscle,

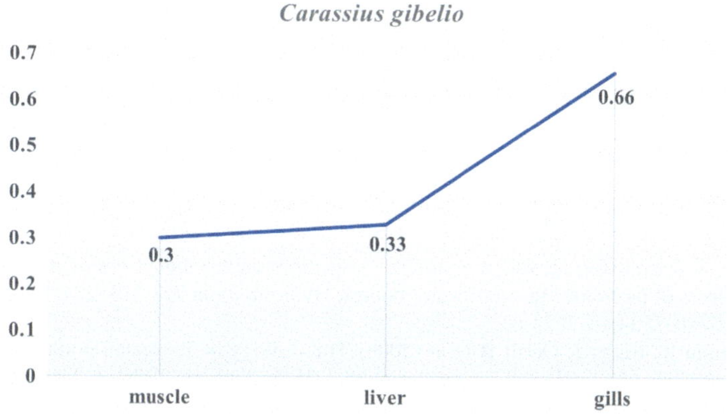

Fig. 19.2 Metal pollution index (MPI) values of the total element accumulation levels in the muscle, liver, and gills tissues in *Carassius gibelio*

liver, and gills of undetermined fish species, so we cannot compare the concentrations of PTEs in our study with their published results. We can solely emphasize that values of PTEs were higher in the study Svirčev et al. [59]. Svirčev et al. [59] excluded the possibility of metal poisoning and assumed that toxic metabolites produced by cyanobacterium *Raphidiopsis* (*Cylindrospermopsis*)*raciborskii* caused the fish mortality in Aleksandrovac Lake.

Aleksandrovac Lake came into the public spotlight after massive mortality of over 1.7 tons of fish. Before this accident, the problems in this small lake were solved occasionally and insufficiently. Recognizing on time the importance of small lakes and ponds is the basis for continuous long-term monitoring and legal regulations which can provide protection of small water bodies. This is a global problem that needs to be solved in the future in order to preserve small water bodies.

19.6 Conclusions

This chapter provides an insight into the pollution status of lakes and ponds in the Western Balkan countries. Analysis of available literature shows that most studies are related to lakes bigger than 50 ha, which leaves a significant knowledge gap for the small lakes and ponds. Presented results of the investigation of PTEs in fish species from small Aleksandrovac Lake in Southern Serbia indicate that fish muscle (meat) was exposed to the lower pressure of PTEs pollution than liver and gills. That suggests, despite massive fish mortality, there was no risk for human health by fish consumption. In the future, more comprehensive studies need to be done globally to conserve small water bodies, especially shallow lakes and ponds.

Acknowledgements The investigation was supported by Grant (Agreement No. 451-03-9/2021-14/200378) funded by the Serbian Ministry of Education, Science and Technological Development.

References

1. Lorenz S, Rasmussen JJ, Süß A et al (2017) Specifics and challenges of assessing exposure and effects of pesticides in small water bodies. Hydrobiologia 793:213–224. https://doi.org/10.1007/s10750-016-2973-6
2. Céréghino R, Biggs J, Oertli B et al (2008) The ecology of European ponds: defining the characteristics of a neglected freshwater habitat. Hydrobiologia 597:1–6. https://doi.org/10.1007/s10750-007-9225-8
3. Downing JA, Prairie YT, Cole JJ et al (2006) The global abundance and size distribution of lakes, ponds, and impoundments. Limnol Oceanogr 51(5):2388–2397. https://doi.org/10.4319/lo.2006.51.5.2388
4. Oertli B, Céréghino R, Hull A (2009) Pond conservation: from science to practice. Hydrobiologia 634:1–9. https://doi.org/10.1007/s10750-009-9891-9
5. Pond Conservation Group (1993) A future for Britain's ponds: an agenda for action. Pond Conservation Group, Oxford
6. Ramsar Convention Secretariat (2013) The Ramsar Convention Manual: a guide to the convention on Wetlands (Ramsar, Iran, 1971), 6th ed. Ramsar Convention Secretariat, Gland, Switzerland
7. Biggs J, von Fumetti S, Quinn-Kelly M (2017) The importance of small waterbodies for biodiversity and ecosystem services: implications for policy makers. Hydrobiologia 793:3–39. https://doi.org/10.1007/s10750-016-3007-0
8. Oertli B, Joye DA, Castella E et al (2002) Does size matter? The relationship between pond area and biodiversity. Biol Conserv 104:59–70. https://doi.org/10.1016/S0006-3207(01)00154-9
9. Søndergaard M, Jeppesen E, Jensen JP (2005) Pond or lake: does it make any difference? Archiv für Hydrobiologie 162:143–165. https://doi.org/10.1127/0003-9136/2005/0162-0143
10. Collinson NH, Biggs J, Corfield A et al (1995) Temporary and permanent ponds: an assessment of the effects of drying out on the conservation value of aquatic macroinvertebrate communities. Biol Conserv 74(2):125–133. https://doi.org/10.1016/0006-3207(95)00021-u
11. Kristensen P, Globevnik L (2014) European small water bodies. Biology and Environment: Proceeding of the Royal Irish Academy 2014. https://doi.org/10.3318/BIOE.2014.13
12. Biggs J, Nicolet P, Mlinaric M et al (2014) Report of the workshop on the protection and management of small water bodies, Brussels, 14th November 2013. The European Environmental Bureau (EEB) and The Freshwater Habitats Trust, p 23
13. Williams P, Whitfield M, Biggs J (2004) Comparative biodiversity of rivers, streams, ditches, and ponds in an agricultural landscape in Southern England. Biol Conserv 115:329–341. https://doi.org/10.1016/S0006-3207(03)00153-8
14. Downing JA (2010) Emerging global role of small lakes and ponds: little things mean a lot. Limnetica 29(1):9–24
15. Bartout P, Touchart L, Terasmaa J et al (2015) A new approach to inventorying bodies of water, from local to global scale. J Geogr Soc Berl 146:245–258. https://doi.org/10.12854/erde-146-20
16. Brönmark C, Hansson L (2002) Environmental issues in lakes and ponds: current state and perspectives. Environ Conserv 29(3):290–307. https://doi.org/10.1017/S0376892902000218
17. Rannap R, Lohmus A, Briggs L (2009) Restoring ponds for amphibians: a success story. Hydrobiologia 634:87–95. https://doi.org/10.1007/s10750-009-9884-8

18. Phillips G, Wilby N, Moss B (2016) Submerged macrophyte decline in shallow lakes: what have we learnt in the last forty years? Aquat Bot 135:37–45. https://doi.org/10.1016/j.aquabot.2016.04.004
19. Hutinec JB, Struna S (2007) A survey of ponds and their loss in Žumberak-Samoborsko gorje Nature Park, northwest Croatia. Natura Croatica: Periodicum Musei Historiae Naturalis Croatici 16(2):121–137
20. Sremački M, Obrovski B, Petrović M et al (2020) Comprehensive environmental monitoring and assessment of protected wetland and lake water quality in Croatia and Serbia. Environ Monit Assess 192(3):1–11. https://doi.org/10.1007/s10661-020-8141-5
21. Ranđelović V, Matejić J, Zlatković B (2007) Flora i vegetacija Batušinačkih bara kod Niša. In: Paper presented at the 9th symposium on flora of southeastern Serbia and neighbouring regions, University of Niš, Niš, 1–3 September 2007
22. Stamenković O, Simić V, Piperac S et al (2021) Direct, water-chemistry mediated, and cascading effects of human-impact intensification on multitrophic biodiversity in ponds. Aquat Ecol 55:187–214. https://doi.org/10.1007/s10452-020-09822-5
23. Stamenković O, Piperac S, Milošević D et al (2019) Anthropogenic pressure explains variations in the biodiversity of pond communities along environmental gradients: a case study in south-eastern Serbia. Hydrobiologia 838:65–83. https://doi.org/10.1007/s10750-019-03978-4
24. Tasevska O, Špoljar M, Gušeska D et al (2017) Summer rotifer assemblages in three reservoirs in the Republic of Macedonia. Acta Zool Bulg 8:77–84
25. Ostojić A (2004) Faunistic compostion of the zooplankton of Lake Bubanj. Kragujevac J Sci 26:107–110
26. Ostojić A, Obradović M, Radoičić D (2012) Faunistic composition of zooplankton in the Šumarice lake (Kragujevac, Serbia). Kragujevac J Sci 34:87–96
27. Ostojić A, Radojević I, Obradović M et al (2012) A checklist of freshwater zooplankton on the territory of Kragujevac city (Serbia) and it's surrounding areas. Water Res Manag 2(3):35–42
28. Špoljar M, Tomljanović T, Lalić I (2011) Eutrophication impact on zooplankton community: a shallow lake approach. Holist Approach Environ 1(4):131–142
29. Špoljar M, Dražina T, Lajtner J et al (2018) Zooplankton assemblage in four temperate shallow waterbodies in association with habitat heterogeneity and alternative states. Limnologica 71:51–61. https://doi.org/10.1016/j.limno.2018.05.004
30. Kuczyńska-Kippen N, Pronin M, Špoljar M (2019) Specificity of zooplankton communities inhabiting small water bodies in two regions of Europe: Poland vs. Croatia. In: Abstracts of Biocenozy wód płynących w kontekście zmian hydrologicznych, Uniwersytet im. Adama Mickiewicza, Szczecin, 25–27 September 2019
31. Mancinelli G, Mali S, Belmonte G (2019) Species richness and taxonomic distinctness of zooplankton in ponds and small lakes from Albania and North Macedonia: the role of bioclimatic factors. Water 11(11):2384. https://doi.org/10.3390/w11112384
32. Ranković B, Simić S, Bogdanović D (2006) Phytoplankton as indicator of water quality of lakes Bubanj and Šumarice during autumn. Prirodno-Matematički Fakultet, Univerzitet u Kragujevcu, Kragujevac
33. Svirčev Z, Simeunović J, Subakov-Simić G et al (2013) Cyanobacterial blooms and their toxicity in Vojvodina lakes, Serbia. Int J Environ Res 7(3):745–758
34. Đorđević N, Simić S (2014) Cyanobacterial blooms in oligosaline and alkaline microaccumulation before and after rehabilitation. Pol J Environ Stud 23(6):1975–1982
35. Simić S, Đorđević N, Milošević Dj (2017) The relationship between the dominance of Cyanobacteria species and environmental variables in different seasons and extreme precipitation. Fund Appl Limnol 190(1):1–11. https://doi.org/10.1127/fal/2017/0975
36. Temunović M, Šerić Jelaska L, Durbešić P (2007) Diversity of water beetles (Hydradephaga, Coleoptera) in temporary ponds of Lonjsko polje Nature Park, Croatia. Entomol Croat 11(1–2):13–24
37. Vilenica M, Pozojević I, Vučković N et al (2020) How suitable are man-made water bodies as habitats for Odonata? Knowl Manag Aquat Ecosyst 421:13. https://doi.org/10.1051/kmae/2020008

38. Pešić S (2000) Weevils (Coleoptera: Curculionoidea) of the lakes of Kragujevac (first communication). Acta Entomol Serbica 5(112):13–28
39. Stanić S, Kekić V, Pavlović-Lučić S (2002) A contribution to knowledge of Drosophilidae (Diptera) fauna in Kragujevac basin. Acta Entomol Serbica 7(1/2):151–154
40. Pešić S (2012) Aquatic weevil (Coleoptera: Curculionoidea) assembly response to the different ecological conditions in artificial lakes in central Serbia. Arch Biol Sci 64(4):1523–1531. https://doi.org/10.2298/ABS1204523P
41. Turić N, Merdić E, Hackenberger Kutuzović B et al (2011) Diversity of aquatic insects (Heteroptera: Nepomorpha, Gerromorpha and Coleoptera: Hydradephaga, Hydrophilidae) in the karst area of Gorski kotar, Croatia. Natura Croatica: Periodicum Musei Historiae Naturalis Croatici 20(1):179–188
42. Belmonte G, Alfonso G, Beadini N et al (2018) An inventory of invertebrate fauna of Albania and Macedonia Lakes. Thalass Salentina 40(2):27–38
43. Pozojević I, Vučković N, Dorić V et al (2021) Contribution to the knowledge of the water mite (Acari: Hydrachnidia) fauna of Croatia-New data and records from a permanent pool in the Dinaric karst region. Persian J Acarol 10(1):1–7. https://doi.org/10.22073/pja.v10i1.62352
44. Pavlović M, Simonović P, Stojković M et al (2015) Analysis of diet of piscivorous fishes in Bovan, Gruža and Šumarice Reservoir, Serbia. Iran J Fish Sci 14(4):908–923
45. Khanom M, Rouf M, Ahsan N et al (2020) Morphological characteristics, growth and age structure of allochthonous fish pumpkinseed, *Lepomis gibbosus* in Bara Lake, Croatia. Bulg J Agric Sci 26(1):213–222
46. Jeppesen E, Søndergaard M, Meerhoff M et al (2007) Shallow lake restoration by nutrient loading reduction—some recent findings and challenges ahead. Hydrobiologia 584:239–252. https://doi.org/10.1007/s10750-007-0596-7
47. Erwin KL (2008) Wetlands and global climate change: the role of wetland restoration in a changing world. Wetl Ecol Manag 17(1):71–84. https://doi.org/10.1007/s11273-008-9119-1
48. EEA ETC/ICM (2010) Climate change impacts on water quality and biodiversity. Background Report for EEA European Environment State and Outlook Report 2010. Report from the EEA European Topic Centre on Inland, Coastal and Marine Waters. http://icm.eionet.europa.eu/ETC_Reports. Accessed 30 March 2021
49. Rasmussen JJ, Wiberg-Larsen P, Baattrup-Pedersen A et al (2012) Stream habitat structure influences macroinvertebrate response to pesticides. Environ Pollut 164:142–149. https://doi.org/10.1016/j.envpol.2012.01.007
50. Beketov MA, Kefford BJ, Schafer RB et al (2013) Pesticides reduce regional biodiversity of stream invertebrates. P Natl Acad Sci USA 110(27):11039–11043. https://doi.org/10.1073/pnas.1305618110
51. Has-Schön E, Bogut I, Vuković R et al (2015) Distribution and age-related bioaccumulation of lead (Pb), mercury (Hg), cadmium (Cd), and arsenic (As) in tissues of common carp (*Cyprinus carpio*) and European catfish (*Silurus glanis*) from the Buško Blato reservoir (Bosnia and Herzegovina). Chemosphere 135:289–296. https://doi.org/10.1016/j.chemosphere.2015.04.015
52. Djikanović V, Skorić S, Spasić S et al (2018) Ecological risk assessment for different macrophytes and fish species in reservoirs using biota-sediment accumulation factors as a useful tool. Environ Pollut 241:1167–1174. https://doi.org/10.1016/j.envpol.2018.06.054
53. Maceda-Veiga A, Monroy M, de Sostoa A (2012) Metal bioaccumulation in the Mediterranean barbel (*Barbus meridionalis*) in a Mediterranean River receiving effluents from urban and industrial wastewater treatment plants. Ecotoxicol Environ 76:93–101. https://doi.org/10.1016/j.ecoenv.2011.09.013
54. Dallinger R, Prosi F, Segner H et al (1987) Contaminated food and uptake of heavy metals by rainbow trout (*Salmo gairdneri*): a field study. Oecologia 73:91–98
55. Smederevac-Lalić M, Višnjić-Jeftić Ž, Pucar M et al (2011) Fishing circumstances on the Danube in Serbia. Water Res Manag 1(4):45–49

56. Van den Roovaart J, van Duijnhoven N, Fribourg-Blanc B et al (2017) Emissions of pollutants to Europe's waters—sources, pathways and trends. Künitzer A (ed) ETC/ICM Technical Report 3/2017, Magdeburg: European Topic Centre on inland, coastal and marine waters, p 96
57. Bartram J, Ballance R, World Health Organization & United Nations Environment Programme (1996) Water quality monitoring: a practical guide to the design and implementation of freshwater quality studies and monitoring programs/edited by Jamie Bartram and Richard Ballance. E & FN Spon, London. https://apps.who.int/iris/handle/10665/41851
58. Đorđević N, Simić S, Ćirić A (2015) First identification of the cylindrospermopsin (CYN) - producing cyanobacterium *Cylindrospermopsis raciborskii* (Woloszyńska) Essnayya & Subra Raju in Serbia. Fresen Environ Bull 24(11a):736–3742
59. Svirčev Z, Obradović V, Codd GA et al (2016) Massive fish mortality and *Cylidrospermopsis raciborskii* bloom in Aleksandrovac Lake. Ecotoxicology 25(7):1353–1363. https://doi.org/10.1007/s10646-016-1687-x
60. Simić S, Komárek J, Đorđević N (2014) The confirmation of the genus *Glaucospira* (Cyanobacteria) and the Occurrence of *Glaucospira laxissima* (G. S. West) comb. nova in Serbia. Cryptogam Algol 35(3):259–267. https://doi.org/10.7872/crya.v35.iss3.2014.259
61. Drobac D, Tokodi N, Lujić J et al (2016) Cyanobacteria and cyanotoxins in fishponds and their effects on fish tissue. Harmful Algae 55:66–76. https://doi.org/10.1016/j.hal.2016.02.007
62. Sakan S, Ostojić B, Đorđević D (2017) Persistent organic pollutants (POPs) in sediments from river and artificial lakes in Serbia. J Geochem Explor 180:91–100. https://doi.org/10.1016/j.gexplo.2017.06.008
63. Romanić S, Vuković G, Klinčić D et al (2018) Organochlorine pesticides (OCPs) and polychlorinated biphenyls (PCBs) in Cyprinidae fish: towards hints of their arrangements using advanced classification methods. Environ Res 165:349–357. https://doi.org/10.1016/j.envres.2018.05.003
64. Grba N, Krčmar D, Maletić S et al (2016) Organic and inorganic priority substances in sediments of Ludaš Lake, a cross-border natural resource on the Ramsar list. Environ Sci Pollut R 24(2):1938–1952. https://doi.org/10.1007/s11356-016-7904-6
65. Dvoršćak M, Fingler S, Mendaš G et al (2019) Distribution of organochlorine pesticide and polychlorinated biphenyl residues in lake sediment cores from the Plitvice Lakes National Park (Croatia). Arch Environ Contam Toxicol 77(4):537–548. https://doi.org/10.1007/s00244-019-00668-z
66. Kljaković-Gašpić Z, Herceg Romanić S, Bituh T et al (2018) Assessment of multiple anthropogenic contaminants and their potential genotoxicity in the aquatic environment of Plitvice Lakes National Park, Croatia. Environ Monit Assess 190(11). https://doi.org/10.1007/s10661-018-7028-1
67. Sula E, Aliko V, Barceló D et al (2020) Combined effects of moderate hypoxia, pesticide and PCBs upon crucian carp fish, *Carassius carassius*, from a freshwater lake—in situ ecophysiological approach. Aquat Toxicol 105644. https://doi.org/10.1016/j.aquatox.2020.105644
68. Malaj E, Rousseau DPL, Du Laing G et al (2011) Near-shore distribution of heavy metals in the Albanian part of Lake Ohrid. Environ Monit Assess 184(4):1823–1839. https://doi.org/10.1007/s10661-011-2081-z
69. Dalo E, Sadikaj R, Sahiti H (2019) Assessment of accumulation of heavy metals and lipid peroxidation in common reed (*Phragmites australis*) in the Albanian Part of Lake Ohrid. J Ecol Eng 20(4):114–120. https://doi.org/10.12911/22998993/102794
70. Has-Schön E, Bogut I, Rajković V et al (2008) Heavy metal distribution in tissues of six fish species included in human diet, inhabiting freshwaters of the Nature Park "Hutovo Blato" (Bosnia and Herzegovina). Arch Environ Con Tox 54(1):75–83. https://doi.org/10.1007/s00244-007-9008-2
71. Has-Schön E, Bogut I, Kralik G et al (2008) Heavy metal concentration in fish tissues inhabiting waters of "Buško Blato" reservoar (Bosnia and Herzegovina). Environ Monit Assess 144(1–3):15–22. https://doi.org/10.1007/s10661-007-9627-0

72. Popović M, Nedić D, Pećanac B et al (2019) The toxic element concentration in fish tissues from Saničani Lake, an urban environment, in Bosnia and Herzegovina. Biol Trace Elem Res 197:271–278. https://doi.org/10.1007/s12011-019-01982-2
73. Ivanković A, Jurković J, Habul EV et al (2016) Contribution to the knowledge of the Blidinje Lake sediment (Bosnia and Herzegovina). Austin J Earth Sci 2(3):1016
74. Vukosav P, Mlakar M, Cukrov N et al (2013) Heavy metal contents in water, sediment and fish in a karst aquatic ecosystem of the Plitvice Lakes National Park (Croatia). Environ Sci Pollut R 21(5):3826–3839. https://doi.org/10.1007/s11356-013-2377-3
75. Horvatinčić N, Briansó JL, Obelić B et al (2006) Study of pollution of the Plitvice Lakes by water and sediment analyses. Water Air Soil Pollut Focus 6(5–6):475–485. https://doi.org/10.1007/s11267-006-9031-8
76. Gashi F, Frančišković-Bilinski S, Bilinski H et al (2017) Assessing the distribution of trace elements in water from Batllava Lake (Kosovo). Sustain Water Resour Manag 3(1):1–12. https://doi.org/10.1007/s40899-017-0077-8
77. Malsiu A, Shehu I, Stafilov T et al (2020) Water quality and sediment contamination assessment of the Batllava Lake in Kosovo using fractionation methods and pollution indicators. Arab J Geosci 13:412. https://doi.org/10.1007/s12517-020-05408-5
78. Sahiti H, Bislimi K, Dalo E et al (2018) Effect of water quality in hematological and biochemical parameters in blood of common carp (*Cyprinus carpio*) in two lakes of Kosovo. Nat Eng Sci 3(3):323–332. https://doi.org/10.28978/nesciences.468987
79. Latifi F, Haziri I, Rama A et al (2017) Assessment of heavy metal contamination in some organs of the common carp growing in lakes in Kosovo. Veterinaria 66(3):139–144
80. Vrhovnik P, Arrebola JP, Serafimovski T et al (2013) Potentially toxic contamination of sediments, water and two animal species in Lake Kalimanci, FYR Macedonia: relevance to human health. Environ Pollut 180:92–100. https://doi.org/10.1016/j.envpol.2013.05.004
81. Vrhovnik P, Šmuc NR, Dolenec T et al (2013) An evaluation of trace metal distribution and environmental risk in sediments from the Lake Kalimanci (FYR Macedonia). Environ Earth Sci 70(2):761–775. https://doi.org/10.1007/s12665-012-2166-1
82. Vrhovnik P, Šmuc NR, Dolenec T et al (2013) Impact of Pb–Zn mining activity on surficial sediments of Lake Kalimanci (FYR Macedonia). Turk J Earth Sci 22:996–1009. https://doi.org/10.3906/yer-1205-1
83. Tziritis EP (2014) Environmental monitoring of Micro Prespa Lake basin (Western Macedonia, Greece): hydrogeochemical characteristics of water resources and quality trends. Enviro Monit Assess 186(7):4553–4568. https://doi.org/10.1007/s10661-014-3719-4
84. Petrović D, Jančić D, Furdek M et al (2016) Aquatic plant *Trapa natans* L. as bioindicator of trace metal contamination in a freshwater lake (Skadar Lake, Montenegro). Acta Bot Croat 75(2):236–243. https://doi.org/10.1515/botcro-2016-0031
85. Kastratović V, Krivokapić S, Đurović D et al (2013) Seasonal changes in metal accumulation and distribution in the organs of *Phragmites australis* (common reed) from Lake Skadar, Montenegro. J Serb Chem Soc 78(8):1241–1258. https://doi.org/10.2298/JSC121026153K
86. Kastratović V, Bigović M, Jaćimović Ž et al (2018) Levels and distribution of cobalt and nickel in the aquatic macrophytes found in Skadar Lake, Montenegro. Environ Sci Pollut Res 25(27):26823–26830. https://doi.org/10.1007/s11356-018-1388-5
87. Rakocevic J, Sukovic D, Maric D (2018) Distribution and relationships of eleven trace elements in muscle of six fish species from Skadar Lake (Montenegro). Turk J Fish Aquat Sci 18(5):647–657. https://doi.org/10.4194/1303-2712-v18_5_01
88. Vukašinović-Pešić V, Blagojević N, Vukanović S et al (2017) Heavy metal concentrations in different tissues of the snail *Viviparus mamillatus* (Küster, 1852) from lacustrine and riverine environments in Montenegro. Turk J Fish Aquat Sci 17(3):557–563. https://doi.org/10.4194/1303-2712-v17_3_12
89. Vukašinović-Pešić V, Blagojević N (2018) Metal pollution: evidences from plants, aquatic invertebrates and fish from Lake Skadar. In: Pešić V, Karaman G, Kostianoy A (ed) The Skadar/Shkodra Lake environment. The handbook of environmental chemistry, vol 80. Springer International Publishing, Springer Nature Switzerland AG, pp 141–151

90. Branković S, Pavlović-Muratspahić D, Topuzović M et al (2010) Concentration of some heavy metals in aquatic macrophytes in reservoir near city Kragujevac (Serbia). Biotechnol Biotechnol Equip 24:223–227. https://doi.org/10.1080/13102818.2010.10817840
91. Branković S, Pavlović-Muratspahić D, Topuzović M et al (2011) Environmental study of some metals on several aquatic macrophytes. Afr J Biotech 10(56):11956–11965. https://doi.org/10.5897/AJB10.2655
92. Milošković A, Branković S, Simić V et al (2013) The accumulation and distribution of metals in water, sediment, aquatic macrophytes and fishes of the Gruža reservoir, Serbia. B Environ Contam Tox 90(5):563–569. https://doi.org/10.1007/s00128-013-0969-8
93. Milošković A, Dojčinović B, Simić S et al (2014) Heavy metal and trace element bioaccumulation in target tissues of three edible predatory fish species from Bovan reservoir (Serbia). Fresen Environ Bull 23(8a):1884–1891
94. Milošković A, Simić V (2015) Arsenic and other trace elements in five edible fish species in relation to fish size and weight and potential health risks for human consumption. Pol J Environ Stud 24(1):199–206. https://doi.org/10.15244/pjoes/24929
95. Sunjog K, Kolarević S, Kračun-Kolarević M et al (2016) Assessment of status of three water bodies in Serbia based on tissue metal and metalloid concentration (ICP-OES) and genotoxicity (comet assay). Environ Pollut 213:600–607. https://doi.org/10.1016/j.envpol.2016.03.008
96. Jaćimović M, Lenhardt M, Višnjić-Jeftić Ž et al (2015) Elemental concentrations in different tissues of European perch and bullhead from Sava Lake (Serbia). Slov Vet Res 52(2):57–65
97. Rašković B, Poleksić V, Skorić S et al (2018) Effects of mine tailing and mixed contamination on metals, trace elements accumulation and histopathology of the chub (*Squalius cephalus*) tissues: evidence from three differently contaminated sites in Serbia. Ecotox Environ Safe 153:238–247. https://doi.org/10.1016/j.ecoenv.2018.01.058
98. Đikanović V, Skorić S, Gačić Z (2016) Concentrations of metals and trace elements in different tissues of nine fish species from the Međuvršje Reservoir (West Morava River Basin, Serbia). Arch Biol Sci 68(4):811–819. https://doi.org/10.2298/ABS151104069D
99. Đikanović V, Skorić S, Jarić I et al (2016) Age-specific metal and accumulation patterns in different tissues of nase (*Chodrostoma nasus*) from the Medjuvršje Reservoir. Sci Total Environ 566–567:185–190. https://doi.org/10.1016/j.scitotenv.2016.05.072
100. Milošković A, Milošević Đ, Radojković N et al (2018) Potentially toxic elements in freshwater (*Alburnus* spp.) and marine (*Sardina pilchardus*) sardines from the Western Balkan Peninsula: an assessment of human health risk and management. Sci Total Environ 644:899–906. https://doi.org/10.1016/j.scitotenv.2018.07.041
101. Nikolić D, Skorić S, Lenhardt M et al (2019) Risk assessment of using fish from different types of reservoirs as human food—a study on European perch (*Perca fluviatilis*). Environ Pollut 113586. https://doi.org/10.1016/j.envpol.2019.11358
102. Milošković A, Dojčinović B, Kovačević S et al (2016) Spatial monitoring of heavy metals in the inland waters of Serbia: a multispecies approach based on commercial fish. Environ Sci Pollut R 23(10):9918–9933. https://doi.org/10.1007/s11356-016-6207-2
103. EU (2006) Commission Regulation (EC) No. 1881/2006 of 19 December 2006 setting maximum levels for certain contaminants in foodstuffs (Text with EEA relevance). Off J Eur Union No. 1881/2006 364:5–24. https://eurlex.europa.eu/eli/reg/200-6/1881/oj. Accessed 23 Feb 2021
104. FAO (1983) Compilation of legal limits for hazardous substances in fish and fishery products. FAO Fish Circ 464:5–100
105. Official Gazzette of RS (2011) Regulation on quantity of pesticides, metals, metalloids, and other toxic substances, chemotherapeutics, anabolics, and other substances which can be found in food. Official Gazzette of RS No 28/2011. www.pravno.informacioni.sistem.rs/SlGlasnikPortal/prilozi/5.html&doctype=reg&abc=cba&eli=true&eliActId=427071&regactid=427071
106. Usero J, González-Regalad E, Gracia I (1997) Trace metals in the bivalve mollusks *Ruditapes decussates* and *Ruditapes philippinarum* from the Atlantic Coast of Southern Spain. Environ Int 23(3):291–298. https://doi.org/10.1016/S0160-4120(97)00030-5

Chapter 20
Conclusions: Small Water Bodies of the Western Balkans—Values and Threats

Vladimir Pešić, Marko Miliša, and Đurađ Milošević

Abstract The small waters bodies such as headwater streams, springs, ditches, small lakes and ponds are critical to maintaining freshwater biodiversity and ecosystem services. This is especially true for Dinaric karst where they are often the only water bodies present. This concluding chapter briefly describes the main conclusions presented in the book "Small Water Bodies of the Western Balkan" which has 19 chapters and is addressed to different types of specialists, from managers and policy makers to hydrologists and ecologists. There has been no recent publication that provides such comprehensive knowledge on small water bodies in the Western Balkans. To address this research gap, our aim was to draw together information on different aspects of these important habitats in one place. The book highlights recent advances and current gaps in our knowledge of these important but still neglected freshwater ecosystems, and promotes their protection in a socio-ecological framework taking into account the contrasting climate and economic development across regions.

Keywords Small water bodies · Ponds · Springs · Streams · Balkan

V. Pešić (✉)
Faculty of Sciences, University of Montenegro, Džordža Vašingtona bb., 81000 Podgorica, Montenegro

M. Miliša
Faculty of Science, Division of Biology, University of Zagreb, Rooseveltov trg 6, 10000 Zagreb, Croatia
e-mail: marko.milisa@biol.pmf.hr

Đ. Milošević
Faculty of Sciences and Mathematics, Department of Biology and Ecology, University of Niš, Višegradska 33, 18000 Niš, Serbia
e-mail: djuradj.milosevic@pmf.edu.rs

V. Pešić et al. (eds.), *Small Water Bodies of the Western Balkans*, Springer Water,
https://doi.org/10.1007/978-3-030-86478-1_20

20.1 Introduction

Encompassing a wide range of small standing and running freshwater ecosystems, small water bodies (SWBs) are among the most common type of aquatic ecosystems worldwide. As highlighted by recent studies (e.g., [1, 2]), small waterbodies are crucial in providing a variety of ecosystem services and maintaining freshwater biodiversity. The term "small water body" is quite ambiguous because there are no universal scientific or legislative definitions [1]. In the limnological literature, small standing waterbodies (SSWB) are most often defined as small (ranging from 1 m^2 to several hectares, <10 ha) and shallow water bodies, not deeper than 20 m [3, 4]. Among them, ponds which can be temporary or permanent, artificial or naturally created, are usually defined as SSWB of 1 m^2 to approximately 2–5 ha, while small lakes are defined as SSWB with a size of more than 1–5 ha and a limit of 50–100 ha [1, 2, 5]. Small lotic water systems such as "small stream" are even more ambiguous [1] and in the recent limnological literature these water bodies are most commonly defined as zero, first, or second-order flow (for discussion see Biggs et al. [1]).

The main parts of the Balkan Peninsula belong to the Alpine belt (Alpides), a high mountainous part of the continent formed from the late Paleozoic–Mesozoic–Tertiary ocean of Thetis, predominantly during the Alpine orogeny [6]. The southeastern continuation of the Alpine orogenic belt in the Balkans is known as the Dinarides, a mountain range that stretches for 645 km along the Adriatic Sea. The Dinaric area started to emerge in the Cretaceous, and already in the Paleogene it was a large uplifted terrain [6]. This mountainous area dominated by highly karstified rocks covers about 30% (200,000 km^2) of the territory of the Western Balkans [7] and represent the most distinctive landscape feature of the entire area.

The turbulent geological history of the western part of the Balkan Peninsula over the last 30 million years, has resulted in great landscape complexity and habitat diversity and patchiness for both terrestrial and aquatic ecosystems [8]. The spatial and temporal distribution of SWBs in the Western Balkans, particularly in its Dinaric part is mainly determined by the morphological and hydrogeological characteristics of the karst, characterized by strong seasonality and unpredictability of water availability [7], as well as the semi-arid climate that prevails in the karst area. The cumulative impact of these factors has resulted in the scarce development of a permanent hydrographic surface network in karst terrains, where small water bodies often dominate.

Although ecosystem services such as carbon cycles may be dominated by SWBs [1], small water bodies are generally less recognized compared to larger water bodies. Moreover, small water bodies often remain neglected and excluded from water planning and government policies such as the Water Framework Directive (WFD). The latter directive sets a clear requirement to protect only large water bodies with areas >50 ha, while the decision to protect smaller water bodies is left to the individual Member States [8]. In recent decades, however, awareness of the importance and conservation of SWBs at both a regional and a global level has nevertheless increased significantly and has been transferred into several important international initiatives, such as the Ramsar Convention on Wetlands and the development of the EPCN (European Pond Protection Network [5]). These initiatives have highlighted the need for an interdisciplinary approach to research on and protection of SWBs.

As a primary requirement for the better inclusion of small water bodies in existing catchments and landscape management plans, recent studies such as Biggs et al. [1] emphasize the need to implement reliable water monitoring programmes that will enable the mapping of SWBs, and facilitate the development of effective measures to protect their biota and the ecosystem services they have provided [1]. This process in the Western Balkans is however, hampered by a lack of quantitative hydrological data, as small waters are often forgotten in existing biomonitoring programmes committed to the implementation of the EU Water Framework Directive [9]. In addition, strategies and measures for the effective conservation of SWBs in the Western Balkans are largely site-specific, making it difficult to propose policy measures for their better integration into general water planning.

The aim of this chapter is to summarize the main conclusions presented in the book "Small Water Bodies of the Western Balkan", aimed at providing an extensive overview of the various environmental aspects of different types of small water bodies. The chapter emphasizes the current gaps, recent advances, and future prospects of our knowledge of these important aquatic ecosystems in the region of the Western Balkans.

20.2 The SWBs in the Western Balkan are Critical to Freshwater Biota and Ecosystem Service Delivery

The environmental characteristics of SWBs in the Western Balkans are primarily determined by the dominant source of water supply, the water stability of these habitats, as well as the geological and climatic context of the geographical areas in which they are located. The response of aquatic biota and ecosystem delivery services of small water bodies also depends on the degree of their hydromorphological modification [10]. Most of the small water bodies in the Western Balkans have already been altered by human induced environmental changes resulting from an increased demand for drinking water. In Chap. 1, Stevanović et al. [7] emphasized the need to ensure an appropriate balance between abstraction for drinking water supply and water-dependent ecosystem supply.

20.2.1 Small Standing Water Bodies

The great diversity of small water bodies (Fig. 20.1) in the Western Balkans generally belong to the SSWB that were the subject of several chapters in the Book [11–20]. A large number of SWBs such as vernal ponds (also called vernal pools) are fed

Fig. 20.1 Small standing water bodies in regions with contrasting climates: **a** Batušničke Bare ponds in Niš Municipality in Serbia. **b** Small mountain pond on the Durmitor plateau in Montenegro. **c** Temporary pond in the Dugi Otok Island in Croatia. **d** Permanent pond near Cetinje in Montenegro. **e** Pool phase of a small intermittent stream in the vicinity of Skadar/Shkodra Lake in Montenegro. Photos by V. Ranđelović (**a**), V. Pešić (**b**, **d**, **e**) and T. Dražina (**c**)

exclusively by atmospheric deposits. This habitat type is most commonly defined as a seasonal temporary wetland that forms in shallow depressions that can retain water and which lasts until it disappears as a result of rising temperatures [21]. Compared to permanent ponds, vernal ponds are poor in nutrients and show significantly larger daily differences in water temperature and other physicochemical parameters such as pH, dissolved carbon dioxide, and oxygen levels [22].

Although some authors have identified vernal ponds as typical of the Mediterranean area [23], in the Dinaric Karst this habitat type is relatively scarce as a result of the highly karstified substrate that does not allow the pool to slowly lose water by drying. Most ponds formed after heavy rains lose their water after a few days or weeks, which is an insufficient time frame for the development of a particular community that would regenerate each spring and disappear in late autumn. True vernal ponds in the sub-Mediterranean part of the Dinaric karst most often lie on a waterproof layer, such as alluvial deposits of finer material (clay). On the Adriatic islands, small and shallow temporary ponds are often the only freshwater habitats found [14].

Regardless of the generally harsh environment, temporary ponds host many specialized taxa that are rare or absent in other freshwater habitats [24]. These inhabitants of temporary ponds can be divided into two groups based on their strategies in terms of how they tolerate or avoid drought [25]. Completing the life cycle before surface water loss is a strategy for those species that do not have drought-resistant phases, as is the case with water mites [26]. Other inhabitants of temporary basins such as rotifers and entomostracan crustaceans (e.g., cladocerans, ostracods, and copepods) have specific drought survival traits, which include forming dehydration-resistant stages that allow them to survive the arrested physiological phases for several months until the water returns [27, 28].

On the other hand, there are a large number of temporary and permanent water bodies where water input is provided by long-distance drainage [22], such as for example, the numerous ponds along the northern shore of Lake Skadar in Montenegro. All of these temporary and/or permanent ponds get their water from the nearby lotic/lentic reaches, and thus acquire their fauna, which is mostly represented by generalist species that likely do not allow other species to inhabit these habitats. The frequency and duration of the period in which these small waterbodies are connected with nearby lentic/lotic reaches affect the composition and abundance of their biotic communities. Cvijanović [12] mentioned that the greatest macrophyte diversity was found in backwaters with a low connection rate with the main channel, while the largest plant biomass was found in SWBs that had a low or moderate connection with the main channel. Stojković-Piperac et al. [17], who studied the ichtiofauna of the Danube floodplains, found the highest species richness of fish in the side arms of the river that has direct connection with the main river flow.

Permanent ponds are common in floodplains and flood-prone areas [29]. About 12% of the Western Balkans is floodplains [12]: significant floodplains can be found along major river systems such as the Danube, Sava, Drina, South Morava, Vardar, Vjosë/Aoös, Neretva and Trebišnjica in Bosnia and Herzegovina, and the Lim and Zeta in Montenegro [12]. Some of these systems, such as the floodplains and wetlands in the Danube basin, are unique not only in Europe but also globally [12, 17].

Small standing water bodies in the Western Balkans, both temporary and permanent, whether natural or artificial, are characterized by diverse and rich aquatic biota that has been the subject of interest of several chapters in this book.

In Chap. 2, Špoljar et al. [11] give an overview of the biotic assemblages inhabiting 28 SWB sites located in Croatia, Serbia and Albania, with a special interest in the zooplankton component.

In Chap. 3, Cvijanović [12] assessed the habitat diversity, and conservation values, of small fluvial and gravel pit lakes along the river floodplain systems in the Western Balkans with a focus on macrophytes. The results showed that these two types of floodplain SWBs, covering a wide trophic gradient, and characterized by high macrophytes' physical structural heterogeneity, create an environment for diverse aquatic assemblages.

In Chap. 4, Čerba and Hamerlík [13] give an overview of the limnological literature devoted to the biota of city fountains, and provide new data based on recent research conducted in Croatia. This study reveals that this type of man-made ecosystem, which has often been neglected as a SWB in the limnological literature, can support the development of specific invertebrate communities, making them important for maintaining aquatic biodiversity in urban areas.

In Chap. 5, Dražina et al. [14] provide an overview of the temporary ponds on three Adriatic islands (Rab, Dugi otok and Korčula) in Croatia, with a special emphasis on zooplankton assemblages. This study showed that fishless ponds with a mixture of submerged macrophyte stands and open waters support greater invertebrate diversity, suggesting that in future conservation and management plans for small isolated ponds these factors should be taken into account [14].

In Chap. 8, Trbojević and Predojević [15] analyzed the diversity and distribution of the algae of SWBs in Serbia. They stressed that underestimating the algal diversity in SWBs is primarily the result of low taxonomic resolution of available data and/or interest for more comparative studies of algal taxonomy and phylogeny, as well as their insufficient inclusion in ongoing monitoring programmes.

In Chap. 14, Miličić et al. [16] studied the large branchipods of the Ramsar site "Bardača Wetland" situated in the northern part of Bosnia and Herzegovina. Even though the large branchiopod crustaceans represent one of the flagship groups in the inland water and temporary ponds ecosystems, in Bosnia and Herzegovina this group is insufficiently studied. This study provides the first records of five species for the territory of Bosnia and Herzegovina, while two species are new for the Western Balkans [16].

In Chap. 16, Stojković-Piperac et al. [17] provide the first data on the community structure and composition of fish fauna within the Danube floodplain in Serbia and Croatia. Their results showed that river-associated wetlands host a high diversity of fish that use the floodplain SWBs as a refuge for feeding, breeding, and nesting purposes at during at least one phase of their life cycle [17]

In Chap. 17, Crnobrnja-Isailović et al. [18] assessed the importance of small water bodies' conservation for maintaining local amphibian diversity in the Western Balkans. This study revealed that SWBs represent suitable habitats that are used by 97% of the 29 amphibian species listed for the Western Balkans during at least one phase of their life cycle [18].

20.2.2 Springs

The Western Balkans hosts a large number of groundwater-dependent ecosystems, which are explored in several chapters of this volume. Springs are classically defined as spatially restricted inland freshwater ecosystems which occur in places where groundwater reaches the surface [30, 31]. The environmental characteristics of the springs of the Western Balkans, especially in its Dinaric (karst) part, and their importance for maintaining freshwater biodiversity have been the subject of a large number of studies published in recent years (see Dedić et al. [32] for an overview of recent crenobiological research in the Western Balkans).

Due to the characteristic behaviour of the karst and groundwater regime, karstic springs are known to exhibit a large variation of discharge throughout the year (Fig. 20.2). The high discharges of springs and rivers during the winter and/or the spring periods may cause continuous seasonal flooding of the land in depressions and karstic fields (poljes); in the dry season, however, all the water passes underground through numerous ponors, and these fields and depressions become dry [7]. The changes in groundwater discharge affect the deposition of material and nutrients in springs, influencing both the biotic communities and the functioning of the

Fig. 20.2 Spring habitats in regions with contrasting climates: **a** Alipašini izvori springs in Gusinje municipality in Montenegro. **b** Small mountain spring on the mountain Komovi in Montenegro. **c** Spring of the Krupa River in Croatia. **d** Spring "Biljanini Izvori" in Ohrid town in North Macedonia. **e** Spring of Ravnjak stream in Montenegro. Photos by V. Pešić (**a**, **b** and **d**, **e**) and M. Miliša (**c**)

ecosystem [33]. Pešić et al. [33] mentioned that a large number of periodic springs that cease to flow, most often during the dry part of the year, may be an important source of organic matter for adjacent lotic/lentic reaches.

A number of studies conducted in recent years have emphasized the importance of different environmental parameters such as water chemistry, discharges, substrate composition [34–37] and hydromorphological modification [10] to different biotic groups inhabiting the springs of the Western Balkans. These studies have highlighted the generally higher sensitivity of spring habitats to environmental changes compared to other aquatic ecosystems, which is primarily caused by their small size and relatively higher climate and human induced pressure on these ecosystems [10]. Most springs in the Western Balkans are more or less tapped, not only for local drinking water supply, but often for aesthetic reasons [10, 32]. The impact of the ongoing climatic changes is particularly visible in the case of so-called riparian springs bordering streams and rivers that are strongly influenced by flood events [10, 33].

The importance of preserving spring habitats is often highlighted in the context of their significant freshwater biodiversity, which includes a large number of endemics which are often a priority for conservation efforts [10, 32]. These interesting issues are discussed in several chapters [33, 38–43] of this book. The proportion of endemic biota in spring ecosystems depends on the geographical isolation of these habitats and the habitat preferences of their inhabitants, which varies from taxa that occasionally occur in springs (crenoxenes) to taxa that are strictly bound to springs (crenobionts) [44].

In Chap. 6, Pešić et al. [33] gave an overview of the ecological and biotic features of riparian springs that appear along adjacent water bodies that are lotic (streams and rivers) or lentic (lakes and ponds). The latter systems are defined as springs that are flooded by adjacent lotic/lentic reaches at least part of the year [33]. The existence of flood and spring aquatic phases is the main driver that affects changes in the biotic assemblages of riparian springs as the transition between flood and spring aquatic phases results in greater variation in the environmental conditions along both the spatial and temporal gradients [33].

In Chap. 7, Stanić-Koštroman et al. [38] discussed the characteristic features of spring habitats and their biotic communities in Bosnia and Herzegovina, with a special emphasis on endemic and rare species. In Chap. 9, Živić et al. [39] provides an overview of the limnological literature dedicated to the investigations of springs and headwater streams in Serbia.

In Chap. 10, Savić et al. [40] analyzed the macroinvertebrate communities of two groups of springs, in the valley of Niš and the Vlasina plateau, both situated in southeastern Serbia, but characterized by different geological histories. This study showed that the changes in macroinvertebrate assemblages were primarily driven by geological impacts that are associated with changes in water chemistry, but also with biotic factors, such as the presence of invasive species [40].

In Chap. 11, Raković et al. [41] offers a review of the distribution of freshwater gastropod fauna in the springs, small mountain streams (low order streams), and groundwaters of the Western Balkan, listing 229 species and subspecies, many of them point endemics.

In Chap. 12, Marinković et al. [42] gives an overview of the diversity of leeches (Hirudinea) in the Balkans, emphasizing that the Dinaric Alps with its karstic springs represent the centre of diversity of this group in the Balkan Peninsula.

In Chap. 13, Pozojević and Pešić [44], assessed the diversity of water mites (Hydrachnidia) in karst springs, listing 44 crenobiontic (exclusively found in springs) water mite species for the territory of the Western Balkans. Their study reveals that most of the crenobiontic species are inhabitants of helocrene and rheohelocrene springs, and to a lesser extent inhabit rheocrenic springs, while only two species can be considered characteristic representatives of limnocrenic springs [44].

In Chap. 15, Vilenica et al. [43] gives a comprehensive overview of mayfly (Ephemeroptera) diversity in the Western Balkans and emphasizes both recent advances and the current gaps in our knowledge, while also providing recommendations for future research on this important insect group.

20.3 An Integrated Look at the Threats to SWBs in the Western Balkans

Almost all chapters in this volume emphasize that the SWB of the Western Balkans face a number of threats and challenges. Human activities such as the alteration of the hydrological regime, agriculture, land use change in the process of urbanization and land conversion into agriculture have been identified as major threats to SWBs in the Western Balkans. Most chapters highlight the impact of the above-mentioned threats on the biodiversity and ecosystem services (Fig. 20.3) of different types of SWB (e.g. [11, 12, 14, 38–40]) or the impact on communities of particular floristic/faunistic groups (e.g., [15–18, 41–43]). Human induced changes resulting in contamination of SWBs and their biota represent a risk to human welfare, both directly, through increased health risks, and indirectly, through reduced ecosystem services [20].

Some chapters [12, 14, 17, 18, 33] have highlighted the impact of climate change and invasive species on the biotic communities of small water bodies in the Western Balkans. Cvijanović et al. [12] stated that SWBs in the floodplains of the Western Balkans are exposed to the influence of invasive macrophyte species. There are also a significant number of invasive species of fish, whose presence in small waterbodies can lead to drastic changes in the composition and abundance of local communities. Dražina et al. [14] showed that the presence of invasive *Gambusia holbrooki* and

Fig. 20.3 Provisioning ecosystem services: freshwater provision by a small water bodies supports well-being especially in karstic areas. **a** Artificial pond Gornja Osječenica in Montenegro. Regulating ecosystem services of small water bodies is important for water quality regulation in agricultural catchment; for example, uptake by semiaquatic plants reduces inorganic nutrient pollution. **b** Batušničke bare ponds in Serbia. Cultural ecosystem services includes environmental education and scientific knowledge and aesthetic values. **c** Scientific research of the aquatic biota of a small pond in Dugi Otok Island in Croatia. In the Western Balkan there are many bridges that are old and are monuments of culture. **d** Adzija's Bridge (Hadzija is a pilgrim) on the river Sušica in Bjelopavlici region of Montenegro. Photos by V. Pešić (**a**, **d**), V. Ranđelović (**b**) and T. Dražina (**c**)

Carassius auratus in temporary ponds on Croatian islands (which are naturally fish-free) affects zooplankton communities. Indeed, invasive species can dominate in some types of SWB (e.g. *Carassius gibelio* is a representative species for oxbow lakes, and the most common and widespread species in the SWBs of the Danube and Sava floodplains [17]—Fig. 20.4).

Several studies have stressed that small water bodies in the Dinaric karst, especially those that are groundwater dependent, are very sensitive to climatic and human induced environmental changes [7, 33]. These ecosystems are likely to become even more vulnerable in space and time in the future, as most of the Western Balkans region becomes characterized by greater climatic extremity in both temperature and precipitation [48].

Stefanović et al. [7] recognized surface and groundwater abstractions and the building of dams and reservoirs as the major threats to spring ecosystems in the karst area. Over the last decade, the Western Balkan region has become the target for the massive deployment of small-scale hydropower projects, leading to an increased

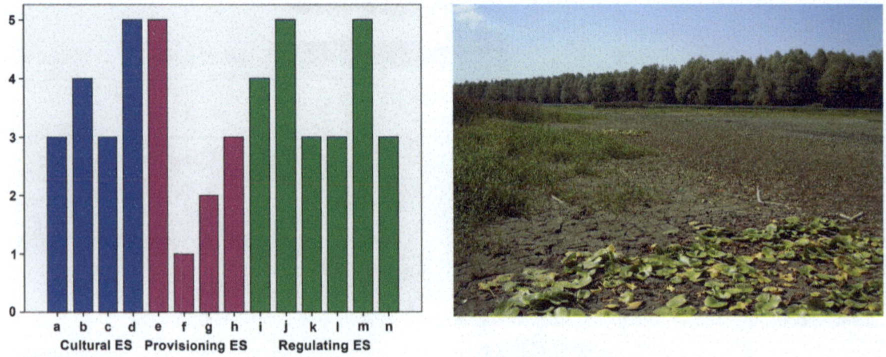

Fig. 20.4 The floodplains and wetlands of the Danube Basin represent a unique and valuable freshwater ecosystem, not only in the Western Balkans, but also in the whole of Europe [45]. Their ecological importance was confirmed by the formation of a Transboundary UNESCO Biosphere Reserve, the so-called "Amazon of Europe", that combines a group of thirteen protected areas within the Mura–Drava–Danube region, which hosts an amazing range of biological diversity and which is a hotspot of rare natural habitats such as large floodplain forests, river islands, gravel and sand banks, side branches and oxbows. Its value is even greater when we remember that floodplains represent a hotspot of ecosystem services (ES), providing goods, supporting biodiversity, regulating the climate and enabling cultural diversity [46]. According to the results of the assessment of provisioning and regulating ecosystem services by stakeholder at the Begečka jama wetland area [47], the most frequently used provisioning ES were game meat, fish, wood and water. This was followed by regulating ES, where the provision of terrestrial habitats and the significant contribution to local climate regulation and air purification were dominant. Within the cultural ES, the most dominant subclass was recreational activity (i.e., hiking, cycling and wildlife viewing). The letters indicate the different ES subclasses: a-water related activity, b-tourism, c-education, d-recreation activity, e-wood, f-animal product, g-wild fish, h-water, i-local climate regulation, j-air purification, k-flood retention, l-nutrition retention, m-provision of terrestrial habitats and n-provision of aquatic habitats and the numbers indicate no (0), very low (1), low (2), medium (3), high (4), and very high (5) provision of ES. Photo by M. Novaković

number of conflicts with the local community over these facilities. As pointed out by Pešić et al. [49] this conflict with the public is "recognized as a consequence of the lack of an established set of equitable sharing principles among the contending parties". From the point of view of biodiversity conservation, this problem is complex and primarily related to providing enough water in the summer months to meet the prescribed minimum of ecological flows in small rivers and streams in the Western Balkans, where small hydropower plants are built [49]. In recent years, the issue of the minimum levels of ecological flows have become a point of contention between water engineers and environmentalists. According to Stevanović et al. [7] the following mitigation measures are recommended: (a) artificial regulation of aquifer regimes with the aim of improving the minimum flow, (b) comprehensive monitoring of abiotic and biotic parameters, (c) reducing the risk of pollution in the karst aquifer, and (d) raising awareness of the sustainable use of water resources.

With regard to freshwater biodiversity, all the chapters emphasized the importance of SWBs in maintaining aquatic biota and their communities and the need for the adoption of mitigation measures which would reduce the pressure on these communities. The importance of preserving small water bodies in impacted areas should not be neglected, even in heavily impacted regions, since even small habitat patches, such as urban ponds, can support diverse and complex biotic communities [50]. As shown in Chap. 4, small man-made water bodies such as city fountains may support diverse invertebrate communities, particularly of insects [13], opening new perspectives on conserving biodiversity in rural areas. In a similar way, the study by Pešić et al. [10] showed that the hydromorphological modification of springs in karst areas does not necessarily lead to a reduction in diversity in their biotic assemblages, provided that the discharge and substrate composition both remain suitable.

Despite ongoing research interest, the effects of anthropogenic pollution on SWBs are poorly investigated in comparison to larger water bodies. In Chap. 19, Milošković et al. [20] analyzed the published studies related to the contamination of small lakes and ponds with potentially toxic elements (PTE) and found that SWB in the Western Balkans were largely neglected in these studies. Understanding the simultaneous effects of multiple stressors on SWB is crucial to understanding the response of aquatic biota and their communities to various human activities [51]. In Chap. 18, Buzhdygan et al. [19] who tested the impact of various co-occurring types of human activities (such as agricultural land use, gravel exploitation, nearby roads, and waste input) on different consumer communities, showed that the intensification of human impacts led to shifts in the trophic composition and induced changes in the diversity of different trophic levels of the tested SWB ecosystems [19]. These results highlight the need to consider the importance of a multistressor and multitrophic perspective in predicting the consequences of human induced changes for both the biotic assemblages and the ecosystem functioning of SWB.

20.4 The Protection of SWBs Within the Socio-Ecological Context

The main goal of this book was to compile the most important scientific knowledge gained in recent years about the small water bodies of the Western Balkans. The geological and climatic history of the Western Balkan that shaped today's physiographic patterns has influenced the spatial and temporal distribution of SWBs in the study area, particularly in its karstic regions. The uniqueness of this area has been and still is a challenge in terms of researching small water bodies as well as in understanding ecosystem functioning, especially in the context of proposing and implementing measures for their protection, and recognizing human dependence on SWBs in the Western Balkan area.

The importance of SWBs in maintaining freshwater biodiversity and providing ecosystem services is particularly noticeable in karst terrains where these types of aquatic ecosystems often dominate. In this volume, special emphasis is placed on the concept of small karst water bodies, with the aim of making the protection of these habitats more manageable and pragmatic. Moreover, by highlighting the different types of SWBs and the ecosystem services they provide, we hope to raise awareness of their socio-ecological contribution to regions with contrasting geological and climatic contexts and developing economies.

Almost all the chapters in the book emphasize that managers and policy makers generally overlook the impact on SWBs in the Western Balkans and that they are not sufficiently taken into account by ongoing monitoring programmes. Environmental policy in the Western Balkans is based on existing EU strategies that have not been adjusted to the specific context of the region. Another problem is that we still lack a common strategy for the protection of SWBs in the EU that could be modified and applied in the national legislation of the Western Balkan countries.

Better recognition of the SWBs in the region of the Western Balkans, as highlighted by Crnobrnja-Isailović et al. [18] is primarily related to their socio-economic status. As increased water demand makes SWBs even more vulnerable, both in space and time, in many regions of the Western Balkans, especially in areas where there are not many large water bodies, the social recognition of these ecosystems is key to motivating the local community to recognize the importance of their protection. This book highlights the need for both scientists and environmental managers to take action to improve knowledge about these important but still neglected freshwater ecosystems and promote their protection in the wider socio-ecological system, one which considers not only their ecological but also their social recognition.

Acknowledgements We are indebted to all the authors of the chapters for their enthusiasm and commitment, and for the insights that have resulted in this book. Moreover, we thank the many peoples who helped us in providing discussions and ideas for this book and that have resulted in this book. Thanks also to our colleagues T. Dražina, V. Ranđelović and M. Novaković for the permission to use their photos.

References

1. Biggs J, Von Fumetti S, Kelly-Quinn M (2017) The importance of small water bodies for biodiversity and ecosystem services: implications for policy makers. Hydrobiologia 793:3–39
2. Céréghino R, Biggs J, Oertli B, Declerck S (2008) The ecology of European ponds: defining the characteristics of a neglected freshwater habitat. Hydrobiologia 597:1–6
3. Rosset V, Angelibert S, Arthaud F et al (2014) Is eutrophication really a major impairment for small waterbody biodiversity? J Appl Ecol 51:415–425
4. Bolpagni R, Poikane S, Laini A et al (2019) Ecological and conservation value of small standing-water ecosystems: a systematic review of current knowledge and future challenges. Water 11:402. https://doi.org/10.3390/w11030402
5. E.P.C.N. (2007) Developing the Pond Manifesto. Ann Limnol Int J Lim 43:221–232
6. Schwan W (1997) Europe. In: Encyclopedia of European and Asian regional geology. Encyclopedia of earth science. Springer, Dordrecht
7. Stevanović Z, Pekaš Ž, Stevanović AM et al Springs as essential water sources for dependent ecosystems in karst. In: Pešić V, Miliša M, Milošević Dj (eds) Small Water Bodies of the Western Balkans. Springer, Cham
8. Popov SV, Rögl F, Rozanov AY et al (2004) Lithological-paleogeographic maps of Paratethys: 10 maps Late Eocene to Pliocene. Cour Forschungsinst Senck 250(1–46):8
9. European Commission (2000) Directive 2000/60/EC, establishing a framework for community action in the field of water policy. Off J Eur Comm Brussels L327:1–71
10. Pešić V, Dmitrović D, Savić A et al (2019) Application of macroinvertebrate multimetrics as a measure of the impact of anthropogenic modification of spring habitats. Aquat Conserv 29:341–352
11. Špoljar M, Shumka S, Tasevska O et al Small standing-water ecosystems in the transitional temperate climate of the Western Balkans. In: Pešić V, Miliša M, Milošević Dj (eds) Small Water Bodies of the Western Balkans. Springer, Cham
12. Cvijanović D Conservation value and habitat diversity of fluvial lakes and gravel pits in river-floodplain systems. In: Pešić V, Miliša M, Milošević Dj (eds) Small Water Bodies of the Western Balkans. Springer, Cham
13. Čerba D, Hamerlík L Fountains—overlooked small water bodies in the urban areas. In: Pešić V, Miliša M, Milošević Dj (eds) Small Water Bodies of the Western Balkans. Springer, Cham
14. Dražina T, Špoljar M, Miliša M Temporary ponds in Mediterranean islands: oases of biodiversity. In: Pešić V, Miliša M, Milošević Dj (eds) Small Water Bodies of the Western Balkans. Springer, Cham
15. Trbojević I, Predojević D Algae in shallow and small water bodies of Serbia: a frame for species and habitat protection. In: Pešić V, Miliša M, Milošević Dj (eds) Small Water Bodies of the Western Balkans. Springer, Cham
16. Miličić D, Šukalo G, Dmitrović D Large branchiopods in small water bodies: a case study of the Ramsar site "Bardača Wetland" (NW Republic of Srpska, Bosnia and Herzegovina). In: Pešić V, Miliša M, Milošević Dj (eds) Small Water Bodies of the Western Balkans. Springer, Cham
17. Stojković Piperac M, Milošević Dj, Čerba D, Simić V Fish communities over the Danube wetlands in Serbia and Croatia. In: Pešić V, Miliša M, Milošević Dj (eds) Small Water Bodies of the Western Balkans. Springer, Cham
18. Crnobrnja-Isailović J, Adrović A, Bego F et al The importance of small water bodies' conservation for maintaining local amphibian diversity in the Western Balkans. In: Pešić V, Miliša M, Milošević Dj (eds) Small Water Bodies of the Western Balkans. Springer, Cham
19. Buzhdygan OY, Stojković Piperac M, Stamenković O et al Human impact induces shifts in trophic composition and diversity of consumer communities in small freshwater ecosystems. In: Pešić V, Miliša M, Milošević Dj (eds) Small Water Bodies of the Western Balkans. Springer, Cham

20. Milošković A, Đuretanović S, Radenković M et al Pollution of small lakes and ponds of the Western Balkans—assessment of levels of potentially toxic elements. In: Pešić V, Miliša M, Milošević Dj (eds) Small Water Bodies of the Western Balkans. Springer, Cham
21. Solomeshch AI, Barbour MG, Holland RF (2007) Vernal pools. In: Barbour MG, Keeler-Wolf T, Schoenherr AA (eds) Terrestrial vegetation of California, 3rd edn. University of California Press, Berkeley
22. Keeley JE, Zedler PH (1998) Characterization and global distribution of vernal pools. In: Witham CW, Bauder ET, Belk D, Ferren WR, Ornduff R (eds) Ecology, conservation, and management of vernal pool ecosystems—proceedings from a 1996 conference. California Native Plant Society, Sacramento, CA, pp 1–14
23. Zacharias I, Dimitriou E, Dekker A, Dorsman E (2007) Overview of temporary ponds in the Mediterranean region: threats, management and conservation issues. J Environ Biol 28:1–9
24. Fonseca LC, Cristo M, Machado M et al (2008) Mediterranean temporary ponds in Southern Portugal: key faunal groups as management tools? Pan Am J Aquat Sci 3:304–320
25. Wigginns GB, Mackay RJ, Smith IM (1980) Evolutionary and ecological stratefies of animals in annual temporary pools. Arch Hydrobiol Suppl 58:97–206
26. Zawal A, Pešić V (2018) The diversity of water mite assemblages (Acari: Parasitengona: Hydrachnidia) of Lake Skadar/Shkodra and its catchment area. In: Pešić V, Karaman GS, Kostianoy AG (eds) The Skadar/Shkodra lake environment. Springer, Berlin, Heidelberg, pp 311–323
27. Williams DD (1998) The role of dormancy in the evolution and structure of temporary water invertebrate communities. Arch Hydrobiol 52:109–124
28. Bilton DT, Freeland JR, Okamura B (2001) Dispersal in freshwater invertebrates. Annu Rev Ecol Syst 32:159–181
29. Keruzoré AA, Willby NJ, Gilvear DJ (2013) The role of lateral connectivity in the maintenance of macrophyte diversity production in large rivers. Aquat Conserv Mar Freshw Ecosyst 23:301–315
30. Odum HT (1971) Fundamentals of ecology. Saunders, Philadelphia
31. Kresic N (2010) Types and classification of springs. In: Kresic N, Stevanovic Z (eds) Groundwater hydrology of springs. Elsevier, Oxford, pp 31–86
32. Dedić A, Gerhardt A, Kelly MG et al (2020) Innovative methods and approaches for WFD: ideas to fill knowledge gaps in science and policy. Water Solut 3:30–42
33. Pešić V, Dmitrović D, Savić A Riparian springs—challenges from a neglected habitat. Riparian springs. In: Pešić V, Miliša M, Milošević Dj (eds) Small Water Bodies of the Western Balkans. Springer, Cham
34. Dmitrović D, Savić D, Pešić V (2016) Discharge, substrate type and temperature as factors affecting gastropod assemblages in springs in northwestern Bosnia and Herzegovina. Arch Biol Sci Belgrade 68:613–621
35. Savić A, Dmitrović D, Pešić V (2017) Ephemeroptera, Plecoptera and Trichoptera assemblages of karst springs in relation to environmental factors: a case study in central Bosnia and Hercegovina. Turk J Zool 41:119–129
36. Savić A, Dmitrović D, Glöer P, Pešić V (2020) Assessing environmental response of gastropod species in karst springs: what species response curves say us about niche characteristic and extinction risk? Biodivers Conserv 29:695–708
37. von Fumetti S, Dmitrović D, Pešić V (2017) The influence of flooding and river connectivity on macroinvertebrate assemblages in rheocrene springs along a third-order river. Fund Appl Limnol 190(3):251–263
38. Stanić-Koštroman S, Kamberović J, Dmitrović D, Dedić D, Škobić D, Lasić A, Gligora Udovič M, Herceg M Ecological characteristics and specifics of spring habitats in Bosnia and Herzegovina. In: Pešić V, Miliša M, Milošević Dj (eds) Small Water Bodies of the Western Balkans. Springer, Cham
39. Živić I, Stojanović K, Marković Z Springs and headwater streams in Serbia: the hidden diversity and ecology of aquatic invertebrates. In: Pešić V, Miliša M, Milošević Dj (eds) Small Water Bodies of the Western Balkans. Springer, Cham

40. Savić A, Đorđević M, Đorđević M et al Springs of Southeastern Serbia with a focus on the Vlasina Plateau: different types of challenges for the macroinvertebrate community. In: Pešić V, Miliša M, Milošević Dj (eds) Small Water Bodies of the Western Balkans. Springer, Cham
41. Raković M, Tomović J, Popović N et al Gastropods in small water bodies of the Western Balkans—endangerments and threats. In: Pešić V, Miliša M, Milošević Dj (eds) Small Water Bodies of the Western Balkans. Springer, Cham
42. Marinković N, Paunović M, Raković M et al Importance of small water bodies for diversity of leeches (Hirudinea) of Western Balkan. In: Pešić V, Miliša M, Milošević Dj (eds) Small Water Bodies of the Western Balkans. Springer, Cham
43. Vilenica M, Petrović A, Rimcheska B et al How important are small lotic habitats of the Western Balkans for local mayflies? In: Pešić V, Miliša M, Milošević Dj (eds) Small Water Bodies of the Western Balkans. Springer, Cham
44. Pozojević I, Pešić V Karst springs: isolated ecosystem ecology from the water mite perspective. In: Pešić V, Miliša M, Milošević Dj (eds) Small Water Bodies of the Western Balkans. Springer, Cham
45. International Commission for the Protection of the Danube River (ICPDR) (2015) The Danube River Basin Management Plan 2015–2021—Factsheet. https://www.icpdr.org/main/sites/default/files/nodes/documents/managementplansbrochure2015.pdf. Accessed 26 Aug 2019
46. Perosa F, Fanger S, Zingraff-Hamed A, Disse M (2021) A meta-analysis of the value of ecosystem services of floodplains for the Danube River Basin. Sci Total Environ 777:146062
47. Danube floodplain Interreg project (2021) Report, database and maps of ecosystem services analysis of the pre-selected pilot areas including a list, description, assessment, and ranking concerning the demands and supplies. http://www.interreg-danube.eu/uploads/media/approved_project_output/0001/44/ca5399ad0f6e1f07de5c080e2d2e72153bef6282.pdf. Accessed 10 June 2021
48. Kostianoy AG, Kostianaia EA, Pešić V (2020) Drainage basins of Montenegro under climate change. In: Pešić V, Paunović M, Kostianoy A (eds) The rivers of Montenegro. The handbook of environmental chemistry, vol 93. Springer, Cham, pp 68–81. https://doi.org/10.1007/698_2020_481
49. Pešić V, Paunović M, Kostianoy AG, Vukašinović-Pešić V (2020) The rivers of Montenegro: from conflicts to science-based management. In: Pešić V, Paunović M, Kostianoy A (eds) The rivers of Montenegro. The handbook of environmental chemistry, vol 93. Springer, Cham, pp 287–301. https://doi.org/10.1007/698_2020_480
50. Hill MJ, Biggs J, Thornhill I et al (2017) Urban ponds as an aquatic biodiversity resource in modified landscapes. Glob Chang Biol 23:986–999
51. Birk S, Chapman D, Carvalho L et al (2020) Synthesizing the impacts of multiple stressors on freshwater biota across scales and ecosystems. Nat Ecol Evol 4:1060–1068. https://doi.org/10.1038/s41559-020-1216-4

The manufacturer's authorised representative in the EU is Springer Nature Customer Service Centre GmbH, Europaplatz 3, 69115 Heidelberg, Germany. If you have any concerns regarding our products, please contact ProductSafety@springernature.com

Printed and bound by CPI Group (UK) Ltd, Croydon, CR0 4YY
30/09/2026
02225695-0015